THE *flavor* THESAURUS

THE *flavor* THESAURUS
Pairings, Recipes and Ideas
for the Creative Cook

NIKI SEGNIT

BLOOMSBURY
NEW YORK · LONDON · NEW DELHI · SYDNEY

Quote from *The Debt to Pleasure* by John Lanchester copyright © 1996 John Lanchester. Reprinted by permission of Henry Holt and Company, LLC. Extracts from *The Bell Jar* by Sylvia Plath copyright © 1971 Harper & Row, Publishers, Inc. Reprinted by permission of HarperCollins Publishers. Extracts from *Venus in the Kitchen* by Norman Douglas reprinted by permission of The Society of Authors as the Literary Representative of the Estate of Norman Douglas. Extracts from *How to Eat* by Nigella Lawson © 1998 Nigella Lawson. Reprinted with permission of John Wiley & Sons, Inc. Every effort has been made to trace the copyright owner of *The Book of New French Cooking* by Paul Reboux. Extracts from *French Provincial Cooking* by Elizabeth David, published by Michael Joseph. Reprinted by permission of the Estate of Elizabeth David. Excerpts from *Everyday Drinking* by Kingsley Amis © 2008 The Literary Estate of Sir Kingsley Amis. Published by Bloomsbury USA.

Published by Bloomsbury USA, New York

All papers used by Bloomsbury USA are natural, recyclable products made from wood grown in well-managed forests. The manufacturing processes conform to the environmental regulations of the country of origin.

LIBRARY OF CONGRESS CONTROL NUMBER: 2010910372

ISBN 978-1-60819-874-0

First U.S. Edition 2010
Revised U.S. Edition 2012

10 9 8 7 6 5 4

Typeset by Hewer Text UK Ltd, Edinburgh
Printed and bound in the U.S.A. by Thomson-Shore Inc., Dexter, Michigan

It seems fitting to dedicate this book to a pair:
my cooking adviser and mother, Marian Stevens,
and my writing adviser and husband, Nat Segnit.

Contents

". . . lamb and apricots are one of those combinations which exist together in a relation that is not just complementary but that seems to partake of a higher order of inevitability—a taste which exists in the mind of God. These combinations have the quality of a logical discovery: bacon and eggs, rice and soy sauce, Sauternes and *foie gras*, white truffles and pasta, *steak-frites*, strawberries and cream, lamb and garlic, Armagnac and prunes, port and Stilton, fish soup and *rouille*, chicken and mushrooms; to the committed explorer of the senses, the first experience of any of them will have an impact comparable with an astronomer's discovery of a new planet."

John Lanchester, *The Debt to Pleasure*

Introduction

I hadn't realized the depth of my dependence on cookbooks until I noticed that my copy of Elizabeth David's *French Provincial Cooking* had fingernail marks running below the recipes. Here was stark evidence of my timidity, an insistence on clinging to a set of instructions, like a handrail in the dark, when after twenty years of cooking I should surely have been well enough versed in the basics to let go and trust my instincts. Had I ever really learned to cook? Or was I just reasonably adept at following instructions? My mother, like her mother before her, is an excellent cook but owns only two recipe books and a scrapbook of clippings, and rarely consults even those. I began to suspect that the dozens of books *I* owned were both a symptom and a cause of my lack of kitchen confidence.

It was at a dinner around the same time that a friend served a dish using two ingredients it would never have occurred to me to pair. How, I wondered, did she know *that* would work? There was something in the air about surprising flavor matches, the kind of audacious combinations pioneered by chefs like Heston Blumenthal, Ferran Adrià and Grant Achatz. What lay at the heart of their approach to food was, as far as I could see, a deeper understanding of the links between flavors. Being an ordinary, if slightly obsessive, home cook, I didn't have the equipment or resources to research these; what I needed was a manual, a primer to help me understand how and why one flavor might go with another, their points in common and their differences. Something like a thesaurus of flavors. But no such book existed and so, with what turned out in hindsight to be almost touching naivete, I thought I might try to compile one myself.

My first task was to draw up the list of flavors. Stopping at 99 was to some extent arbitrary. Nonetheless, a flavor thesaurus that accounted for every single flavor would be as impractical as it would be uncomfortable on the lap. Other than potatoes, the staple carbohydrates have been omitted. The same goes for most common condiments. There are, of course, plenty of interesting things to say about the flavors of rice, pasta, black pepper, vinegar and salt, but their flavor affinities are so wide as to exclude themselves by virtue of sheer compatibility. Other omissions, like zucchini, might strike you as odd: all I can say to the zucchini fan is (a) sorry, and (b) this book makes no claims to be the last word on the subject. Any book on flavor is going to be at least in part subjective, and in writing about the pairings I find most interesting or like to eat the most, I will inevitably have left gaps that come down to nothing other than a matter of taste.

The majority of flavors appear under their own heading. In a few instances, where it seemed to make sense, some very similarly flavored ingredients share a heading. Anise, for example, covers anise seeds, fennel, tarragon, licorice and pastis. Similarly, neither bacon and ham nor Brussels sprouts and cabbage could easily be separated, so they labor slightly uncomfortably under composite categories. When it came to a choice between untidiness and boring the reader with repetitions, I chose untidiness every time.

Then I sorted the flavors into categories. Most of us are familiar with the concept of flavor families, whether we know it or not. Floral, citrus, herbaceous: the sort of descriptors you might encounter on the back of a wine bottle, to help conjure an idea of how something might taste. And it's into these, or adjectival headings like them, that the flavors are divided. The flavors in each family have certain qualities in common; in turn, each family is linked in some way to the one adjacent to it, so that, in sum, they comprise a sort of 360° spectrum, represented opposite as a flavor wheel.

Take the Citrusy family, for example. This covers zesty, citric flavors like orange, lemon and cardamom. Cardamom, in turn, has flavor compounds in common with rosemary, which is the first flavor in the next flavor family, Berry & Bush. At the other end of that family, blackberry leads to the first flavor in the Floral Fruity family: raspberry. And so on around the wheel, flavor leading to flavor, family to family, in a developing sequence of relations you might enter at lemon and leave at blue cheese.

I acknowledge that this methodology has its limitations. Some flavors resisted easy categorization: coriander seed, for instance, ended up under Floral Fruity, but might as easily have sat in Citrusy or Spicy. And how an ingredient is prepared can make all the difference to its character. The flavor of cabbage, for example, is mustardy when raw, sulfurous cooked. The flavor wheel, in short, is by no means intended to be an inarguable, objective framework for understanding flavor—but it does provide a stimulating and intriguing means of navigating your way around the subject.

Next came the pairings. Clearly, dishes often have more than two primary ingredients, but a couple of considerations led me to make pairs of flavors the organizing principle of my *Flavor Thesaurus*. First, sanity (mine). Even restricting myself to 99 flavors, if I had set out to write about flavor *trios* I would have been faced with 156,849 possible combinations; the 4,851 possible pairings seemed more to scale with the sort of book it would be both possible to write and pleasurable to read. Second, clarity. To assess, in the mind's palate, the compatibility of two flavors is exponentially easier than imagining the interplay of three or more. Necessarily, I often discuss a flavor combination in the context of a dish that contains other ingredients (for example, parsley and mint in tabbouleh), but the emphasis is always on the main flavor pairing under discussion.

The entries elaborate on each of these pairings, drawing promiscuously on flavor science, history, culture, chefs' wisdom and personal prejudice— anything that might shed light on why certain flavors work together, what they bring out in each other, how the same flavor pairings are expressed in different cuisines and so on. I've given any recipes in the briefest possible

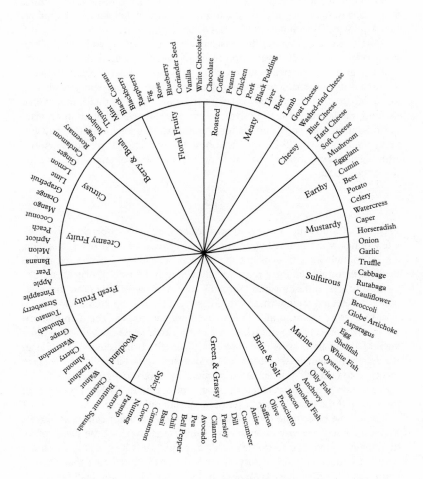

terms, rather in the manner of Victorian cookbooks—compressed in the expectation that you have some experience in the kitchen. If you're tempted to cook anything in the pages that follow, it's best to read the recipe through before you start (see Beet & Pork, page 88, if you need convincing). I've assumed you know that you usually need to add salt to savory dishes, taste them and adjust before you serve, turn off the stove when you're finished, and fish out any ingredients that might choke your loved ones. If something in a recipe isn't clear—stop, think, and if the solution still isn't forthcoming, find a similar recipe and see if that sheds any light.

Most often, I guarantee, the answer will just dawn on you. One of the great satisfactions of discovering more about flavor combinations is the confidence it gives you to strike out on your own. Following the instructions in a recipe is like parroting pre-formed sentences from a phrasebook. Forming an understanding of how flavors work together, on the other hand, is like learning the language: it allows you to express yourself freely, to improvise, to find appropriate substitutions for ingredients, to cook a dish the way you fancy cooking it. You'll be surprised how rarely things go seriously wrong. Although the author takes absolutely no responsibility for anything that ends up in the trash.

Flavor is, of course, notoriously subjective and hard to describe. Still, there are a few things worth noting before you try. As anyone who has been on a wine course will tell you, flavor is not the same as taste. Taste is restricted to five qualities detectable on the tongue and elsewhere in the mouth: sweetness, saltiness, sourness, bitterness and "umami" (or savoriness). Flavor, on the other hand, is detected mainly thanks to our sense of smell, by the olfactory bulb and, to a lesser extent, orally. Pinch your nose and you can tell if an ingredient is sweet or salty, but not what its flavor is. Your sense of taste gives you a back-of-an-envelope sketch of what a particular foodstuff is like: flavor fills in the details. Nonetheless, in its general, broadest use, the term "flavor" tends to incorporate taste, as well as the "trigeminal" qualities of ingredients—that is, the sensation of heat from chili, pepper and mustard, the cooling properties of menthol and the drawstring pucker of tannins in red wine and tea.

Beyond the basic taste elements, to characterize a flavor can be as elusive a task as describing any other sort of feeling. Inasmuch as the flavor of an ingredient is attributable to the chemical compounds it contains, we might with some degree of objectivity observe that two ingredients that share a compound have similar flavors. Holy basil and clove, for instance, both contain a compound called eugenol—and holy basil has a clove-like flavor. But what do we mean by a clove-like flavor? To me, it's a bit like sucking on a sweet, rusty nail. However, no person's taste buds, or olfactory systems, are quite the same, and neither are their faculties for converting sensory input into words.

Where you come from and what you're accustomed to eating are also important determinants of how you sense and describe flavor, and of which flavors you tend to pair with others. I've used expert opinions to lend to my own judgments as robust an edge of objectivity as possible. But there's no escaping the fact everyone's flavor thesaurus would to some extent be different. Flavor is, among other things, a repository of feeling and memory: just as smell is said to be the most redolent sense, so the flavor of a certain dish can transport us back instantaneously to the time and place we first experienced it, or experienced it most memorably. *The Flavor Thesaurus* may look like, and even sometimes read like, a reference book, but for all its factual content it's an inescapably subjective one.

Writing *The Flavor Thesaurus* has taught me many things, not least to take a more open-minded approach to combinations that other cooks, in other cultures, take for granted. But as a naturally untidy person, I'm always looking for patterns, some means of imposing order on unruly reality. And in part I suppose I expected the book to add up, over its length, to something along these lines, a Grand Unifying Flavor Theory that would reconcile the science with the poetry and my mother's thoughts on jam.

It didn't. Or not quite. I *did* learn some broadly applicable principles, like how to use one flavor to disguise, bolster, temper or enliven another. And I'm now far more alert to the importance of balancing tastes—salt, sweet, bitter, sour and umami—and making the most of contrasting textures and temperatures. But what *The Flavor Thesaurus* does add up to, in the end, is a patchwork of facts, connections, impressions and recollections, designed less to tell you exactly what to do than to provide the spark for your own recipe or adaptation. It's there, in short, to get the juices flowing.

Niki Segnit
London, March 2010

ROASTED

Chocolate

Coffee

Peanut

Chocolate

The complex processing undergone by most chocolate explains the huge variation in flavor. The untreated cocoa beans are astringent and bitter but fermentation gives rise to fruity, wine-like or sherry flavors, and the roasting process can introduce a nearly infinite variety of nutty, earthy, woody, flowery and spicy notes. The flavor of good-quality chocolate is best appreciated by pushing a piece to the roof of your mouth and letting it melt. The more sweetened the chocolate, the quicker it will reveal its flavor. As you work your way up the cocoa percentages you'll notice that it takes longer for the flavor to develop, and that there's an increase in bitterness and length—the time the flavor lingers in your mouth. When you get to 99 or 100 percent cocoa content, you may also note that the experience is like running your tongue along the main London–Edinburgh railway. In this section, "chocolate" is taken to cover dark chocolate, milk chocolate and cocoa. White chocolate is dealt with separately (see page 342).

Chocolate & Almond What does parental guilt taste like? Chocolate and almond, the ingredients in the Toblerone your dad grabbed at the airport instead of a pair of maracas or a genuine bear's paw. The secret of its success must have something to do with the genial compatibility of chocolate and almond. A wealth of nutty flavor notes is formed when cocoa beans are roasted during the chocolate-making process. Similarly, the flavor of almonds is intensified by toasting, which helps them stand up to chocolate's strength of flavor. Put this to the test in Christopher Tan's chocolate soup with salted almonds. Melt 1 tbsp unsalted butter in a skillet over a medium-low heat, add 1½ oz sliced almonds and sauté with care for 4–5 minutes, until golden brown. Set aside. Whisk 8 fl oz water, 3 fl oz whipping cream, 1 oz superfine sugar and 1¼ oz sifted cocoa powder in a saucepan over a medium-low heat until the sugar has dissolved. Let the mixture bubble gently for 2–3 minutes, then add 3½ oz chopped dark chocolate containing at least 60 percent cocoa solids. Stir with the whisk until smooth, then pour into little bowls, top with the almonds and sprinkle over a few grains of *fleur de sel*. Divide between 4–6 bowls and serve immediately.

Chocolate & Anise See *Anise & Chocolate, page 180.*
Chocolate & Apricot See *Apricot & Chocolate, page 275.*
Chocolate & Avocado See *Avocado & Chocolate, page 195.*

Chocolate & Bacon Chocolatier Katrina Markoff, founder of Vosges Haut-Chocolat, combines applewood-smoked bacon and smoked salt with a dark milk chocolate in her Mo's Bacon Bar. She says she was inspired by eating chocolate-chip pancakes with maple syrup and bacon when she was six. You might see how the classic combination of savory and sweet would work and be made that bit more unusual by the smoky element in the bacon. Taking the combination a step farther, Tee and Cakes in Boulder, Colorado, makes maple-flavored cupcakes topped with a slightly salted dark chocolate ganache and a sprinkle of chopped-up bacon.

Chocolate & Banana *See Banana & Chocolate, page 272.*

Chocolate & Beet *See Beet & Chocolate, page 87.*

Chocolate & Black Currant Dark and heavy as Finnish Goth poetry, but not quite as popular. A few British brand names (Matchmakers, Jaffa Cakes) have given the pair a whirl, announcing them with a fanfare before ushering them silently out of the back door. The combination is more likely to work with the soothing influence of dairy—say, in a black-currant mousse, parfait or fool with a chocolate sauce, or a chocolate gâteau with a fresh cream and black-currant filling.

Chocolate & Black Pudding *See Black Pudding & Chocolate, page 42.*

Chocolate & Cardamom Like a puppeteer's black velvet curtain, dark chocolate is the perfect smooth background for cardamom to show off its colors. Use the cardamom in sufficient quantities and you can pick out its enigmatic citrus, eucalyptus and warm, woody-floral qualities. I find adding a pinch of ground cardamom can make even the most ordinary dark chocolate taste expensive. This tart is spectacularly delicious and very quick to make but needs a couple of hours in the fridge to set. Prepare and bake a 9-in sweet pastry shell. Slit open 10 cardamom pods, grind up the contents with a mortar and pestle, add to 1¼ cups heavy cream in a pan and scald. Remove from the heat and add 7 oz dark chocolate, broken into pieces, and 2 tbsp unsalted butter. Stir until melted and well mixed. When cooled a little (don't let it set), pour into the pastry shell and place in the fridge for two or three hours. When it has hardened, sift a little cocoa powder over it and serve with a modest dollop of crème fraîche.

Chocolate & Cauliflower *See Cauliflower & Chocolate, page 122.*

Chocolate & Cherry *See Cherry & Chocolate, page 243.*

Chocolate & Chestnut Charles Ranhofer, chef at Delmonico's restaurant in New York in the nineteenth century, used to fashion potatoes out of chestnut ice cream, with almond slivers for eyes, and the whole thing rolled in grated chocolate for an authentic muddy look. Tempting, perhaps, to make these and bury them in a deep soil of finely grated chocolate, then let your guests dig for the potatoes with spoons. For fear of friends thinking you've lost it completely, a chocolate sauce on a chestnut ice cream would be tasty, if less fun. If you're on a chestnut bender, use the egg whites left over from making the ice cream to make a Mont Blanc, the classic dessert of meringue topped with a mound of sweetened chestnut purée, a summit of whipped cream and a light dusting of confectioners' sugar.

Chocolate & Chili One of the original "wow" flavor pairings to have made its way around the world. As chilies turn red, they develop a sweet, fruity flavor that combines harmoniously with bitter chocolate—even more so when the chili is dried and has taken on still sweeter, raisiny, leathery notes. Look out for mulato and ancho dried chilies, which are considered inherently chocolatey themselves. Besides flavor compatibility, the fattiness of chocolate offsets some of the chili heat, as in a chili-rich Mexican *mole*. *Mole* simply means "sauce," and there are many different types. Most of them contain dried chilies, but as a rule chocolate turns up only in "red" or "black" *moles*. As well as chili and chocolate, these contain various dried fruits, bread, nuts, tomato, onion, garlic, seeds, dried and fresh herbs, spices, oil, lard and stock. As you might imagine, the result is a complex, sweet-piquant sauce that requires much pounding, grinding and toasting to prepare. Incidentally, the meat is either browned and added to the sauce to finish cooking, or cooked (usually roasted) separately and served with the sauce draped over it. Fresh *moles* are primarily confined to special occasions. If you fancy making one to an authentic recipe but don't have Mexican chocolate (which is coarse, dark and often blended with cinnamon and vanilla), the cookbook writer Rick Bayless suggests using a third as much unsweetened cocoa powder instead. Aside from *moles*, American spice shops sell prepared blends of cocoa, chipotle and paprika to add to chili con carne, stews or even cakes and brownies. You might also try adding a few pinches of dried red chili flakes to chocolate cornflake clusters—I call these mini Krakatoa cakes. The corn flavor is very harmonious with the chili and chocolate, and the crunchy texture adds to the fun. Get the basic recipe from a five-year-old. But best not to serve him or her the results.

Chocolate & Cinnamon *See Cinnamon & Chocolate, page 213.*

Chocolate & Coconut Just as government health departments warn that using marijuana can lead to harder drugs, so sweet tobacco led to my addiction to cigarettes. You could, with rice paper, make a rollie out of these strands of cocoa-flavored coconut. "Don't let's ask for the moon," I puffed at the dog, à la Bette Davis. From there it was only a few fake hacking coughs to the truly rank taste of real cigs. Crikey, I thought, as I inhaled my first. These are *terrible*. If, as I'd been led to believe, the cigarette companies were so dastardly, why didn't they learn something from the candy companies and make their products irresistible? I soon learned that they had, but it had nothing to do with the taste. Years later, I was back on the candy again, this time in the form of fancy chocolate with notes of tobacco and smoke. If you're looking for a hit, try Pralus's Tanzanie (tobacco, treacle, molasses, raisin) or their Vanuatu (smoke, spice, licorice). For a chocolate with notes of tobacco and coconut, try Michel Cluizel's Mangaro Lait 50%, a milk chocolate that, as the name suggests, contains a whopping 50 percent cocoa.

Chocolate & Coffee *See Coffee & Chocolate, page 24.*
Chocolate & Fig *See Fig & Chocolate, page 332.*
Chocolate & Ginger *See Ginger & Chocolate, page 303.*
Chocolate & Goat Cheese *See Goat Cheese & Chocolate, page 58.*

Chocolate & Hazelnut We have the scarcity of cocoa in late-nineteenth-century Piedmont to thank for the popularity of this heavenly combination. The bulking out of chocolate with ground hazelnuts led (eventually) to the invention of Nutella, although it was originally sold as a solid loaf and called *pasta gianduja*. *Gianduja*, which means something along the lines of "John the wandering man," is a carnival character representing the typical Piedmontese, and still the generic term for the sweet paste made from chocolate and hazelnut. In the 1940s mothers would cut a slice off the loaf, put it between slices of bread and give it to their children, who were smart enough to throw away the bread and just eat the chocolate. Piedmontese ducks must have been very fantastically plump mid-century. In 1951 a technique was developed to soften the mixture, and the product was renamed *Supercrema Gianduja* and sold by the jar. Finally, in 1964, its name was changed to the more internationally pronounceable Nutella, and today it outsells peanut butter worldwide. If you find Nutella too sweet, you might like to get your *gianduja* fix from a Ferrero Rocher or from Baci—or, if you prefer something a little more unusual, try Valrhona's Caraibe Noisettes or Amedei's milk chocolate with Piedmont hazelnuts. See also Nutmeg & Walnut, page 219.

Chocolate & Lemon *See Lemon & Chocolate, page 298.*

Chocolate & Lime Chocolate limes are a classic British sweet. In your mouth, the lime candy falls away in sharp, slatey layers to reveal a dry, crumbly chocolate center. Sadly the combination rarely crops up in other forms, although I once ate a spectacular dark chocolate tart with a sharp lime sorbet at one of Terence Conran's restaurants.

Chocolate & Mint *See Mint & Chocolate, page 321.*

Chocolate & Nutmeg Few recipes call for milk chocolate. It's more difficult to work with than dark chocolate, usually doesn't have as much cocoa flavor and is, in most cases, achingly rich. If you genuinely can't bear dark chocolate, consider a milk chocolate and nutmeg tart. The nutmeg boosts the flavor of the chocolate and freshens its cloying sweetness (nutmeg has a similar effect in creamy custard tarts and eggnog (see Egg & Nutmeg, page 133). The milk chocolate needs to be at least 30 percent cocoa solids. Follow the recipe in Chocolate & Cardamom, page 8, but scald the cream with ¼ of a whole nutmeg grated into it instead of the cardamom. Cool the chocolate a little before grating in an additional ¼ nutmeg. Taste for strength, then pour into the pastry shell and leave to set in the fridge. Grate over a little more nutmeg before serving.

Chocolate & Orange *See Orange & Chocolate, page 288.*

Chocolate & Peanut According to Alexandre Dumas, the Spanish called peanuts *cacohuette* because of their resemblance in flavor to cocoa. He goes on to note that they took advantage of this flavor harmony by mixing small amounts of expensive cocoa into a peanut mixture to make a sort of cheap chocolate. Fifty years later, in 1912, the Goo Goo Cluster, a mixture of chocolate, peanuts, caramel and marshmallow, became the first combination chocolate bar in the U.S. By the close of the 1920s, Reese's and Mars had respectively launched Peanut Butter Cups and Snickers, the latter becoming America's favorite chocolate bar, a position it holds to this day. Unroasted peanuts actually don't taste very good with chocolate, since (being legumes) they have a greenish, vegetal taste; the success of most peanut-chocolate combinations is down to the formation of pyrazines during the roasting process, which are harmonious with roasted notes in the chocolate. Use the combination at home for diner-style treats like a sundae made with vanilla ice cream topped with chopped, roasted peanuts and chocolate sauce, or a milkshake made with liquefied peanut butter and chocolate ice cream. Chef Paul Heathcote uses dark chocolate instead of the usual milk in his salty chocolate, caramel and peanut tart.

Chocolate & Pear A little chocolate will highlight pear's sweetness; too much and you swamp the fruit's flavor. *Poires Belle Hélène*—poached pears with chocolate sauce—is frequently a case in point. Too often a thick blanket of chocolate overpowers the dish, so use it sparingly and be sure to poach the pears in vanilla syrup to create a connection between the two flavors. Nuts fulfill a similar bridging role; pear and chocolate both love hazelnut, and the trio make a great cake. Or follow Nigel Slater's decadent tip of stirring broken-up florentines into whipped cream and spooning it into the cored cavities of poached pears. Use bought florentines or the recipe in Ginger & Chocolate, page 303.

Chocolate & Pineapple *See Pineapple & Chocolate, page 261.*

Chocolate & Raspberry Raspberry is a reflex pairing for chocolate tarts and puddings. Too often, in my humble opinion, berries are strewn on a chocolate dessert plate for no better reason than to pretty it up. All very well if there's enough cream to smooth the transition between the two, but if there's not, or if the raspberries aren't perfectly ripe, the combination is like being offered a soothing cuddle only to be pinched hard on the fleshy underside of your arm. A more balanced chocolate-raspberry experience is to be had in chocolate with strong raspberry notes, such as Valrhona's gorgeous Manjari or Amano's Madagascar.

Chocolate & Rose *See Rose & Chocolate, page 334.*
Chocolate & Rosemary *See Rosemary & Chocolate, page 309.*

Chocolate & Strawberry Not all it's cracked up to be. Strawberry's heartlike shape and color have seen it unimaginatively match-made with that default love token, chocolate. But doesn't a strawberry dipped in chocolate just look like a fruit wearing big underpants? And aren't they the sort of thing corporate raiders feed to call girls in cream-colored hotel rooms? I'd take chocolate and hazelnut over these two any day.

Chocolate & Thyme *See Thyme & Chocolate, page 317.*

Chocolate & Tomato A hint of chocolate flavor in spicy tomato recipes such as chili con carne, caponata, ketchup or meatballs is recommended by the American food historian Alice Arndt. Mexican cooks think of cocoa/dark chocolate as spices as well as sweet ingredients; for them, chocolate is a flavoring that, used in moderation, adds richness and depth to savory dishes and smooths the raw edges of sharp ingredients such as tomato.

Chocolate & Vanilla *See Vanilla & Chocolate, page 340.*

Chocolate & Walnut A classic in brownies. It's also worth throwing a handful of walnuts into a chocolate bread-and-butter pudding. Or add caramel to chocolate and walnuts to make what is sometimes called "turtle" flavor. Turtles are a popular candy in Canada and the United States. The name comes from the shape: a small pile of nuts (usually pecan or walnut) is held together with caramel and covered with a smooth shell of chocolate under which some of the nuts stick out like the head and legs of a turtle. Add a swirl of caramel to the recipe in Chocolate & Almond (see page 17) and you might call it turtle soup.

Chocolate & Watermelon *See Watermelon & Chocolate, page 245.*
Chocolate & White Chocolate *See White Chocolate & Chocolate, page 343.*

Coffee

Like chocolate, coffee goes through multiple stages before it reaches the cup, which accounts for its complexity. More than 800 aroma compounds have been identified in the roasted bean. Raw, the green seed is a relatively unfragrant little thing. Roasting expands it by 50 to 100 percent and, in turning it brown, releases its flavor. As a rule, lighter-brown beans (roasted for 9–11 minutes) are truer to their origins—i.e., they reveal more about the variety of bean and its growing conditions. When a coffee is darker (12–13 minutes' roasting time) and oily on its surface, roasted flavors will have begun to dominate, and it'll be spicier and more chocolatey. Coffee might contain notes of black currant, coriander seed, clove, vanilla, chocolate and nuts, all of which make harmonious matches, and are used to flavor the kinds of syrup you find in coffee bars. Tia Maria and Kahlua are both coffee-flavored liqueurs but have markedly different characters.

Coffee & Almond Once roasted, coffee is one of the most complex of all flavors, and one of its most desirable characteristics is "nutty"—which might explain why it's so frequently paired with nuts. Walnut (see Coffee & Walnut, page 25) is the most common nut note identified in coffee but almond (or marzipan) is also typical. Even if you don't detect almonds in your cup of joe, it'll still work a treat with a plump almond croissant covered with a generous dusting of confectioners' sugar. French women seem to be able to eat these without looking as if they've been caught in a snowstorm. I can't.

Coffee & Avocado *See Avocado & Coffee, page 196.*
Coffee & Banana *See Banana & Coffee, page 272.*

Coffee & Beef Caffeinated red meat. Something to serve your most militantly health-conscious friends. Why not add a garnish of lit cigarettes? There's a well-reported flavor overlap between roasted coffee and cooked beef, which might account for the combination turning up in fancier restaurants in recent years. Safe to assume cowboys eschew molecular gastronomy, but they have traditionally enriched their gravy with leftover joe. And coffee has long been a popular marinade or rub for meat in the American South. When I tried something similar at home, I thought the coffee gave my steak an overpoweringly gamy flavor and concluded the two were best left at least one course apart at dinner. Years later, in Lisbon, Portugal, I came across a dish called *bife à café*, popular in homely restaurants and neighborhood canteens, and had to revise my opinion. Thin-cut beef steak was served in a creamy, coffee-laced sauce, as if the chef had spilt his espresso in a stroganoff. It was pretty good—the cream placating the fiercer demeanors of coffee and beef—even if it took an effort of will to suppress the association with coffee-flavored cake frosting.

Coffee & Black Currant A mysteriously good pairing that often crops up in wine tasting notes. Once vinified, the rare Lagrein black grape, native to the Italian Alps, captures both flavors. I encountered them just over the border in Haute-Savoie in a heavenly *vacherin glace*: layers of meringue, black currant sorbet, whipped cream and coffee ice cream with a sprinkling of toasted almonds. It's in the running for the most delicious sweet thing I have ever put in my mouth. The coffee flavor had the fresh fragrance of just-ground beans and the black currant that hint of muskiness that processed fruit can't help but lose by oversweetening. Worth trying in a variant of pavlova (coffee-flavored meringue with cream and a just-sweet-enough black currant compote), or even black currant jam in a coffee gâteau.

Coffee & Cardamom The Bedouin way of preparing Arabic coffee. In Morocco, Algeria, or the Algerian Coffee Stores in Soho, London, for that matter, they'll grind together coffee beans and cardamom in equal measure (although obviously adjust proportions according to your preference). Drunk in tiny cups with or without sugar, maybe with a drop or two of orange-flower water, the aromatic spice rounds out the tartness of the coffee. I pair the flavors in this coffee-iced cardamom cake. The cardamom makes for a soft, fragrant sponge, while coffee icing lends a zingy contrast. Put ⅞ cup sugar, 1¼ cups self-rising flour, 1 stick softened butter, the seeds of 12 cardamom pods, ground to a powder, 1¼ tsp baking powder, 4 tbsp milk and 2 eggs in a large bowl and beat well for 2–3 minutes. Transfer to a greased and lined round 8-in springform cake pan and bake at 350°F for 50–60 minutes. Remove and leave to cool. Dissolve 2 tsp instant coffee in 2 tbsp hot water with a few drops of vanilla extract. Mix into 1 cup confectioners' sugar. The icing should be translucently thin, so add the liquid drop by drop until you've achieved an easily spreadable consistency. Spread the icing over the cooled cake. Note that the cardamom tastes stronger the next day.

Coffee & Cherry While the world fretted over who killed Laura Palmer in David Lynch's *Twin Peaks*, all I wanted was to slide up and join Agent Cooper in the Double R Diner for damn fine coffee and a slice of cherry pie.

Coffee & Chocolate Matari, or mocha, coffee beans come from the Yemeni city of Moka. They have a rich, mellow chocolate aftertaste and have lent their name to the various concoctions that attempt to reproduce their flavor by adding chocolate to less exotic coffees. Commonly this involves a shot of espresso in a cup of steamed milk and cocoa. Personally, I don't like this pairing in drinks. Part of the pleasure of coffee is feeling the sort of new-dawn determination I associate with Melanie Griffith taking the streets of Manhattan in her stride in the opening titles of *Working Girl*. Whereas hot chocolate I associate with Meg Ryan, in an oversize sweater, clutching a massive mug with both hands as if in supplication. Hard to be determined *and* vulnerable. Forget hot drinks. Coffee and chocolate work much better together in mousses, truffles and cakes. Or use them as uncredited flavor boosters. A little coffee flavor in chocolate dishes can make them taste more chocolatey, and vice versa.

Coffee & Cinnamon Cinnamon has the strength and sweetness to round out coffee flavor in baking. In cafés in Mexico they sometimes give you a stick of cinnamon to stir your coffee. Tastes good and saves on the washing up. One of Thomas Keller's more famous dishes pairs doughnuts dusted in cinnamon sugar with a coffee cup of cappuccino-flavored semifreddo.

Coffee & Clove See *Clove & Coffee, page 216*.
Coffee & Coriander Seed See *Coriander Seed & Coffee, page 337*.

Coffee & Ginger In the late seventeenth century, English coffeehouses served their brews black with optional additions of ginger, clove, cinnamon or spearmint. To this day in the Yemen, ginger is a popular flavoring for a tea brewed with coffee husks. The combination is called *qishr*, and is short and golden, with a bite to make you wince over your *bint al sahn*, or honey cake. An espresso lover's alternative to the milky gingerbread concoctions the coffee chains serve at Christmas time. See Ginger & Cinnamon, page 303.

Coffee & Goat Cheese See *Goat Cheese & Coffee, page 58*.

Coffee & Hazelnut If you find yourself at a good-quality ice cream parlor in France or Italy and you suffer an attack of selection anxiety, remember this: coffee and hazelnut, coffee and hazelnut, coffee and hazelnut.

Peanut & Apple An apple grated or sliced into a peanut butter sandwich is the worthy parent's alternative to jam or jelly. Righteousness aside, a sharp Granny Smith can provide a more refreshing contrast than sugary jam. Apple is particularly good with the cinnamon-raisin swirl peanut butter made by Peanut Butter & Co., and not at all bad with their maple-syrup-flavored one either. You can buy it in a jar, or try it at their peanut butter café in Greenwich Village, New York.

Peanut & Asparagus *See Asparagus & Peanut, page 129.*
Peanut & Banana *See Banana & Peanut, page 272.*

Peanut & Beef In common with Indonesian satay, in Peru the term *anticucho* refers to the process of cooking skewered meat over a grill rather than to the ingredients used. That said, ox heart is the usual meat of choice, although beef is becoming more prevalent. Either way, the meat is marinated in vinegar, garlic, chili, cumin and oregano, as it is in Bolivian *anticuchos*, but with the addition of thick peanut and chili sauce on the side. In both versions there may also be a few baby potatoes threaded onto the skewer. The combination of beef, peanut and potatoes may put you in mind of Thai *mussaman* curry. David Thompson says that this peanut-enriched sauce, thick with potatoes and meat (not always beef—lamb and duck are frequent alternatives), is the most laborious of all the Thai curries—and the most delicious.

Peanut & Black Currant *See Black Currant & Peanut, page 325.*

Peanut & Broccoli Crunchy peanut butter was made for broccoli, which catches the kernel crumbs in its unopened flowers. Put this to the test in a salad dressed like the one in Peanut & Coconut on page 27. Or, if you're good at chopping the florets into neat pieces, give them the *kung pao* treatment—see Peanut & Chicken, below.

Peanut & Carrot Nigella Lawson gives a recipe for "The Rainbow Room's Carrot and Peanut Salad," named in honor of the restaurant where her mother ate something similar. The mix of ingredients might, she admits, sound odd, but it works, particularly if you're brave with the vinegar: its astringency cuts through the oiliness of the nuts and, in combination with sweet carrot, gives the whole thing a mouthwatering quality that I associate with lime-juicy Asian salads. Coarsely grate 4 carrots and mix them with 3 oz salted peanuts, 2 tbsp red wine vinegar, 2 tbsp peanut oil and a few drops of sesame oil. Eat immediately.

Peanut & Celery *See Celery & Peanut, page 97.*

Peanut & Chicken The obsessively tidy may find comfort in the Sichuan dish of *kung pao* chicken, in which peanuts are paired with chicken, chilies and scallions cut into peanut-sized pieces. Stir-fried, everything bar the chilies takes on a uniform golden color, belying the contrast in the mouth of bland, soft chicken against crunchy, rich nuts in their spicy sauce. Cut 2 chicken breasts

Coffee & Orange Breakfast companions. If you can find it, San Matteo of Sicily makes a heavenly orange and coffee marmalade. I once had burnt orange and coffee ice cream, bitter as a custody battle but resolved by the sweetness of the cream. Orange and coffee tiramisu is also nicer than it sounds. You could even make it with this orange and coffee-bean liqueur. The recipe is adapted from one by Patricia Wells, who originally used eau de vie. I rather like the way, with marvelously arbitrary bossiness, it calls for 44 coffee beans, no more and no less. Take a large orange and make 44 slits in it. Put a coffee bean in each. It will now look like a medieval weapon or tribal fetish. Put 44 sugar cubes in a jar. Position the orange on top and pour over 2 cups brandy, rum or vodka. Leave it to steep for 44 days, then squeeze the juice out of the orange, mix it back into the alcohol, strain and pour into a sterilized bottle. Alternatively, put it somewhere dark and cool, forget about it completely, find it covered in dust something like 444 days later, try it skeptically, and realize on your second sip that it's absolutely delicious without the addition of the juice. Perfectly balanced, not too sweet, and with a complex lingering coffee/orange flavor that proves to be as good at rounding off a day as it is at starting one.

Coffee & Rose Bitter flavors can be used as an antidote to the overpowering sweetness of florals such as rose. I wouldn't normally relish a Turkish coffee, as thick and black as liquid tarmac, but a cube of rose-flavored Turkish Delight brings it into balance.

Coffee & Vanilla *See Vanilla & Coffee, page 340.*

Coffee & Walnut I've always thought that walnuts had a slightly nicotine character, which might explain why coffee and walnut is such a natural combination. Barring a ban on eating walnuts in public places—someone with a nut allergy might be nearby, after all—a partial substitute for a cigarette and coffee is a large slice of coffee and walnut cake. But it's hard to find a good one. The coffee sponge must be moist, the icing not too sweet, and the walnuts plenteous.

Coffee & White Chocolate *See White Chocolate & Coffee, page 343.*

Peanut

Raw peanuts taste beany. Roasted or fried, they take on a sweet flavor, with hints of chocolate and meat, plus a vegetal undertone. It's a highly complex and satisfying flavor experience that works well with rich meats, sweet shellfish and sharp fruits such as apple and lime. Ground into peanut butter, the flavor becomes sweeter and saltier with a pleasing edge of bitterness. Some say bacon goes with everything, but peanut is surely even more marriageable and, unlike bacon, it has huge international and cross-cultural appeal.

into half-inch squares and place in a mixture of 1 tbsp light soy sauce, 2 tsp Shaoxing wine, 2 tsp cornstarch and ½ tsp salt. Leave to marinate while you slice the white parts of 6 scallions into half-inch pieces, then do the same for 6 dried chilies (removing the seeds if you want to contain the heat). Combine 1 tbsp sugar, 1 tbsp black rice vinegar, 2 tbsp water, 1 tsp cornstarch, 1 tsp light soy sauce, 1 tsp oyster sauce and 1 tsp sesame oil, then set aside. Heat some peanut oil in a wok and add the chilies, taking care not to burn them. Add the chicken, brown it, then add 2 sliced garlic cloves, an inch of fresh ginger, finely chopped, and the scallions. Fry until the chicken is cooked through. Add the sauce and cook until thick and shiny. Then add a generous handful of unsalted roasted peanuts, give a quick stir and serve. You could use cashews instead, but peanuts are the more authentic choice. Chicken and cashew nuts has a special place in my heart—it was the first Chinese food I ever ate. When I came to describe it to my mother, my nine-year-old mind struggled to evoke the sheer unfamiliarity of it. How to describe a bean sprout? Or a water chestnut, with its odd, raw-potato crispness? Or the soft, salty gloop they came in? I'd never seen these things before. I *still* hadn't seen them, in fact, as we'd eaten our takeout in a car, in a parking lot, after dark, the windows steamed up with egg-fried mist. "The rice had peas in it," I managed. See also Peanut & Lamb, page 28.

Peanut & Chili *See Chili & Peanut, page 206.*
Peanut & Chocolate *See Chocolate & Peanut, page 20.*
Peanut & Cilantro *See Cilantro & Peanut, page 194.*

Peanut & Cinnamon Castries Crème is a peanut liqueur with a strong peanut-butter start, according to its maker, giving way to subtle notes of spicy cinnamon and brown sugar. See also Peanut & Apple, page 26.

Peanut & Coconut Frequently paired in Indonesian cooking. Used to pep up rice dishes, *seroendeng* is toasted grated coconut cooked with onion, garlic and spices, then mixed with peanuts. *Rempeyek kacang* is a deep-fried snack made with chopped peanuts folded into a spicy coconut milk and rice flour batter. Most famously, peanuts and coconut are combined in a sauce for satay, and in the dressing for *gado gado*, the classic Indonesian salad. Not to undervalue *gado gado* in any way, but it's a great recipe to have up your sleeve when you have vegetable odds and ends to use up. Roast 1⅔ cups skinned peanuts in the oven at 375°F for 6–8 minutes. Cool, then whiz in a food processor until finely ground. Add 3 tbsp soy sauce, 2 tbsp brown sugar, the juice of ½ lime, a couple of crushed garlic cloves (fried with 2–3 chopped shallots) and chili to taste. Whiz until smooth, add 1½ cups coconut milk and whiz again until all is well combined. Use this to dress a combination of cooked, blanched and raw vegetables. Cooled cooked potato, blanched green beans and raw bean sprouts are a common combination, variously supplemented with carrot, scallions, cabbage, cucumber or lettuce. Boiled eggs, shrimp chips, fried onion or tofu are used as a garnish. The dressing can be frozen.

Peanut & Cucumber In India, *khamang kakadi* combines peeled, diced cucumber, roasted crushed peanuts, finely chopped fresh green chili and grated coconut. It's dressed with lemon juice, salt, sugar and cumin, or sometimes an oil containing mustard seeds that's been heated until the seeds pop and release their flavor. Serve as a chutney.

Peanut & Grape *See Grape & Peanut, page 248.*

Peanut & Lamb In Bolivia they pair lamb with peanut in a soup or stew. In West Africa peanut and lamb might turn up in a *mafe* (see Peanut & Tomato, below) and in Thailand in a *mussaman* curry (see Peanut & Beef, page 26). Small pieces of lamb might be threaded onto a skewer, grilled and served with a peanut sauce in a Southeast Asian satay. The term "satay" refers to the technique of cooking food on a stick over charcoal, and doesn't imply any particular flavor combination. In the West, we have come to expect satay accompanied by a peanut sauce, but in Indonesia it might come served with *kecap manis* or a mixture of tomato and chili. As to the satay itself, anything goes: lamb, goat, chicken, beef, seafood, minced duck, ox offal, water buffalo, turtle, tofu, or pretty much anything you can thread on a stick.

Peanut & Lime Packets of peanuts flavored with lime, or lime and chili, are popular in Mexico, where you can also buy them fresh from a street vendor, fried in their skins in pork fat and scooped still hot into a cup with a squeeze of lime juice. Peanut and lime is also a very common pairing in Thailand and Vietnam, where both are used to garnish noodle dishes, soups and salads. Like the vinegar in Nigella Lawson's "Rainbow Room" salad (see Peanut & Carrot, page 26), the astringent juice provides a great counterpoint to the fattiness of the nuts.

Peanut & Mint *See Mint & Peanut, page 323.*
Peanut & Pork *See Pork & Peanut, page 40.*
Peanut & Potato *See Potato & Peanut, page 93.*

Peanut & Shellfish Satay sauce can turn a skewer of lean mussels or large shrimp into a rich treat. In the intensely flavored, hearty Brazilian dish, *vatapá*, shellfish is stewed in a peanut and coconut sauce thickened with bread. When combined, the shrimp and peanuts share a nutty sweetness that sits just on the satisfying side of overwhelming.

Peanut & Tomato Combined in the popular West African stew *mafe*. A *mafe* might be made with chicken, goat or beef, and a variety of vegetables, but the peanut and tomato sauce is a constant. The following is more of a rough guide than a recipe, but you'll get the idea. Cook a chopped large onion in oil until softened, then throw in a couple of finely chopped garlic cloves and some chopped red chilies. Add about 2 lb meat, jointed or cut into chunks, and allow to brown. Empty in a can of tomatoes, 2 tbsp tomato purée, a bay leaf, and ½ cup peanut butter whisked into 2 cups hot water or stock. Simmer for about an hour, then add a green bell pepper and some root vegetables (carrot, sweet potato, squash, yam—say about 4 large carrots' worth), chopped into bite-sized pieces. Continue to simmer until they're just tender. Give it a good stir and serve on rice, couscous or the more traditional millet.

Peanut & Vanilla I had a happy childhood, but it might have been a deal happier if snow-white, vanilla-flavored Marshmallow Fluff had been available in Hampshire. I could have demanded, or even made, fluffernutter sandwiches. One slice of (white) bread is spread with the fluff, the other with peanut butter, before they're pressed together and sliced. In season four of *The Sopranos*, Christopher Moltisanti asks his mother to make one and she refuses. Look how *he* ends up.

MEATY

Chicken

Pork

Black Pudding

Liver

Beef

Lamb

Chicken

Chicken has a reputation for being bland—the magnolia of foods—and yet standing up to 40 cloves of garlic (see Garlic & Chicken, page 112), or to big flavors like rosemary, thyme and lemon, takes some serious meatiness. The well-exercised joints—legs, thighs—are the tastiest, even more so when cooked skin-on and bone-in. It's the skinless, boneless breast meat, especially from intensively farmed birds, that has earned chicken its pale reputation. It's like a sort of dry tofu for carnivores. The best that can be said of it is that it adds bite to dishes and doesn't get in the way of more interesting flavors in a sauce—salty, sweet, nutty, fruity, spicy, even fishy. This chapter also touches on turkey, goose, quail and the odd game bird. And swan.

Chicken & Almond *See Almond & Chicken, page 238.*
Chicken & Anise *See Anise & Chicken, page 180.*

Chicken & Avocado Good together, if a little blandly healthy, like those smug couples you see jogging in the park. Give the chicken a smoke and things could start to look up. Or throw them some toasted pine nuts and a handful of raisins, toss through some leaves and dress with something sharp.

Chicken & Bacon *See Bacon & Chicken, page 165.*
Chicken & Banana *See Banana & Chicken, page 271.*
Chicken & Basil *See Basil & Chicken, page 209.*

Chicken & Bell Pepper One of the easiest, most foolproof combinations in this book. Seed 6–8 peppers (red, yellow or orange, not green), chop them into generous chunks and put in a large non-stick saucepan with 8 chicken thighs, skin-on and preferably bone-in too. Leave over a medium heat. Keep an eye on it for the first ten minutes, giving it the odd stir to prevent it sticking. Then all of a sudden the peppers release their juices and you can leave it alone. Put a lid on and cook over a low-medium heat for 30 minutes, or until the pan is half-full of sweet, oily, autumn-colored stock. It's a bit of a miracle, this—you can hardly believe the rich complexity of the sauce comes from just two ingredients. Season and serve with rice, couscous or French bread, whichever you prefer to mop up with.

Chicken & Blue Cheese *See Blue Cheese & Chicken, page 64.*
Chicken & Cabbage *See Cabbage & Chicken, page 118.*

Chicken & Caviar In Sylvia Plath's *The Bell Jar*, Esther Greenwood attends a smart luncheon where she hatches a plan to monopolize an entire bowl of caviar. If, she observes, you carry yourself with a certain arrogance when you do something incorrect at the table, people will think you're original rather than bad-mannered: "Under cover of the clinking of water goblets and silverware and bone china, I paved my plate with chicken slices. Then I covered the chicken slices with caviar thickly as if I were spreading peanut-butter on a piece of bread. Then I picked up the chicken slices in my fingers one by one, rolled them so the caviar wouldn't ooze off and ate them."

Chicken & Celery *See Celery & Chicken, page 96.*
Chicken & Chestnut *See Chestnut & Chicken, page 229.*

Chicken & Chili The Portuguese went to Mozambique and came back with chicken peri peri (or piri piri), a simple dish of flame-grilled chicken marinated in oil, chili, salt and citrus juice. *Peri peri* is a generic African word for chili, but usually refers to the hot, simply flavored bird's eye variety. Having taken to the dish themselves, the Portuguese exported it to their colonies, including Goa, where it's particularly popular. The peri peri diaspora has been accelerated in recent years by the South African chain, Nando's, which, spotting the mass-market appeal of chicken that can make tears run down your cheeks, has opened restaurants in five continents since 1987. See also Ginger & Chili, page 302, and Peanut & Chicken, page 26.

Chicken & Cilantro Cilantro is widely used in Thai chicken dishes such as green curry, and in Vietnam, *rau ram*, or "hot mint," is included in chicken salads and summer rolls. Unrelated botanically to cilantro, *rau ram* nonetheless has a similar, if slightly more peppery, citrusy flavor. In Malaysia it's known as the "laksa herb," after the noodle soup, called *laksa lemak*, that it's often used to garnish.

Chicken & Coconut *See Coconut & Chicken, page 280.*
Chicken & Egg *See Egg & Chicken, page 132.*
Chicken & Garlic *See Garlic & Chicken, page 112.*
Chicken & Grape *See Grape & Chicken, page 248.*

Chicken & Hard Cheese In the 1980s there was something of a vogue for chicken Cordon Bleu, a somewhat unbalanced dish of skinless, boneless chicken breasts stuffed with slices of Gruyère and ham. This recipe for *poulet au Comté* is an improvement, not least because the skin-on, bone-in roasted chicken has enough character to take on the weight and fruity, nutty, caramelized flavors of the cheese. Joint a chicken into 4 pieces and lightly dust with seasoned flour. Brown the pieces in butter, remove from the pan and keep warm. Deglaze the pan with 1¼ cups dry white wine and 2 tbsp strong mustard. Pour this sauce over the chicken in an ovenproof dish and bake for 40 minutes at 400°F, turning a few times. Sprinkle with 4 oz finely grated Comté and put back in the oven for about 5 minutes, until the cheese starts to brown. Serve with boiled potatoes or rice.

Chicken & Hazelnut *See Hazelnut & Chicken, page 235.*
Chicken & Lemon *See Lemon & Chicken, page 297.*

Chicken & Lime Citrus fruits are paired with chicken in cuisines the world over. I love the tang of lime juice in spicy chicken soups such as the famous *sopa de lima* of the Yucatán peninsula. Shredded chicken, chili and strips of tortilla are served in a chicken and tomato broth seasoned with cinnamon, garlic, allspice, and black peppercorns and finished with a generous squeeze of lime and some cilantro.

Chicken & Mushroom *Grifola frondosa*, or hen of the woods, is a species of mushroom named for its resemblance to a chicken ruffling its feathers. Chicken of the woods, *Laetiporus sulphureus*, which looks more like a flattened chicken nugget than anything, is perhaps the closest to actual chicken in texture, but opinion is divided as to whether the flavor bears much comparison. Add a handful or two of mushrooms to the pot with your chicken and they will contribute a gamy flavor that makes the bird taste as if it really did come from the woods, as opposed to the middle shelf of your refrigerator. Add them to a braising pheasant or partridge and you'll almost be able to hear the twigs snap underfoot. The pairing of morels with chicken in a cream sauce is altogether less rustic; morels are often said to be closer to truffles in their complexity and refinement of flavor. Like truffles, they come in white and black forms, both of which the late American food writer Richard Olney thought were "exquisite." He added that while dried morels have their uses in sauces and terrines, they can never quite measure up to fresh. The general feeling is that the drying process robs morels of some of their honeyed sweetness. Fresh or dried, black or white, morels must be cooked.

Chicken & Onion Brillat-Savarin wrote that "poultry is to the kitchen what canvas is to the artist." In its neutrality, the chicken finds common ground with the leek, a native not of Wales but, appropriately enough, of Switzerland, according to the *Encyclopaedia of Domestic Economy* (1855). Offend no one (but the vegetarians) with a cock-a-leekie soup—a traditional Scottish preparation of leeks, prunes and chicken stock.

Chicken & Oyster *See Oyster & Chicken, page 149.*

Chicken & Parsnip Roasted parsnips make a welcome side dish to roast chicken, and they're essential with the roast turkey at Christmas. Some cooks swear by the use of parsnip to flavor a really good chicken broth, although if you don't have a parsnip at hand, chef Robert Reid says a pinch of curry powder will improve your stock at a subliminal level, while some mushroom peelings will give it a meatier quality. As will chicken feet, which also give the finished article a pleasantly gelatinous texture. Easy and cheap to put to the test if you have a Chinese supermarket nearby.

Chicken & Pea *See Pea & Chicken, page 199.*
Chicken & Peanut *See Peanut & Chicken, page 26.*

Chicken & Pear Chicken and pear might not sound like much to sing about, but when braised together in good stock with bacon and shallots, partridge and pear is. The pear pieces absorb the rich cooking liquid but retain their subtle, fruity sweetness. For a Christmastime dinner, heat 1 tbsp oil in a hefty casserole and brown 4 small partridges. Set the birds aside and in the same pan cook about 5 oz chopped smoked bacon with 2 tbsp butter, 20 peeled whole shallots, 4 peeled, cored and quartered pears and a chopped garlic clove. When the shallots are golden and softened, return the partridges to the pan and pour over ⅔ cup hot chicken stock. Season, cover, and cook in the oven at 325°F for 20–25 minutes. You can serve this straight from the pot, but it's better to remove the partridges and rest them under foil while you put the casserole back on the heat and stir in 7 oz cooked, peeled chestnuts until they're warmed through. Five gold rings in the form of hot apple fritters would make the perfect pudding—see Vanilla & Apple, page 339.

Chicken & Potato In Antibes, on the Côte d'Azur, I lost my heart not to a lifeguard in a stripy swimsuit but to a humble, freestanding rotisserie. Strolling along the rue Aubernon, I was stopped in my tracks by a bizarrely beautiful contraption of black iron and brass, like one of Jean Tinguely's creaking kinetic sculptures, except it smelled of roast chicken. Rows of birds levitated in different stages of readiness, goosebump-raw to bronzed, reaching the top of their elliptical cycle before gravity turned the skewer and the birds adjusted to the downward part of their journey, with a jolt that shook free molten droplets of fat. The machine had a similarly hypnotic effect to the Penny Falls in amusement arcades, as if by pushing a raw chicken into a slot at the top a cooked one might be displaced into the tray at the bottom. In fact the tray at the bottom was full of potato chips. Could there be anything more delicious in the world? Chicken and roast potatoes, maybe, or Portuguese-style chicken and fries, but there was something particularly irresistible about the chips from the bottom of the rotisserie, chewy and glisteningly coated in the fat dripping from the chickens above.

Chicken & Rose Chicken served with rose petals, or rosewater, was popular in Moghul, Moorish and medieval English kitchens. In Laura Esquivel's novel *Like Water for Chocolate*, the heroine, Tita, brings her sister, Gertrudis, to an orgasmic boiling point with a dish made from rose petals and quail. Brillat-Savarin would *not* have approved. In his opinion, the flavor of quail was the most exquisite, but the most fugitive, of all game, and therefore to serve it any way other than plainly roasted or *en papillote* was nothing short of barbaric. If you subscribe to this view, you might find chicken an acceptable substitute in quail dishes that call for sauce; many cooks do. The psychologists Hollingworth and Poffenberger claim that with "tactual" qualities eliminated, most people fail to distinguish between the flavors of chicken, turkey and quail. But tactual qualities are not to be sniffed at. Would Gertrudis have had such a good time chomping through a skinless chicken breast as gnawing on a sticky little quail's leg?

Chicken & Saffron *See Saffron & Chicken, page 177.*

Chicken & Sage Sage, usually paired with onion in a stuffing or a sauce, bolsters the savory quality of chicken, although gamier turkey is better able to roll with the pungent herb's punches. Sage is also good with goose, as the bird has more fat, something with which sage has a particular affinity, as the writer Harold McGee points out. Goose stuffed with sage and onion and served with applesauce was the classic Christmas dish in the UK from the reign of Elizabeth I to the Second World War, although by the end of the Victorian era most people, particularly in the south of England, had switched to turkey. Victoria herself didn't care for either bird, preferring beef or a bit of roast swan. If you're curious, Peter Gladwin, once chef to Elizabeth II, compares swan's dark, tough meat to undernourished goose.

Chicken & Shellfish *See Shellfish & Chicken, page 138.*

Chicken & Thyme Thyme is often used to flavor roast chicken, pushed under the skin or into the cavity. Brining a chicken before roasting it, however, makes the meat juicier and lend an extra intensity of flavor, not only because the salt penetrates the meat but because the brine can be flavored with herbs, spices and/or vegetables. Put about 2½ oz sea salt and ¼ cup sugar in a pan with 2 cups of water and a dozen sprigs of thyme (or 1 tbsp dried thyme), heat gently till dissolved, then leave to cool. Add 6 cups of cold water to the salt water and refrigerate. When the brine is good and cold, wash the chicken and put it in a large roasting bag, big enough both to hold it and to immerse it totally in brine. Pour in the brine, then seal the bag, smoothing out as much of the air as possible. Leave in the fridge for 4–8 hours, moving it around from time to time. Give the chicken a thorough rinse in cold water before patting it dry. It can then be roasted as normal, immediately or after a day or two left covered in the fridge.

Chicken & Tomato Hard to get that excited about, once you've grown out of drenching your dinosaur-shaped chicken bites in tomato ketchup. Tomato and chicken are the controlling partnership in chicken tikka masala and in chicken cacciatore, or hunter's stew—which is not, sadly, the invention of pockmarked Sicilian peasants, returning home with a brace of feral chickens slung over their waistcoats, but an English recipe from the 1950s, taught to nice girls by their mothers in the hope they'd bag the sort of chap who'd be neither too unadventurous nor too suspiciously cosmopolitan to object to a lightly herbed slop of chicken in tomato sauce.

Chicken & Truffle *See Truffle & Chicken, page 116.*

Chicken & Walnut Ground nuts make an excellent basis on which to build a stew, contributing their own light, buttery background flavor as well as absorbing the rich flavors of meat and spice to make a thick, luxurious sauce. The popular kormas of northern India (and every curry house in the world) are based on ground almonds, cashews or coconut, an idea dating from Moghul times. The same principle applies to the Turkish dish of Circassian chicken. Poached chicken is shredded and served at room temperature in a sauce made with onions, garlic, ground walnuts, soaked bread and maybe some ground coriander or cinnamon. In Georgia, *satsivi* is made with chicken, fish or vegetables cooked in a mixture of walnuts and a long list of spices, including cinnamon, cloves, coriander, paprika and cayenne. Unlike Circassian chicken, no bread is used to supplement the nuts, but a sour flavoring such as vinegar or pomegranate juice will be added for balance—as it is in Iranian *fesenjan*, sometimes made with chicken but more often with duck. The combined tannic punch of walnut and pomegranate counters duck's fattiness deliciously. See also Almond & Chicken, page 238.

Chicken & Watercress *See Watercress & Chicken, page 99.*

Pork

Despite its prohibition by two of the major religions, pork is the world's most consumed meat. It tastes less salty and somewhat sweeter than beef. When roasted, good pork contains a tantalizing combination of woodland and farmyard flavors, which can be emphasized with garlic, mushrooms, cabbage, potatoes and robust herbs or contrasted with the sharp flavor of apple. Sweet anise, white pepper and cooked onions all tease out its savory meatiness. Remember that the color of meat is an indicator of how much exercise the animal has been allowed; when raw, fully flavored pork will therefore be a healthy rose pink. If it's the color of a student's dishcloth, it will undoubtedly have seen as much action.

Pork & Anise How long have you got? Three hours? Braise a 4 lb piece of pork belly in the aromatic stock ingredients given in Anise & White Fish, page 183: put the pork (in one piece) in an ovenproof pot with a tight-fitting lid, add 2 quarts of water, bring slowly to the boil and, at a low simmer, skim the scum until it desists. Stir in all the other ingredients, cover and transfer to the oven. Cook at 275°F for 2–3 hours. It is ready when the meat yields easily to a skewer. Set the meat aside while you simmer the sauce on the stovetop to reduce it. Serve the meat in pieces with white rice, the sauce and some sort of freshening garnish like shredded scallions or cilantro. One hour? Stud a pork tenderloin with garlic, brush with olive oil, sprinkle over 1 tsp crushed fennel seeds and roast in the oven at 425°F for about 30 minutes, maybe with some sliced fennel bulb underneath the meat. Half an hour? Make a pasta sauce of delicious sweet Italian sausages—see Pork & Tomato, page 41. Five minutes? Unwrap the lacy, rose-pink slices of fennel-dotted *finocchiona* that you bought to lay on home-made pizza and fold them into different shapes as you eat them all, one by one.

Pork & Apple *See Apple & Pork, page 267.*
Pork & Apricot *See Apricot & Pork, page 276.*

Pork & Bacon Bacon is like pork's older, more experienced brother. They team up in the full English breakfast (bacon, sausage), in French *choucroute garnie* (back bacon, pork knuckle, pork shoulder, salt pork and Frankfurt,

Strasbourg and Montbéliard sausages) and in the astonishingly delicious Asturian bean stew, *fabada*—see Black Pudding & Pork, page 43. Ruddy-faced bacon lends pale pork color, as well as a salty, brothy flavor. The pork in your pork pie would have a rather insipid city-dweller complexion if it wasn't mixed with bacon to give it its rosy flush: 1 part bacon to 3 parts pork shoulder works well. See also Pork & Egg, page 38.

Pork & Beef In *Goodfellas*, Vinnie tells us that for authentic meatballs three different types of ground meat are needed. "You got to have pork. That's the flavor." But beef's flavor too, albeit a darker, ferrous note against the lighter, sweeter background of pork. Veal is there for texture. In the United States, you can buy the trio premixed, often in equal measures, in the form of meatloaf or meatball mix. The flavor and texture of a beef stew can be enriched by the addition of pork rind—hold a little back when you're roasting a pork joint, even if that means less crackling. And pork lard is hard to beat when it comes to browning meat for a pot roast.

Pork & Beet See Beet & Pork, page 88.
Pork & Black Pudding See Black Pudding & Pork, page 43.

Pork & Broccoli Broccoli raab (or rabe), the broccoli lover's broccoli, is hugely popular in southern Italy. Some people say it's more mustardy and spicy than standard broccoli; I'd add that it has a ferrous tang and a salty hint of licorice. And with those hot and anise-like qualities, it's just perfect with Italian pork sausages (often flavored with chili and fennel seeds themselves). There are several ways to cook this dish. Sometimes the broccoli is cooked with the pasta, or everything is thrown into the same pan together. Some recipes stipulate that the sausages should be kept whole, others that the meat should be liberated from the casings, others still that the sausages be sliced into bite-sized pieces. I favor the last method, cooking the pieces slowly in olive oil. When some of the fat has rendered, I add chopped broccoli and cook it with the sausage pieces for 15–20 minutes. A tablespoon of hot pasta cooking water can be added to the sauce now and then to create a brothy juice that's soaked up by the florets. Serve with orecchiette pasta, so that, properly mixed into the sauce, the "little ears" of pasta fill up with the broccoli's tiny green flower buds like miniature salad bowls.

Pork & Butternut Squash See Butternut Squash & Pork, page 227.
Pork & Cabbage See Cabbage & Pork, page 119.

Pork & Celery Some people detect an anise or fennel quality in celery, and it certainly shares those flavors' affinity for pork. Celery seed is used extensively in charcuterie, liver sausages and meatloaf. The stalks of celery are braised with cubed pork, onion, white wine and stock, and thickened by *avgolémono* (see Egg & Lemon, page 133) in the simple, savory Greek stew, *hoirino me selino*. See also Prosciutto & Celery, page 169.

Pork & Chestnut *See Chestnut & Pork, page 229.*
Pork & Chili *See Chili & Pork, page 207.*

Pork & Cilantro In Portugal, cilantro is the most widely used herb, and the word *coentrada* on the menu means that the dish has been cooked with lots of the stuff. A popular recipe is pig's ears with cilantro and garlic, tossed in oil and vinegar and served cold. If pig's ears sound a little on the chewy side, consider cilantro's knack of cutting through fatty Asian dishes such as *roujiamo*, a Chinese hot meat sandwich usually served as a street snack. Originally from Shaanxi Province, it can involve various ingredients, but a common combination is braised pork stuffed into wheat flatbread with lots of cilantro and some bell pepper.

Pork & Cinnamon *See Cinnamon & Pork, page 214.*
Pork & Clove *See Clove & Pork, page 216.*

Pork & Coconut The Vietnamese dish *thit heo kho tieu* consists of caramelized pork slow-cooked in coconut water (not milk) and fish sauce. Boiled eggs are added toward the end. The Melbourne chef Raymond Capaldi offers a less rustic take on the combination—cold coconut noodles served with a neat brick of hot, gelatinous pork belly. The noodles aren't rice noodles cooked in coconut milk, as you might expect, but are *made* of coconut milk, chili oil and palm sugar, set with agar-agar. They're as thick as licorice bootlaces and white as fresh paint. The dish is garnished with typically Vietnamese herbs such as mint and cilantro and served with a *laksa* vinaigrette.

Pork & Coriander Seed *See Coriander Seed & Pork, page 338.*
Pork & Cucumber *See Cucumber & Pork, page 185.*

Pork & Cumin As lamb is often suited to flavors redolent of its habitat—grassy, herbal, *maquis*—so pork is complemented by the earthy flavor of cumin. Sprinkle ground cumin on an oiled pork tenderloin or chops before cooking, or let them bask in this glorious marinade. Mix 1 tbsp honey, 2 tsp ground cumin, 5 tbsp red wine, 2 tbsp olive oil and 1 tbsp red wine vinegar in a medium-sized freezer bag. Add the pork and muddle it around, then seal it and leave to rest in the fridge for a few hours before removing and cooking.

Pork & Dill *See Dill & Pork, page 188.*

Pork & Egg Come together in an English fried breakfast or a Sausage McMuffin, neither of which is a patch, visually, on a slice of gala pie. That slender keyline of hot-water pastry enclosing another of shimmering aspic, the mottled-pink rectangle of minced pork and bacon, the ovoid cross section of boiled egg white, the yellow circle of yolk. Maybe it's the Pop Art neatness that appeals, as I feel the same way about Scotch eggs. Good examples of either are hard to find. Best to make your own. For 4 Scotch eggs, mince together about 7 oz pork shoulder and 3 oz unsmoked streaky bacon and season. Divide into 4 and, with wet hands, form into balls and then into cups. Dust 4 peeled boiled eggs with flour and sit the fat end of each in a "cup." Work the meat around the egg, making sure there are no holes. Dip in beaten egg, then breadcrumbs and deep-fry for 7–8 minutes, turning a few times for even color. Little Scotch quail's eggs are a brilliant variation, but only if someone else is peeling the shells off. See also Black Pudding & Egg, page 42.

Pork & Garlic Pair pork with shameless amounts of garlic: they're made for each other. *Adobo*, a Filipino stew that has a claim to be the national dish, combines loads of garlic with meat—usually fatty pork—vinegar, soy sauce, bay leaves and peppercorns. You can make a quick, no-fuss *adobo* by piling all the ingredients into a pot, bringing to the boil, and simmering, covered, until the meat is tender—about 1½ hours. The following method involves a little more work, but the results will be considerably more flavorful. Marinate, for anything between a few and 24 hours, 1 lb pork shoulder or belly, cut into chunks, all but 4 cloves of a garlic bulb, crushed, 4 tbsp soy sauce, ½ cup rice vinegar, 1 bay leaf and 1 tsp freshly ground black pepper (for a slightly different but still delicious flavor, you can use inexpensive balsamic or Chinese black vinegar instead of the rice vinegar). Transfer to a pan, add enough water to cover, then bring to the boil, cover and simmer until the meat is tender. Strain, pour the broth back into the pan and boil until reduced to a thick gravy. Meanwhile, heat some peanut oil in a frying pan and fry the pork pieces, adding them back to the broth when they're crisp. Finally, crush the remaining 4 cloves of garlic, fry them until golden and add to the pot. Simmer for a few minutes, then serve with rice or, even better, *sinagong*—garlic fried rice. Simply fry cold cooked rice with crushed garlic, finely chopped shallots and a little soy sauce.

Pork & Ginger *See Ginger & Pork, page 304.*
Pork & Globe Artichoke *See Globe Artichoke & Pork, page 127.*
Pork & Grape *See Grape & Pork, page 248.*

Pork & Grapefruit Some years ago I was lying on the beach in Antigua irritated at the sunshine, the cool breeze in the palms, the white sand fine as castor sugar, and the volcano on distant Montserrat, still puffing after its eruption six months earlier, for the expectation they embodied that I was in paradise. And then a waft of jerk pork drifted over and I was. I turned onto my stomach, partly to stop it rumbling. A woman had set up an oil-drum barbecue in the shade of the beach's fringe of trees. Twenty minutes later I had a fresh roti in one hand, stuffed with hot pork fiery with allspice and Scotch bonnet peppers, and a can of ice-cold grapefruit Ting in the other.

Pork & Juniper *See Juniper & Pork, page 316.*
Pork & Mushroom *See Mushroom & Pork, page 81.*

Pork & Oily Fish "The eel goes very well with pork, because it is among fish what the pig is among quadrupeds," writes Norman Douglas in *Venus in the Kitchen, or Love's Cookery Book*, before giving a recipe for suckling pig stuffed with eels. As preparing a romantic dinner goes, wrestling an eel into a piglet has to give Annie Hall and Alvy's lobsters a run for their money. Douglas suggests stuffing a gutted 10–15-day-old suckling pig with thick pieces of boned, vinegar-washed eel, peppercorns, cloves and sage.

Pork & Onion *See Onion & Pork, page 110.*
Pork & Oyster *See Oyster & Pork, page 150.*
Pork & Parsnip *See Parsnip & Pork, page 220.*
Pork & Pea *See Pea & Pork, page 200.*

Pork & Peanut Peanut and pork are often combined to make the American take on *dan dan* noodles, but peanut is not part of the authentic Sichuan version. Mine is an inauthentic version of the inauthentic version, so should probably be called spicy pork and peanut noodles. Roughly chop 4 oz salted roasted peanuts. Heat 1 tbsp peanut oil in a frying pan and fry sliced dried red chilies to taste. Don't let them burn. Add 10 oz minced pork, mixed with 1 crushed garlic clove, and fry slowly until browned and cooked through. Drain off any fat and, back on the heat, sprinkle over 1 tbsp soy sauce and 2 tbsp light brown sugar. Stir in most of the chopped peanuts. In a separate pan, cook 4 sheets of fine or medium egg noodles in 3 cups of chicken stock laced with 2 tbsp soy sauce and 1 tsp sesame oil. Put a little chili oil in the bottom of 4 soup bowls. Using a slotted spoon, divide the noodles between them, add a few tablespoons each of stock and top with the pork mixture. Garnish with the rest of the crushed peanuts and some rings of fresh red chili.

Pork & Pear *See Pear & Pork, page 269.*
Pork & Pineapple *See Pineapple & Pork, page 262.*

Pork & Potato A salty, earthy pairing that gives us sausages and mash, frankfurters with potato salad, and a hearty Tuscan roast—see Rosemary & Pork, page 311. In Peru, they meet in *carapulcra*, a stew made with dried potato, chilies and peanuts. And in Korea, a soup called *gamjatang* combines pork backbone with potatoes and lots of spice, and is often served as a late-night corrective to a few too many beers.

Pork & Rhubarb Rhubarb chutney is a familiar accompaniment to pork chops but more recently the pair has been turning up in barbecue and Asian-style recipes. Rhubarb's sour fruitiness is apt to absorb complex spice mixtures and can cut through sweet, salty sauces.

Pork & Rosemary *See Rosemary & Pork, page 311.*

Pork & Rutabaga Rutabaga makes most sense mashed under a hunk of braised or roasted meat steeped in gravy. There's no point trying to make this dish refined; it's as coarse as a farmer cursing in his long johns. Try it with a hand of pork (the fatty shoulder—sweeter than the leg) or the pig's cheeks. Both are usually used for making sausages but also reward a long braising.

Pork & Sage *See Sage & Pork, page 314.*

Pork & Shellfish *See Shellfish & Pork, page 141.*

Pork & Thyme *See Thyme & Pork, page 319.*

Pork & Tomato The acidity of tomato makes a delicious contrast to pork's fatty sweetness. Cut into pieces and cooked in tomato, good Italian sausages (with fennel, if you can get them) will disintegrate into the liquid, giving you a *ragù* of a richness and depth you normally have to drum your fingers for two hours to achieve. Soften a little garlic in olive oil, then add 4 sliced sausages, allowing them to brown a little before adding a can of peeled plum tomatoes. Break the tomatoes up with a spoon and, while you're at it, use the tomato liquid to help scrape off the tasty bits of pork stuck to the bottom of the pan. Season and allow to simmer for about 20 minutes. Serve with pasta, polenta, bread or rice. In Thailand, tomatoes add a sweet element to *nam prik ong*, a cooked dip made with a fried paste of chili, lemongrass, shallot, garlic, shrimp paste, coarsely chopped tomato and minced pork, seasoned with fish sauce and palm sugar. It's eaten with raw or boiled vegetables, steamed rice or fried pork skin.

Pork & Truffle *See Truffle & Pork, page 117.*

Pork & Watercress *See Watercress & Pork, page 100.*

Pork & Watermelon *See Watermelon & Pork, page 246.*

Black Pudding

A good black pudding will have a velvety texture and a rich, mellow sweetness that should please fans of foie gras. Black pudding from Britain, *boudin noir* from France, *morcilla* from Spain or Argentina, *sanguinaccio* from Italy, *Rotwurst* from Germany and *kishka* from Eastern Europe may all be made with fresh blood, but the additional ingredients can vary from country to country, region to region, sausage maker to sausage maker. The blood in question is often pig's, which has a particularly fine flavor, but lamb's blood is also used, and in *The Odyssey* a feast is prepared of she-goat blood cooked in the animal's intestine. Oats, barley, cubed fat, rice, pine nuts, chestnuts, almonds, cream and combinations thereof are used to thicken blood puddings, and a blood sausage may also include meat or offal. The mixture is seasoned according to the sausage maker's blend of choice, stuffed into lengths of intestine or synthetic casings, then poached. Black pudding rubs along particularly well with autumnal flavors such as apple and sweet root vegetables, and is good used as a seasoning ingredient, enhancing the flavor of meat dishes.

Black Pudding & Apple *See Apple & Black Pudding, page 264.*

Black Pudding & Bacon There's more to the partnership of bacon and black pudding than their contribution to a breakfast fry-up. They can be used in a salad or a pilaf, or combined with pork in a pie or stew. See also Black Pudding & Pork, page 43.

Black Pudding & Chocolate A mixture of chocolate and cream is combined with blood to make the Italian black pudding, *sanguinaccio*. If that doesn't sound rich enough to begin with, it's often embellished with sugar, candied fruit, cinnamon or vanilla. *Sanguinaccio* is sometimes made into a sausage form, like other black puddings, but is also eaten (or drunk) while still in its creamier liquid state. This may bring to mind recipes for jugged hare, in which the animal is cooked in its own blood with dark chocolate.

Black Pudding & Egg Terence Stamp and Julie Christie. Very English. And, like Terry, black pudding consorts with the best of them these days, having left the confines of the greasy spoon for the swank expanses of Michelin-starred restaurants, where it hangs out with scallops and fancy salad leaves. If you're in a mood to make something fiddly, it's fun to fill ravioli with black pudding and a soft egg yolk that breaks when you cut into it. Serve with sage butter. For a more down-to-earth take on the combination, mix 1 part black pudding with 2 parts sausage meat for deliciously dark Scotch eggs. See Pork & Egg, page 38, for the Scotch egg recipe.

Black Pudding & Lamb A handsome, unpretentious dinner-party dish. Unroll a 1½ lb boned loin of lamb. Uncase a black pudding and carefully lay it the length of the meat, as if it were Cleopatra rolling herself up in a carpet. Roll it up, tie with string and present to Caesar. Or season and roast at 350°F for approximately 1 hour. Worth trying with pork too.

Black Pudding & Liver Grilled black pudding with foie gras has been on the menu at Andrew Pern's The Star in Yorkshire for over a decade. The foie gras is sandwiched between two slices of black pudding, topped with a slice of caramelized apple and served with a salad of watercress, apple and vanilla chutney and a scrumpy reduction. Pern named his 2008 book after the bestselling combination.

Black Pudding & Mint A few years ago I decided I'd learn more about cooking if I concentrated on the cuisine of one country a month. Breakfast, lunch and dinner, I'd cook nothing but Indian, or French, or Japanese food. Spain came first and *The Moro Cookbook* by Sam and Sam Clark saw me through most of a hot but cloudy June—weather that found its perfect match in *habas con morcilla*, fava beans with blood pudding. I associate black pudding with cold weather and hearty dishes, perky fava beans with bright, summery sweetness. Fresh mint has a foot in both camps. It's a summery herb that yearns for the cold; the flavor of shade. Fry ½ lb *morcilla*, cut into thick rounds, in 3 tbsp olive oil over a medium heat, until the contents burst from their casings like biceps from a bodybuilder's T-shirt. Set them aside and, in the same pan, cook a couple of thinly sliced garlic cloves and ½ tsp fennel seeds. When the garlic starts to turn golden, add 1 lb fava beans and ½ cup water. Cook for 3–5 minutes, until the beans are tender. Finally, return the *morcilla* to the pan to reheat and add a good handful of roughly chopped mint. Season and serve on toast.

Black Pudding & Onion Combined with potato, black pudding and leek make delicious patties. Boil 4 large potatoes until soft, then drain and mash them with seasoning but no butter or cream. While the potatoes are cooking, soften 3 or 4 thinly sliced leeks in butter. Fold the leeks into the potato with a generous pinch of mustard powder and a shake of white pepper. Finally mix in about 6 oz crumbled black pudding. Wet your hands and shape the mixture into 6–8 patties. If there's time, it's best to let them rest in the fridge for at least half an hour—this stops them breaking up in the pan. Lightly dust the patties with flour, then shallow-fry till they are thoroughly heated through (the pudding is already cooked) and tantalizingly browned on each side. Serve with spinach salad in a mustardy dressing, plus a glass of ale.

Black Pudding & Pork The next time a friend goes to Spain, tell them to forget about the straw donkey and bring you back a bag of *fabes*. *Fabes* are big white beans—bigger than butter or lima beans. Bigger, in fact, than a woodsman's thumbs. They're heaven, each a perfect mouthful of soft, beany purée flavored with whatever rich stock they've been cooked in. If your friend forgets, butter beans or limas will do fine for this Asturian dish called *fabada*. Soak 1 lb large, dried white beans in cold water overnight. The next day, put them in a roomy pan with a sliced Spanish onion and 3 chopped garlic cloves, cover with water and bring to the boil, skimming as necessary. Then add ½ lb whole *morcilla* and ½ lb chorizo sausages (having pricked them a few times to prevent them bursting), ½ lb pork belly slices and 3 smoked bacon slices. Pour in extra boiling water to cover and, once it starts bubbling, reduce to a low heat and simmer for about 3 hours. Top up the water if it drops below the ingredient line (which it will: this is a thirsty stew). Agitate the pan now and then to stop things sticking—stirring will break up your precious ingredients. They'll break up anyway, to an extent, but you don't want this too mushy. In the last hour, taste from time to time and add salt if necessary. Cut the meat into bite-sized portions and serve.

Black Pudding & Potato In Ireland you might eat black pudding with potato pancakes or with champ, a dish of buttery mashed potatoes and scallions. In a similar vein, German *Himmel und Erde* ("heaven and earth"), a combination of apple and potato, sometimes mashed, sometimes in pieces, is often served with black pudding. See also Black Pudding & Onion, above.

Black Pudding & Rhubarb Like apple, rhubarb's sharpness can cut through the richness of black pudding. A tart, quick-to-make rhubarb chutney is a case in point—see the recipe at Rhubarb & Oily Fish, page 251. Try this in a *morcipan*, grilled *morcilla* served on a roll, as sold from street carts in Argentina.

Black Pudding & Shellfish A modern classic. In fancy restaurants, pale scallop is often found perched, trembling like an ingénue, on filthy old black pudding's knee.

Liver

Liver has a recognizable flavor across animal species. Both its intensity of flavor and its level of tenderness or toughness depend to some extent on the animal's age, which is why calf's and lamb's liver are generally favored over ox liver. That said, a well-reared, grass-fed cow will usually furnish superior liver to that of an ill-kept calf pumped full of antibiotics. The liver is the clearinghouse for the body's toxins, and an animal that has experienced a lot of stress will have had to metabolize more compounds that detract from the ultimate eating quality of the meat.

Liver & Apple It's blood that gives liver its characteristic flavor. The liver is one of the few places in a mammal's or bird's body where a large amount of liquid blood collects. So it's perhaps not surprising that apple, which goes so well with rich, delicate blood sausage, should go well with liver too. In mainland Europe, poultry liver and apple are combined in mousses, pâtés and terrines.

Liver & Bacon I wonder if the key to this classic combination might be that liver is low in fat while bacon often has some to spare. Fry your slices of bacon until crisp and set them aside somewhere hot while you quickly fry the liver in the bacon fat. Calf's liver is often served with bacon in restaurants, but lamb's liver can be excellent too. If you're worried that the liver might be too strongly flavored, you can soak it in milk (as some do with anchovies) for an hour or so before cooking, remembering to pat it dry. And bear in mind that, like cabbage and eggs, liver's flavor intensifies with longer cooking.

Liver & Beef *See Beef & Liver, page 48.*

Liver & Beet Nutritionists say that beets are good for your liver. I say it's even better for the liver on your plate. Fergus Henderson notes that when he first tried venison liver he had expected it to be rather bitter and taste strongly of iron. In fact, he found it particularly sweet and delicate, and suggests it as a great match for roast beets.

Liver & Black Pudding *See Black Pudding & Liver, page 42.*

Liver & Chili Liver needs robust partners to set off its bold iron flavor. Try marinating chicken livers in an Indian sauce of hot paprika, mustard oil and yogurt, then grilling them over charcoal. Chef and writer Mridula Baljekar notes that while it's rare to find liver on the menus of Indian restaurants, it's frequently cooked in Indian homes. She coats chicken livers in a mixture of flour, chili, cumin and garam masala, then fries them with garlic in oil, serving them with browned fried onions and a tomato and cilantro sauce.

Liver & Fig *See Fig & Liver, page 332.*

Liver & Garlic As every parent knows, children are inherently conservative. As a kid, I implored my mother not to put egg sandwiches in my packed lunch, in order to avoid the merciless mime festival of pinched noses and dry retching that attended the release of their sulfurous odor, intensified after a morning brooding eggily in a plastic lunchbox. Okay: no egg sandwiches. Then, one day, I suppose after my parents had had a dinner party, I peeled back the corner of the lunchbox and filled the classroom with a heady mix of chicken liver, garlic, brandy and thyme, as if a gourmet giant had galumphed in and belched. I nibbled my sandwiches amid howls of derision from my classmates.

Liver & Oily Fish Red mullet is sometimes known as the "woodcock of the sea" as, like the woodcock, its liver is highly prized. It might be served separately as a pâté or pounded up into a sauce to accompany the fish. In Provence, it's sometimes used in the saffron and garlic rouille that accompanies a bouillabaisse. *Ankimo*, or monkfish liver, is one of the most revered *chinmi*, or rare delicacies, of Japan. Its velvety, creamy texture has a greater claim than even red mullet liver to be the foie gras of the sea.

Liver & Onion The sweetness of onion contrasts with liver's bitter, savory flavor. The combination is a classic all over the world. In Poland they fry pig's liver in breadcrumbed strips and serve them on a bed of fried onion. In the Philippines pig's liver is commonly fried with garlic, onions and a little pork. The English sauté lamb's liver with onions, or braise ox liver in a rich, mahogany-colored stew. The Indian dish of tandoori chicken livers, cooked on a skewer and served on a sizzling plate of onions, is in contention for the best take on the pairing, but for me the finest liver and onion dish of all is the Venetian *fegato alla veneziana*—lightly fried strips of calf's liver and meltingly soft, slow-cooked onions, served with polenta or rice.

Liver & Sage *See Sage & Liver, page 314.*
Liver & Truffle *See Truffle & Liver, page 116.*

Beef

The taste of beef is predominantly salty and umami-ish, with some sweetness and sourness (and bitterness if the meat is rare). Its flavor is clean, yeasty and meaty, with a slight metallic edge, and little of the animalic character so apparent in most pork and lamb. As with all meat, the flavor will to some extent depend on the species, how the animal has been reared, the cut and the cooking method. Grass-fed beef is fuller flavored than grain-fed. Most beef benefits from hanging, which gives it a deeper, gamier flavor. Beef works well with vegetables and shellfish, as do other meats, but it has a particular affinity for sharp flavors, notably horseradish and mustard.

Beef & Anchovy *See Anchovy & Beef, page 159.*
Beef & Anise *See Anise & Beef, page 179.*

Beef & Bacon Even bullish beef can benefit from the bolstering flavor of bacon. Cured meat brings a generic meatiness to a dish, not to mention some very welcome fat. The leanness of many beef cuts has always invited a good larding, and in French butcher's shops you can still see cuts of meat with the fat neatly stitched onto them. Roughly the same principle obtains when pancetta is added to spaghetti bolognese, or lardoons to beef bourguignon. In Germany, bacon is rolled between thin slices of beef with mustard and pickles to make *Rouladen*. Similarly, in Italy *saltimbocca* is made by laying prosciutto and sage leaves on very lean veal before cooking.

Beef & Beet In New England, red flannel hash is prepared with beets, corned beef, onion and cooked potatoes (diced or mashed). Fry it up for breakfast, with a fried or poached egg on top. Add some pickles on the side and you could call it *labskaus*, a dish that in various guises was common to northern European port cities such as Hamburg and Liverpool. In Hamburg, the dish might have salted herring mashed into it too, or come served with rolled herring as a garnish along with the egg and pickles. The related British dish, lobscouse, is more of a simple meat and potato stew than a hash, and piquant additions like pickled beets, red cabbage or onions are served on the side. See also Caper & Beet, page 102.

Beef & Bell Pepper *See Bell Pepper & Beef, page 202.*
Beef & Blackberry *See Blackberry & Beef, page 325.*
Beef & Blue Cheese *See Blue Cheese & Beef, page 63.*
Beef & Broccoli *See Broccoli & Beef, page 124.*
Beef & Cabbage *See Cabbage & Beef, page 118.*

Beef & Caper Not surprising that beef should have an affinity for caper, as the latter is a tropical relative of the mustard family and contains isothiocyanate, or mustard oil. Capers are known as *mostacilla* (little mustard) in Cuba, and *jeerba* (herb) mustard in Aruba and Curaçao. Make a dressing of olive oil, lemon juice, chopped parsley and drained rinsed capers for cold roast beef. See also Beef & Egg, page 48.

Beef & Carrot In the nineteenth century the English upper classes ate their carrots with rib of roast beef. The working classes made do with cheaper beef cuts such as brisket, which they ate salted and boiled. Carrots were thrown into the pot toward the end of the lengthy cooking process. A similar Jewish dish, sweetened with honey, syrup or dried fruit, is called *tsimmes*. In the 1850s, Thackeray acknowledged both the deliciousness and the popularity of the combination in his essay "Great and Little Dinners." Sixty years later, it was still popular enough to merit a music-hall song, "Boiled Beef and Carrots" by Harry Champion.

Beef & Celery *See Celery & Beef, page 95.*

Beef & Chili The *carne* in chili con carne is usually beef, although a combination of pork and beef is common too. The dish is said to have originated in San Antonio, Texas, in the late nineteenth century, when it was sold from cauldrons under lamplight by the famous "chili queens." Health and Safety did for them in the 1940s, but at heart chili (or at least a "bowl of red") still belongs to Texas. There is a predictably high level of debate about what constitutes an authentic chili, but most recipes are agreed on a spice mix of dried powdered chili, cumin and oregano. The pepper of choice is the smoky, dried ancho, with a blast of extra heat supplied by the infernal cayenne. An altogether different take on this pairing comes in the form of weeping tiger salad from Thailand, in which seared beef steak is served on a crisp salad with a chili, lime and fish sauce dressing.

Beef & Cinnamon Cinnamon, which is used in a lot of Greek meat dishes, is a key ingredient in *pastitsio*, a pasta dish somewhere between moussaka and *lasagne al forno*. Layers of macaroni are alternated with a spiced meat and tomato sauce, then topped with a thick béchamel. In Italy the term *pasticcio*, which translates (like *pastitsio*) as "mess," or "muddle," is used to describe any pie that contains a mixture of sweet and savory ingredients. Elizabeth David gives a recipe for *pasticcio* in which cooked spaghetti is layered with a beef ragù flavored with orange zest and cinnamon, then baked in a double crust of sweet pastry (the use of sweet dough isn't that unusual for savory pies in southern Italy). The Italian practice of spicing beef with cinnamon goes back to Roman times, and to this day is a feature of braised oxtail, *brasata di coda di bue*. They might put a little chocolate in it too.

Beef & Clove *See Clove & Beef, page 216.*

Beef & Coconut The Blue Elephant's *beef penang* was unforgettable, but The Fatty Crab's *beef rendang* knocked me sideways. Beef *penang* and *rendang* are both salty, sweet, slow-cooked coconut-based stews from Southeast Asia. *Rendang*, from Indonesia, is the more concentrated, all the water in the coconut milk having been evaporated until the meat is left to fry in the residual coconut oil, which makes the dish very intensely flavored. If cooking is halted while the dish is still wet, it's called *kalio*. This makes it more like the *penang*, which is a curry with plenty of sauce. Both dishes include shallots, garlic, ginger or galangal, chili and lemongrass, to which the *penang* adds lime leaves, cilantro root, fish sauce and lots of peanuts.

Beef & Coffee *See Coffee & Beef, page 23.*
Beef & Dill *See Dill & Beef, page 187.*

Beef & Egg Pending the inevitable advent of Health & Safety Directive 9/24675(F): Beef Tartare (Severe Danger of Death), you must order one *now*, this minute, before it's too late. Raw beef has a mild ammonia flavor with a hint of pepperiness and the gentlest hint of fish, as if at some point in its life the cow had yearned for the sea. The raw egg yolk beautifully emphasizes the meat flavor. How exposed your palate will be to these flavors will depend on the amounts and relative proportions of chopped anchovy, caper, shallot, parsley and mustard you choose to mix in. Even if you don't quite fancy the idea, it'll be something to tell the grandchildren, along with the one about driving without seatbelts and smoking in bars.

Beef & Garlic Garlic brings out the butch character in beef. Pair a rib-sticking roast beef with garlicky mashed potatoes, or make a sauce for thin slices of teriyaki beef by softening a fistful of peeled garlic cloves in beef stock and then puréeing it. Or simply push slivers into the nicks you've made in a joint, very much like an inverse of the game Operation, without the burden of that pesky rubber band.

Beef & Ginger *See Ginger & Beef, page 301.*
Beef & Hard Cheese *See Hard Cheese & Beef, page 67.*

Beef & Horseradish This most English of combinations may well have originated in Germany. In sixteenth-century England, horseradish was used for medicinal but not culinary purposes, although the contemporary botanist John Gerard noted that in Germany the root was used to accompany fish and meat dishes in the way that the English used mustard. The beef-on-weck sandwich, a specialty of Buffalo, New York, also has German roots—*Weck* is a German dialect word for "roll," and the specific type used is called *Kummelweck*, topped with a mixture of kosher salt and caraway seeds. The roll is stuffed with thin slices of rare roast beef and horseradish. Horseradish and beef can also be combined in a stew. Horseradish loses its pungency when cooked, so is added toward the end; used in the form of a creamed sauce, it gives a stroganoff-like effect, its sharp, slightly sour bite giving an edge to the rich beef sauce.

Beef & Juniper The Corsican dish *premonata* (or *prebonata*) involves beef, kid or goat served in a rich braised sauce of wine, tomatoes, peppers and juniper berries. As sinister as the backstreets of Calvi at midnight.

Beef & Lemon *See Lemon & Beef, page 297.*
Beef & Lime *See Lime & Beef, page 294.*

Beef & Liver Less fancy versions of beef Wellington prevail, but the real deal is fillet steak slathered in *pâté de foie gras* and shavings of fresh truffle, then wrapped in puff pastry. Putting it together is simple, cooking it less so. Left in the oven for too long, the meat begins to steam and shrink. Lose your nerve and take it out too soon and you'll have soggy pastry and undercooked beef; exactly what happens to Doris Scheldt in Saul Bellow's *Humboldt's Gift*, when she makes beef Wellington for her boyfriend, Charlie Citrine. Charlie's next girlfriend, Renata, plays it safe and sticks to serving champagne cocktails while wearing feathers and a G-string. Worth bearing in mind if you're having an off day in the kitchen.

Beef & Mint *See Mint & Beef, page 320.*

Beef & Mushroom A garnish of mushrooms is a steakhouse standard. Beef and mushroom are also paired in a stroganoff, or in a pie with a thick, booze-laced gravy. A few years ago, at the Fence Gate Inn in Burnley, Lancashire, the chef, Spencer Burge, caused a fuss by making a beef and mushroom pie that cost £1,000 (about $1,500) a slice. It consisted of 5 lb wagyu beef, 3 lb rare, cinnamon-scented Japanese *matsutake* mushrooms (so precious they were picked under the protection of an armed guard), black truffle, and that favorite ingredient of the honest British pie, gold leaf. Most connoisseurs recommend eating wagyu raw or rare truly to appreciate its quality. If stuffing it in a pie isn't enough to make them weep, the two bottles of Chateau Mouton Rothschild '82, reduced to half their volume for the sauce, should do it.

Beef & Oily Fish The famous Italian summer dish *vitello tonnato* consists of thin slices of cold cooked veal served with a tuna mayonnaise sauce. Marcella Hazan sounds a warning note: canned tuna is essential; fresh won't work. Chicken can be substituted for veal—Elizabeth David gives a recipe for *pollo tonnato*, so that makes it okay. The chicken might seem like a compromise but in flavor terms it has a strong harmony with canned tuna, as does cooked beef. According to *Sensory-Directed Flavor Analysis* by Ray Marsili, 2-methyl-3-furanthiol has one of the meatiest characters of all flavor compounds, and is very pronounced in cooked beef, chicken broth and canned tuna.

Beef & Olive *See Olive & Beef, page 173.*

Beef & Onion The precursor of the Philly cheesesteak sandwich, which was devised by Philadelphians Pat and Harry Olivieri as something a little different to offer on their hot dog cart, consisted of just finely sliced steak and onions in a soft white roll. The addition of provolone cheese came some years later. In the Japanese *gyudon*, steak and onions are likewise thinly sliced and fried, then a little soy, rice wine and water are thrown in to deglaze the pan and provide a sauce. It's served on rice. In the Lyonnaise dish of *grillade des mariniers*, thinly sliced rump steak is marinated in olive oil, red wine vinegar, bay leaf, orange rind and cloves. When it's had time to infuse, the meat is layered in a heavy pot with oodles of sliced onions and garlic, sprinkled with the marinade and braised for hours. To lend piquancy, a little anchovy and garlic, mixed to a paste with some of the cooking juices, is sometimes added in the last 15 minutes of cooking. And if you've ever tasted French onion soup made with veal stock, you'll never want it any other way.

Beef & Orange *See Orange & Beef, page 288.*
Beef & Oyster *See Oyster & Beef, page 148.*

Beef & Parsley St. John restaurant in London is famous for its dish of marrowbone with parsley salad. Middle veal marrowbone is roasted until the marrow is loose but not melted away, then served with a salad made with lots of parsley, a little sliced shallot and some tiny capers, dressed with olive oil and lemon. They give you some toast and a hillock of sea salt too. I once had a first date with a guy who ordered it. I quite fancied him to begin with, but the sight of him frowning at a bone, turning it over, and then over again, picking at it, smearing it on toast and adding pinches of salt like a geriatric pharmacist cast an irreversible pall over the evening. We might have been better off eating Argentinian: charred steak with *chimichurri* sauce is not only delicious but can be eaten with a knife and fork, and is a cinch to make at home. Finely chop a large bunch of flat-leaf parsley and mix it with 5 tbsp olive oil, 2 tbsp red wine vinegar, a crushed garlic clove (or two) and some seasoning. Serve on the side of a steak cooked as you like it.

Beef & Parsnip *See Parsnip & Beef, page 220.*
Beef & Pea *See Pea & Beef, page 198.*
Beef & Peanut *See Peanut & Beef, page 26.*

Beef & Pear Nashi pear and beef are combined in two popular Korean dishes, one with raw beef and one with cooked. *Yuk hwe* is a special-occasion dish of finely sliced raw beef tenderloin marinated in soy, sesame, garlic, scallions and chili, served with shredded Asian pear and maybe some pine nuts. As in beef tartare, a raw egg yolk might be stirred into the beef. *Pulgogi* consists of thin strips of beef marinated in grated Nashi pear, lemon juice, rice wine, sesame oil, sesame seeds, garlic, soy and sugar for a few hours and then fried quickly. It's eaten wrapped in small, crisp lettuce leaves with raw vegetables such as carrot, cucumber and radish.

Beef & Pork *See Pork & Beef, page 37.*
Beef & Potato *See Potato & Beef, page 89.*

Beef & Rutabaga One wintry afternoon we inserted the car into the herring-bone pattern of trucks and made our way under the low lintel to the striplit interior of the building. At high, Formica-topped tables drivers sat hunkered over their food, staring at us like dogs drowsily protective of their bones. There was no menu, and as the staff didn't seem to speak much English—or speak much at all—there was nothing for it but to point at what we wanted. It wasn't long before we, too, were hunched over our food, just as jealous of it as the wordless *routiers* patting flakes of pastry off their stomachs. Thin layers of rutabaga, turnip and potato lined the bottom of the casing; above that, a generous pile of tender beef, seasoned as it should be with white pepper. And just below the pastry lid lay a seam of soft onions, whose juice had seeped through the meat, tenderizing it all the way through to the strata of vegetables, so the whole pie was steeped in a tangy, earthy sweetness that set off the robust savoriness of the meat. So simple, yet so sensitive to the ingredients. Were we at a simply wonderful, if (thankfully!) little-known, Relais Routiers somewhere between Normandy and Burgundy? No, we were in a gas station off the A390 in Cornwall. We followed our Cornish pasties with the most marvelous pouch of Maltesers.

Beef & Shellfish *See Shellfish & Beef, page 138.*
Beef & Thyme *See Thyme & Beef, page 317.*

Beef & Tomato With a keen fisherman for a father, a great cook for a mother, a granny with her own miniature fruit orchard, and the cold hands of natural pastry makers, it was inevitable that my sister and I should be filled with lust for processed food. We nagged for orange-crumbed fish fingers, level as mantelpieces, and sawdusty cakes in lurid icing. When the Pot Noodle (like Cup Noodles in the United States) launched in the 1970s, I pestered and pestered until my mother gave in. One Beef and Tomato Pot Noodle was procured from the corner shop and the entire family huddled round, as a generation back they might have round an early television, to peer into the cup and witness the alchemical transformation from fish food into a hot meal. Mixing up the noodles, soy-based meat substitute, freeze-dried vegetables and bouillon powder with a sachet of tomato ketchup beat my mother's spaghetti bolognese hands down for entertainment value but offered an eating experience whose weirdly one-dimensional inauthenticity I dimly understood was half the point: a sort of slippery soup that tasted of barbecue-beef-flavor chips.

Beef & Truffle The French and Italian ways of serving truffles with beef say a lot about their respective cuisines. Tournedos Rossini calls for perfectly cooked steak and a slice of foie gras, served on a crouton with Madeira sauce. Black truffle is shaved over the top and usually added to the sauce too. The Italians might shave a little white truffle—which doesn't tolerate being cooked—over a plate of carpaccio. No frills or furbelows: they just let the ingredients speak for themselves.

Beef & Walnut Pickled walnuts became fashionable in England in the eighteenth century. They're preserved in vinegar when the nut is still green, before the shell has formed. If you're making your own, this means gathering the nuts in summer. They're lovely things: black as crude oil, with a slight beet-like resistance to the teeth, a mild piquancy and a flavor only subtly reminiscent of unpickled walnuts. They're frequently paired with Christmas foods—cold roast beef, blue cheese or leftover turkey. Chef Fergus Henderson makes a beef stew with pickled walnuts, cooked in red wine with lots of red onion, garlic, herbs and rich stock.

Beef & Watercress *See Watercress & Beef, page 99.*

Lamb

In the United States, lamb comes bottom of the flavor-preference table, behind beef, chicken, fish, pork, turkey and veal. Per capita consumption is only 1 percent that of beef or chicken. The flavor, even of young lamb, is considered by many to be too sheepy and gamy, and the use of flavor descriptors like "sweaty-sour" is enough to make anyone pause before taking a fork to a heavenly, subsiding lamb shank. The characteristic flavor is primarily in the fat—in blind tastings, many people find it hard to distinguish the flavors of very lean lamb and beef—and improves with hanging, even in the case of young lamb. Mutton, for its part, not only develops a deeper flavor when hung for two or three weeks but becomes tender enough to roast. One of the great pleasures of full-flavored lamb is its ability to stand up to other strong flavors—robust herbs such as rosemary, for example, or bold spices in curries and stews.

Lamb & Almond A luxurious pairing, fit for a feast, or a delicious meal for two. Scale down the Moroccan tradition of stuffing whole lambs with rice, spices and almonds to your humbler requirements by using a boned shoulder of lamb. Alternatively, pair lamb and almonds in a slow-cooked tagine. Or try my Moroccan-style lamb meatballs, cooked with spiced rice and garnished with toasted almonds, which boasts similarly fine flavors to a tagine but can be made in minutes rather than hours. Slowly soften an onion in oil and a little butter with a cinnamon stick. Meanwhile, mix 2 tsp allspice into 1 lb minced lamb. Season and form into walnut-sized balls. Steep a pinch of saffron threads in hot water. When the onion is soft, add 3 cups hot water, 2 tsp honey, 2 tsp pomegranate molasses, 2 tsp allspice, 1 tsp salt and the saffron water, then bring to the boil. Add the meatballs. When the water starts to bubble again, add 1 cup basmati rice, stir and simmer, covered, for 10 minutes. Check for seasoning and the rice's texture—you don't want it to turn mushy. When the rice is just cooked, stir in 6 tbsp chopped parsley, 3 tbsp chopped mint and a drained can of chickpeas. Heat through and serve with lots of toasted almond slivers and a little more chopped parsley or mint. This is, in essence, a thick soup, so needs no other accompaniment.

Lamb & Anchovy This has to be one of my favorite combinations. Anchovy works as a flavor enhancer for the meat. All you have to do is stick a knife into a shoulder or leg joint at various intervals, then push drained salted anchovies into the slits. I tend to use 8–12 anchovies for a 4 lb joint. You can always make more slits and press in some garlic and/or rosemary too— all feisty flavors that rub along very nicely together. Roast the lamb in the usual way. The anchovy melts into the meat and intensifies the flavor with a mouthwateringly rich, savory saltiness, and the gravy should be terrific.

Lamb & Anise Although the tarragon-sharpened flavor of Béarnaise sauce is most strongly associated with beef, it's perfectly suited to roast and grilled lamb too. And Pernod gives a warm breeze of anise to this summery lamb dish. Soften a diced onion and a little garlic in olive oil. Add 8 lamb loin chops dusted with seasoned flour and cook until browned. Add 3 tbsp Pernod and allow to cook over a high heat for 1 minute. Then add 3 small zucchini, cut into coins, and a can of peeled plum tomatoes, chopping them up in the pan with a spoon. Stir in some seasoning, 1 tsp herbes de Provence and 1 tbsp tomato purée. Bring to the boil, then simmer, covered, for 45 minutes, giving the odd stir. Serve with saffron rice or couscous.

Lamb & Apricot Go way back. *The Baghdad Cookery Book*, dating from the thirteenth century, features a slow-cooked lamb and apricot tagine called *mishmishiya*, which means "apricoty." Lamb and apricot both have an affinity with sweet spices, and while their sharpness cuts through the lamb's fattiness, the intense sweetness of the dried apricots throws the lamb, spices and almonds, all very sweet themselves, into a far more savory light, making the meat taste meatier. Meanwhile, the musky, rich apricots plump up with lamb stock, cinnamon, coriander and cumin, so by the time the dish is served they're neither too cloying nor too fruity. You might also pair apricot with lamb in a similarly spiced pilaf, or stuff lamb with a mixture of chopped dried apricots, onion, almonds and rice or couscous.

Lamb & Black Pudding *See Black Pudding & Lamb, page 42.*

Lamb & Cabbage *Farikal* is a popular Norwegian dish traditionally eaten around September to mark the passing of summer. It's simply mutton (or lamb) on the bone, simmered with cabbage, served with boiled potatoes and washed down with beer or aquavit. If this sounds stark, remember that lamb and mutton make boldly flavored, fatty stock. In Venice a (slightly) more festive combination of salted, smoked mutton and Savoy cabbage is called *castradina*. It's served on November 21 to celebrate the *Festa Madonna della Salute*, when locals offer up thanks for their health. The lamb and cabbage stand for the food sent across the Adriatic by their Dalmatian neighbors when its isolation during the plague threatened the city with starvation.

Lamb & Caper *See Caper & Lamb, page 102.*
Lamb & Cardamom *See Cardamom & Lamb, page 307.*
Lamb & Celery *See Celery & Lamb, page 96.*

Lamb & Cherry Broadly speaking, cherry flavor divides into two groups: sweet and sour. The sour cherry is too sharp to eat from the tree but has more flavor than the sweet variety, making it better for cooking. Morello (called *griottes* in France) and Montmorency are famous sour cherries, cropping up in lots of Middle Eastern, Russian and Eastern European dishes. Claudia Roden gives a recipe for shoulder of lamb with rice stuffing and sour cherry sauce, remarking that in the West the cherry has yet to become as popular a pairing for lamb as apricot. A rich mutton and potato stew with sour cherries is made in Azerbaijan, while in Turkey you might find sour cherries and lamb in a pilaf accompanied by any combination of onions, saffron, almonds, pomegranate, feta, mint, parsley and pistachios.

Lamb & Chestnut *See Chestnut & Lamb, page 229.*

Lamb & Cilantro *See Cilantro & Lamb, page 192.*

Lamb & Cinnamon *See Cinnamon & Lamb, page 213.*

Lamb & Cumin In *Shark's Fin and Sichuan Pepper*, Fuchsia Dunlop writes about the Uighur migrants of Chengdu in southwestern China, who sell salty lamb kebabs flavored with chili and cumin, cooked on portable grills they set up on the street. They sell hash, too, and you could tell when the police were having a crackdown, she recalls, because the fragrance of sizzling lamb and cumin was abruptly conspicuous by its absence from the streets.

Lamb & Dill *See Dill & Lamb, page 187.*

Lamb & Eggplant Eggplant was, I can only imagine, designed with lamb in mind. Its kitchen-towel propensity to soak up fat and juices suggests as much. Moussaka is the most obvious application, but there's also *hünkar beğendi*, a Turkish dish of cubed lamb served with an eggplant purée, and *patlican kebabi*, wherein lamb and eggplant are simply cubed, threaded on sticks and cooked over the grill—or, if you're lucky, whole eggplants are stuffed with minced lamb, cooked until the skin is wrinkled and the flesh yieldingly soft, then served in slices with thick white yogurt on the side.

Lamb & Garlic *See Garlic & Lamb, page 113.*

Lamb & Globe Artichoke *See Globe Artichoke & Lamb, page 127.*

Lamb & Goat Cheese Goat cheese and lamb have, unsurprisingly perhaps, a great flavor affinity, and have defied what might seem their unfashionable level of richness to become an increasingly popular combination. They're paired in warm flatbread wraps and two-bite filo pastry parcels. Goat cheese and feta are crumbled over lamb pasta sauces and pushed inside lamb burgers. And moussaka made with a goat cheese béchamel is a rich treat, although for light relief you might serve it with baby spinach leaves tossed in a lemony dressing.

Lamb & Lemon *See Lemon & Lamb, page 299.*

Lamb & Mint The French say *bof* to the Brits' love of mint with lamb, and they might have a point when it comes to the brutally vinegary strains of mint sauce. In 1747 Hannah Glasse wrote that a roasted, skinned hindquarter of pork will eat like lamb if served with mint sauce, which must have more to do with the overpowering nature of the sauce than any true similarity between the meats. But mint as a partner for lamb should not be dismissed wholesale. Lamb has a natural affinity for herbal flavors and, like citrus, mint's cleansing properties serve the useful purpose of deodorizing some of lamb's funkier notes. Consider, for example, the lamb and mint ravioli served at Mario Batali's Babbo restaurant in New York. Or *sauce paloise*, which is like Béarnaise but swaps the tarragon for mint and is served with roast or grilled lamb. And in Azerbaijan a minted soup called *dusbara* is served with teeny lamb-filled tortellini bobbing in it. It's most often garnished with sour cream and garlic, although some prefer vinegar and garlic, which sort of takes us back to where we started.

Lamb & Nutmeg *See Nutmeg & Lamb, page 218.*
Lamb & Onion *See Onion & Lamb, page 109.*

Lamb & Pea Peas act as apt reminders that lamb, not yet at its best by Easter, starts to develop its true lamby flavor in early summer, when the peas will be straining from their pods like glimpses of belly in an undersize shirt. Lamb and fresh peas, grassy as the fields the lambs have nibbled, are a gorgeous combination, whether you roast the lamb and boil the peas with earthy new potatoes or make *agnello alla romagnola*, lamb cut into pieces and stewed in butter with pancetta, tomato and peas, Emilia Romagna style. For minced lamb, try a spicy, aromatic Indian *keema* curry, polka-dotted with peas and served with rice or chapatis.

Lamb & Peanut *See Peanut & Lamb, page 28.*
Lamb & Potato *See Potato & Lamb, page 91.*
Lamb & Rhubarb *See Rhubarb & Lamb, page 250.*
Lamb & Rosemary *See Rosemary & Lamb, page 310.*

Lamb & Rutabaga Walk into a room where a haggis is being cooked and the first thing to hit you will be the aroma of lamb. Of all meats, lamb is the easiest to identify by cooking fragrance alone, as the meat contains chains of fatty acids not found in chicken, beef or pork. And haggis is intensely lamby. It's made from the "pluck"—the animal's liver, heart and lungs—mixed with oatmeal, suet, pepper, allspice, clove and nutmeg. Neeps (rutabaga) and tatties (mashed potatoes) are the essential side dish. The peppery sweetness of the neeps chimes with the haggis's own spiciness, and makes the meal worthy of its arrival to a flourish of bagpipes. See also Potato & Rutabaga, page 93.

Lamb & Saffron Iranian saffron is sweeter in character than the Spanish variety. Iran is the world's biggest producer of saffron by some way, and so it's no wonder that its national dish, *chelow kebab*, consists of lamb or chicken kebabs and saffron rice, sometimes accompanied by charred tomatoes and maybe a raw egg yolk and a sprinkle of citrusy sumac. The steaming, fragrant rice is usually flavored with a pat of saffron butter that melts into it like solid sunlight. Make some by grinding 30 saffron threads with a mortar and pestle, then adding it to 1 tsp hot water and putting it aside while you zest a lemon. Mix the lemon zest, saffron water and a little lemon juice to taste with 1 stick softened butter. Shape into a log, wrap in cling wrap, place in the fridge and use in slices as and when needed.

Lamb & Shellfish *See Shellfish & Lamb, page 140.*
Lamb & Thyme *See Thyme & Lamb, page 318.*
Lamb & Tomato *See Tomato & Lamb, page 255.*

CHEESY

Goat Cheese

Washed-rind Cheese

Blue Cheese

Hard Cheese

Soft Cheese

Goat Cheese

The flavors in goat cheese range from the lightest and most citric to the fullest Billy Goat Gruff. Its sharpness works well with sweet, dense ingredients like beets, butternut squash, good bread and honey. Traditionally feta and halloumi are made with sheep's milk, or a combination of sheep and goat's milk. They are included in this chapter because they share many flavor affinities with goat cheese. Like all cheeses, the flavor of goat cheese is markedly influenced by what the animal has eaten. In a sensory evaluation study conducted in 2001, over two-thirds of the tasting panel correctly identified which one-day-old goat cheese had come from pasture-grazed goats and which from animals fed on hay and concentrate. For the twenty-day-old cheese, the figure rose to 100 percent.

Goat Cheese & Anise *See Anise & Goat Cheese, page 181.*

Goat Cheese & Apricot A great stand-up kitchen snack. Stuff a plump dried apricot with a modestly pungent goat cheese and note just how meaty the combination is. The sweet, perfumed fruitiness of apricot emphasizes the savoriness of the cheese, and the whole somehow recalls lamb. Goat cheese, in common with cooked lamb and mutton, contains caprylic acid, which may account for the similarity in flavor.

Goat Cheese & Basil *See Basil & Goat Cheese, page 210.*
Goat Cheese & Beet *See Beet & Goat Cheese, page 88.*

Goat Cheese & Blackberry *Banon*, or *banon à la feuille*, is a French goat cheese that's dipped in brandy before being wrapped in chestnut leaves. Once it's wrapped, it will be ready to eat in about three weeks, but leave it a little longer and it will develop rich, fruity, woody notes that complement blackberry right down to the hairs between its drupelets.

Goat Cheese & Butternut Squash *See Butternut Squash & Goat Cheese, page 227.*

Goat Cheese & Caper Australian agricultural writer E. A. Weiss writes that the flavor of caper is mainly attributable to capric acid, which develops after the buds are pickled. In its pure form, capric acid is pungently goaty, but the tiny amount in capers is modified by the pickling process. Capric acid also makes a significant contribution to the flavor of goat cheese—a soft goat cheese will work well in the recipe suggestion under Caper & Soft Cheese, page 103.

Goat Cheese & Cherry *See Cherry & Goat Cheese, page 244.*

Goat Cheese & Chili The Spanish habit of giving everything a liberal dusting of heat extends to the Canary Islands, where one of the three versions of the local Majorero cheese comes rubbed with *pimentón*. Texturally similar to Manchego (the popular Spanish ewe's milk cheese from La Mancha), Majorero is a firm, white cheese made with fatty goat's milk from Fuerteventura. Arico, from neighboring Tenerife, is rubbed with a combination of *pimentón* and *gofio*, a toasted cereal, and was declared supreme champion at the World Cheese Awards in 2008.

Goat Cheese & Chocolate At a tasting of cheese and chocolate pairings held by chocolatier Paul A. Young, we kicked off with a ten-day-old Cerney, a goat cheese from Gloucestershire, paired with an Amedei 63 percent cocoa dark chocolate, moving on to a matured version of the same cheese with a 64 percent Valrhona Madagascan Manjari. Chocolate with cheese may sound an unlikely, even offputting, proposition, until it strikes you how comparable it is to chocolate and milk (or cream) and, further, how the flavor notes in some good chocolate—spicy berry fruits, dried fruits, caramels—are natural partners for cheese. We also tried Colston Bassett Stilton with Valrhona 70 percent, following the precedent of Young's Stilton and port truffles, which were originally conceived as a seasonal special but proved so popular that they're now on the menu year-round. But the revelation of the night was the pairing of Milleen's, a deliciously piquant, floral washed-rind Irish cheese made from cow's milk, first rind-on with an 85 percent Valrhona African, then rind-off with a Valrhona 40 percent Java milk chocolate. Try adding a few types of chocolate to your cheeseboard at the end of a meal—if your guests really can't face combining chocolate and cheese, they can always eat them separately. And both will work with port.

Goat Cheese & Cilantro In Mexico the fresh, citrusy flavor of cilantro helps to cut through the fattiness of cheese in the ubiquitous enchiladas and quesadillas. Mexican cheeses are hard to find beyond the Americas, but Rick Bayless says that *queso fresco* is not unlike fresh goat cheese, if a little drier, saltier and crumblier. Parmesan or (even better) pecorino romano can be substituted for *queso añejo*. And for the Chihuahua (the cheese, not the dog) that's used in quesadillas, try mild Cheddar or a Monterey Jack. Or you could follow Bayless's suggestion and mix cream cheese with goat cheese, scallions, a little salsa and chopped cilantro, and spread it onto bread with some sliced tomatoes.

Goat Cheese & Coffee Coffee with cheese might sound even less promising than chocolate and cheese; but then Norwegian *ekte gjetost* is only cheese in a manner of speaking. It's made with whey left over from the cheese-making process, cooked until the lactose caramelizes, then poured into rectangular molds. Left to cool, the smooth, sweet, fudge-colored, caramel-flavored results are eaten in thin slices on toast or crispbread with morning coffee. It's also served with fruitcake or used in a sauce for game. A cow's-milk *gjetost* is popular too, but goat's-milk is the more traditional version.

Goat Cheese & Coriander Seed *See Coriander Seed & Goat Cheese, page 337.*

Goat Cheese & Cucumber *See Cucumber & Goat Cheese, page 184.*

Goat Cheese & Fig *See Fig & Goat Cheese, page 332.*

Goat Cheese & Garlic A goat cheese and garlic pizza has been on the menu at London's Orso restaurant for over twenty years. The sweet flavor of the garlic takes the edge off the sharpness of the cheese. It's quite a mouthful, but the pizza base is light and crisp and the whole thing small enough not entirely to preclude a *secondo*. Make something similar by spreading a small amount of the tomato sauce in Garlic & Basil, page 112, on a pizza base and decorating it with thin slices of goat cheese. While the oven is warming up, put some unpeeled garlic cloves in it on a roasting tray; by the time the oven is hot, they'll be lightly cooked. Scatter them on the pizza and bake it at 450°F for 10 minutes. If your tolerance for outlandishly pungent flavors is high, you may enjoy *Foudjou*, a potted French combination of fresh and older grated goat cheese with garlic, herbs, brandy and olive oil. It's kept in a crockery jar and left to mature for a few months, after which it's eaten spread on French bread or baked potatoes.

Goat Cheese & Lamb *See Lamb & Goat Cheese, page 54.*

Goat Cheese & Lemon *See Lemon & Goat Cheese, page 299.*

Goat Cheese & Mint Authentic halloumi cheese is made from a mixture of goat's and sheep's milk. Mint is often added at the dry-salting stage; you may taste a hint of it, in addition to a citrusy flavor, in the end result. Like halloumi, feta is a brined cheese, whose saltiness benefits from the cooling contrast of mint. Mash feta with chopped mint, snipped chives and a grinding of black pepper, fold into little filo parcels and bake. On a sweeter note, in Ibiza a cheesecake-like pudding called *flaó* is traditionally made with sweetened fresh goat cheese, mint, anise and honey.

Goat Cheese & Mushroom *See Mushroom & Goat Cheese, page 79.*

Goat Cheese & Olive *See Olive & Goat Cheese, page 173.*

Goat Cheese & Pear Grazalema is a goat (or sheep) cheese from the sunbaked province of Cádiz. It's a "wolf in sheep's clothing," according to *The Murray's Cheese Handbook*, which finds "sweet ripe pear and nectarine notes that remind us of chocolate-dipped fruit." Which should be enough to sell you the cheese *and* the book. A buttery Chardonnay is the recommended wine pairing. Or you might try a manzanilla sherry.

Goat Cheese & Raspberry Raspberries combine well with young cheeses that still have their milky, lactic tang and a streak of citrus. Try them together in a fool. Crush 2 cups fresh raspberries. Whip ¾ cup heavy cream to soft peaks. Whisk 5 oz soft, fresh goat cheese with 1 tbsp confectioners' sugar and a squeeze of lemon until soft. Fold into the cream with the raspberries. Check for sweetness, and divide between 4 bowls.

Goat Cheese & Rosemary Goat cheese and lamb tend to share flavor affinities, including rosemary. Perroche, from Herefordshire, is a soft, lemony, unpasteurized goat cheese that comes rolled in either rosemary, tarragon or dill. It's worth trying all three, but rosemary with gently citrusy cheese is a particularly winning combination. Or make this goat cheese and rosemary tart. Line an 8-in flan pie pan with shortcrust pastry and blind-bake it. Cook 3 or 4 sliced leeks in olive oil until soft, season, then leave to cool a little. Spread them over the pastry. Whisk ⅓ cup half-and-half with 4 oz soft, rindless goat cheese. When well combined, beat in 2 eggs, 1 egg yolk and 1 tsp very finely chopped rosemary, and season. Pour over the leeks, top with milk (about ½ cup), and bake at 375°F for 25–30 minutes.

Goat Cheese & Thyme *See Thyme & Goat Cheese, page 318.*

Goat Cheese & Walnut There's a Persian place called Patogh off the Edgware Road in London that just might be my favorite restaurant anywhere. It serves *chelow* kebabs (see Lamb & Saffron, page 55) and enormous, cratered moonscapes of bread, hot from the oven and scattered with toasted sesame seeds. Walk in the door and the aroma of charring meat and baking bread will make the temptation of a starter irresistible. *Paneer* is a white tablet of feta, as smooth as a bar of Ivory soap and usually scattered with crisp walnuts. It's generally accompanied by *sabzi*, a thicket of fresh herbs, to offset its richness. There'll be plenty of mint, plus tarragon and dill, bulbous scallions and, nestled somewhere among the sprigs and leaves, little radishes, like baby robins in their nest. Quite apart from the heavenly contrast of freshness and salt, the chance to forage and eat with your hands makes this a fun and very easy appetizer or lunch to serve at home.

Goat Cheese & Watercress *See Watercress & Goat Cheese, page 100.*

Goat Cheese & Watermelon Beautiful combined in a salad. Beets are perhaps a more familiar partner to briny feta, but watermelon is a refreshing pretender to beet's earthier, more substantial sweetness, and has, in fact, been a common fixture in Greek salads since time immemorial. Bear in mind that both the salty cheese and olives, if you're using them, will draw juice out of the watermelon, so add the fruit just before serving to prevent the salad from getting soggy. You might also consider how good goat cheese would be with (or in?) the barbecued watermelon in Rosemary & Watermelon, page 312.

Washed-rind Cheese

Washing cheese in brine and other solutions dates back to the Middle Ages, when European monks discovered that the practice stimulated the growth of bacteria, which gave the cheese less acidic, more pungent flavors, thus making abstention from meat a lot more bearable. In research conducted at Cranfield University in the UK, the beery washed-rind Vieux Boulogne was declared the world's smelliest cheese, outstinking even brandy-washed Époisses, which is banned on public transportation in France. As a rule, washed-rind cheeses are best left in the quiet company of bread or crackers, but they also go well with ingredients that share their piquant spiciness (like cumin, sharp apples or raw onion) or their soft earthiness (like potato). This chapter covers Stinking Bishop, Munster, Pont L'Évêque, Langres, Livarot, Vacherin and Celtic Promise.

Washed-rind Cheese & Anise Washed-rind cheeses such as Munster or Stinking Bishop are simply too opinionated to be paired with many flavors, but anise (sweet, wonderful, beloved pet flavor of *The Flavor Thesaurus*) is up to the challenge. Slice or scoop your cheese onto these thin fennel-seed crackers. Sift 1 cup plain flour into a bowl with ½ tsp baking powder, ½ tsp salt and 2 tsp fennel seeds. Add 1½ tbsp olive oil and ½ cup water in increments until the mixture is wet enough for a dough to be formed. Knead for 5 minutes, roll out to roughly ¼ in thick, and press out your crackers with a biscuit cutter. You'll get about 24 crackers of 2-in diameter. Place on a greased baking sheet, brush with water and cook for 25 minutes at 325°F. Substitute whole wheat flour if you prefer. Also try celery, cumin or caraway seeds.

Washed-rind Cheese & Apple A crisp apple works well with gentler-flavored washed-rind cheeses. Slice and core a tart green apple and eat it with a fruity, grassy Pont l'Évêque. Chef Pierre Gagnaire makes a chantilly of Pont l'Évêque and serves it with apple sorbet. Hard-core washed-rinds prefer their apple in the form of cider or the apple brandy, calvados—Celtic Promise, for example, is one of several cheeses washed in cider to help the development of its flavor and, other than a hunk of bread, a glass of cider is all the accompaniment it needs. I was so taken with Celtic Promise that I wanted to share it with the guests at my wedding, until I drove one back to London from Wales and it became all too apparent that two dozen of them in a hot room might have made my big day memorable in a way I'd want to forget.

Washed-rind Cheese & Bacon Langres from the Champagne region of France is a salty little cheese with a flavor that's often compared to bacon. It looks like one of those unctuous yellow Portuguese custard tarts and is every bit as rich. Each cheese has a hole in the top, into which you pour a little champagne or marc: quite a flavor sensation, and very handy if you've forgotten to pack glasses for your picnic. Langres isn't the only cheese that has a meaty quality. Some detect a beefiness in Italy's forceful taleggio. Émile Zola thought that Camembert smelled like venison. The French called Livarot from Normandy "the meat of the poor." I find some fresh goat cheeses taste like a butcher's shop smells. And then there's the steak-fat blue cheese discussed in Blue Cheese & Beef, page 63. See also Potato & Washed-rind Cheese, page 95.

Washed-rind Cheese & Cumin *See Cumin & Washed-rind Cheese, page 86.*

Washed-rind Cheese & Garlic Vacherin du Haut-Doubs, or Vacherin Mont d'Or, is a (very) soft cheese from Franche-Comté or the Swiss canton of Fribourg that frequently makes it onto lists of the world's top cheeses. Peel back the rumpled, tarpaulin-like rind and you'll discover an ivory-colored, fondue-like liquid, with a milky, salty, slightly fruity flavor, and a whiff of spruce from the box in the case of the (always unpasteurized) Haut-Doubs version from the French side of the border. Some people eat it straight from the box with a spoon, but I find a finger sufficient. It heats well too: take the lid off the box, wrap the box in foil, prick the surface of the cheese, insert some slivers of garlic, pour over ⅓ cup dry white wine, then bake it at 350°F for 20 minutes. Scoop from the box with good bread.

Washed-rind Cheese & Pear Livarot is a gamy, buttery cheese made from the fatty milk of Normandy cows. It has a rather spicy quality to its flavor, which makes it good with pears; no coincidence, given the abundance of orchards in Normandy. Combine them in a tart or simply sliced on a plate, served with a glass of creamy pear hard cider, otherwise known as perry, or *poiré* in France. Calvados is a brandy traditionally made with apples, but the Domfrontais variety contains at least 30 percent pear and is a recommended companion for Livarot. Stinking Bishop cheese, incidentally, is washed with perry made from the Stinking Bishop pear.

Washed-rind Cheese & Potato *See Potato & Washed-rind Cheese, page 95.*

Washed-rind Cheese & Walnut Walnut is a fine companion to all cheeses, and it's one of the few flavors that can rise to the challenge of a pungent washed-rind. Its bitter, tannic quality is apt to cut through the fatty headiness of the cheese, and its sweetness peeps through too. Walnut bread, walnut crackers, or fresh "wet" walnuts when they're in season are all very good simply paired with Livarot.

Blue Cheese

What makes blue cheese blue is a powdered blue-green mold called *Penicillium roqueforti*. The dominant flavors in blue cheese are the fruity, spicy notes attributable to the ketone 2-heptanone, and a green, fatty, metallic note from 2-nonanone. Other than that, blue cheeses vary as widely in flavor as they do in provenance, depending on the animal's diet, whether its milk has been pasteurized, the kind of starter and secondary cultures used, and the length of time and conditions under which it's stored. Roquefort, famously, is aged in caves—its authenticity depends on it having been stored in the natural caves of Cambalou near Roquefort-sur-Soulzon in the Aveyron. Stilton is aged in cellars, where cool air can penetrate the tiny holes made in the cheese to help the mold spores do their work. A notable exception to the *Penicillium roqueforti* rule is a rare blue cheese called Bleu de Termignon, made in tiny quantities in the French Alps with milk from pasture-grazed cattle, and allowed to blue naturally. Other blue cheeses covered in this chapter include Gorgonzola, Cabrales, Rogue River Blue, Shepherd's Purse, Beenleigh Blue and Fourme d'Ambert.

Blue Cheese & Avocado Missing out on avocado and bacon is one of the drawbacks of vegetarianism, but a deep-flavored, salty blue cheese is some compensation. Pair them seventies-style, the fruit's cavity brimming with blue cheese dressing, or spread a toasted slice of brioche with avocado mashed with a little lemon juice and top with crumbled blue cheese.

Blue Cheese & Bacon See *Bacon & Blue Cheese, page 165.*

Blue Cheese & Beef It's the tang of full-flavored, properly aged beef that accounts for its successful pairing with blue cheese, according to writer Peter Graham. Steak au Roquefort is a classic, and in Spain a similar dish is made using *queso de Cabrales*, a boisterous blue from Asturias. Stilton will make an opinionated cheeseburger to shake up sluggish taste buds at a hungover brunch. A handful of crumbled blue cheese is sometimes added, at the end of cooking, to a *carbonnade de boeuf*, melting into the stock and taking on its beery meatiness. Devotees of the combination should seek out Shepherd's Purse Buffalo Blue Cheese, made in North Yorkshire with locally reared buffalo's milk, which is higher in saturated fat than cow's milk. It has a luxurious flavor redolent of the fat on a really good, aged rump steak, with a similar melting texture. Kills two flavors with one stone—and it's suitable for vegetarians.

Blue Cheese & Blueberry See *Blueberry & Blue Cheese, page 335.*

Blue Cheese & Broccoli A threat or a promise, depending where you stand on pungent flavors. Combine them in a soup, in macaroni and cheese or, for more timid palates, a salad of chopped raw broccoli with a mild blue cheese dressing.

Blue Cheese & Butternut Squash Pumpkin ravioli are often served with a Gorgonzola sauce (see Blue Cheese & Sage, page 65). The complex, salty cheese makes a striking contrast to the dumb sweetness of the squash, particularly if you use a bumptious *piccante* Gorgonzola rather than the milder variants sometimes referred to as *dolce*. Alternatively try roasting small cubes of butternut squash and piling them on doorsteps of toast. Dot with Gorgonzola and blast under the grill. Serve with a glass of German Spätlese, a wine with a fair amount of residual sugar, which would work well with the caramelized squash and contrast nicely with the salty sharpness of the cheese.

Blue Cheese & Cabbage The recipe for blue cheese dressing under Blue Cheese & Chicken, below, works a treat on coleslaw. But it's not the sort of thing to yank you out of bed in the morning. For that you need the kimchi butter and sweet Gorgonzola croissant from the Momofuku Bakery in NYC. Best to hold the coffee with that.

Blue Cheese & Celery *See Celery & Blue Cheese, page 96.*

Blue Cheese & Chicken Paired with Buffalo chicken wings—see Celery & Blue Cheese, page 96—but I think the richness of the combination is better diluted in a salad. Prepare a blue cheese dressing by thoroughly mixing ⅓ cup mayonnaise, ½ cup sour cream, 2 oz chopped blue cheese, 1 tbsp lemon juice, 2 tbsp parsley, 1 crushed garlic clove and some seasoning. Cool in the fridge while you slice 2 celery stalks into matchsticks. Tear Cos lettuce leaves as you would for a Caesar salad. Cut 4 boneless, skinless chicken breasts into slices, dip in seasoned flour and fry in a mixture of peanut oil and butter. When the chicken is cooked, toss while still hot in a warmed mixture of 2 tbsp butter and 4 tbsp of hot sauce—Frank's RedHot if you can get it, but anything nice and fiery will do. In a large bowl, toss the leaves and celery in the cold dressing and pile the chicken onto a platter. Serve with a glass of IPA or a German wheat beer.

Blue Cheese & Fig While brown dried figs are reminiscent of sweet Pedro Ximénez sherry, black-purple mission figs are more suggestive of port and, appropriately enough, make a really heavenly match for Stilton. Combine them in these Stilton and fig straws and serve with chilled tawny port. It'll make a change to savor this pair of flavors at the start of an evening. In a food processor, make a dough of 1 cup plain flour, 1⅔ cups crumbled Stilton, 4 tbsp butter, a pinch of salt and 1–2 tbsp cold milk. Roll out into a rectangle roughly 12 in x 8 in. Using scissors, snip about 8 dried mission figs into thin strips and lightly press into the dough. Along its width, fold one half of the dough over the other, with the fig on the inside. You will now have a rectangle roughly 6 in by 8 in. Roll it out to about ¼ in thick and cut into straws, using a palette knife to transfer them to a greased baking tray. Bake at 350°F for about 15 minutes. You might alternatively try blue cheese and fig in a salad with sturdy leaves and a port dressing. Boil ¾ cup port until reduced by about half, add 1 tsp honey and leave to cool. Whisk in 3 tbsp extra virgin olive oil, 2 tbsp balsamic vinegar and season.

Blue Cheese & Grape *See Grape & Blue Cheese, page 248.*

Blue Cheese & Grapefruit Modern in an obsolete way, like the Niagara Falls Skylon or flying cars. Combine them in a salad with red onion, beets and some crisp, bitter lettuce. If you can get it, grapefruit marmalade makes a terrific sandwich combined with blue cheese.

Blue Cheese & Mushroom *See Mushroom & Blue Cheese, page 78.*

Blue Cheese & Peach Peach is a great match for Gorgonzola. Both have a fruity, creamy quality. They're often paired in salads and on bruschetta, but all they really need is a plate and a knife. The nineteenth-century explorer, F. W. Burbidge, described the flavor of the Asian fruit durian as "a combination of cornstarch, rotten cheese, nectarines, crushed filberts, a dash of pineapple, a spoonful of old dry sherry, thick cream, apricot pulp and a soupçon of garlic all reduced to a consistency of a rich custard." To this day in Singapore they're banned from many hotel rooms—no drugs, no firearms, no durian.

Blue Cheese & Pear The gentler flavors of pear can be lost with some blue cheeses, but Fourme d'Ambert is at the milder end of the scale. It's sweet and milky, with a mild mustiness that comes from its aging in caves, and is injected with a sweet wine just detectable to the palate. You might otherwise find blue cheese and pear combined in a salad (see Pear & Walnut, page 269), or in a blue cheese fondue served with slices of pear, figs and walnut bread.

Blue Cheese & Pineapple Martin Lersch has a blog called Khymos that runs a regular TGRWT (They Go Really Well Together) recipe challenge; blue cheese and pineapple was the subject of TGRWT #10. Interestingly, Lersch notes that he was unable to find an overlap in the reported flavor odorants in these ingredients, but that this is the case with many common pairings. Other TGWRTs have investigated apple with rose, chanterelle mushrooms with apricot, Parmesan with cocoa, and banana with parsley.

Blue Cheese & Sage Salty-sweet blue cheese craves bitterness and sage can provide it. Melt 2 tbsp butter, add 3 large sage leaves and stir for 30 seconds. Over a low to medium heat, add ½ cup crumbled Gorgonzola and ½ cup heavy cream, stirring while it melts and melds. Remove the sage, adjust the seasoning and serve on pasta or gnocchi.

Blue Cheese & Truffle Blue cheese and truffle are commonly paired in a creamy sauce to serve with *filet mignon*. The food writer Jenifer Harvey Lang gives a recipe for choux pastries stuffed with a mixture of blue cheese, cream cheese and cream, which she tops with shavings of black truffle. I once had my head turned by the combination of truffled honey and Stilton, but even without the addition of cream, there's something about the pair that feels like wearing a low-cut top and a short skirt.

Blue Cheese & Walnut Toasted walnuts can develop something of a blue cheese flavor. Take it as a hint. Milky, sweet, bitter-skinned walnut works wonders with all types of blue cheese. It's a classic partnership, especially crumbled Roquefort and walnuts in a salad with chicory. Good dressed with extra virgin olive oil, cider vinegar and cream, 5:3:2, shaken up with seasoning. Try walnuts on a cheeseboard with Beenleigh Blue, made in Devon from ewe's milk, which when mature has an almost fudgy texture and intensely peppery veins. Or with *queso de Cabrales* from Asturias in Spain, which has legendary strength: this cheese could fight bulls.

Blue Cheese & Watercress *See Watercress & Blue Cheese, page 99.*

Hard Cheese

This covers a wide range of cheeses, including Cheddar, Parmigiano-Reggiano, Manchego, Comté, Gruyère, pecorino, Lincolnshire Poacher, Berkswell and Mahón. Flavors vary according to milk type, milk quality, the cheesemaker's recipe, microflora and the cheese's age. A good cheesemonger will be able to furnish you with a tasting flight of the same cheese at different stages of maturation, so that you can experience at first hand the effect aging has on flavor. Many hard cheeses are sweet, sour and salty, and gain crystals of umami as they age. It's this array of tastes, in combination with its high fat content, that makes hard cheese both a satisfying eat in itself and a complement to other ingredients, enhancing their flavors with its rich roundness. Red wine, raw onion, watercress and walnuts often work well with hard cheese, as they add a balancing touch of bitterness. And sweet partners like dried fruit, cooked tomato and cooked onion can bring out the cheese's savory side. When tasting hard cheese, look out for creamy, buttery, coconut, caramel, fruity (especially pineapple), sulfurous (chopped boiled egg), cooked, roasted and nutty flavors.

Hard Cheese & Almond Both Keen's and Montgomery Cheddar contain almond notes, and although walnuts are the more common nut pairing, there's no reason why almonds shouldn't work as well with a good, nutty Cheddar as they do with a Manchego. Authentic Manchego is made with sheep's milk in La Mancha, central Spain. It's as widespread in its popularity as in the redolence of its flavors. Young Manchegos have a fresh, grassy quality evocative of the cool pastures of Galicia in the north. Older specimens exhibit the hot, dry saltiness of the Andalusian coast. Both would be good cut into thin slices and served with a bowl of toasted, salted almonds. If you're lucky, they'll be heart-shaped Spanish Marcona almonds, prized for their luscious, milky flavor.

Hard Cheese & Anchovy *See Anchovy & Hard Cheese, page 160.*

Hard Cheese & Anise Mahón, a hard cheese from Menorca, is made with pasteurized cow's milk and has a salty, lemony flavor. It's traditionally eaten in thin slices with fresh tarragon leaves, olive oil and black pepper. Leonie Glass writes about Salers, a semi-hard cheese from the mountainous Cantal *département* in south-central France, where the cows have a richly aromatic diet of licorice, arnica, gentian and anemone, all of whose flavors are discernible in the milk. Serve with the fennel crackers described in Washed-rind Cheese & Anise, page 61.

Hard Cheese & Apple *See Apple & Hard Cheese, page 265.*
Hard Cheese & Apricot *See Apricot & Hard Cheese, page 276.*

Hard Cheese & Asparagus People get superstitious about asparagus. Otherwise sane cooks will insist on cooking it in special pans, or turning the spears three times counterclockwise but never under the light of a full moon. Cast aside cabbalistic practices by simply roasting asparagus in the oven, then serving with grated Parmesan, which has a harmoniously sulfurous character. Make sure to scatter lots of cheese on the tips, where it will catch deliciously between the bracts.

Hard Cheese & Bacon *See Bacon & Hard Cheese, page 166.*

Hard Cheese & Banana Comté is made in the Alps with raw cow's milk and is the most popular cheese in France. When young, it has flavors of fresh hazelnut, dried apricot, soft caramel and boiled milk; when fully mature, according to the Comté cheese assocation, it has "rich, persistent walnut, hazelnut, chestnut, grilled almond, melted butter and spice flavors softened by hints of matured cream or citrus fruits." Notes of leather, white chocolate and prune may also be apparent. Try the Comté representative's recommendation of white bread topped with sliced banana, sliced Comté and a pinch of Espelette pepper (more about that in Oily Fish & Chili, page 154), grilled for a few minutes until the cheese has melted. The banana, they say, helps bring out the cheese's multitude of flavors, whereas the chili adds "a final hot, persistent touch."

Hard Cheese & Basil If there's such a thing as a cult restaurant, La Merenda is it. The Michelin-decorated chef Dominique Le Stanc used to cook at the Hotel Negresco on the promenade in Nice. Like a disillusioned sheriff, he handed back his star and headed into the Old Town to cook in the less rarefied confines of his tiny, bead-curtained, ten-table joint scarcely bigger than a kebab shop. There's no fancy stemware, no flower arrangements, no flattering lighting, no wine list, no backs to the chairs and, advance planners should note, no telephone. The menu, chalked on a blackboard you prop on the table or your knee, is invariably Niçoise in provenance, and usually includes a *tagliatelle au pistou*. *Pistou* is essentially a variation on pesto, trickled around the coast from its home in Liguria, discarding a vowel on the way and picking up some new ones like souvenirs. It primarily differs from pesto in lacking pine nuts, and is sometimes made with Emmental in place of Parmesan. Le Stanc serves his *pistou* with fresh spinach pasta.

Hard Cheese & Beef Trained tasting panels might use the term "brothy" to describe the flavor of hard cheese, meaning it has similar properties to a beef stock cube. As any cheeseburger fan can testify, many hard cheeses make a harmonious match for beef. When I tire of reading about wonderful organic produce markets, or clod-booted laments for the good old days of subsistence farming, I like to pick up *The $100 Hamburger* by John F. Purner. It's a guide to cafés and restaurants near airfields, rating both the quality of the ribs or cheeseburgers and the runways that'll take you to them; like a gastronomic *Earth from Above*, where the aerial vantage is given on places like the Kanab Muni airfield in Utah, for example, where great burgers are available at the nearby Houston Trails End Café, or the grass airstrip at Gaston's in Arkansas, where you can stay in a cabin and eat fresh trout from the lake. Now flying, burgers and eating fish are frowned upon, reading the book is like looking at footage of glamorous smokers in bars: nostalgia we weren't expecting to come so soon.

Hard Cheese & Broccoli Broccoli begs for a hard, under-the-counter-strength cheese. Parmesan is ready to eat, having matured for a relatively lengthy eighteen months. If you're buying it cut fresh from a deli, look for the date stamped on the wheel. Older Parmesans, matured for three or four years, are sometimes called *stavecchio* (or at their oldest, *stravecchio*), and are saltier, spicier and generally more intense. They're what you really want on your extra-bitter broccoli raab. Combine them in a pasta or risotto.

Hard Cheese & Cauliflower *See Cauliflower & Hard Cheese, page 122.*
Hard Cheese & Chicken *See Chicken & Hard Cheese, page 32.*

Hard Cheese & Chili Italian pecorino can be pretty spicy to begin with, but if you want it properly hot you'll need to buy it *con peperoncino*—flecked with little pieces of chopped-up, medium-hot, dried red pepper. In the United States, jalapeño Jack works on the same principle, and can be used to make a fondue-style dip for tortilla chips: deconstructed nachos. The flavor of corn is particularly harmonious with hard cheese and chili; try tossing freshly popped kernels with chili powder, paprika, melted butter, grated Parmesan and salt. See also Butternut Squash & Chili, page 226.

Hard Cheese & Clove Friese Nagelkaas is a Gouda-style cheese from Friesland in Holland. *Nagel* means "nail," referring to the nail-shaped cloves that are shattered and added to the cheese with a little cumin. If you've ever tasted Green & Black's Maya Gold chocolate, this is the cheese equivalent. The cloves and cumin contribute a heady spiciness whose hints of orange and lemon peel lend the cheese a Christmassy flavor.

Hard Cheese & Fig The *consorzio*, or safeguarding consortium, responsible for Parmigiano-Reggiano recommends dried fig, hazelnuts, walnuts and prunes as pairings for their twenty-four- to twenty-eight-month-old Parmesans. These will have lost some of the milkiness of their youth and started to develop notes of fruit and nut, alongside a melted-butter flavor—hence the pairings. As the cheese continues to age, the nut flavors get more pronounced, and spicy notes (nutmeg in particular) become detectable.

Hard Cheese & Globe Artichoke In Italy, raw baby artichokes are sliced paper-thin and tossed with olive oil, lemon juice and wisps of Parmesan to make a salad that's bitter, sharp and salty, and an excellent appetizer. Strip your artichokes of all their dark external leaves. Lop ½–1 in off the top of each, cut most of the stem off and peel what's left of the stem with a potato peeler. Rub with lemon juice, then cut in half lengthways, checking there's no choke—if there is, remove it with a teaspoon. Slice each half into very fine strips, cutting from the top of the leaf to the stem. As you go, put the strips into water acidulated with lemon juice to prevent them browning. Once you've chopped all your stems, strain off the lemon water, pat dry and toss in olive oil. Season, and garnish with thin parings of Parmesan.

Hard Cheese & Grape Grapes have a generic fruitiness that makes them a safe bet on the cheeseboard. Whether it was for want of fresh grapes that someone first paired tangy, slightly lemony Wensleydale cheese with a slice of fruitcake I don't know, but they make a very cozy couple. Of course, the affinity of cheese with grape is clear in countless cheese-and-wine pairings. "Buy with crackers, sell with cheese," they say in the wine trade: crackers cleanse the canny buyer's palate, whereas cheese coats the taste buds with fats and proteins that can attenuate a wine's harsher, tannic qualities.

Hard Cheese & Juniper *See Juniper & Hard Cheese, page 315.*
Hard Cheese & Mushroom *See Mushroom & Hard Cheese, page 80.*
Hard Cheese & Nutmeg *See Nutmeg & Hard Cheese, page 218.*

Hard Cheese & Onion Cheese and onion was the first flavored chip introduced in England and Ireland. But there are countless variations on the theme. Stinky German Limburger is paired with raw onion, rye bread and mustard and served with strong dark beer. In Wales, leeks and salty, white, acidic Caerphilly are shaped into vegetarian Glamorgan sausages and coated with breadcrumbs and herbs. Classic French onion soup is *gratiné* with Gruyère, giving a lovely fruity twang to the rich onion flavor and making it nearly impossible to eat without infinitely extendable arms. A red onion tart will often include an atoll of goat cheese, while Berkswell, a hard sheep's cheese made in the West Midlands, kills two birds by having strong hints of caramelized onion itself.

Hard Cheese & Orange *See Orange & Hard Cheese, page 289.*

Hard Cheese & Parsnip Parsnip has a serious appetite for salty foods that emphasize its sweetness. Parmesan is a popular pairing—typically, parboiled parsnips are tossed in flour and grated Parmesan before being roasted. Or make a Parmesan cream garnish for parsnip soup by folding 1 tbsp grated Parmesan and 2 tbsp snipped chives into 4 tbsp whipped heavy cream. Mark Bittman gives a recipe for parsnip gnocchi, which he recommends serving with Parmesan, butter and sage.

Hard Cheese & Pea *See Pea & Hard Cheese, page 199.*
Hard Cheese & Pear *See Pear & Hard Cheese, page 269.*

Hard Cheese & Pineapple Outsnoot snobs sneering at your sticks of cheese and pineapple. The flavors can be naturally harmonious: the pineapple note of ethyl caproate is present in some of the world's finest cheeses, including Comté, Lincolnshire Poacher and Parmesan. Ethyl caproate also occurs naturally in clove, figs and wines.

Hard Cheese & Potato Once upon a time there was a coarse-skinned baked potato that had a lovely smell of malt but a messy thatch of Cheddar on the top, so it had to be eaten in the scullery. Everybody saw its goodness and beauty, but it was never invited to the ball. One evening, when the potato was all alone, a fairy godmother appeared and asked the poor, calloused, muddy vegetable its wish. And the potato said it would like nothing better than to be rid of its rough raiments and be made smooth and silky and acceptable to people who dined in the best restaurants of France. The fairy godmother waved her wand. The potato fell into a swoon and, in the winking of one of its many eyes, awoke in a restaurant called L'Ambassade d'Auvergne, right in the very heart of Paris, as part of the richest, silkiest mashed potato all the customers of the restaurant had ever seen. It was called *aligot*, and, beside potato, was made with rich Laguiole cheese, garlic, cream and butter, all stirred up together until the *aligot* was so elastic it could be whipped out of the pan to a height of three or even four feet without breaking. The following night the fairy godmother appeared again and asked the potato what its next wish was. The potato was feeling a bit queasy from the night before and fancied something cozy but sophisticated. The fairy waved her wand and the potato found itself lying sliced in warm, creamy layers with Gruyère cheese and his old friends, cream and garlic, in a potato dauphinoise. The potato loved to be free of its tatty jacket but wasn't so sure about the roguish garlic and cream. The next evening, when the fairy godmother appeared and asked the potato what its third wish was, it has to be said that the potato came across as a little jaded. How about bathing in a sticky, fruity *raclette*, the fairy godmother suggested—perhaps with a gaggle of cornichons? Or turning into smooth, pebble-like gnocchi smothered in fontina sauce? "Actually," said the potato, "what I'd like most is to stay at home with a bottle of beer. And some Cheddar. As long as it's Keen's, mind you, or Westcombe's, because if I've learned anything it's that I'm too good for that greasy, one-dimensional pap I used to get saddled with." "Get you," said the fairy godmother, and vanished in a puff of smoke.

Hard Cheese & Sage Derby is a pressed cow's milk cheese, softer and more delicately flavored than Cheddar. Sage Derby is the flavored variant. Avoid the lurid green blocks found on some deli counters and go for the sort of quality cheese made by Fowlers Forest Dairy, which scatters chopped sage leaves through the center of its Derbys. Slice it thin and pair with prosciutto in a sandwich, or use for cheese on toast topped with a fried egg. And if good

Soft Cheese & Grape Grapes and young, buttery Brie are paid one of the highest culinary compliments in British cuisine: they're paired in a sandwich. Stronger than Brie, Arômes au Gène de Marc is steeped in grape brandy for a month with the pips, skins and stalks that are left over from grape pressing. Sold still freckled with this debris, the cheeses look as if they've been rolled along an autumn footpath on the way to market.

Soft Cheese & Mushroom Tasting a ripe Camembert at room temperature is like sitting on a bale of fresh straw next to a basket of just-picked mushrooms while eating a truffled fried egg. The *Penicillium camembertii* that gives it its white kidskin rind is also responsible for the characteristic mushroom note. Capitalize on this flavor harmony by taking the rind "lid" off a Camembert or Brie, scattering just-cooked wild mushrooms over it and placing it in the oven so the cheese melts a little before serving.

Soft Cheese & Smoked Fish In North America the terms "smoked salmon" and "lox" have become interchangeable, though strictly speaking lox isn't smoked but cured in brine, which helped it survive its long journey to market. When Russian and Eastern European immigrants arrived in America at the end of the nineteenth century, they found that salmon, a luxury back home, was in plentiful and affordable supply, and started to eat much more of it. Cream cheese, which began to be sold on a mass scale at around the same time, had a similar taste to the dairy products familiar from the old countries, and as it softened the extreme saltiness of the lox, the combination soon became a staple of Jewish-American cuisine. Some delis will gouge a trench in the bottom half of the bagel to accommodate a more satisfying depth of cheese. See also Onion & Smoked Fish, page 110.

Soft Cheese & Strawberry *See Strawberry & Soft Cheese, page 259.*

Soft Cheese & Tomato Mozzarella is classically paired with tomato in a salad or on a pizza. Your best ripe tomatoes will be well served by real buffalo mozzarella or, even better, burrata. Imagine a mozzarella shaped like a drawstring money-bag, filled with a mixture of thick cream and off-cuts of mozzarella. Cut into it and its center oozes like a slow groan of pleasure. The lactic freshness mingles with the sweet-sour tomato juice to make an unforgettable dressing. Enhance it with extra virgin olive oil, fresh basil and seasoning. Burrata is originally from Apulia, the stiletto heel of Italy, but cheesemakers in the United States have caught on to its appeal.

Soft Cheese & Truffle The fragrance of truffles is often compared to garlic and cheese; conversely, cheeses like Brie de Meaux and Saint Marcellin are frequently said to have a truffly quality about them. Take a good Brie and slice it in half, so you have two rounds. Cover the bottom half with thin slices of truffle, replace the top half, wrap in cling wrap and leave in the fridge for 24 hours before serving at room temperature. Brie can be bought already truffled, as can the soft triple-cream Brillat-Savarin.

sage cheese isn't available, you can always pair a strong hard cheese with sage in scones, and eat them warm with butter.

Hard Cheese & Shellfish *See Shellfish & Hard Cheese, page 140.*

Hard Cheese & Tomato That tomatoes and cheese can transform a thin bread crust or plate of pasta into something transcendent is testament to the umami-rich splendors of the combination—although, as any Italian will tell you, the ingredients need to be good for the magic to happen. Given a high-quality English cheese, by contrast, many people will advise against cooking with it, thinking it should be preserved for the cheeseboard. The makers of Lincolnshire Poacher (a Cheddar-like cheese with a bold, nutty flavor) beg to differ, on the parallel principle that you should apply the same standards to the wine you cook with as you do to the stuff you drink: "cooking" wine, like "cooking" cheese, will be as inferior in your dish as it is in the glass/on the board. I use a decent Côtes du Rhône for this hearty dish, which I call Lincolnshire Poacher's Pot. Imagine that ratatouille went to a cheese and wine party, got drunk and lost eggplant and zucchini along the way. Soften a chopped large onion in oil over a medium heat for 5 minutes, then add a finely chopped garlic clove and a chunkily chopped green or red pepper (green brings a pleasing freshness). Cook slowly for another 5 minutes. Add a can of good-quality plum tomatoes and break them up with a spoon. Then add a scant tsp dried mixed herbs, $\frac{1}{3}$ cup good red wine, 2 tbsp water, a pinch of sugar and some seasoning. Bring to the boil and simmer for 20–30 minutes. Taste and adjust the seasoning as necessary. Before serving, cut 6–8 oz Lincolnshire Poacher into half-inch cubes, add to the pan, still on the heat, then stir, giving them a minute or two to warm through. Decant into 2 earthenware bowls. Eat it before the cheese melts entirely. Crusty bread is essential.

Hard Cheese & Walnut *See Walnut & Hard Cheese, page 233.*
Hard Cheese & White Fish *See White Fish & Hard Cheese, page 145.*

Soft Cheese

Many of the soft cheeses covered in this chapter are eaten young, and retain fresh dairy flavors—the clean milkiness of mozzarella and cottage cheese, for example, or the slightly richer creaminess of Brillat-Savarin and Corsican brocciu. Bloomy cheeses like Brie and Camembert have a buttery taste when young but become more pungent, earthy and vegetal as they age, recalling the farmyard more than the dairy. While fresher, younger soft cheeses work well with salty and fruity (particularly berry) flavors, the more aged types are particularly harmonious with other earthy ingredients like mushrooms and truffles.

Soft Cheese & Anchovy Nigella Lawson describes a sandwich-bar lunch of cottage cheese on white bread, no butter, "but with anchovies; the saltiness, the aggressive and indelicate invasiveness of those cheap and unsoaked tin-corroded fish made me feel, after it was finished, that something actually had been eaten." In Naples, *mozzarella in carrozza* consists of mozzarella and anchovy sandwiched between two slices of white bread, dipped in flour and egg, then fried. A few high-quality, pink anchovy fillets can be served with a ball of mozzarella for a simple lunch. If you prepare this for a solitary meal, I think you should be allowed the entire ball. Not for gluttony's sake, but simply for the pleasure of taking the whole thing in your hand and biting into it like a juicy apple. I love the initial resistance, followed by the absolute give. Chase with a sliver of anchovy.

Soft Cheese & Apple Eat young Brie or Camembert cut into thin pieces with corresponding slivers of apple—it's like eating apples with cream or slices of butter. Or buy a whole cheese in a box, remove the wax paper, then put the cheese back in the box, prick the top and pour over some apple brandy. Bake for 20 minutes at 400°F. Serve with slices of apple. The sweet-toothed might prefer to drizzle a whole baked Brie with caramel, sprinkle it with walnuts and serve with wedges of sharp apple. I also like to serve apple with older Camembert. The cheese takes on something of a cooked-cabbage flavor as it ages, and this works very well with the fruit.

Soft Cheese & Avocado *See Avocado & Soft Cheese, page 197.*
Soft Cheese & Basil *See Basil & Soft Cheese, page 211.*

Soft Cheese & Bell Pepper In Corsica it's difficult to eat a meal without the local cheese, brocciu, a ricotta-like sheep's cheese, sometimes with a bit of goat's milk mixed in. Brocciu is served with fruit and jam for breakfast, with charcuterie for lunch and in cannelloni for dinner. You'll see the sheep and goats it comes from hanging around on the hairpin bends of Corsica's heart-stopping coastal-road system (they'll quite possibly be the last things you see). We stopped in Calvi to buy a beach picnic for a walk along the coast path. We had stocked up on charcuterie, bread and tomatoes when the shop owner insisted that we also try some miniature red bell peppers stuffed with brocciu. A couple of hours later we were clambering over rocks in search of a lunch spot. As we unpacked the picnic, we were joined by first one wasp, then another, then nine more, until our *al fresco* paradise began to look like the set of a B-movie. As I lifted a stuffed pepper to my lips, a wasp heli-skied into the cheese, and in my flustered hand movement shooing it away I launched a piece of prosciutto onto a rock. One wasp chased after it, and then another, quickly followed by the others, and, ingenious by accident, we got on with our picnic in peace. Which begs the question: do wasps particularly like prosciutto or are they sick to the mandibles of brocciu? I'd come round to it myself, its sweet milkiness at once complemented and offset by the sweet smokiness of the peppers.

Soft Cheese & Black Currant Sharp, bitter black currants offset th[e] ing creaminess of cheesecake. If you don't have time to make your o[wn] don't want a huge one lurking in the fridge, spread a digestive biscui[t] cream cheese and top it with black currant jam: cheatscake. Don't b[e] measly with the cream cheese—you need to feel your teeth sink into it.

Soft Cheese & Caper *See Caper & Soft Cheese, page 103.*
Soft Cheese & Caviar *See Caviar & Soft Cheese, page 152.*

Soft Cheese & Celery Celery might not be as prized as it was in the ni[ne]teenth century, when it was presented at table in special glass or silver vas[es] but it does have a refreshing, bitter quality, with a hint of anise, that shou[ld] earn it a place on the cheeseboard. Unlikely as it may seem, celery has flav[or] traits in common with walnut, also a classic pairing for cheese—see Walnu[t] & Celery, page 231. In his book *Eggs*, the chef Michel Roux gives a recip[e] for Camembert ice cream, which he serves with tender celery leaves, littl[e] radishes and crackers for cheese.

Soft Cheese & Cinnamon *See Cinnamon & Soft Cheese, page 214.*
Soft Cheese & Eggplant *See Eggplant & Soft Cheese, page 84.*

Soft Cheese & Fig In Syria, where figs are plentiful, they're eaten fresh for breakfast with dazzling-white soft cheese. They might also be turned into a rough jam, which you can make by adding 1 lb roughly chopped fresh figs to 1 cup sugar dissolved in 3 cups hot water. Bring to the boil and simmer gently until good and thick, making sure it doesn't catch on the bottom. Cool and keep in a lidded pot in the fridge—it will last a week. Use the same mixture, but perhaps with some vanilla or orange zest and Cointreau added, for a compote to serve with *coeurs à la crème*. Blend 12 oz cottage cheese, 8 oz cream cheese and 1 cup heavy cream until smooth, then divide the mixture between 4 heart-shaped molds and leave to drain overnight. If you don't have heart-shaped molds, use cheesecloth-lined flowerpots (the kind with drainage holes in the bottom). You can add a few tablespoons of confectioners' sugar if you prefer your *coeurs* sweeter but a little tartness is a welcome contrast to the super-sweet fig. See als[o] Fig & Anise, page 331.

Soft Cheese & Garlic There are worse things you can do, when you[r] options are limited, than reach for a garlic and herb Boursin, a baguette an[d] a bottle of Beaujolais, as in the ad: *du pain, du vin, du Boursin*. The inspira[a]tion for the product, launched in the late 1950s by François Boursin, was t[he] long-standing custom of serving soft cheese with a mix-your-own selection [of] fresh herbs. At the time of writing, it's still made with Normandy cow's m[ilk] and cream, using the original production process. Presented with a smea[r on] a slice of baguette, my husband gave a deep, Gallic shrug. "Tastes like g[arlic] bread." And so it does. Which is no bad thing.

Soft Cheese & Walnut New-season walnuts are at their very best in autumn, which makes it as good a time as any to prepare labna, a soft yogurt cheese whose lovely lactic tang makes walnuts seem very sweet. Line a colander or large sieve with clean, damp cheesecloth and place over a pan or bowl high enough for some of the liquid to drain off under the colander. Mix 1 tsp salt into 4 cups natural yogurt and transfer to the lined colander. Leave to drain at room temperature for about 8 hours. Serve with walnut oil, chopped walnuts, maybe some runny honey, and warm brown bread. In Syria and Lebanon, labna is eaten for breakfast with walnuts and a few dried figs.

EARTHY

Mushroom

Eggplant

Cumin

Beet

Potato

Celery

Mushroom

This section covers, among others, button mushrooms, morels, porcini and chanterelles, but not truffles, which have their own section (see page 115). All of them contain a flavor-identifying (or "character-impact") compound called 1-octen-3-one. Mushrooms vary greatly in texture and this has more bearing on how they're cooked than on their individual flavors. There are mushrooms that taste of almonds, shellfish, meat, anise, garlic, carrots and rotting flesh, but generally speaking they all prefer the same flavor partners, primarily those that enhance their own flavor—garlic, bacon, Parmesan.

Mushroom & Anise Tarragon brings a welcome fresh, grassy note of anise to all types of mushroom and shares their love of cream. Sour cream works particularly well with tarragon and mushroom in a stroganoff-style dish. In a large frying pan, fry some garlic in a combination of butter and oil over a medium heat, adding mushrooms (chopped if necessary) when the garlic begins to color. Season, and before all the mushroom juices have evaporated, add a dash of brandy. When that's all but gone too, remove the pan from the heat and stir in chopped tarragon and just enough sour cream to make a sauce. Heat through gently and serve on white rice. *Pleurotus euosmus*—the tarragon oyster mushroom—is a close relative of the plain oyster mushroom, but identifiable by its strong tarragon aroma. An anise note is also found in the aroma of the fleshy white horse mushroom, *Agaricus arvensis*, and in its flavor when young and unopened.

Mushroom & Apricot Mushroom and apricot are paired in stuffings for venison, hare and quail and added to beef or lamb in the sort of towering pie that makes table legs buckle at banquets. The combination will remind mushroom foragers of the legendary apricot aroma of chanterelles (also called girolles), which the more experienced among them can apparently nose on the air. Chanterelles are very popular with chefs for their peppery, fruity flavor, which they nonetheless have only when fresh: chanterelles don't really survive drying.

Mushroom & Asparagus *See Asparagus & Mushroom, page 129.*

Mushroom & Bacon Dried morels have a smoky, meaty flavor subtly redolent of bacon. Their complexity of flavor (and their expense) lends weight to the argument that they should be the sole focus of a meal; nonetheless, like all fungi, they take well to the flavor enhancement offered by bacon, whose salty fattiness infiltrates their many wrinkles and crevices. Morel caps look like homemade beanies, knitted with big needles after one too many parsnip wines. In the nineteenth century a dish *à la forestière* signified a garnish of morels and diced bacon; today it is more likely to mean button mushrooms. Not that button mushrooms and bacon aren't good in an omelette or crêpe, or even outstanding in a double-crust pie, when the mushroomy roux has turned to jellified umami. An old recipe worth reviving is the *croûte baron*: a savory of grilled mushrooms and bacon on toast, covered with beef bone marrow and breadcrumbs, given another flash under the grill, then garnished with parsley. You might use a little olive oil or clarified butter if bone marrow isn't available.

Mushroom & Beef *See Beef & Mushroom, page 49.*
Mushroom & Blueberry *See Blueberry & Mushroom, page 336.*

Mushroom & Blue Cheese Blue cheeses are made blue by fungi, so it comes as no surprise that many of them have hints, or in the case of Gorgonzola, clear enunciations, of mushroom, with which they share important flavor compounds. Polenta with Gorgonzola and porcini is a popular dish in the Trentino region of Italy; blue cheese can be stirred into wild mushroom risotto before serving; and mushrooms and blue cheese make a delicious soup with leeks.

Mushroom & Butternut Squash Toadstools and pumpkins. A fairy-tale combination. Pumpkin and squash may be sweet but they have an earthy side that makes them highly compatible with the woodsy mushroom. Chanterelles, in which you may detect a slightly fruity, pumpkin-like quality, are an ideal match. Philip Howard, chef at The Square in London, serves pumpkin purée, chanterelles, leeks and black truffle with scallops (he also makes a dish of langoustine with pumpkin purée, rings of trompette mushrooms, a field mushroom purée, Parmesan gnocchi and a potato and truffle emulsion). At Kitchen W8, the bistro Howard part-owns, a simpler combination of butternut squash and chanterelles is offered in a red wine risotto, topped with a soft poached egg.

Mushroom & Chestnut Chestnut mushrooms are merely the brown form of the common mushrooms you find in those little plastic boxes on supermarket shelves. Japanese shiitake mushrooms take their name from *shii*, a species of chestnut tree, and *take*, meaning mushroom. The chestnut association is strong in mushrooms because of their symbiosis in the wild: mushrooms grow under chestnut trees. In northern Italy a tagliatelle is made with chestnut flour to serve with earthy-sweet dried porcini. Mushrooms and chestnuts are cooked together bourguignon-style, with shallots, bacon and red wine, and served in a pastry or suet crust. And the food writer Richard Mabey combines porcini and chestnuts in a soup. It can be made with fresh, but he pronounces vacuum-packed pre-cooked chestnuts excellent. Simmer 8 oz vacuum-packed chestnuts in just enough water to cover for 40 minutes. Meanwhile, rehydrate 1 oz dried porcini mushrooms in just enough hot water to cover them for 30 minutes. Cook 1 diced onion and 4 chopped bacon slices in a bit more butter than you ought and add to the cooked chestnuts with the mushrooms and their soaking water. Simmer for 15 minutes, then purée in batches. Reheat and season to taste, adding a squeeze of lemon and a schooner of fino sherry before serving. Apply some rouge to your cheeks and a little dab of cocoa under the fingernails, and you can claim to have foraged for the ingredients beyond the back of the cupboard.

Mushroom & Chicken *See Chicken & Mushroom, page 33.*

Mushroom & Dill In Russia they call the porcini the tsar of mushrooms. Pair it with pine-scented dill and you'll have a dish to make you yearn for the forests of Siberia. Chopped mushrooms are stewed with dill, salt and pepper, mixed with butter and sour cream and stuffed into Eastern European dumplings called *pirozhki* or *pierogi*. Or they're cooked in the style described in Mushroom & Anise, page 77, and served with rice or boiled potatoes and a glass of chilled vodka. According to David Thompson, dill is paired with mushrooms in curries from northeastern Thailand, near the border with the dill-loving Laotians.

Mushroom & Egg The giant puffball, *Calvatia gigantea*, has a particularly mushroomy flavor. If you like mushroom fritters, try dipping piano-key slices of giant puffball in egg and breadcrumbs before frying them. Alternatively, treat them like French toast, dipping them in egg, then frying them in butter for breakfast. The Chinese dish *mu shu* is a little more elaborate: inspired by the forest floor, its pancake base is strewn with stir-fried lily buds and delicately flavored wood ear mushrooms and dotted with little "flowers" of scrambled egg. Strips of pork are (sometimes) added too. See also Asparagus & Mushroom, page 129.

Mushroom & Garlic Even the blandest mushrooms take on some of their wild cousins' intensity of flavor under the influence of garlic, but garlic and shiitake mushrooms enjoy an extra-special relationship. Shiitake contain a compound called lenthionine, chemically similar to the sulfides found in alliums such as garlic and onion, and, like them, they are prized for their flavor-enhancing properties. Lenthionine content is maximized by drying and rehydration. You won't get the same results with fresh shiitake, but you can tuck slivers of garlic into their gills and give them a drizzle of olive oil and a sprinkle of salt before grilling or frying them.

Mushroom & Goat Cheese A young *Agaricus bisporus* is the squeaky white button mushroom. Older, browner specimens are sold as chestnut, crimini or baby bella mushrooms, and have more flavor than their white siblings. Six or seven days later, when they open fully to display their inky gills, they become portabellas. Until the 1980s they were considered unsalable, and mushroom farm workers took them home as perks. Then it dawned on someone that their pronounced flavor might, after all, be marketable, especially if they had a fancy Italian name. So they made one up. The portabella owes its success partly to its size and shape: it's at once an instant veggie burger and a mushroom that life isn't too short to stuff. Its rusticity is particularly well paired with goat cheese, which adds a welcome tanginess. Place 6 portabellas, cap down, on an oiled baking tray. The black gills look like lined-up 45s seen through the window of a jukebox. Scrape them out, but leave the stalks. Mash ½ lb goat cheese with 1 tbsp olive oil, a handful of chopped parsley and plenty of salt and pepper. Stuff the mixture into the mushroom caps, taking care not to overfill them, as they will shrink a little in the oven. Bake at 400°F for 15 minutes, then serve sprinkled with more chopped parsley.

Mushroom & Hard Cheese The species of mushroom that we commonly eat contain no salt, which is why you must either add some or pair mushrooms with salty ingredients to realize their flavor fully. Parmesan provides some saltiness in a mushroom risotto, pasta or bruschetta, while Gruyère is delicious with mushrooms on toast. And grated pecorino romano with finely chopped mushrooms is the beginning of a pesto-like sauce. Put some cooked mushrooms in a food processor with grated pecorino, toasted walnuts, garlic, parsley (or basil, or both) and olive oil—all particularly good partners for mushroom—and pulse until you've achieved your desired texture. Balance the flavors to taste. Great on pasta, clearly, but also spread on a baguette for a steak or sausage sandwich.

Mushroom & Mint In Tuscany, porcini mushrooms are often sold alongside, and served with, the herb nepitella, or *Calamintha nepetha*. Nepitella has a woody flavor comparable to mint, which can be substituted for it. It might be worth checking your garden to see if you have some—it's a popular decorative shrub. Try it chopped and scattered over thinly sliced fresh porcini, with thin parings of pecorino cheese and your best olive oil.

Mushroom & Oily Fish The forest-floor mustiness of mushroom particularly suits the earthy flavors of freshwater oily fish. A finely chopped crumb of shiitake brings out the best in a salmon fillet. Salmon is also paired with chopped mushrooms (and rice) in coulibiac, a French puff pastry dish descended from *kulebjaka*, a Russian pie made with yeast pastry (although not always containing fish or mushrooms). Antonio Carluccio considers chanterelles particularly good with red mullet. He marinates the fish fillets in olive oil, lime juice and seasoning before frying them, skin-side first. If you're trying this at home, serve the mushrooms on the side, cooking them as per the recipe in Mushroom & Anise, page 77, but using shallots and parsley instead of garlic and tarragon. I'd serve mackerel fillets the same way, except I'd leave out the cream and give both fish and mushrooms a quick squeeze of lemon.

Mushroom & Onion Warm, soft and as inviting as a pair of sheepskin slippers. *Duxelles* is not, as it sounds, a kind of mushroom but a combination of finely chopped mushroom and shallot (or onion) that has been slowly sautéed in butter. Try a ratio of about 7:1 (in weight). In the largest frying pan you can lay your hands on, soften the shallots over a low heat, without browning them, then add the mushrooms, continuing to cook until all their juices have evaporated and the mixture is dark and soft. *Duxelles* can be used as a sauce—for example, with fish or chicken—as a stuffing (some include it in beef Wellington), in scrambled eggs and omelettes or simply on toast. Alternatively, make a mushroom and shallot dressing by cooking the mixture for a shorter time—i.e., until the mushroom juices have reduced to almost nothing—adding a little red wine vinegar to deglaze, then transferring to a dish to cool before whisking in olive oil. Wonderful with globe artichokes.

Mushroom & Oyster *See Oyster & Mushroom, page 150.*

Mushroom & Parsley Parsley lends a lovely, grassy note to rich, earthy, autumnal mushrooms fried in olive oil or butter, maybe with a little garlic. Inhale that damp, turfy aroma and you could almost believe you were up at the crack of dawn, strolling through the dewy grass with a basket hooked over your forearm.

Mushroom & Pork *Boletus edulis* are better known by the Italian term, porcini, which means "little pigs"—perhaps after the coarse appearance of the mushroom stalks, which look (but don't feel) like bristly pigskin. Porcini also share pork's reputation for needing to be cooked thoroughly: eaten raw, they can cause stomach upsets. But their flavor is magnificent, and not only survives drying but is thought by many to be improved by it. It's the mushroom for the late-rising city dweller, who need only make a slipper-shod shuffle to the fridge for a wax-paper bag of pork and porcini sausages. You can also buy the mushroom in powdered and stock-cube form, handy for pork and porcini pasta sauces or aromatic noodle broths.

Mushroom & Potato *See Potato & Mushroom, page 92.*
Mushroom & Rosemary *See Rosemary & Mushroom, page 310.*

Mushroom & Shellfish The soft sweetness of scallops is a great contrast to the deep, earthy intensity of cooked mushrooms—posh porcini and roasted crimini alike. In Thailand, shrimp is matched with similarly nutty straw mushrooms in the coconut milk and lemongrass soup *tom yang gung*. Japanese shiitake and shrimp *gyoza* dumplings give a delicious double hit of umami. In France, mussels, mushrooms and sometimes oysters are combined in *sauce normande*, which is served with fish, especially sole. In the autumn, foragers can feast on clams and chanterelles, brought together by their mutual love of garlic, wine and parsley. Shellfish is yet another of the flavors identified in mushrooms: *Russula xerampelina*, the shrimp mushroom, or crab brittlegill, is found in the coniferous forests of northern Europe and America, and some say it imparts a shellfish- or crab-like flavor to dishes.

Mushroom & Soft Cheese *See Soft Cheese & Mushroom, page 74.*

Mushroom & Thyme Although mushrooms feel at home with grassy herbs such as tarragon and parsley, the woody herbs are their true kindred spirits. The piney, smoky nature of thyme makes a harmonious match with the thick, earthy flavor of mushrooms, especially dried ones. Combine them in a risotto or a rich white bean stew, or simply on toast.

Mushroom & Tomato Tomatoes and mushrooms garnish a steakhouse steak and, alongside sausages, bacon and eggs, are part of a full English breakfast. They sit on the plate as separate as boys and girls at their first dance. Never an entirely easy partnership, but can work together in sauces for pasta or fish. Moti Mahal, an Indian restaurant in London, serves a mushroom *shorba* (soup) with tandoori bread and tomato chutney.

Mushroom & Truffle Kissing cousins. Truffles are not mushrooms but they are fungi. Truffle oil, paste or butter is often used to enhance the flavor of mushroom dishes; like a culinary push-up bra, the aim is to give the more ordinary fungi the full, in-your-face sexiness of truffle. It works, but it's a little obvious. Classier to use them together but keep them distinct, as at Carlos' restaurant in Illinois, where they serve a mushroom soup topped with truffle foam and sprinkled with porcini powder, cappuccino-style.

Mushroom & Walnut Both ligneous. Mushrooms tend to the sort of heavy, damp woodiness you greedily inhale on a forest walk in autumn; walnuts to the warmly sweet, dry fragrance of timber in a DIY store. Mushrooms cooked in walnut oil are characteristic of southwestern French cuisine, while in many parts of Europe and North America mushrooms and walnuts are frequently paired in sauces, soups and salads. Raw button or crimini mushrooms have a subtle flavor and make for a great texture contrast with crisp toasted nuts. Toss them with a walnut oil and sherry vinegar dressing, and consider adding some goat cheese or blue cheese. *Mousserons* you may know better as fairy ring mushrooms, although lawn owners might as well call them turf herpes: they're virtually impossible to eradicate. On the plus side, they're delicious cooked, with faint flavors of anise and almond. Dried, they take on a sweeter, nutty character that some people compare to walnut, to the extent that they recommend trying them in cookies.

Mushroom & White Fish Italian chef Giorgio Locatelli combines turbot (or brill) with porcini, although he says his grandfather would "turn in his grave" at the idea. In his experience, parsley acts as a bridge between the two ingredients. The mushroom expert John Wright singles out the rich, buttery horn of plenty mushroom as a "happy companion" to fish, particularly white fish. Beech mushrooms are also frequently paired with fish: they have a nutty, some say shellfish-like, flavor, and retain their crunchy texture when cooked. Two reasons to dust off your recipe for sole *bonne femme*—sole in a white wine, butter and mushroom sauce.

Eggplant

When raw, a good eggplant tastes like a bland, sweet apple; cooked, it's transformed into something very savory. Frying eggplants lends them a wonderful creaminess that's particularly lovely sprinkled with sweet, warming spices. Stewed or roasted, they take on a musky, mushroomy quality that works well with salty ingredients. Short of taking a surreptitious nibble, the best way to check if an eggplant has the requisite flavor and texture is to test it for tautness. Ideally an eggplant should be as tight and shiny as dolphin skin. Similarly, they squeak when you pinch them.

Eggplant & Bell Pepper Of all Turkey's many eggplant dishes, *patlican biber* is one of the most popular, so much so that in early evening the smell of eggplants and green peppers frying in olive oil fills the air the length and breadth of the country. Once they've cooled, they are served with two simple sauces, one of cooked tomatoes and garlic and another of thick yogurt mixed with salt and more garlic. See also Garlic & Thyme, page 114.

Eggplant & Chili There's a Sichuan delicacy called fish-fragrant eggplants, which is an awful lot nicer than it sounds. "Fish-fragrant" refers not to fish itself, which isn't used in the recipe, but to the seasoning, which is more often applied to fish in Sichuan cuisine. Not all versions include pork, but this one does. Take 1½ lb small, slim eggplants and cut them lengthways into quarters. Heat 2 cups oil in a wok and deep-fry the eggplant pieces a few at a time until golden and tender, then drain on kitchen paper. Discard all but a few tablespoons of the oil, get the wok nice and hot again, then add 1–2 tbsp Sichuan chili bean paste, muddling it into the oil. Add 2 tbsp each finely chopped fresh ginger and garlic, 1 lb minced pork and 3 tbsp sliced scallions. Stir-fry for 30 seconds, then add 3 tbsp rice wine (or sherry), 3 tbsp black rice vinegar (or cheap balsamic), 2 tbsp sugar, 1 tbsp roasted and crushed Sichuan peppercorns, and 2 tsp ground red chili. Cook over a high heat for 2 minutes, add ½ cup chicken stock and simmer for a further 3 minutes. Finally add the eggplants and simmer for 3 minutes. If you can't bear to deep-fry your eggplants, shallow-fry them instead. You don't get the same texture, but the sauce is so good I'm tempted to say it hardly matters.

Eggplant & Garlic We're all either radiators or drains. Radiators are outgoing, effusive, participatory; drains suck the energy out of the room. Couples often comprise one of each. Your eagerness to see *x*, the radiator, is tempered by your dread of being stuck next to *y*, the drain. In this partnership, garlic is the radiator. It's a beguiling extrovert. Eggplant is the drain: unpredictable, often bitter, and needing a lot of attention (or an unhealthy amount of lubrication) to cajole it into a companionable mood. Together they make *baba ghanoush*—mix the flesh of roasted or grilled eggplants with raw garlic, tahini, olive oil, lemon juice and parsley.

Eggplant & Ginger Japanese eggplants are milder than ordinary ones, if that's imaginable, thinner-skinned, and particularly worth seeking out. If you can't find them, there's no reason not to give an ordinary eggplant the Japanese treatment. They love to soak up the flavor of miso, or this ginger and soy broth. Cut a couple of eggplants into bite-sized pieces, sprinkle with salt and leave for 20–30 minutes. Rinse, squeeze gently and pat dry. Fry in peanut oil until just golden, then add 1 tbsp grated fresh ginger, 2 tbsp soy sauce, 1 tbsp sugar and just enough water to cover. Simmer for 20–30 minutes with a foil or greaseproof paper lid on top of the mixture (not on top of the pan). Garnish with the green part of scallions, thinly sliced, and/or a scattering of sesame seeds. Serve with boiled rice.

Eggplant & Lamb *See Lamb & Eggplant, page 54.*

Eggplant & Nutmeg Freshly grated nutmeg puts an extra "ahhh" into eggplants. There should be a global chain selling paper cones of nutmeggy fried eggplant slices. (The EggPlant™. I'm rich!) Evelyn Rose writes that if you deep-fry them they absorb less fat than when they're shallow-fried because the surface becomes "sealed." Alternatively you can salt the slices first to draw out their moisture, thus making them less fat-absorbent, before shallow-frying. Either way, don't be tempted to use ready-ground nutmeg—it has to be freshly grated to order.

Eggplant & Prosciutto *See Prosciutto & Eggplant, page 169.*

Eggplant & Soft Cheese According to Elizabeth David, eggplant and cheese are a less than ideal combination. If you've ever wrapped a soft stole of char-grilled eggplant around the quivering white shoulders of delicate mozzarella, you may well beg to differ. See also Tomato & Eggplant, page 254.

Eggplant & Tomato *See Tomato & Eggplant, page 254.*
Eggplant & Walnut *See Walnut & Eggplant, page 232.*

Cumin

Cumin seeds are too harsh and unfriendly to nibble straight out of the pot, as you might anise or coriander seeds. They're dry, woody and musty—inhale from the jar and you may recall the note in bought curry powder that smells like the inside of a secondhand wardrobe. Fortunately, they are transformed by cooking. Roasted and crushed, they release nutty, lemon notes; fried in oil, they lend dishes like dhal a lively, piquant bite.

Cumin & Apricot There's so much going on in a tagine that, even if it's cumin-scented and full of plump apricots, you're not likely to notice how the sun-baked, floral, woody notes of apricot withstand the earthy, dusty pungency of cumin. Their reciprocity is clearer in this Armenian apricot soup recipe from David Ansel. Apricots originated in Armenia, where fruit (especially cherry) soups are popular. Dice an onion and a couple of carrots, sauté in olive oil for 10 minutes, then add 2 tsp ground cumin. Lower the heat, cover and sweat for 10 minutes. Add 1½ cups red lentils and as much of 5 cups water as you need to cover them. Bring to a simmer and cook for 20 minutes, adding more of the water if necessary as the lentils expand. Remove from the heat and stir in 6 oz chopped dried apricots, some salt and the rest of the water. Blend roughly, in batches if necessary. Cumin is also excellent in an apricot conserve to eat with Camembert.

Cumin & Beet Aside from their shared earthiness, beets and cumin couldn't be more different. The sweetness of beets is enlivened by cumin's smoky, citric edge. They make a complex soup, with a sharp swirl of crème fraîche or sour cream. Or combine them with chickpeas, harissa (Middle Eastern hot chili paste) and lemon to make a beet and cumin hummus.

Cumin & Carrot *See Carrot & Cumin, page 224.*

Cumin & Cauliflower Roasted together, cumin and cauliflower take on a nutty sweetness without losing their essential characters. Cut a cauliflower into small florets and toss in oil before shaking over 1 tbsp ground cumin. Roast at 350°F for about 30 minutes, until soft, stirring once or twice. Sprinkle with salt and serve warm. Some people blanch or steam the florets first but it's not necessary. They're even more irresistible when slightly charred around the edge. I first came across this combination rolled into a falafel in Amsterdam, which was so intriguing I had to unravel it to investigate the identity of the mystery ingredient.

Cumin & Cilantro *See Cilantro & Cumin, page 192.*

Cumin & Coriander Seed More likely to be found together than apart in Indian, Middle Eastern and North African cooking. In India they're even sold together, ground or whole. In Morocco they might be added to a hot harissa, to give it an extra aromatic edge, or shaken over deep-fried chickpeas, sold in cones by street vendors. In Egypt, cumin and coriander seed are mixed with sesame seeds, chopped hazelnuts, salt and pepper to make the famous *dukkah*, eaten with olive oil and bread. Food writer Glynn Christian believes the underlying orange flavor of coriander seed and the lick of lemon in cumin explains their powerful affinity. I think they're (beneficially) opposed in some ways, too: coriander is bright and perfumed, whereas cumin is rather murky and gruff.

Cumin & Cucumber In *Taste*, Sybil Kapoor writes that cumin is very good combined with other bitter ingredients such as cucumber, eggplant and cauliflower—paradoxically highlighting the second ingredient's natural sweetness by deepening its bitterness. The obvious way to pair these two is by using cumin to spice up a yogurt-based cucumber soup or raita. But for something a little different, see Peanut & Cucumber, page 28.

Cumin & Egg *See Egg & Cumin, page 133.*
Cumin & Lamb *See Lamb & Cumin, page 54.*

Cumin & Lemon Citral, a key compound in lemon flavor, is often used in cleaning fluids and furniture polish. The flavor of cumin is frequently compared to dirty socks. But don't let that put you off. To make a lovely marinade for a couple of fish fillets or some lamb chops, mix the zest of a lemon with ½ tsp ground cumin and 2 tbsp olive oil. Or make this glorious dhal. Soak 9 oz chana dhal in water for 2 hours, then drain and tip it into a pan with 2 cups cold water. Bring to a simmer, skim off the scum and add 1 tbsp chopped fresh ginger, ¼ tsp turmeric and chili to taste. Simmer for about 45 minutes, partly covered, stirring now and then. Add a little boiling water if it dries out. When the dhal is almost cooked, heat some peanut oil in a pan and fry a sliced large onion until golden, adding 2 tsp cumin seeds and 1 tsp garam masala toward the end. Stir this mixture into the dhal with the zest of ½ lemon and 1–2 tbsp lemon juice.

Cumin & Lime *See Lime & Cumin, page 295.*
Cumin & Mango *See Mango & Cumin, page 284.*

Cumin & Mint Inhale cumin and dried mint and you could be in Cairo. Use them together to season lamb kebabs or burgers. They're also delicious with fava beans and with soft, tangy cheeses. In India, a lassi might be flavored with a few pinches of each and some salt.

Cumin & Oily Fish Cumin is a great pairing for tuna because it's potent enough to stand up to the rich, oily fish without swamping it. Rub tuna fillets with olive oil, give them a hearty shake of ground cumin and seasoning, then fry them quickly (about 1 minute each side for a ½-in-thick steak). Let them rest for a few minutes, then cut them into strips. Pile them into warm corn tortillas or tacos with plenty of lime-tossed shredded lettuce or cabbage and some tangy mango or tomato salsa. Garnish with a little cilantro.

Cumin & Pork *See Pork & Cumin, page 38.*

Cumin & Potato Cumin has something of a musty flavor and shares an earthy quality with potato. Combine them and you might expect something reminiscent of a trudge around a ruined castle on a damp Sunday after-noon—yet they taste anything but gloomy in the Indian potato dish, *jeera aloo*. When cooked, potatoes and cumin take on a new sweetness, and the latter becomes a deal more aromatic. Boil unpeeled new potatoes until just tender, then drain, dry and cut into halves or quarters. Fry in oil with cumin and salt until browned. Garnish with chopped fresh cilantro.

Cumin & Shellfish *See Shellfish & Cumin, page 139.*

Cumin & Washed-rind Cheese Cumin seeds are the classic partnership for Munster cheese. So classic, in fact, that you can buy Munster already encrusted or riddled with the seeds, and there are bakeries in France that make cumin bread specifically to go with it. In its homeland of Alsace, Munster is served with boiled potatoes and a pile of roasted cumin seeds—a very macho trio, redeemed by the company of a delicately feminine (and local) Gewürztraminer white wine. Try other washed-rind cheeses with cumin, too. At Galvin at Windows in London, they serve Stinking Bishop with a Jersey Royal potato salad and a cumin tuile. Or follow the recipe for crackers in Washed-rind Cheese & Anise, page 61, using the same amount of cumin seeds in place of fennel.

Beet

An unlikely sort of vegetable: dense, bluntly sweet, needing two hours' boiling, with more than a hint of the garden shed in its flavor and a habit of bleeding over everything. And yet neither golden beets nor the pretty pink and white tie-dyed Chioggia varieties have posed much threat to the prevalence of traditional beets, crimson as a Russian doll's cheeks. The secret of beet's success is its strange combination of sweetness and earthiness, which sets off ingredients that are predominantly sour, salty, or both, like goat cheese. The flavor of beet is also found in its leaves, which can be used in salads or cooked like spinach.

Beet & Anchovy Sweet beets are more than happy to take on salty ingredients like goat cheese, capers and piquant anchovy. In the South of France, beets are diced and mixed with lots of anchovies, garlic and olive oil. *Uhlemann's Chef's Companion* notes that salad of beets, anchovies, small crayfish and lettuce is famous among epicures, and is named after the writer Alexandre Dumas.

Beet & Apple *See Apple & Beet, page 263.*
Beet & Beef *See Beef & Beet, page 46.*
Beet & Caper *See Caper & Beet, page 102.*

Beet & Chocolate There's a popular, or at least widespread, cake recipe that pairs chocolate and beet. Its champions can hardly *believe* the lusciousness and chocolatiness of the combination. I couldn't either, and having tried it I still don't. Carrot works in cakes because it is sweet, floral and spicy, and the grated pieces create a lovely rickety-rough texture. In chocolate beet cake, the cocoa almost entirely overwhelms the beet flavor, leaving nothing but a hint of its earthiness, which makes the cake taste like a cheap chocolate cake that's been dropped in a flowerbed. And the raw cake mixture was so unpleasant that no one wanted to scrape the bowl clean. Case closed, at least in my kitchen.

Beet & Coconut *See Coconut & Beet, page 280.*
Beet & Cumin *See Cumin & Beet, page 84.*

Beet & Dill The Italian gastronome Pellegrino Artusi, writing in the nineteenth century, noted that although the Florentines used lots of herbs in their cooking they were missing a trick with dill, especially dill mixed with beets. He had tried the combination in Romagna, where beets and dill were sold bundled together in the market. The ingredients have traditionally been more a feature of northern and eastern European cooking. Dill is often used to flavor borscht, and is essential for the cold beet soup from Lithuania called *saltibarsciai*. Whisk a little water into a sour dairy base (kefir cheese, buttermilk or sour cream) to thin it a little, then add plenty of grated cucumber, grated cooked beets, chopped boiled eggs, dill and chives. Chill, then serve with cold boiled potatoes and more dill.

Beet & Egg *See Egg & Beet, page 131.*

Beet & Goat Cheese A lively, stinging goat cheese is the perfect foil to beet's sweetness. Just as well, as this has surely been the signature partnership of the last ten years, in the same way roasted peppers were inseparable from tomatoes in the 1990s. Goat cheese soufflé with beet ice cream. Horseradish and beet tart with a goat cheese meringue. Goat cheese panna cotta with beet caviar. Primp them as you will, to me they're at their best muddled with warm green beans and a few crushed walnuts. And this risotto is good too, if rather pink. Soften a finely chopped small onion in olive oil, then add ¾ cup risotto rice, stirring until the grains are coated in oil. Add a sherry glass of white wine and cook until evaporated. Mix in 4 medium diced or grated cooked beets, then add 3 cups hot vegetable stock, one ladle at a time, stirring constantly, until the rice is cooked to your liking. Add a few tablespoons of finely grated Parmesan and some seasoning. Divide between 2 plates and serve, topped with a scattering of goat cheese cut into small cubes. Garnish with lemon thyme leaves, if you have them.

Beet & Horseradish *See Horseradish & Beet, page 104.*
Beet & Liver *See Liver & Beet, page 44.*

Beet & Oily Fish In Scandinavian and Baltic countries beet is commonly paired with salty fish, especially herring. The fish and beets are mixed with onion, potato and apple and dressed with vinegar or maybe a mustardy mayonnaise. The dish is called *sillsallad* in Sweden and *rosolje* in Estonia. On the Danish island of Bornholm, they eat salt-fried herring on dark rye bread with beets and hot mustard. The London smokers H. Forman & Son sell a beet-cured salmon.

Beet & Onion *See Onion & Beet, page 107.*
Beet & Orange *See Orange & Beet, page 288.*

Beet & Pork From a recipe for *barszcz* (i.e., borscht) in Louis Eustache Audot's *French Domestic Cookery*, published in 1846: "Put into a stockpot eight pounds of beef, two pounds of smoked ribs of pork, half a pound of ham, thirty morels, onions, and leeks, and some beet juice. Make the whole into bouillon; strain it, and add to it a hare, a roasted fowl and a duck; then, again, a quantity of beet liquor. Let it boil a quarter of an hour, strain the bouillon afresh, add a few whites of eggs beaten up with a little water; boil it up and strain it again; cut up the boiled viands, and serve them with the bouillon garnished with morels, onions, slices of beef intermixed with celery and sprigs of parsley, the whole stewed beforehand; together with fennel, broiled sausages, and balls of *godiveau* [a veal stuffing]." I suggest you make this while singing "Louis Audot made some borscht" to the tune of "Old MacDonald." Incidentally, Audot goes on to give a recipe for beet juice, which is often cooked separately from the meat in borscht in order to keep it a vibrant crimson. It's not until right at the end that he adds, "This soup can be made with much less meat than directed in the last receipt, and will be *barszez* if it merely contain the juice of beetroot," which only goes to show that you should always read the recipe through before starting to cook. See also Onion & Beet, page 107.

Beet & Potato *See Potato & Beet, page 90.*
Beet & Walnut *See Walnut & Beet, page 231.*

Beet & Watercress Earth and iron. Like a Zola novel with a happy ending. The love of rustic, rosy-cheeked beet for outspoken watercress *is* requited. Combine lots of finely chopped watercress with sour cream and use to dress cooked beets. Serve with oily fish, liver or a rare steak.

Potato

Potatoes are sweet and slightly bitter (if the bitterness is pronounced, and the skin green, they're fit only for the compost bin). Good potatoes are characterized by combinations of buttery, creamy, nutty and earthy flavors. When baked, the skin of old potatoes can have a malty, dusty cocoa character. Cooks tend initially to divide potatoes by texture rather than flavor, as the waxiness, flouriness or firm bite dictates the ideal cooking method. Waxy potatoes usually have a more concentrated flavor, while floury are often described as bland or light tasting. A trained testing panel coordinated by the Scottish Crop Research Institute recently found a direct correlation between levels of umami-forming compounds in potato varieties and their intensity of flavor. The best-performing cultivars were derived from the "phureja" strain—look out for Mayan Gold, whose flesh is butter-yellow, rich in flavor and makes fries so tasty you'll be tempted to skip the ketchup. The potato's sweetness is particularly heavenly when contrasted with salty foods such as fish, hard cheese, caviar and salted butter.

Potato & Anchovy *See Anchovy & Potato, page 161.*
Potato & Asparagus *See Asparagus & Potato, page 130.*

Potato & Bacon Driving past the Farmer's Market Café on the A12 in Suffolk, England, I saw a sign outside that read, in huge letters, Ham Hock Hash. Nothing else. No other food, no opening times, nothing. Just three little words that launched a thousand U-turns.

Potato & Beef Harold McGee writes that the flavor of maincrop (i.e., old) potatoes intensifies when they're stored in the dark at a temperature between 44 and 50°F, and slow enzyme action creates floral, fruity and fatty notes. At too low a temperature, the starch turns to sugar, so that when the potatoes are cooked they begin to caramelize, resulting in a dark-brown French fry with an unpleasantly bittersweet flavor. Which won't do. The partnership of potato and beef is too important, whether in *steak frites* or burger and fries, the fanciest rib of beef with roast potatoes or the everyday cottage pie.

Potato & Beet Potato and beets make a very pink mash that is both sweet and earthy. Beet's potting-shed flavor is attributable to a compound called geosmin, which can also be detected in the smell of just-caught, bottom-feeding freshwater fish such as carp, and in "petrichor," a term coined by two Australian researchers for the distinctive aroma released by rain on earth or concrete following a dry spell. Potato's earthy flavor comes from a different compound. If you like earthy aromas and flavors, they're also to be found in mushrooms, truffles, cooked onion and garlic, some cheeses, and aged Bordeaux and Burgundy. See also Beef & Beet, page 46.

Potato & Black Pudding See Black Pudding & Potato, page 43.

Potato & Cabbage In Ireland, mashed potato + kale or cabbage = colcannon. The Portuguese make a similarly unadorned dish, a soup called *caldo verde*. This is a rustic blend of potatoes and onions with cabbage, often made even less decorative by the addition of a ragged hunk of chorizo. Originally from the Minho region in the north, *caldo verde* is now something of a national dish and is equally popular in Brazil. Try it and you'll see why: high-quality potatoes are used, and the Galician cabbage, with its wide, kale-like leaves, gives the soup a dark, serious depth. Try making it anywhere else and you'd be advised to use floury, rather than waxy potatoes—they need to fall apart and thicken the broth. Galician cabbage can be hard to come by outside Portugal, and collard greens, spring greens or *kai-lan* (Chinese broccoli) can be used instead. Peel 2 lb floury potatoes, cut into chunks and bring to the boil in 6 cups salted water. Simmer until tender, then roughly mash them into the water and bring back to the boil. Add ½ lb finely chopped greens and cook for about 10 minutes. If you like the idea of the sausage, put it in with the potatoes, remove it at the mashing stage and cut it into chunks, then return it to the soup with the greens. See also the introduction to Broccoli, page 124, and Cabbage & Onion, page 119.

Potato & Caper See Caper & Potato, page 103.
Potato & Cauliflower See Cauliflower & Potato, page 123.

Potato & Caviar Simon Hopkinson and Lindsay Bareham write that caviar and truffles are probably better paired with potatoes than with anything else. Potato and truffle share a warm earthiness that makes for a harmonious combination, whereas caviar and potato are a contrasting pair; the sweet, bland softness of potato is pitched against the salty complexity of the taut fish eggs. They're popularly paired in a canapé of warm roasted baby potatoes topped with cold sour cream and caviar, the temperature differential creating an additional pleasing contrast. See also Potato & Truffle, page 94.

Potato & Celery See Celery & Potato, page 97.
Potato & Chicken See Chicken & Potato, page 34.

Potato & Chili There's a sharpness to the flavor of paprika that, along with its smoky quality, makes a successful partnership with potato. Paprika potato chips are by far the most popular flavor in Germany. And potato and paprika make great chili fries too. Mix 4 tbsp olive oil and 4 tsp paprika together in a freezer bag. Using a mandoline or food processor disc, cut 4 large potatoes into strips roughly ½ in square in cross section. Pat them dry, then toss them in the bag until well covered in spicy oil. Transfer them from the bag to a baking tray and roast in the oven at 425°F for 20–25 minutes, tossing once or twice. A pinch or two of cayenne in the oil mix will give them a bit more kick. See also Chili & Tomato, page 207.

Potato & Cilantro *See Cilantro & Potato, page 194.*
Potato & Cumin *See Cumin & Potato, page 86.*
Potato & Dill *See Dill & Potato, page 188.*

Potato & Egg The simplest, cheapest proof that you needn't be an oligarch to eat like a king. A fleshy, fluffy French fry dipped deep in runny yolk. Or a fried egg on a pillow of potato purée. Add a little onion and you've got all you need for a Spanish tortilla: sweet, cakey omelette, soft, earthy potato, bitter-sweet caramelized onion. Eat one fresh from the pan, yellow as a Euro on a mid-afternoon sidewalk, or the next day, by which time it will have taken on a deeper, more savory flavor, plus a grayish tinge that won't matter in the slight-est if you slip it into a crusty white roll for an eggier take on the classic English delicacy, the chip butty (or fried-potato sandwich). A bit rustic, perhaps, but nothing on the *huevos con patates* we ate late one evening in Spain. We ordered it thinking it would be a tortilla, or a Brit-pandering plate of egg and fries. What arrived was a dainty saucer piled with plain potato chips that had been knocked about the pan with a couple of eggs, not quite scrambled, not quite fried. A self-respecting two-year-old would have thrown it away and started again. But it was after midnight and nowhere else was open. Once the chewy slices of potato, bound into clusters by buttery, lacy egg, had proved defini-tively resistant to division by fork, we rolled back our sleeves and ate with our fingers. It was quite delicious. We finished it down to the very last scrap, etch-ing our wineglasses with greasy fingerprints. See also Ginger & Egg, page 303.

Potato & Garlic *Skordalia* is a Greek dish made by beating pounded raw garlic into mashed potatoes with olive oil and a little white wine vinegar or lemon juice until the required consistency is achieved. An egg yolk is added sometimes too. Ground almonds, bread or mashed beans may be used instead of potato or to supplement it. You get the picture. It's white, it's garlicky and it's good with fish. Or simply scoop it up on strips of warm pita bread.

Potato & Globe Artichoke *See Globe Artichoke & Potato, page 128.*
Potato & Hard Cheese *See Hard Cheese & Potato, page 70.*
Potato & Horseradish *See Horseradish & Potato, page 105.*

Potato & Lamb Should get a room. They're all over each other. Stoic beef keeps its distance, even squeezed up close to potato in a cottage pie. The fattiness of lamb, on the other hand, seeps into potato in shepherd's pie, thick, spicy Indian *gosht aloo*, or Lancashire's famous hotpot. John Thornton, the mill owner in Elizabeth Gaskell's *North and South*, is surely in contention for one of the earliest bourgeois enthusiasts of peasant food. Of the hotpot he eats with his workers, he says, "I have never made a better dinner in my life." Years later, conceivably in homage to Mrs. Gaskell, factory boss Mike Baldwin raved over Betty's hotpot in the British soap opera *Coronation Street*. The real secret of a great hotpot is not the use of tasty neck-end of lamb (or mutton), or browning the meat in dripping, or lots of layers of meat, onion and potato, or even adding a few kidneys under the final roof of sliced potatoes. It is time, the one thing the bustling industrialist lacks.

Potato & Lemon *See Lemon & Potato, page 300.*
Potato & Mint *See Mint & Potato, page 323.*

Potato & Mushroom Just as a good fish soup should have a slightly disturbing oceanic depth, so a proper wild mushroom soup should live up to its name: a wolf in soup's clothing. The trick is to use a mixture of different fungi to layer the flavors, as you might use a variety of fish in a bouillabaisse. I like the Italian chef Gennaro Contaldo's recipe, not just for its mixture of fresh and dried mushrooms but for the addition of potato, which thickens the soup while contributing its own earthy flavor. Soften a chopped onion in 4 tbsp olive oil, then add 1 lb chopped wild mushrooms and sauté for 5 minutes. Add 1 quart vegetable stock, 1 oz reconstituted dried porcini (with their soaking water) and a peeled and finely chopped potato. Bring to the boil, turn down the heat and simmer for 20 minutes. Cool slightly and blend until smooth. Reheat, season to taste and serve with crusty bread and the door firmly bolted.

Potato & Nutmeg Nutmeg is used to obscure some of potato's ruder, earthier flavors. It's often given the same purpose with pumpkin or spinach. A little grating should do the job. But then nutmeg is so lovely with all these ingredients, why restrain yourself?

Potato & Oily Fish *See Oily Fish & Potato, page 156.*
Potato & Olive *See Olive & Potato, page 174.*

Potato & Onion Such dependable ingredients that you forget what magic they can work together. When I was a kid, there was always a sack of potatoes and a sack of onions in the garage: they had the same status as logs. Living in a poky flat in the middle of the city, as I do now, the car is parked in the street and has become a garage itself—internalizing what would have housed it. Because it's dark and usually larder-cool, I keep my potatoes and onions in the trunk. And yet within half an hour, these utilitarian vegetables, nestled amid the outdoor gear and plastic bottles of brake fluid, might be baked in milk, or sautéed together, or combined to make a creamy onion mash, or grated and fried in a rösti, or chopped in a simultaneously comforting and zestful potato salad, or layered with cheese for a golden-brown pan haggerty or, most transcendently of all, combined with a few eggs in a tortilla, so much more than the sum of its parts.

Potato & Parsley *See Parsley & Potato, page 190.*
Potato & Parsnip *See Parsnip & Potato, page 220.*

Potato & Pea The presiding culinary spirits of uncontrollable children, acne, acrimonious divorce and mass unemployment. Nothing wrong with French fries and peas *per se*, of course, but in too many pubs, canteens and cafés, depressing slabs of undercooked frozen potato blunder onto the plate, backed up by an entourage of hard, polyhedral peas. Best to close your eyes and think of a verdant pea and potato soup, or Indian *aloo matar*, in which the pea and potato are submerged in a sauce of fresh ginger, garlic and ground spices.

Potato & Peanut Peanut has a meatiness that inevitably sees it paired with potato. In the city of Popayán in Colombia, they make small pasties stuffed with fried potato and peanuts called *empanada de pipián*, which are served with a peanut sauce. Boiled potatoes are usually part of an Indonesian *gado gado* salad—see Peanut & Coconut, page 27. Even in Thailand, the potato gets a rare culinary outing in the peanut-laced *mussaman* curry. Lastly, there's one of my husband's signature dishes, the Dalston Dinner, akin to a fish supper but with the added advantage that you don't need to leave the pub to eat it. Simply empty a packet of salted peanuts into a bag of salt and vinegar chips, clench the bag shut and shake. Mysteriously more delicious than it should be. Good with lager.

Potato & Pork *See Pork & Potato, page 40.*

Potato & Rosemary Rosemary's affinity for potato makes it irresistible on a spookily bloodless Roman *pizza bianca*, topped with potato, garlic and not even so much as a rumor of tomato. A sturdily flavored, waxy new potato such as Yukon Gold is usually specified for this recipe, or you might try Duke of York or La Ratte. Parboil, then slice them very thinly and arrange, just overlapping, on an oiled (uncooked) pizza base sprinkled with a clove or two of finely chopped garlic. Drizzle with more oil, season well and scatter with finely snipped rosemary. You can pep up the flavor, if not the color, with cheese (Parmesan, mozzarella, Asiago) or onion.

Potato & Rutabaga A combination of mashed potato and rutabaga is called clapshot. Mix 1:1 and add plenty of butter, plus either chives or some crisp fried onion. Conjoined but not combined, "neeps and tatties" are the essential side dish for haggis. The tatties, or potatoes, are mashed with butter, and cozy up to the haggis's mellow sheepishness, while the neeps, or rutabaga, are bolstered in their natural spiciness by nutmeg, answering the spicing of the meat.

Potato & Saffron *See Saffron & Potato, page 178.*

Potato & Shellfish Whole cultures have been founded on this combination: chowder in New England, *moules frites* in Belgium, potato gnocchi with spider crab sauce in Venice. At Romerijo's in El Puerto de Santa María, between Jerez and Cádiz, you can tuck into net-fresh *mariscos* or *pescados fritos* with crisp, golden fries. Romerijo takes up more than its fair share of space in this lively seaside town—two branches face each other across a perpetually crowded narrow street, one specializing in fried seafood, the other in boiled. Vast glass counters house every variety of shellfish you've ever heard of, and plenty you haven't; from minuscule *camarones*, hardly bigger than the commas on the menu, to lobsters meaty enough to have rowed themselves ashore and smacked a longshoreman in the chops for his trouble. Sea-salty fried potatoes are served in paper cones printed with the restaurant's blue and white logo. Customers sit outside at Formica tables, shouting over the din as they decapitate their *langostinos*, or peel off their delicate orange armor and toss it in plastic buckets already brimming with crab claws and winkle shells as dark and shiny as ceremonial helmets. It's like a scene by Brueghel's more optimistic brother: Brueghel the Happier. Round the corner on Calle Misericordia, you'll find plenty of places for an *aperitivo*, and the trick is to drink enough oaky white Rioja to order without fear from the range of bivalved and tentacular monstrosities twitching on the ice, but not so much that you arrive at Romerijo when all that's left is a mound of *percebes*—a crustacean whose appearance can be compared only to the armor-plated foot of a miniature aquatic pig. In English-speaking countries, *percebes* are known as goose barnacles, and were once believed to metamorphose into barnacle geese—an altogether more appropriate fate, in my book, than going anywhere near anyone's mouth. Just have the fries.

Potato & Smoked Fish *See Smoked Fish & Potato, page 164.*

Potato & Tomato Let's not beat around the bush—fries and ketchup. Arguably the most popular partnership in the Western world. The exact origins of *frites* are obscure but the term "French-fried potatoes" began to appear in the mid-nineteenth century, shortly before the introduction of Heinz tomato ketchup in 1876. The singular thing about ketchup is its strength in all five basic taste categories: it is sweet, sour, bitter, salty and rich in umami. Tomato naturally contains umami, as does potato; that's deliciousness squared. Elizabeth David gives a Greek recipe for mashed potato mixed with skinned, finely chopped tomato, sliced scallion, parsley, melted butter and flour, shaped into patties, then fried or baked. She also mentions a French recipe for a cream of tomato and potato soup: in brief, you soften the chopped whites of 2 leeks in butter, then add ½ lb roughly chopped tomatoes and cook them until they ooze their juice. Next add ¾ lb diced peeled potatoes, some salt, a little sugar and 3 cups water. Bring to the boil and simmer for 25 minutes. Liquefy and sieve, then return to a clean pan and add ⅔ cup cream. Warm through and serve with parsley or chervil. On Pantelleria, an island off Sicily, *insalata pantesca* combines cooked new potatoes with chunks of raw tomato, olives, red onion and capers in olive oil and vinegar. And, of course, potato and tomato provide the background to a New York clam chowder. See also Chili & Tomato, page 207.

Potato & Truffle Friends of the earth. Encountering potatoes for the first time in Colombia, sixteenth-century Spanish explorers attributed truffle-like qualities to them, even naming them "earth truffles." Infuse a standard potato dish with truffle and their harmoniousness is clear; for example, in truffled mashed potatoes, potato gratin, or a mayonnaise made with truffle oil for potato salad. Giorgio Locatelli serves black truffles on potato gnocchi, because he believes they have a great affinity, although for pasta dishes and risotto he prefers white truffles.

Potato & Washed-rind Cheese Let warm, fruity Vacherin Mont d'Or (see Washed-rind Cheese & Garlic, page 62) ooze slowly into the accommodating flesh of a jacket potato. In tartiflette, a dish from the Savoie region of France, nutty-flavored, whey-washed Reblochon cheese is melted over potatoes, bacon and onion. See also Hard Cheese & Potato, page 70, and Cumin & Washed-rind Cheese, page 86.

Potato & Watercress See Watercress & Potato, page 100.
Potato & White Fish See White Fish & Potato, page 146.

Celery

This section covers celery stalks, celeriac (the swollen base of a celery relative) and celery seeds, which are harvested from smallage, or wild celery, and crushed to make celery salt. All share with the herb lovage a pair of compounds that give them their characteristic celery flavor. This is most potent in celery seeds, which are warm and bitter with a complex herbal, citrus quality; some are more lemony than others. Celery seeds can be useful when you don't have any stalks, or when they prove unwieldy. The stalks have a rather more salty, anise character, whereas celeriac is mild and adds root-vegetable sweetness and earthiness to the celery flavor. The seeds, stems and leaves of lovage are all used, and an alcoholic cordial is made of it to drink with brandy. Celery flavor is particularly savory, recalling stocks and broths; accordingly it's used in soups and stews, where it emphasizes the sweetness of meat and seafood.

Celery & Apple In *American Psycho*, Patrick Bateman's girlfriend, Evelyn, breaks down when her Waldorf salad turns out to be gross. No need to make the same mistake. Chop 3 unpeeled, cored apples along with a stalk or two of celery and bind them to half a cup of walnuts with a tablespoon or so of mayonnaise—but don't drown it. Serve at Christmas with a thick slice of ham and a glass of tawny port.

Celery & Beef Cooked slowly, celery takes on a sweeter, brothier character, and the stalks (and sometimes the seeds) are used to add depth to braises and stews. The same use is made of the celery-flavored herb, lovage, which in Germany is sometimes called *Maggikraut*, in reference to its meaty, yeasty similarity to Maggi stock cubes. Besides dropping them in a pot, you can cross-hatch celery stalks and use them as an edible rack for your beef joint, giving a deliciously deep, savory basis for gravy. And if you're in New York, make sure you try a Dr. Brown's Cel-Ray soda—probably the only celery-flavored pop in the world—with your salt beef or pastrami sandwich.

Celery & Blue Cheese A staple at the Christmas table. Or in a Boxing Day soup as pale and fatty as I feel after three days mainlining After Eights. Also paired in a garnish for spicy Buffalo wings—chicken wings deep-fried, then tossed in margarine and hot sauce and served on a platter with a blue cheese dip and little sticks of celery. I love the idea of all those strong flavors and the contrasting textures and temperatures, but the wings part leaves me cold. Eating chicken wings, I feel like a hamster nibbling on a pencil. Or like one of those crab-eating sea otters that die of starvation because the calories expended getting at the meat outnumber the calories gained when they do. For a more satisfying take on the trio, see Blue Cheese & Chicken, page 64.

Celery & Carrot See Carrot & Celery, page 223.
Celery & Chestnut See Chestnut & Celery, page 228.

Celery & Chicken Celery makes an excellent companion to chicken, even if it's not as popular as it once was. Throughout the nineteenth century, British and American cookbook writers stipulated celery sauce as the natural accompaniment to boiled fowl—indeed, celery sauce was traditionally served with an oyster-stuffed turkey at the English Christmas dinner. Hannah Glasse gives several recipes for it. One involves cutting celery stalks into pieces and simmering them in a little water. When the celery is soft, add mace, nutmeg and seasoning, then thicken the cooking liquid with butter and flour. Other versions are made with veal stock, cream or both. In recent years, scientific studies have shown that celery contains volatile compounds that, although not individually distinguishable to the human palate, significantly enhance sweet and umami notes in chicken stock.

Celery & Egg See Egg & Celery, page 132.

Celery & Horseradish Early recipes for the Bloody Mary omit both celery and horseradish. As did Jacques Petiot, the cocktail's self-proclaimed inventor, in the recipe he gave in the *New Yorker* in 1964—vodka, tomato, cayenne, lemon, black pepper, salt, Worcestershire sauce. But haven't they made themselves indispensable since? Like a couple at an okay party, who turn up late and spice things up; the horseradish makes your sinuses fizz, the celery leaves tickle your cheeks, and the stalk, with the runnels of tomato juice in its furrows, makes an ideal instrument of emphasis in drunken conversations. And of course, they make the drink a meal in itself, so your needs are pretty much taken care of.

Celery & Lamb In Persian cooking, celery and lamb are combined in a *khoresh* much like the one given in Rhubarb & Lamb, page 250, except that the celery is fried with the herbs before being added to the meat. In Turkey and Greece, celery and lamb are stewed in a lemon sauce. At his restaurant in

Langen, Germany, chef Juan Amador makes a dish of Aragon lamb cooked with celeriac, coffee and walnuts.

Celery & Nutmeg *See Nutmeg & Celery, page 217.*
Celery & Onion *See Onion & Celery, page 108.*
Celery & Oyster *See Oyster & Celery, page 149.*

Celery & Peanut "Ants on a log" is an American snack in which the concave groove of a celery stalk is filled with peanut butter and lined with marching raisins. Actually the combination is rather good, beyond the kitschy look of the thing—the crisp, slightly bitter bite of celery balances the salty fattiness of peanut and the sweet grapey-ness of raisin. You might take the elements and recast them in a more grown-up, Thai-influenced salad. Cut 4 tender celery stalks into matchsticks, mix with half a handful of raisins and toss with the dressing in Lime & Anchovy, page 293. Roughly crush half a handful of roasted peanuts, stir half into the celery and scatter the rest over the top.

Celery & Pork *See Pork & Celery, page 37.*

Celery & Potato Crushed, freshly roasted celery seed stirred into your potato salad is a great way to add a little savory spike to it. Potato and celeriac mash is good too; to avoid excessive wetness, it's best to cook the vegetables separately. Be sure to drain both vegetables well, and dry them out over a low heat before adding the butter. See also Truffle & Celery, page 115.

Celery & Prosciutto *See Prosciutto & Celery, page 169.*

Celery & Shellfish The New England lobster roll is one of those legendary sandwiches that sound simple enough but are the cause of multiple disputes over the exact manner of their construction. Everyone agrees on the basics—that it should consist of a generous pile of lobster meat in a soft white hot dog bun—but there are different schools of thought on the presence of lettuce and/or celery and whether to robe the meat in melted butter or mayo. Waking from a coma in season six of *The Sopranos*, the first thing Tony asks for is a lobster roll from the Pearl Oyster Bar in the West Village. If you've ever wondered why mobsters are fat, you might like to note that these contain melted butter *and* mayonnaise. Mix lobster meat, a little finely chopped celery, Hellmann's mayonnaise, a squeeze of lemon and seasoning, and leave in the fridge while you open out hot dog buns like books and brown the insides in a pan of melted butter. Stuff the lobster mix into the bun. Eat lying back on a sun lounger, thinking of New England.

Celery & Soft Cheese *See Soft Cheese & Celery, page 73.*
Celery & Truffle *See Truffle & Celery, page 115.*
Celery & Walnut *See Walnut & Celery, page 231.*
Celery & White Fish *See White Fish & Celery, page 144.*

MUSTARDY

Watercress

Caper

Horseradish

Watercress

With its bittersweet, peppery, mineral freshness, watercress used to be a popular garnish for roast meats. All very well but it deserves a starrier role than culinary spear carrier. It is great paired with a salty and a sweet ingredient—blitzed into a salty stock with sweet milk or cream for a sublime soup, or pressed between slices of sweet bread and salty butter in a sandwich. Liquefied with sour cream and a pinch of salt, watercress makes a sauce that is as refreshing as dandling your feet in the river on a hot afternoon.

Watercress & Anchovy *See Anchovy & Watercress, page 162.*

Watercress & Beef *Tagliata* is a Tuscan dish of beef steak that's seared, then thinly sliced and laid on uncooked arugula, which is dressed by the meat's cooking juices. It's a lovely means of showcasing good steak—with only the mouthwateringly peppery herb for company, you can really taste the meat. Why not anglicize the dish by substituting watercress for the arugula? A change from the more familiar practice of cramming the leaves in a thick sandwich with roast beef. Don't be tempted to serve it with potatoes, which would add a floury heaviness when the delight in *tagliata* is its lightness, in substance if not in flavor. Restrict any accompaniments to a dollop of Dijon mustard.

Watercress & Beet *See Beet & Watercress, page 89.*

Watercress & Blue Cheese The sweet-saltiness of Stilton contrasts nicely with the bitter pepper flavor of watercress. You might also detect a faint metallic tang in them both, as if you'd let the tines of your fork linger in your mouth a moment too long. By all means combine them in a salad with pear and walnut, or in a soup, soufflé or tart. But there's nothing quite like spreading bread with buttery Stilton, deep enough to leave a pleasing impression of your teeth when you bite into it, and scattering it with watercress leaves. See also Parsnip & Watercress, page 221.

Watercress & Chicken I think of roast chicken and watercress as a warm-weather counterpart to roast beef and horseradish. The sweetness of the meat is emphasized by the hot, peppery kick of watercress, but the leaf's refreshing greenness simultaneously lightens the combination. In France, watercress is the classic garnish for roast chicken, and works particularly well when the sweetness of the flesh and the bitterness of the leaves are balanced by crisp, salty skin—although a few olives, or a Thai dressing made with fish sauce and lime juice, do the same trick if you like to eat only the lean white meat. A warm watercress sauce for chicken can be made very quickly. Soften a couple of shallots in butter, then add about ⅓ cup white wine and simmer until reduced to about 1 tbsp. Add 1⅔ cups hot chicken or vegetable stock, simmer for 5 minutes and add ⅔ cup cream. When heated through, add 8 oz chopped watercress, cook for 1–2 minutes, check the seasoning and blend.

Watercress & Egg Garden cress (sometimes called peppergrass) has a rather shyer bite of mustard oil than its close relative, watercress, but both work in pleasingly prickly contrast to the cozy comforts of egg. There's no finer contrast for a sandwich to eat with your little finger in the air. The Ritz Hotel in London serves egg mayonnaise and cress bridge rolls (like miniature hot dog buns) as part of its afternoon tea. Watercress also makes a good omelette, or a salad with a soft poached egg and pieces of chorizo.

Watercress & Goat Cheese As an intensely blue sky sharpens the objects against it, so the bitterness of watercress gives vibrant goat cheese a cleaner, more defined edge. And both tend to a similarly ringing minerality. Cooked watercress loses its kick but is still well matched with the cheese, particularly in a watercress soup with a goat cheese garnish. Or pair them raw in a salad, dressed with walnut oil and sherry vinegar, as long as you're prepared, as a goat would be, to give the leaves a long, ruminative chew.

Watercress & Grapefruit *See Grapefruit & Watercress, page 293.*

Watercress & Oily Fish Trout and cress; not so much a pairing as a reunification. Trout feed on the more tender leaves of watercress but they're really after the sowbugs, tiny crustaceans that live in its thickets. As with other natural pairings, trout and watercress are sufficiently harmonious to need a minimum of preparation. To serve two, fry a trout each in clarified butter for about 5 minutes on each side, and make a watercress sauce by blending a bunch with ⅔ cup sour cream, a squeeze of lemon and a pinch of salt and sugar. The Japanese, who didn't come to watercress until the late nineteenth century, make a cooked salad called *o-hitashi* with it. Blanch a bunch, plunge it into iced water, drain, chop into edible lengths, dress with dashi (dried tuna stock), a little rice wine and soy, leave to marinate for a while, then eat cold.

Watercress & Orange *See Orange & Watercress, page 291.*
Watercress & Parsnip *See Parsnip & Watercress, page 221.*

Watercress & Pork In southern parts of China, watercress is slowly simmered with pork ribs to make a simple soup. It might be flavored with fresh ginger or with jujubes, a fruit that looks and tastes like date.

Watercress & Potato The Chinese like their watercress soup in (often pork-based) broth form. In France and Britain, the taste is for the thicker potage style, achieved with the use of cream, potatoes or both. I like all watercress, but the stuff sold in bunches seems to have a cleaner, mineral fragrance, as opposed to the bagged watercress that gets all pondy if you don't use it straight away. If you use a lot of watercress, you might consider growing land cress, which has a similar flavor and grows easily through the winter under a cloche. Some say land cress is stronger flavored, and that you should use less of it in a soup than you would watercress. I say use more stock and potato and make more soup.

Watercress & Shellfish The pepperiness of watercress tempers the overblown richness of shellfish. Nobu serves a salad of watercress with lobster and black sesame seeds, Alain Ducasse a watercress and scallop soup that's the bottomless green of a Chesterfield sofa in a gentleman's club.

Watercress & Smoked Fish Rich, salty smoked fish and the hot pepperiness of watercress make a fine match but need some sweet relief: eggs and cream in a watercress and smoked trout tart, for instance, beets in a hot smoked salmon and watercress salad, or potatoes in fishcakes with a watercress sauce. See also Watercress & Chicken, page 99.

Watercress & Walnut Watercress sandwiches are unlikely to be met with much enthusiasm, however neatly you cut the crusts off. If the leaves are trapped between slices of homemade walnut bread, it's another story. You might add a little smoked salmon, or thin slices of Brie, but don't underestimate the ability of watercress to draw out the flavor of the bread. In fact it makes rather good bread itself: in Mark Miller and Andrew MacLauchlan's *Flavored Breads*, a recipe is given for an arugula and watercress flatbread, plus a variation using watercress, cilantro and mint. Dip in warm walnut oil.

Caper

Capers, the buds of the caper bush, can revitalize a bland meat dish or an old-fashioned seafood cocktail like a new accessory perks up an old dress. They add a fresh nip, a quirky flavor and a splash of briny liquid that's particularly pleasing with fish. Salted capers retain more of their interesting flavor than brined; soak them for 15 minutes before using and the herbal-mustard character will be more apparent. Some chefs marinate the rinsed capers in white wine and herbs before using them in salads or sauces. Caperberries, which are the fruit rather than the buds of the bush, are also covered in this chapter.

Caper & Anchovy Just a little caper and anchovy can transform bland or oily dishes. We should carry them around in a little envelope, like a sewing kit, for culinary emergencies. Lemon mellows them somewhat; they're a good trio for a flavored butter to use on grilled salmon, tuna steaks or lamb chops. Pound 4 anchovy fillets with 1 tsp lemon juice and ½ tsp lemon zest. Mix into 1 stick softened butter with 3–4 tsp small capers and season. Transfer to a square of cling wrap and shape into a cylinder. Chill, then slice into discs to use.

Caper & Beef *See Beef & Caper, page 46.*

Caper & Beet Sugary beets and mustardy capers make for a kind of honey-mustard combination. They're mixed into minced beef, cooked potato and onion and fried in patties in the popular Swedish dish, *biff à la Lindström* (not unlike the *labskaus* or red flannel hash discussed in Beef & Beet, page 46). Add rinsed capers to an olive oil and red wine vinegar dressing for beet and goat cheese salad.

Caper & Cauliflower *See Cauliflower & Caper, page 121.*

Caper & Cucumber Cornichons, small cultivars of the cucumber family, are usually picked at 1–1½ in long. Their thin, knobbly skin and crunchy flesh make them ideal for pickling. Capers and pickled cornichons are combined in tartare sauce, with herbs and hard-boiled egg in *sauce gribiche*, and in Liptauer, a soft, spicy cheese from central Europe, flavored with paprika, mustard and chives.

Caper & Goat Cheese *See Goat Cheese & Caper, page 57.*

Caper & Lamb Capers' salty bitterness smartly chaperones the sweetness of lamb. The best capers contain notes of thyme and onion, both of which go well with lamb too. This classic recipe is from Keith Floyd. He says it serves six, so you might want to make more sauce if your leg is feeding more than that. Submerge a leg of mutton, or *gigot*, in an oval-shaped pot full of water and bring very slowly to the boil. Skim the fat from the surface and add 6 leeks, 2 rutabagas, 6 carrots and 4 turnips, all chopped into hearty chunks. Simmer for about 2 hours. Just before the cooking time is up, melt 1 oz butter in a saucepan, add 1 oz plain flour and stir into a creamy paste. Pour in ⅓ cup warm milk and whisk until smooth. Then add ⅔ cup stock from the mutton pan and simmer gently for about 20 minutes, until you have a velvety, luxurious sauce. Stir in 3–4 tbsp rinsed, drained capers, check the seasoning and pour into a jug. Remove the mutton from the stock, place on a serving plate surrounded by the vegetables and pour over the sauce. See also Goat Cheese & Caper, page 57.

Caper & Lemon Could wake the dead. Stir them into mayonnaise or combine them in a dressing for smoked salmon or fried fish. Or make more of a fuss of their salty acidity by cooking lemon and caper spaghetti. Follow the recipe in Lemon & Basil, page 297, substituting 1 tbsp rinsed capers for the basil. See also Caper & Anchovy, page 101.

Caper & Oily Fish A significant flavor compound in capers, methyl isothiocyanate, has a strong mustard character, and is also found in horseradish—a famous partner for oily fish. Capers cut through the fattiness of fish in much the way horseradish does. Oily fish and capers are especially good paired on pizza. Spread a base with tomato sauce (see Garlic & Basil, page 112, for a good one), then scatter over thin slices of red onion, canned tuna, anchovies, olives and capers. Bake for 10 minutes at 450°F.

Caper & Olive *See Olive & Caper, page 173.*
Caper & Parsley *See Parsley & Caper, page 189.*

Caper & Potato Capers like to razz up mild flavors and cut through fat. They're good in a Mediterranean potato salad (see Potato & Tomato, page 94), can be added to *skordalia* (see Potato & Garlic, page 91) or served with hot (especially sautéed) potatoes. It's worth remembering to add the buds to the cold oil and bring them up to heat. Tossed into hot fat, they're in danger of becoming tiny, briny incendiary devices. Note that long cooking intensifies the flavor of capers. In Greece, the leaves of the caper bush are eaten with potatoes and with fish. They have a more mustardy-thyme flavor than the buds. See also Saffron & Potato, page 178.

Caper & Shellfish Capers work well with all seafood. The pickled variety, whose flavor, unlike salted capers, is somewhat masked by vinegar, are typically used in pickled shrimp, a dish popular in the American South. It's like a ceviche—cold, piquant, with lots of citrus juice—but the shellfish is cooked. See also Caper & White Fish, page 103.

Caper & Smoked Fish Capers make a classic garnish for smoked salmon, cutting through the fish's fattiness and, in their extreme saltiness, making it seem sweeter. Caperberries can be put to the same use, and are excellent with cured meats too. They're the size of a small olive and come with a stalk that pulls off with a satisfying *tock*. In contrast to the baggy texture of capers, caperberries are firm and full of seeds that flood your mouth when you bite into them, like a briny, coarse mustard. The flavor is similar to capers but a little milder. Serve with smoked salmon pâté on whole-wheat toast.

Caper & Soft Cheese Famously paired with smoked salmon, but they're quite delicious together without the fish. Mix drained whole capers into a thick, rich, ivory-colored cream cheese. Use French nonpareil capers, if you can. They're the really small ones that look like green peppercorns and are highly regarded for their finer, radishy, oniony flavor. Spread the mix on crackers or rye bread and brace yourself for the little shocks of caper in each bite. The culinary equivalent of walking barefoot along a stony beach. See also Caper & Cucumber, page 102.

Caper & Tomato *See Tomato & Caper, page 254.*

Caper & White Fish Skate with capers in black butter is a deservedly classic dish but, as a critically endangered species, skate is off the menu. Try the sauce on scallops instead. When skate was cheap and plentiful, crafty fishmongers used to press "scallops" out of skate wings; some believe the sweet, delicate flavors are similar. Taking care not to let it burn, heat 6 tbsp butter in a pan until deeply golden. Add 1 tsp white wine vinegar and 1–2 tbsp drained rinsed capers. Stir, then pour over cooked scallops, any white fish, or the sauce's other classic partner, ox brains. Capers are, of course, also an essential element of tartare sauce.

Horseradish

Horseradish is a bruiser with a gentle side. Cut or grate it and its natural defense mechanism releases hot, bitter compounds to make your eyes run and your nose burn. For that reason, it's usually paired with the sort of pugnacious flavors that can stand up to it—smoked foods, roast beef, piquant cheeses and spicy, boozy tomato juice. Treated gently, however, in small doses, it can tease out the delicate flavors of raw seafood. In common with other hot ingredients such as chili and mustard, horseradish is a showcaser, drawing attention to what's in your mouth. It has a slight freshening quality, too, knocking a little fishiness off uncooked seafood, or the earthiness from potato and beet. This chapter also covers wasabi.

Horseradish & Apple *See Apple & Horseradish, page 266.*

Horseradish & Bacon Bacon and tomato find each other irresistible. Salty, sweet, sour: how could they not? Which is why so many swear by tomato sauce on their bacon sandwiches. I understand the force of habit in these cases, but everyone should try horseradish sauce at least once, especially when smoked bacon is at stake. If it's good smoked bacon, I'd go so far as to say that horseradish is just plain better, complementing the bacon's charry brininess where ketchup puts up a fight. Mix 1 tbsp prepared horseradish with 4 tbsp mayo. Spread more than you should on toasted brown bread, lay the bacon on top, then some sickle-shapes of sliced avocado and a few leaves of crisp Cos lettuce. Sink your teeth in and feel your eyeballs begin to turn upwards. Note how the horseradish gets behind the bacon and gives it a nip on the backside.

Horseradish & Beef *See Beef & Horseradish, page 48.*

Horseradish & Beet Mellow beets talk down headstrong horseradish. Pair in a salad, or in a relish such as the Ukrainian *tsvikili*, a 6:1 mixture of grated cooked beets and grated fresh horseradish, seasoned with salt, pepper, sugar and vinegar to taste. The sweetened Jewish sauce, red chrain, works on similar principles and is served with gefilte fish. Alternatively, pay homage to the couple's Russian roots with a crimson borscht, lifted by a swirl of horseradish mixed with sour cream.

Horseradish & Celery *See Celery & Horseradish, page 96.*

Horseradish & Oily Fish Horseradish is served as a sauce with simply cooked oily fish; chef Richard Corrigan cures herring with a horseradish mixture; and, in the form of wasabi, it provides a hot contrast to raw tuna or salmon sushi and sashimi.

Horseradish & Oyster In New Orleans, oysters are simply dressed with horseradish, or a mixture of tomato ketchup and horseradish. The thrill of cold oyster and nose-tingling horseradish enacts a sequence of shocks, like the lime wedge after your tequila shot. New York chef David Burke pairs cooked oysters and horseradish in a risotto.

Horseradish & Pea *See Pea & Horseradish, page 200.*

Horseradish & Potato Season mash to taste with frisky, fibrous horseradish for a fresher thatch on your cottage pie. Or use horseradish mayonnaise—see Horseradish & Bacon, page 104—in a potato salad to accompany smoked salmon, mackerel or trout.

Horseradish & Smoked Fish *See Smoked Fish & Horseradish, page 163.*
Horseradish & Tomato *See Tomato and Horseradish, page 255.*

Horseradish & White Fish Purists may disapprove, but I love wasabi with my sushi, pushed into the rice like a lump of green explosive. Wasabi and horseradish are close relatives. Both have a pungent, metallic flavor that comes from volatile sulfur compounds liberated as a defensive measure when the plant is damaged—i.e., grated. Horseradish has the more radishy, watercress-like flavor. Most wasabi served in Western restaurants isn't wasabi at all but horseradish dyed green; the same goes for many store-bought varieties. Mind you, any more than a lentil's worth and you'll be weeping hot tears through your nose, so in the quantities it's appropriate to use you'd be hard pushed to notice the difference in flavor. If you do overdo it, the secret, according to Harold McGee, is to breathe out through the mouth, so the wasabi fumes don't irritate your nasal passages, and in through the nose, to bypass the residue in your mouth. Or you could just weep and wait like the rest of us.

SULFUROUS

Onion

Garlic

Truffle

Cabbage

Rutabaga

Cauliflower

Broccoli

Globe Artichoke

Asparagus

Egg

Onion

The hardest workers in the food business. Across the species that comprise the *Allium* genus, onion contributes a range of distinct flavors, from the light, herbal freshness of chive to the delicate, perfumed flavor of shallots, the tear-jerking boisterousness of the bulb onion, and the more vegetal, green-tinged earthiness of leeks and scallions. Raw, onions lend a sharp, crisp edge to dips and salads; roasted or braised, they become sweet and succulent; fried until black-edged, they add a bittersweet dimension to a hot dog.

Onion & Anchovy On a rainy day in Venice with friends, we went hunting for lunch in the backstreets of Dorsoduro. We chose a café with a short menu and no English spoken. Dorsoduro is a little less touristy than San Marco, the other side of the Grand Canal, and the proprietor and his wife seemed delighted to entertain some *inglesi* in their restaurant, as if the very idea of foreign visitors to Venice was a novelty. We wanted to try the local dish of bigoli pasta with anchovies and onions. A platter so enormous a gondolier might have punted it onto our table arrived, along with several pitchers of Soave. Bigoli is like whole-wheat spaghetti but thicker, with a nutty, rugged character. It makes an ideal carrier for the sweet and salty mixture of anchovies and onions. You can make this with whole-wheat spaghetti if you can't get the real thing. For two people, soften 3 thinly sliced large onions in 2–3 tbsp olive oil over a medium heat for about 20 minutes, without letting them color. Once they've had about 10 minutes, add 4 or 5 chopped anchovies. Mix them in, break them up a bit and season, bearing in mind how salty the anchovy is to begin with. Cook the pasta until it is *al dente*. Drain it, reserving about a tablespoon of water, and return both to the pan. Place back on the heat and stir in the sauce. For an extra touch of sweetness, add a tablespoon of currants to the onion mix. You could try some of anchovy's other matches in this too: a little rosemary, perhaps, some blanched chopped broccoli or a sprinkling of capers.

Onion & Bacon *See Bacon & Onion, page 167.*
Onion & Beef *See Beef & Onion, page 49.*

Onion & Beet Beet's sweetness is offset to great effect by raw onion. When cooked, onion takes on a sweetness of its own, which can be balanced out by vinegar in this beet and onion chutney to serve with homemade sausage rolls or in cheese sandwiches. Simmer 1½ lb chopped onions with 1 lb peeled, cored and diced eating apples in 1¼ cups red wine vinegar until tender—this takes about 20 minutes. Add 1½ lb diced cooked beets, another 1 cup vinegar, 2 cups sugar, 1 tsp salt and 2 tsp ground ginger. Boil for 30 minutes more. Spoon into sterilized jam jars while still hot and seal. Makes about 5 standard (1 pound) jars' worth.

Onion & Bell Pepper Stray into a residential area in Spain or Portugal in the early evening and the air will be sweet with onions and peppers being softened for supper. If the combination always smells better than it does at home, I put this down to better ingredients, the cheering late-day sunshine, and the simple fact of being on vacation. But when I came to learn more about Spanish cooking I discovered that, above all, patience is the secret ingredient. An onion and pepper mixture cooked for 20 minutes is, it turns out, actually four times better than one cooked for 5. So much so, in fact, that you can pile it straight from the pan onto coarse white bread and eat it for supper. Whether cooking them for 40 minutes would be twice as good as 20 remains to be seen. I've got *some* patience, but I'm no Job.

Onion & Black Pudding *See Black Pudding & Onion, page 43.*
Onion & Cabbage *See Cabbage & Onion, page 119.*
Onion & Carrot *See Carrot & Onion, page 224.*

Onion & Celery Dice carrot, celery and onion and you have the aromatic base for many stocks, soups and stews known by chefs as mirepoix. Add some salty bacon or cured fat for a *mirepoix au gras* and it's a bit like being dealt three of the same-numbered cards in a hand of poker. You'd be unlucky not to end up with something winning. If you're cooking something that needs less sweetness or a fresher, more herbaceous base, skip the carrot. If you're making, say, a pale broth, and don't want carrot color seeping into it, use a parsnip instead. The mirepoix-like combination of onion, celery and green pepper is called the holy trinity in Cajun cooking. A standard mirepoix calls for two parts onion to two parts carrot to one part celery by volume, whereas the holy trinity is 1:1:1.

Onion & Chicken *See Chicken & Onion, page 33.*

Onion & Clove Pity the clove-studded onion. In bread sauce it sticks with the project all the way through, only to be discarded at the end. Bread swans in at the last minute and takes all the credit. But it's the tang of the onion, softened by the aromatic, fireside warmth of the clove, that gives the sauce its special, irresistible depth. It's the Ugg boot of sauces. You might also try the combination in a clove-infused chicken and onion sandwich. For two baguette sandwiches of about 6 in each, thinly slice a Spanish onion, then cook it nice and slowly in 1 tbsp peanut oil, a dab of butter and ½ tsp ground cloves. Warm the bread, slice some hot, cooked chicken. When the onion is meltingly soft, you could add a few tablespoons of cream, then warm the mixture through. Spread it onto the bread before laying on the chicken slices. Great with a glass of Pinot Noir.

Onion & Cucumber According to the Chinese, watery, cooling cucumber is very yin, warming, bright, strong, dry onion yang. Their yininess and yanginess is never more apparent than in pancakes filled with thick, sweet hoisin sauce, crunchy-soft shreds of crispy aromatic duck, and julienned cucumber and scallion.

Onion & Egg *See Egg & Onion, page 134.*

Onion & Garlic *See Garlic & Onion, page 113.*

Onion & Ginger Forget Fred; scallions are the perfect partner for ginger. A staple combination in Chinese cooking, and so versatile and delicious it can even turn tofu into a savory feast. Fuchsia Dunlop (who learned no fewer than nine ways to cut a scallion at chef school in Sichuan) explains how they're used to temper unsavory flavors in meat and seafood and, more tantalizingly, as a seasoning, shredded and scattered over steamed fish, then drizzled with hot oil to awaken their flavor before a dash of dark soy is added. For me, ginger and scallion is never better than when stir-fried with fresh crab, a dish that can take up to an hour to dismantle with your fingers but is fortunately still delicious when cold.

Onion & Hard Cheese *See Hard Cheese & Onion, page 69.*

Onion & Lamb Onion sauce is a classic partner for roast lamb or mutton. Food writer Charles Campion recalls this pairing was the "banker" recipe his mother always fell back on. Largely, he recalls, because of the outstanding sauce she made with lots of onions softly fried in butter, then simmered in milk with nutmeg and black pepper and thickened with potato flour and cream. The movies are less kind to messy eating habits than books, and James Bond's fondness for lamb and onion kebabs is a notable omission from the film version of *From Russia with Love*: while his Turkish contact, Darko, tucks into what sounds like *kibbeh nayeh* (minced raw lamb, finely ground with chives and peppers), Bond plumps for particularly young, charcoal-grilled lamb with savory rice and lots of onions. No wonder he doesn't get any action that night.

Onion & Liver *See Liver & Onion, page 45.*

Onion & Mint As parsley is to garlic breath, mint is to onion. Remember the 1980s ad for Wrigley's Doublemint, set in a restaurant, in which Girl A, speculating about where the boys might be taking them tonight, asks Girl B whether her hair looks okay? Girl B responds that if she were Girl A she'd be less worried about her hair than her onion breath. Why didn't Girl A gently put down her fork and punch Girl B on the nose? And why was Girl B so worried about Girl A's breath, when Girl A was having the skimpiest side salad, of which the onions formed part, for her main course? Was this why they were eating *before they met the boys?* Because Girl A and Girl B were locked in a toxic cycle of narcissistic and mutually destructive sexual competitiveness? Far from encouraging me to buy gum, it instilled in me a determination to date boys in front of whom I could eat the most lingeringly pungent foods. If he still fancied me after an onion-bhaji-scented kiss, with a trailing after-note of minty raita, I would have found my match. See also Chili & Mint, page 206.

Onion & Mushroom *See Mushroom & Onion, page 80.*

Onion & Nutmeg Often thought of as primarily a sweet spice, nutmeg also has a bitter streak, which provides a useful counterpoint to onion's sweetness. It's good in a sauce made with cooked, puréed onions, added to a béchamel made with milk. Or cook onions very slowly in butter until soft, purée and sprinkle liberally with nutmeg. Loosen with a little chicken stock or cream if necessary. Both these sauces can be served with roast pork, lamb or duck.

Onion & Oily Fish *See Oily Fish & Onion, page 156.*

Onion & Orange Thin sliced rounds of both can make a lovely, crisp salad. Look out for sweet onion varieties such as Vidalia, Empire and Supasweet. The higher sugar levels in sweeter onions come at the expense of pyruvic acid, the defensive chemical responsible for stronger onions' pungency, aftertaste and the teardrops on your cutting board. If you can't get naturally sweet onions, you could try giving your cut onions a rinse in cold water, which arrests some of the stronger sulfur compounds released when you damage their flesh. Blood orange and red onion make a pretty pair on the plate, and red onion is often (though not always) on the sweeter, milder side.

Onion & Oyster *See Oyster & Onion, page 150.*
Onion & Pea *See Pea & Onion, page 200.*

Onion & Pork Unpretentious pork gets along with all the onion family—garlic with roast pork, chives with Chinese pork dumplings, a tangle of onion gravy on sausages, or, best of all, pork and leek sausages. Leeks offer a creamy combination of onion and cabbage flavors—and cabbage is really pork's very best partner of all.

Onion & Potato *See Potato & Onion, page 92.*
Onion & Rosemary *See Rosemary & Onion, page 311.*

Onion & Sage Sage and onion is the classic stuffing for a reason: sweet, herbal, deep and mulchy under a crunchy crust. Perhaps mainly for associative reasons, the combination is redolent of meat and is a knockout with cannellini beans in this bruschetta—a sort of Tuscan beans on toast. Soften a finely chopped onion in olive oil, then add a drained 14-oz can of cannellini beans and a couple of finely chopped sage leaves. Cook over a low heat for 5–10 minutes, then semi-mash, season and serve on rounds of toasted French bread. See also Chicken & Sage, page 34.

Onion & Smoked Fish Thin rings of raw red onion cut through the fattiness of smoked salmon yet match its boisterous flavor—perfect on a warm bagel spread with cream cheese. Instead of smoked salmon, you might choose lox—not the same stuff, even if the terms are often used interchangeably.

Lox is salmon cured in brine, sometimes with onion and spices, but never smoked. The proper stuff is as salty as a fisherman's moustache. Russ and Daughters in New York cures its own lox and serves it with a wine or cream sauce with pickled onion. See also Soft Cheese & Smoked Fish, page 74.

Onion & Thyme *See Thyme & Onion, page 318.*
Onion & Tomato *See Tomato & Onion, page 255.*

Garlic

Adding a small amount of garlic to meat, seafood, green vegetables and even truffles is like drawing a keyline around their flavor—everything gains a sharper definition. Garlic also adds something of a succulent quality. For a mild garlic flavor, infuse warm cooking oil with a whole, unchopped clove; for something more potent, use it raw and crushed to a paste. There are two types of garlic—softneck and hardneck. Softneck is the type you're most likely to find in the supermarket or plaited into a braid, as it's easier to grow, but hardnecks tend to have the better flavor. Try a side-by-side garlic tasting by making garlic bread or aioli with a range of different bulbs, looking out for their different hot, sweet, earthy, metallic, fruity, nutty, rubbery and floral (lily) characters. Elephant garlic is milder and considered part of the leek rather than the true garlic family. Garlic can be bought in powdered, flaked, paste and salt forms, or ready chopped in jars, although I find these prepared products often have a piercing quality and miss the earthy, volatile perfume of fresh cloves. Garlic's main use is in heightening the flavor of savory dishes, but it makes a fantastic primary flavor on a sweet, bland background like bread or pasta, or, in the case of a roasted garlic risotto, rice.

Garlic & Almond This version of the classic Spanish cold soup, *ajo blanco*, involves blending almonds with raw garlic in a food processor—a surefire way of experiencing garlic flavor at its most potent. Garlic contains a sulfox-ide called alliin, which is converted into allicin when the garlic is sliced or crushed. Allicin, in turn, is converted into the sulfide compounds responsible for crushed garlic's characteristic aroma. The more brutally you rupture your garlic, the more alliin is converted. The almond in *ajo blanco* takes the edge off the pungency, as does serving it cold, so you shouldn't be tempted to underplay the garlic. Soak a handful of crustless stale white bread in water or milk. In a food processor place 2 cups ground blanched almonds, the bread (squeezed dry) and 2 peeled, crushed garlic cloves. Pulse until you have a paste. With the motor running, add 3 tbsp olive oil slowly through the feed tube followed by 3 cups iced water. Season and stir in sherry vinegar (1–3 tbsp) to taste. Chill mercilessly and serve with a garnish of grapes, seeded if necessary. See also Almond & Melon, page 240.

Garlic & Anchovy *See Anchovy & Garlic, page 160.*

Garlic & Basil The effect that a few garlic cloves and some basil leaves have on canned tomatoes is nothing short of miraculous. Essentially they pull the flavors of the tomato in opposite directions, stretching it out to its fullest potential. At one end of the spectrum, garlic picks up on the strong dimethyl sulfide flavor of canned tomato, taking it further in a savory vegetal direction; at the other, basil replaces the green, slightly grassy flavors that tomatoes lose in the canning process, lightening and freshening the sauce. Roughly chop 4 or 5 garlic cloves and soften them in olive oil. When they've taken on a little color, add 4 cans of whole plum tomatoes and a generous handful of roughly torn basil leaves. Break up the tomatoes, season and bring to the boil. Remove from the heat and pass through a sieve, working the mixture with a wooden spoon. Discard the garlicky, basilly tomato pulp and pour the sauce back in the pan. Bring to a simmer and let it reduce until it starts plopping like a hot swamp—it should be roughly the consistency of ketchup. Use for pasta or pizza. Gets really heavenly if you keep any excess in the fridge overnight with a couple of fresh basil leaves in it.

Garlic & Beef *See Beef & Garlic, page 48.*
Garlic & Broccoli *See Broccoli & Garlic, page 125.*
Garlic & Cabbage *See Cabbage & Garlic, page 118.*
Garlic & Cauliflower *See Cauliflower & Garlic, page 122.*

Garlic & Chicken A drawback of chicken Kiev is having to dodge the geyser of molten garlic butter as it arcs over your shoulder. Safer to stick to the old Provençal recipe for chicken with 40 cloves of garlic. As the name suggests, you simply roast your chicken with 40 garlic cloves, unpeeled and with their pointy noses lopped off in order that the hot, sweet garlic paste can be squeezed from the skin directly onto toasted rounds of French bread. The garlic goes through a complete personality change, losing its aggressive pungency and turning caramel-sweet, with a chestnutty flavor. But for all its apparent mellowness, roasted garlic lingers no less obstinately on the breath. Last time I made the recipe, I took a cab the following morning and the driver kept giving me dirty looks in the rearview mirror.

Garlic & Chili *See Chili & Garlic, page 205.*
Garlic & Cilantro *See Cilantro & Garlic, page 192.*
Garlic & Coriander Seed *See Coriander Seed & Garlic, page 337.*

Garlic & Cucumber Ken Hom writes about a pickled cucumber salad he enjoyed in Shanghai. The original was made with pungent raw garlic, but he tamps things down by using fried in his version. In tsatsiki, perhaps garlic and cucumber's most famous collaboration, yogurt moderates matters even further, but the hot ardency of garlic still persists. Seed, then coarsely grate or dice half a cucumber, squeeze as much water out of it as you can, and mix with 1 full cup yogurt, a garlic clove crushed with salt, and a tablespoon or two of chopped mint (you might try using yogurt made with sheep's or goat's milk, the flavors of which go particularly well with cucumber). Serve with fatty lamb chops and crusty bread. Or thin with light stock to make a refreshing cold soup, remembering that cucumber will yield more of its water as the flavors meld in the fridge.

Garlic & Eggplant *See Eggplant & Garlic, page 83.*

Garlic & Ginger Good cop/bad cop. Ginger's the good cop: fresh, with a light touch, and a knack for teasing out the sweetness in meat, seafood and greens. Garlic, on the other hand, is rude, coarse, and leaves a firm, sulfurous impression. Indian chefs make a ginger-garlic paste to use as a marinade for meat or as part of a sauce. Process equal quantities by weight of peeled garlic and ginger into a paste, adding about 1 tbsp water per 3 oz. Will keep for a couple of weeks in the fridge or longer in the freezer.

Garlic & Goat Cheese *See Goat Cheese & Garlic, page 59.*
Garlic & Hazelnut *See Hazelnut & Garlic, page 235.*

Garlic & Lamb In Barcelona there's an old inn with an interior like a working forge. A grill the size of a pool table roars and spits beneath immense cuts of lamb on the bone and bundles of artichokes and leek-like *calçots*. A waiter brings you a basket of thick slabs of rough white bread, whole tomatoes and garlic cloves, with which you make your own *pa amb tomàquet*. I followed this with a stack of glistening, salty lamb chops served with a bowl of *aioli*, whose fierceness was perfectly pitched to the sweet fattiness of the meat. When the sting of garlic subsided, each mouthful of lamb tasted sweeter. When the main course had been cleared, a man at the next table leaned over and insisted we order the tangerine sorbet. We were glad he did. It tasted as if the very essence of tangerininess, every perfumed nuance, had been cryogenically preserved.

Garlic & Liver *See Liver & Garlic, page 45.*
Garlic & Mint *See Mint & Garlic, page 322.*
Garlic & Mushroom *See Mushroom & Garlic, page 79.*
Garlic & Oily Fish *See Oily Fish & Garlic, page 155.*
Garlic & Olive *See Olive & Garlic, page 173.*

Garlic & Onion The air our kitchens breathe. Chinese Buddhist monks, however, abstain, classing onions, garlic, shallots, chives and leeks as *wu hun*, or the "fetid" or "forbidden five," on the basis that they excite the senses. A similar injunction against onion and garlic exists in Hindu Brahmin culture. The Jains, for their part, avoid onion and garlic as well as root vegetables, for fear that they might contain living things. As a substitute, both Brahmins and Jains might use the resinous gum called asafetida, which has a pungent, alliaceous flavor. That it's also known as both "devil's dung" and "food of the gods" might recall the equivocal gorgeous/gruesome qualities ascribed to truffles. Raw asafetida resin smells sickeningly sulfurous (the powdered form much less so), but cooked it imbues food with an earthy, savory tang. Although most commonly associated with Indian cuisine, it's native to Iran, where it's sometimes rubbed onto serving plates for meat. Asafetida was so treasured in Roman times that it forms part of half the recipes in Apicius. If you're quite into having your senses excited, you might try five-onion soup, made with garlic, onion, scallions, leek and chives.

Garlic & Parsley *See Parsley & Garlic, page 189.*
Garlic & Pork *See Pork & Garlic, page 39.*
Garlic & Potato *See Potato & Garlic, page 91.*

Garlic & Rosemary The gutsy flavors of garlic and rosemary are an ideal combination for outdoor food. Use them on your barbecued lamb, pork or rabbit, to enrich roasted new potatoes and for an aromatic garlic bread. Or relive the summer on a cold evening by using them in a simple pasta in the "olio, aglio" style—see Chili & Garlic, page 205. Warm some sliced garlic in olive oil with finely snipped rosemary (½ tsp per serving) and mix into cooked spaghetti with plenty of grated Parmesan.

Garlic & Shellfish *See Shellfish & Garlic, page 139.*
Garlic & Soft Cheese *See Soft Cheese & Garlic, page 73.*

Garlic & Thyme Kinder than garlic and rosemary, garlic and thyme work wonders on just about anything savory: vegetables, lamb, chicken, olives. Richard Olney thinks that plenty of thyme and garlic is vital to a good ratatouille. The following (foolproof) method follows his advice and will mean that you'll never again have to consign an insipid attempt to the trash. Chop an eggplant, a red onion, a red bell pepper and 3 small zucchini into small chunks (not dice; think about half a wine cork) and put in a solid ovenproof dish with a can of peeled plum tomatoes. Break up the tomatoes with a spoon and mix in 8 thinly sliced garlic cloves, 8 sprigs of thyme, a few generous glugs of olive oil and some seasoning. Roast at 375°F for an hour, stirring once or twice. Some add a tablespoon of red wine vinegar to perk it up before serving. Really good with Gruyère in buckwheat pancakes.

Garlic & Tomato *See Tomato & Garlic, page 254.*
Garlic & Truffle *See Truffle & Garlic, page 116.*
Garlic & Walnut *See Walnut & Garlic, page 232.*
Garlic & Washed-rind Cheese *See Washed-rind Cheese & Garlic, page 62.*

Garlic & White Fish Aioli turns ordinary white fish into a bit of a feast, especially in the case of *bourride*, an aioli-thickened fish stew from Provence. *Skordalia* is a Greek dip of garlicky mashed potato that's delicious served with grilled or poached fish fillets—see Potato & Garlic, page 91.

Truffle

The price of fresh truffles usually means that they're used sparingly and allowed to dominate without other ingredients getting in the way of their musky, pungent complexity. "Truffle flavor" has, of course, become increasingly ubiquitous. The flavor of white truffles has been successfully synthesized in the form of a compound called bis(methylthio)methane, which is used to make the truffle oil drizzled, not always advisedly, over thousands of restaurant dishes, and sometimes injected into inferior, flavorless truffles for shaving tableside. Fungal fraud aside, there's nothing wrong with synthetic truffle oil, which can work wonders pepping up mashed potato, cauliflower, cabbage or macaroni and cheese, and will certainly give you an idea of what truffle tastes like if you've never had the real thing. In the same way that reading the study guide for *Anna Karenina*, rather than the actual novel, will give you an idea of the book.

Truffle & Asparagus *See Asparagus & Truffle, page 130.*

Truffle & Bacon Everything tastes better with bacon, so the saying goes: even truffles. From an 1833 recipe for "Truffles with Champagne" by Richard Dolby: "Take ten or twelve well-cleaned truffles; put them into a stewpan on slices of bacon, add a bay leaf, a seasoned bouquet, a little grated bacon, some stock, a slice or two of ham, and a bottle of champagne; cover them with a piece of buttered paper, put on the lid, and set the stewpan on hot ashes; put fire on the top and let them stew for an hour. When done, drain them on a clean cloth, and serve on a folded napkin."

Truffle & Beef *See Beef & Truffle, page 51.*
Truffle & Blue Cheese *See Blue Cheese & Truffle, page 65.*

Truffle & Cabbage The characteristic aroma of truffle is attributable to a number of chemical compounds, but it's primarily dimethyl sulfide (DMS) that truffle-hunting pigs and dogs are trained to detect. Truffle grower Gareth Renowden writes that at high concentrations DMS smells like cooked cabbage—a similarity you might note in some oils made with truffle flavoring as opposed to the real thing. Marco Pierre White exploits the harmony of the two ingredients in his truffled cabbage soup, and at Nobu they're paired with beef in truffled cabbage steak.

Truffle & Cauliflower *See Cauliflower & Truffle, page 123.*

Truffle & Celery Cookbook author Elisabeth Luard recommends mixing celeriac purée with an equal amount of mashed potato, lots of cream and some grated truffle—although not before noting that truffles are an acquired taste, redolent of "old socks, the locker-room after a rugby match, unwashed underpants, methylated spirits, gas-pump on a wet Saturday."

Truffle & Chicken In the early twentieth century, Lyons was unusual in that its restaurants were dominated by female chefs, or *cuisinières*, the greatest of whom was the legendary Mère Fillioux. At her tiny restaurant on the rue Duquesne, customers would be served an unchanging menu of charcuterie, quenelles of pike, globe artichokes with foie gras, and *poularde en demi-deuil*—chicken in half-mourning—in which the golden skin of a plump bird was darkened with slices of black truffle slipped underneath. The Ivy in London pairs chicken and truffle in roast *poulet des Landes*. A. A. Gill, who wrote the restaurant's cookbook, claims to have done so in order to learn the recipe. It's a deal more fiddly than the *demi-deuil*, as it involves stuffing the chicken's partially boned legs with a mixture of shallots, mushrooms, breadcrumbs, parsley and foie gras, before they are poached and then roasted with the rest of the chicken. The leg is then served with the breast meat, *gratin dauphinoise* and a thick gravy made of a mixture of chicken and dark meat stocks, black truffle and Madeira.

Truffle & Egg It is a truth universally acknowledged that anyone in possession of a fresh truffle must put it in a lidded container with some eggs. When the truffle has had a day or two to infuse the fatty egg yolk, make a simple omelette, scrambled egg or fried egg dish. You won't be sorry. Fat is essential for releasing and carrying the flavor of truffle; butter and goose fat give particularly good results. Some cooks use garlic to intensify the flavor of truffle served with eggs, by rubbing a cut clove either on a slice of toast or on the inside of the receptacle that the eggs are beaten in before scrambling.

Truffle & Garlic Garlic is often invoked as a flavor descriptor for truffle. Brillat-Savarin reports that he and some "men of unimpeachable integrity" agreed that white truffles from Piedmont had a "taste of garlic, which mars their perfection not at all." Black truffles are less garlicky than white and a little sweeter and mustier, with a more obvious note of mushroom in the spectrum of forest flavors. Both white and black truffles combine easily with garlic and could probably turn Styrofoam into a rich, sticky feast.

Truffle & Globe Artichoke See *Globe Artichoke & Truffle, page 128.*

Truffle & Liver Foie gras studded with pieces of black truffle is about as understated a combination as a gold-plated Ferrari. Maguelonne Toussaint-Samat writes that integral pieces of truffle detract from the singular quality of foie gras; better, in her opinion, to wrap a bacon-larded whole black truffle in paper, cook it in a low oven and serve in thin wisps over sliced foie gras. The two are often paired in dishes *à la Rossini*, most famously Tournedos Rossini, in which the fillet of beef creates an effective bridge between the ingredients, as it does in the burger stuffed with foie gras and truffle and served on a Parmesan bun at Daniel Boulud's Bistro Moderne in New York.

Truffle & Mushroom See *Mushroom & Truffle, page 82.*

Truffle & Pork Rescued from marauding pirates, all but one of the Swiss Family Robinson decide to stay on the island, having recently discovered a wild boar and some truffles. I could sympathize, as long as there was a nice barrel of Barolo washed up somewhere. As Father says, the truffles are "very different from the tough leathery things I remember from Europe," reasoning that he'd only ever tasted truffles robbed of their freshness during long transportation. Truffles start to oxidize as soon as they're unearthed, at first becoming quite pungent, then fading to a bland bitterness. At a charity auction in 2004, a consortium forked out £28,000 (roughly $42,000) for the second-largest white truffle ever found, which was then locked up in a refrigerated safe at the London restaurant Zafferano. Then the restaurant manager went away for four days with the key in his pocket. By the time he returned, the truffle was ruined. There's simply no effective way to keep truffles fresh—a problem the Swiss Family might have solved by making truffled wild boar salami, mortadella or pancetta—or their home country's truffled *cervelas* sausage.

Truffle & Potato See *Potato & Truffle, page 94.*

Truffle & Shellfish Truffle and lobster enjoy an especially harmonious relationship. A key flavor compound of white truffle, bis(methylthio)-methane, which has a sulfurous, garlicky, spicy, mushroom quality, also naturally occurs in lobster (and in shiitake mushrooms and some pungent cheeses such as Camembert). Other crustaceans and scallops are also often paired with truffle, and cabbage forms a common trio.

Truffle & Soft Cheese See *Soft Cheese & Truffle, page 74.*

Cabbage

This section includes various types of cabbage, including preserved cabbage and Brussels sprouts, and a wide range of flavors, from the fresh spiciness of raw cabbage to the corridor-filling sulfurousness of well-cooked common cabbage. Somewhere between these two extremes lie pickled sauerkraut and *kimchi*. Cooked cabbage pairs brilliantly with other pungent ingredients such as garlic, onion, truffle and mustard. Salty bacon or anchovy knocks back some of the bitter flavors—think how bacon tames sprouts. Savoy and Chinese (or Napa) cabbages are revered for their fine flavors, and their crinkly leaves are ideal for absorbing sauces in stir-fries. Intriguingly, cabbages have a clear affinity for apple, and sprouts are delectable with both dried cranberries and red currant jelly.

Cabbage & Apple See *Apple & Cabbage, page 264.*

Cabbage & Bacon A big, brazen, unrefined pairing that resists gentrification. Mrs. Beeton noted that a good stomach was needed to digest it. This warm autumn dish, inspired by panzanella, is so rustic it should come in a John Deere salad bowl. Tear half a slightly stale ciabatta loaf into chunks and grill them. Roast 1 lb halved small Brussels sprouts in olive oil at 350°F for 15 minutes. Meanwhile, slowly fry a chopped garlic clove, 5 scallions cut into ½ in pieces and 6 oz bacon lardoons in 2 tbsp olive oil until the lardoons have browned and a good deal of their fat has rendered into the pan. Remove with a slotted spoon and add to the sprouts with the toasted bread. Toss the mixture, put it back in the oven and turn off the heat. Deglaze the bacon and onion pan with 2 tbsp red wine vinegar, add 1–2 tbsp water or chicken stock and cook for 1 minute. Remove the sprout mixture from the oven and drizzle it with the contents of the pan. Season, add 2 tbsp dried cranberries and toss. Serve while still warm.

Cabbage & Beef Irish-Americans may claim corned beef and cabbage as their own, but it's not a traditional dish in Ireland, where cabbage was more likely to accompany a thick wedge of boiled ham. During the great wave of Irish immigration in the mid-nineteenth century, beef was far more plentiful, and cheaper, in America than it was in Ireland, and the newly arrived took to curing and cooking it as they would a ham, serving it with the same accompaniments of cabbage and potatoes. (In Jewish neighborhoods a similar practice was to brine joints of brisket with spices, then smoke them to make pastrami.)

Cabbage & Blue Cheese *See Blue Cheese & Cabbage, page 64.*
Cabbage & Carrot *See Carrot & Cabbage, page 223.*
Cabbage & Chestnut *See Chestnut & Cabbage, page 228.*

Cabbage & Chicken Some might see the presence of the Brussels sprout on the Christmas table as a Scrooge-like corrective to all the fun, but its role is essential. The bitterness of sprouts, which they owe to their high levels of glucosilonates, offsets the hefty turkey, stuffing, bread sauce, roast potatoes, parsnips, chipolatas (skinny sausages) and chestnuts. If you find their bitterness unpleasant, you might try growing Thompson & Morgan's Trafalgar F1 sprouts from seed. They're bred to be sweeter, and the company claims they'll have children coming "back for seconds," even offering a money-back guarantee if you don't think they're the best you've ever tasted.

Cabbage & Chili *See Chili & Cabbage, page 204.*
Cabbage & Egg *See Egg & Cabbage, page 132.*

Cabbage & Garlic If, as Mark Twain has it in *Pudd'nhead Wilson*, "cauliflower is nothing but a cabbage with a college education," cavolo nero is a cabbage with a vacation home in Tuscany. It's not, however, too refined to submit to simple home cooking, as in this lovely bruschetta. Trim off the stalks from the cavolo nero and blanch the leaves for a few minutes. Drain, to be livened up a bit, which is where capers come in. You don't so much add capers to cauliflower—you set them on it. Cook a chopped onion in olive oil until soft, add some chili flakes and chopped garlic and stir for a few seconds. Add some blanched cauliflower florets, breadcrumbs and raisins and cook until the breadcrumbs are browned. Finally add capers and parsley, warm them through, then serve the mixture tossed with rigatoni.

Cauliflower & Caviar *See Caviar & Cauliflower, page 151.*

Cauliflower & Chili Deep-frying transforms cauliflower. It takes on a sweet, almost musky taste and a creamy texture. I like to give it the salt-and-pepper-squid treatment and serve it with a chili sauce. Cut a cauliflower into florets the size of button mushrooms and dredge with a mixture of cornstarch, salt and black pepper. Deep-fry the florets in peanut or sunflower oil and serve with a sweet chili dipping sauce.

Cauliflower & Chocolate When Heston Blumenthal wanted to show cauliflower how much he loved it, he came bearing chocolate. The result was a cauliflower risotto with a carpaccio of cauliflower and chocolate jelly. The idea was that each component would release its flavor in sequence, culminating in a burst of bitterness from the specially encapsulated chocolate that Blumenthal compared to an espresso at the end of a meal. To prepare the dish, he made a cauliflower stock, a cauliflower cream, cauliflower discs, dried cauliflower, a cauliflower velouté, chocolate jelly cubes and chocolate jelly discs—and *then* he made the risotto.

Cauliflower & Cumin *See Cumin & Cauliflower, page 85.*

Cauliflower & Garlic If you've ever tasted cauliflower and garlic soup or purée, you'll know that the cooking and pulverizing processes bring out their flavors in the strongest possible way. It puts me in mind of cauliflower cheese made with a particularly ripe Camembert. Roast the cauliflower with whole garlic cloves and you'll still enjoy a rich result, with sweet nutty flavors but without the farmyard air.

Cauliflower & Hard Cheese In New York I saw a stall in the Greenmarket on Union Square decorated with an arrangement of differently colored cauliflowers: ivory, purple, lime green and a pale-orange variety called the cheddar cauliflower that I'd never seen before. If I was being pedantic, in color terms it might have more accurately been called the Double Gloucester, but I kept this to myself and simply asked the stallholder what it tasted like. "Cauliflower," he said, as if I was dumb to ask, but I bought one anyway, as by that point cauliflower cheese was a foregone conclusion for that evening's supper. I might equally, in fact, have made it with Double Gloucester, as its rich butteriness and hints of citrus and onion make a delicious cheese sauce. But there was none to be had, so I used a heady Gruyère, whose combination of potency and nuttiness is a match for the cauliflower's.

cool and squeeze them dry, then chop and fry in garlicky olive oil. Season and serve on toasted, robust white bread. I add a pinch of chili flakes and maybe a short, sharp grating of Parmesan. Kale will do fine if you can't get the fancy stuff.

Cabbage & Ginger *See Ginger & Cabbage, page 302.*
Cabbage & Juniper *See Juniper & Cabbage, page 315.*
Cabbage & Lamb *See Lamb & Cabbage, page 53.*
Cabbage & Nutmeg *See Nutmeg & Cabbage, page 217.*

Cabbage & Onion Cabbage (or sprouts) and onion add flavor to cooked potatoes in bubble and squeak, the UK's favorite leftover dish. In the nineteenth century, bubble and squeak was a fried combination of leftover beef and cabbage, but the dish is now invariably meatless. I love it so much that I often cook more potatoes than we can possibly eat with the roast chicken on Sunday. Whether premeditated surplus actually counts as leftovers is, however, a taxing philosophical problem. Processed-food companies now sell ready-made bubble and squeak, i.e., leftovers with no main meal from which they've been left over, which raises a number of dizzying questions. Can, for example, leftovers left over from nothing actually be said to exist? And what do you call it if you have any left over?

Cabbage & Pork Cabbage puts pork's fattiness to good use—for example in *choucroute garnie* from Alsace, the traditional English side dish of buttered cabbage with roast pork, and the crisp, shredded raw cabbage that's served on the side of the Japanese fried pork cutlet, *tonkatsu*. My favorite use for this super-compatible pair, however, is a French *chou farci*, or stuffed cabbage. You can stuff individual leaves, arranging them in rows like a parade of toads, or layer leaves lasagna-fashion with the stuffing mixture. But for my money the best kind of *chou farci* is the size and weight of a bowling ball, served cut into Edam-like wedges that showcase the stuffing. For the force-meat, cook 2 finely chopped shallots and 3 finely chopped garlic cloves in oil until softened. Soak 1 oz crustless bread in a little milk. Mince 10 oz pork shoulder, 10 oz beef steak and 5 oz pork belly. Add the shallot mix and disintegrated bread to the raw meat with 1 egg, 1 tbsp each chopped parsley and chives, ½ tsp dried thyme, some seasoning and, most importantly, ½ tsp of the world's most beautiful spice mixture—and the very making of this dish—*quatre-épices*. *Quatre-épices* is a hot, sweet, potent blend of white pepper, nutmeg, clove and ginger. Parboil a whole Savoy cabbage for 10 minutes, allow it to cool, then fold back the leaves, taking care not to snap them off. When you're left with a cabbage center about the size of a small tangerine, slice it out and replace with an equal volume of stuffing. Replace the leaves layer by layer, cramming a measure of stuffing between each, until your cabbage is fully reassembled. Tie with string to secure. Finally, melt a few tablespoons of butter in a lidded pot big enough to take the cabbage, add a chopped small onion and carrot and cook until soft. Throw in a bouquet garni and some seasoning, add 1 cup chicken or vegetable stock and lower in your cabbage. Cover and simmer for 1½ hours. When it's done, take the cabbage out and keep it warm while you strain the cooking liquid into a jug to serve as a sauce. See also Clove & Ginger, page 216.

Cabbage & Potato *See Potato & Cabbage, page 90.*

Cabbage & Shellfish Dimethyl sulfide (DMS) is an important component in the flavor of cabbage, with an odor variously described as cabbage-like and slightly oceanic—fittingly enough, as it's largely responsible for the odor of green (and some brown) seaweeds. It's also present in seafood: the primary flavor of scallops comes from a precursor that releases DMS (a precursor is a compound involved in a chemical reaction that releases another compound). DMS has been used, not always successfully, to re-create more authentic seafood flavors. More importantly from the home cook's perspective, it proves the deep affinity cabbage has with shellfish—if you needed proof having taken a single bite of a minced shrimp and Chinese cabbage dumpling. See also Truffle & Cabbage, page 115.

Cabbage & Smoked Fish *See Smoked Fish & Cabbage, page 162.*
Cabbage & Truffle *See Truffle & Cabbage, page 115.*

Rutabaga

The hot, peppery sweetness and dense flesh of rutabaga can be sensational when you play up to its natural spiciness, as in Scotland where it's often seasoned with nutmeg. Sweet, spicy star anise is another obvious partner, and so are sweet, earthy flavors such as root vegetables and roasted garlic. Unlike parsnip and potato, rutabaga has a rather good flavor when raw—it's hot and sweet like a radish.

Rutabaga & Anise Combine rutabaga with star anise in dark, Asian braised dishes or, as Hugh Fearnley-Whittingstall does, with leftover goose in pasties. The intense sweetness of the anise accentuates rutabaga's savoriness, while its deeply aromatic character draws attention away from the vegetable's somewhat vulgar edge.

Rutabaga & Beef *See Beef & Rutabaga, page 50.*

Rutabaga & Carrot Served cooked as a duo of fat, waterlogged sticks in British schools. We would invariably eat the carrots and leave the rutabaga on the side of the plate. Scientists now believe that the aversion to bitter vegetables, indeed all bitter flavors, is genetically determined, and boils down to sensitivity to the compound 6-n-propylthiouracil, known in flavor-scientist circles as "prop." If you can strongly detect the taste of prop in broccoli, for instance, you may well be one of the 25 percent of the population kno "super-tasters," which is more of a hindrance than it sounds. Super-ta who are more likely to be female, find brassicas, grapefruit and black unbearably intense. Another quarter of the population can't detect pr all. I might have put my childhood dislike of rutabaga down to being a su taster if it hadn't vanished overnight in my late teens, like acne or the d to wear purple eye shadow. Carrot sweetens rutabaga, and they're excel mashed together with lots of butter and white pepper.

Rutabaga & Lamb *See Lamb & Rutabaga, page 55.*

Rutabaga & Nutmeg Rutabaga is thought to be a hybrid of cabbage ar turnip, and has inherited its parents' weakness for nutmeg. In Scotland nutmeg is often used to season buttery mashed rutabaga, known as neeps Farther north, the Finnish do something similar in a dish called *lanttulaa tikko*, except the mash is creamier and it's served with ham or pork. Neithe version is better than the other, unless you're playing Scrabble. See also Potato & Rutabaga, page 93, and Lamb & Rutabaga, page 55.

Rutabaga & Pork *See Pork & Rutabaga, page 40.*
Rutabaga & Potato *See Potato & Rutabaga, page 93.*

Cauliflower

The palest member of the brassica family in color, but not necessarily in flavor. Cut into large florets and quickly steamed, cauliflower can be the essence of *cucina bianca*: gentle tending to bland. Roasted, fried or puréed, however, it reaches its full musky, earthy potential, and makes a great match for other bold, spicy flavors. Strong cheeses, chili, cumin and garlic all work well, and serve to emphasize cauliflower's sweetness.

Cauliflower & Almond Anthony Flinn, who has worked at El Bulli and now cooks at his own restaurant, Anthony's, in Leeds, created a cauliflower trifle—a purée of cauliflower and cream with grape jelly and brioche. He's also conceived a cauliflower and almond crème caramel. This consists of a layer of salted caramel under a crème caramel made with cauliflower purée, topped with warm almond cream.

Cauliflower & Anchovy *See Anchovy & Cauliflower, page 159.*
Cauliflower & Broccoli *See Broccoli & Cauliflower, page 124.*

Cauliflower & Caper Cauliflower is broccoli that can't be bothered. Where its dark-green cruciferous cousin is frisky, iron-deep and complex, cauli flower is keener on the quiet life, snug under its blanket of cheese. It need

Cauliflower & Nutmeg It was Louis XIV who popularized cauliflower in France. He liked it boiled in stock, seasoned with nutmeg and served with melted butter.

Cauliflower & Potato Like all cruciferous vegetables, the more you cut cauliflower, the more sulfurous it tastes. Puréed in a soup, it can go from meek to funky quicker than a Sunday-school teacher on the hooch. But its ruder notes can be tamped down with cream or potato. Choose the latter to make a sweet base for this *aloo gobi* soup. *Aloo gobi* is a popular Indian dish of spiced cauliflower and potato, served wet or dry. Spicy vegetable soups are too often let down by a curried muddiness, whereas this one captures the vivid, fresh spiciness of the dish that inspired it. Cook a finely chopped onion in peanut oil until soft, then stir in 1 tsp very finely chopped fresh ginger, 1 chopped seeded green chili, ¼ tsp ground turmeric, ½ tsp ground coriander and 1 tsp ground cumin. Add 6 oz diced peeled potato and ½ cauliflower, chopped into florets. Stir until they are coated in the spices, then add 3 cups cold water and bring to the boil. Simmer for 15–20 minutes, until the vegetables are soft. Cool a little, then blend until smooth. Serve garnished with chopped, fresh cilantro.

Cauliflower & Saffron See Saffron & Cauliflower, page 177.

Cauliflower & Shellfish Cauliflower purée is often served with seared scallops. Its bitter flavor contrasts with the sweetness of the shellfish and chimes with the slight bitterness on its caramelized crust. On a different note, the Japanese food writer Machiko Chiba makes a salad of blanched cauliflower and broccoli mixed with fresh crabmeat and dressed with soy, rice wine, sesame oil, rice wine and sugar. She recommends a Halbtrocken Riesling as the ideal wine match.

Cauliflower & Truffle Truffle and cooked cauliflower have a flavor overlap. David Rosengarten suggests enhancing the subtle truffle notes already present in a cauliflower risotto by shaving over fresh white truffle before serving. He describes the flavor of truffle as a combination of cheese, garlic, cauliflower and sex. Like sleeping with the greengrocer, in other words.

Cauliflower & Walnut I used to go to a little café for lunch where they made the most delicious raw cauliflower, walnut and date salad. I only wish I'd asked for the recipe, as reconstructing the dressing has proved difficult. I'm there, pretty much, but there's still something missing (if you think you know what it is, please let me know). The white cabbage flavor of raw cauliflower combined with the dried fruit might put you in mind of a coleslaw with raisins, but the chunkier texture, the extreme sweetness of the dates, and the sour cream make it discernibly different. Cut a cauliflower into small florets about 1 in long, ½ in across. For each 4 oz cauliflower florets, chop up 1 Medjool date and break 5 or 6 walnut halves in half again. Mix in a bowl and (for around 8–12 oz cauliflower) dress with a combination of ⅔ cup sour cream, 2 tsp lemon juice, 2 tsp superfine sugar, ¼ tsp salt and a grind of black pepper. Mix thoroughly and it's ready to eat.

Broccoli

Different varieties of broccoli balance varying proportions of bitterness and sweetness. Calabrese, the familiar bushy variety, is toward the sweeter end of the scale, while purple sprouting has a more pungent depth of flavor and a bitterness taken to extremes by broccoli raab. *Kai-lan*, or Chinese broccoli, is closely related to calabrese and is similarly sweet, but has lots of stalk and leaf and not much head, which makes it juicier. Joy Larkcom remarks on its deliciousness and how easy it is to grow; she also notes its botanical similarity to the rightly celebrated Portuguese Tronchuda cabbage. All varieties have a great predilection for salty ingredients—hence their frequent pairing with anchovies and Parmesan in Italian cooking, blue cheese in Britain, or soy sauce and black beans in China.

Broccoli & Anchovy *See Anchovy & Broccoli, page 159.*

Broccoli & Bacon Broccoli and pancetta make a delicious bittersweet, salty combination. Add sun-dried tomatoes, Parmesan, pine nuts and chili for an Italian combination that's almost Asian in its extremes of sweet, sour, salty, hot and umami. Toast 2 oz pine nuts and set them aside. In a large skillet, over a medium heat, flavor 3 tbsp olive oil with 10 dried chilies and 6 sliced garlic cloves, removing the garlic when golden along with the chilies. Put 14 oz linguine on to cook, meanwhile adding about 6 oz chopped pancetta to the garlicky oil. Fry until crisp, then toss in 14 oz broccoli cut into small florets, making sure they are well coated in the oil. Cook for 4 minutes, add 3 oz sliced sun-dried tomatoes and cook for a minute longer. Add 3–5 tbsp of the pasta cooking water to loosen the mixture. Drain the pasta and add it to the frying pan. Turn the heat off, season, and mix thoroughly. Divide between 4 plates and scatter generously with the pine nuts and some grated Parmesan.

Broccoli & Beef A partnership forged on their shared ferrous tang as much as the bittersweet contrast. Paired in a popular Chinese-American stir-fry with salty oyster sauce, ginger and garlic. If, however, you're feeling wan and actively craving the bitter iron flavor of rare meat, a quick-seared steak should hit the spot, served with broccoli cooked with garlic and anchovies.

Broccoli & Blue Cheese *See Blue Cheese & Broccoli, page 63.*

Broccoli & Cauliflower Bushy calabrese broccoli and cauliflower both belong to the *Brassica oleracea* family. They have many flavor compounds in common, which is evident when you nibble them raw. Cooking clarifies their differences—broccoli's bitter depths of iron versus cauliflower's thick sulfurousness. A hybrid called broccoflower looks, as you might expect it to, like a queasy cauliflower and tastes like weak broccoli. In my experience people tend to love or hate broccoli, and the point of a vegetable that's just like it, except less so, is lost on me. The other variety of green cauliflower is perhaps better known as Romanesco, in equal parts famous and frightening for its custard-green coloring and fractal swirls that give it the appearance of an ancient Thai pagoda. Or a prog-rock album cover. It has a milder, less sulfurous flavor than cauliflower or broccoli.

Broccoli & Chili Like cauliflower, broccoli takes on a richer, sweeter flavor when roasted. Red chili (dried and smoky or fresh and sweet) makes an ideal companion. See also Anchovy & Broccoli, page 159.

Broccoli & Garlic The most fun you can have eating healthily. Stir-fry them with ginger in oyster sauce or cook them, with or without anchovies, in the classic style outlined in Anchovy & Broccoli on page 159, and serve with pasta. Or try this Thai-inspired noodle dish. This makes one serving. Rehydrate a sheet of fine egg noodles as per the instructions on the packet, drain them and pat dry. Heat 1 tbsp peanut oil in a wok and fry 3 garlic cloves, cut into quarters. When golden, remove with a slotted spoon and set aside. Add the noodles and a cup small broccoli florets to the oil and fry for 3 minutes, stirring constantly. Scramble a lightly beaten egg into the pan, mixing it thoroughly with the noodles, then add a sauce made with 2 tsp light soy, 1 tsp dark soy, 1 tsp oyster sauce, 1 tsp sugar and 1 tbsp water. Mix well, heat through for a minute, sprinkle lightly with white pepper and serve garnished with the garlic pieces.

Broccoli & Hard Cheese *See Hard Cheese & Broccoli, page 68.*

Broccoli & Lemon For all broccoli's good points, it does lose its heat very quickly after cooking. The Italian restaurant Orso in London has the answer: serve it warm, not hot, with a squeeze of lemon. If it just turned up like that at the table without any explanation, you might be disappointed, but with a cunning bit of neuro-linguistic programming they forewarn you that it comes warm, which makes it as miraculously delicious as the slow-roast pork with crackling it would be foolish not to order it with.

Broccoli & Peanut *See Peanut & Broccoli, page 26.*
Broccoli & Pork *See Pork & Broccoli, page 37.*
Broccoli & Walnut *See Walnut & Broccoli, page 231.*

Globe Artichoke

The humdrum origins of the artichoke—it's the flower of a thistle—belie its wondrous and complex flavor. See Globe Artichoke & Lamb, page 127, for a full description. Globe artichokes contain a phenolic compound called cynarin, which has the peculiar effect of making anything you eat directly afterwards taste sweet. It temporarily inhibits the sweet receptors in your taste buds, so that when you follow a bite of artichoke with, say, a sip of water, flushing the compound off your tongue, the receptors start working again and the abrupt contrast fools the brain into thinking you've just swallowed a mouthful of sugar solution. This makes for a diverting, if swiftly tedious, party game—sweet radicchio!—but it's bad news for wine. And the enemy of wine is my enemy. The problem can be minimized by using ingredients that create a bridge between the wine and the artichoke (or simply taking a bite of something else before you take a sip of wine). Or you could ditch the wine altogether and drink Cynar, an artichoke-flavored liqueur from Italy.

Globe Artichoke & Bacon In Lazio, a boyfriend and I were speeding through a landscape of fairy-tale castles, well on our way to not living happily ever after. We had been arguing with such uninterrupted intensity that it was only a promising road sign that reminded us that it was well past lunchtime and we were hungry. Our motherly Italian hostess, perhaps picking up on the friction between us, took pity and led us to a table under an olive tree. Mercifully soon, she brought a label-less bottle of cold, dry white wine, an enormous spoon and a terracotta dish of something covered in breadcrumbs and cheese, molten bubbles popping on the surface like the meniscus of a volcano. My boyfriend, or ex-boyfriend, or whatever he was at that moment, took the spoon and, breaking through the crust, emerged with a steaming heap of rigatoni, pancetta and artichokes, in a rich béchamel savory-sweet with Parmesan. They say hunger is the best sauce, but if that lunch under the olive tree is anything to go by, the point in a relationship where *it doesn't matter any more* runs it a close second. We smiled at each other. I topped up our glasses. He piled the pasta on our plates. The bitter, nutty greenness of the artichoke cut through the richness of pancetta and cheese. It was by far the best last date I've ever had. If *your* relationship is on the rocks, get 8 oz rigatoni on to cook. Soften a finely chopped onion and 2 garlic cloves in olive oil with 3 oz sliced pancetta. Add 4–6 cooked artichoke bottoms (good jar ones will do), sliced into sixths. In a bowl, mix ½ cup milk with ⅔ cup heavy cream and 2 oz grated Parmesan. The pasta should be *al dente* by now. Drain it, empty back into the pan and add the milky, creamy, cheesy mixture and the onion and artichokes. Stir and check for seasoning, then transfer to a baking dish. Cut a ball of mozzarella into slices and lay them on top. Cover with a mixture of 2 oz breadcrumbs and 1 oz grated Parmesan and bake for 30 minutes at 400°F, covering it with foil if it looks in danger of burning. Serve with a bottle of cold, cheap Italian white.

Globe Artichoke & Hard Cheese *See Hard Cheese & Globe Artichoke, page 69.*

Globe Artichoke & Lamb Describing the flavor of globe artichokes could be a parlor game. Some say it's a little like asparagus, but I'd say asparagus has a greener flavor, and artichoke bottoms taste more like calabrese broccoli stalks, cooked in a mushroomy vegetable stock until soft, then smothered in butter, with a hint of pewter. The flavor is wonderful, but the magic really resides in the texture—dense, yielding and velvety, it's the foie gras of vegetables. Artichokes are a spring vegetable in Italy and Spain, and make a seasonal pairing with lamb, especially in stews. Lamb seems to mellow the vegetable's bitterness in a way that beef and pork don't. The only drawback to lamb and artichoke stew is having to peel and chop globe artichokes. You end up looking as if you've been playing patty-cake with Edward Scissorhands. You *could* use canned or frozen artichokes, but don't. They're no substitute in either flavor or textural terms. Like beauty, flavor is pain.

Globe Artichoke & Lemon Fly to Rome at Easter time and buy yourself a paper bag of *carciofi alla guida*—deep-fried whole artichokes. Find some lemon wedges and a wodge of napkins and eat them under a tree while they're still hot. Boiled or steamed artichokes can be a tough match for wine, but fried they'd go nicely with a fresh, acidic, dry Prosecco, if you can handle the social stigma of being slumped under a tree with a bottle of booze and a brown paper bag.

Globe Artichoke & Mint The seriousness of globe artichoke is lightened by mint. The two ingredients famously come together in *carciofi alla romana*, in which the artichokes are trimmed, have chopped mint and garlic stuffed between their leaves, and are then simmered, stem end up, in a combination of water, lemon juice, oil and more mint. Lamb and artichoke stews are also garnished with a generous scattering of mint.

Globe Artichoke & Oyster In an early-seventeenth-century manuscript called *The Fruit, Herbs and Vegetables of Italy*, Giacomo Castelvetro cites the pairing of small cooking artichokes with oysters and beef marrow in little pies. These days the pairing of artichoke and oyster is most popular in a soup or bisque in oyster-mad Louisiana.

Globe Artichoke & Pea *See Pea & Globe Artichoke, page 199.*

Globe Artichoke & Pork In France and Italy, globe artichokes are stuffed with minced pork or sausage meat, pushed between their trimmed bracts. It's a little fussy for both cook and diner, but the flavor combination is terrific and, allowed to infiltrate its leaves, pork lends artichoke an irresistible sticky saltiness. I combine them in a pie. Drain and rinse a jar of artichoke hearts, cut them each in half and pat dry. Line an 8-in pie pan with shortcrust pastry and spread ½ lb sausage meat over it. Arrange the artichokes on top, then cover with another ½ lb sausage meat. Cover with a pastry lid, seal the edges and make a little hole in the center. Bake for an hour at 350°F. Best eaten cold with plenty of salad and pickles.

Globe Artichoke & Potato The Jerusalem artichoke is not related to the globe artichoke but was named after it because of its flavor, which many would say was a cross between globe artichokes and top-quality potatoes. Jerusalem artichokes would arguably be more popular if they weren't so famously difficult to digest. A combination of globe artichoke and potato is kinder on the stomach. In Provence, they're thinly sliced and baked together in olive oil and garlic (Escoffier ritzed the dish up by substituting truffle for garlic and butter for olive oil). Or you might pair them cold in a salad with a mayonnaise deliciously enhanced with flesh scraped from the cooked leaves. In Italy the two are combined in soup and, according to *The Silver Spoon*, in a pie.

Globe Artichoke & Prosciutto *See Prosciutto & Globe Artichoke, page 169.*

Globe Artichoke & Shellfish Globe artichokes contain cynarin, a chemical that gives them a strange and insincere sweetness. Shellfish is a good pairing because it has a natural, more pleasing sweetness and its saltiness points up artichoke's tastier qualities. You might remove a cooked globe artichoke's tough outer leaves, purple inner leaves and hairy choke to create a "cup" in which to serve crab, lobster or shrimp in mayonnaise or vinaigrette. Or make a hot dip by mixing thin slices of cooked artichoke heart with crabmeat, mayonnaise, finely chopped shallot and grated Parmesan and baking under a herb crumb crust. Eat fresh from the oven with crackers. Hearty enough for an iron smelter's cocktail party.

Globe Artichoke & Truffle In 1891 Anton Chekhov wrote to his brother complaining about the restaurants in Monte Carlo. "They fleece one frightfully and feed one magnificently . . . Every morsel is rigged out with lots of artichokes, truffles and nightingales' tongues of all sorts." That artichokes have so frequently been paired with the most expensive ingredients—truffles and foie gras—is testament to the thistle's extraordinary attraction. Even today they're served together in the best restaurants. Three-star French chef Guy Savoy has a signature dish of the earthy, if not very down-to-earth, soup *d'artichaut à la truffe noire*, served with a brioche layered with mushrooms and truffle butter.

Asparagus

Salty dairy ingredients are a heavenly match for asparagus: butter, Parmesan and hollandaise sauce simultaneously contrast with and enhance its sweet, sulfurous vegetable flavor. Other sulfurous-tasting foods, such as eggs, crustaceans and garlic, also make harmonious partners. Thin spears of asparagus are called sprue, and tend to have a more piquant flavor than the full-grown variety. White asparagus is grown from the same seed as green but is covered with soil to inhibit the production of chlorophyll. It's pretty much as flavorless as it is lacking in color, and while its apologists describe it as subtle, mild or "dainty," I can't find a whole lot to like in a slimy, anemic tube free of the rich nuttiness that makes proper green asparagus such a treat. For some reason, the Spanish, who have so much to recommend them gastronomically, put it in salads, where it lies hidden among the leaves like a chef's long-forgotten finger.

Asparagus & Almond *See Almond & Asparagus, page 237.*

Asparagus & Anise Tarragon isn't quite as flimsy as its slight green leaves suggest—think how it flavors vinegars, mustards and pickles. Asparagus, none too shyly flavored itself, is often paired with tarragon. The spears might be served with the tarragon-and-shallot-flavored Béarnaise sauce, while asparagus, tarragon and egg make a fine trio in an omelette, baked en cocotte or in a warm tart in which asparagus forms the spokes.

Asparagus & Egg *See Egg & Asparagus, page 131.*
Asparagus & Hard Cheese *See Hard Cheese & Asparagus, page 67.*
Asparagus & Lemon *See Lemon & Asparagus, page 297.*

Asparagus & Mint American chef Daniel Boulud steams asparagus over a little water with fresh mint and lemon zest added to it, then dresses it with extra virgin olive oil, lemon juice and more mint. If you're trying this yourself, go easy on the mint. You don't want to go turning your asparagus spear into a toothbrush.

Asparagus & Mushroom Morels are a great seasonal match for asparagus. They like to grow on charred ground, which might explain their slightly smoky character when fresh, and their love for sulfurous flavors like asparagus. Chef David Waltuck of Chanterelle serves an asparagus flan with sautéed morels in a creamy sauce made with oysters and Madeira. Egg works exceptionally well with both flavors and often forms a trio with them—the classic *oeufs Jessica* pairs baked eggs with minced morels, asparagus and a little quality meat stock.

Asparagus & Oily Fish *See Oily Fish & Asparagus, page 153.*
Asparagus & Orange *See Orange & Asparagus, page 288.*
Asparagus & Pea *See Pea & Asparagus, page 198.*

Asparagus & Peanut This might seem as incongruous as playing darts in a ballgown but the rich, meaty flavor of asparagus is, in fact, very good with peanuts, especially when given an Asian inflection. Steam your spears and serve them with this peanut dressing: mix 3 tbsp sunflower oil, 3 tbsp lemon juice, 2 tbsp light soy sauce, a pinch of sugar and some seasoning, then stir in a scant cup chopped roasted unsalted peanuts.

Asparagus & Potato Share an earthy, nutty character. Jane Grigson recommends boiling new potatoes with asparagus and serving them for lunch with a soft-boiled egg, homemade bread, butter and a Loire white wine. Sancerres from the Loire and Sauvignon Blancs from New Zealand are made from the same grape and have an asparagus quality, which makes them an ideal match for the difficult vegetable.

Asparagus & Prosciutto *See Prosciutto & Asparagus, page 168.*
Asparagus & Shellfish *See Shellfish & Asparagus, page 137.*

Asparagus & Truffle In his book *Aphrodisiacs*, Peter Levene notes that there is a diuretic in asparagus that stimulates the kidneys and "excites the urinary passages." If for any reason this fails to put you in the mood for love, try shaving black truffle, or drizzling a little truffle oil, over your asparagus soup. When cooked, asparagus develops a strong sulfurous-sweet characteristic that goes particularly well with black truffle. Analysis of truffle extract reveals that it contains traces of male-pig sex pheromones, which is thought to account for the sow's happy acquiescence when dragged about the woods all day snuffling in the undergrowth. By all means, make this for your date, but if they like it, you might ask yourself what this says about them.

Asparagus & White Fish *See White Fish & Asparagus, page 143.*

Egg

Side-by-side tastings reveal that hen eggs vary noticeably in their flavor characteristics. If you're lucky, you'll find one with a buttery, naturally salty yolk that makes a mild but very satisfying dipping sauce when soft boiled. Different egg flavors are mainly attributable to their age and storage conditions, although they can also be due to variations in the bird's diet. Some claim that the rare and expensive gull's egg has a fishy flavor (as, apparently, does the penguin's), and that goose and pheasant eggs are gamier, although I can't say I've ever been able to detect either of these. Quail and duck eggs are often described as creamy, but this is mainly attributable to their higher yolk-to-white ratio.

Egg & Anchovy If you see anchovies lying prostrate on a fried egg, there's probably a Wiener schnitzel underneath. A garnish of anchovy, egg and sometimes caper makes for Wiener schnitzel *à la Holstein*, named after the Prussian diplomat Friedrich von Holstein, who supposedly ate them for breakfast. There are several variations on the dish, some including caviar, salmon, beets, pickles and lobster. Scotch woodcock is an old English recipe that sounds as if it should be a breakfast dish—scrambled eggs on toast with a garnish of anchovies—but was in fact served in genteel Victorian households as the culmination of a six-course meal. If you're going to try this, I recommend (a) not eating five courses beforehand and (b) soaking the anchovies in milk to soften their flavor. *Nasi lemak* is a Malaysian dish of coconut rice garnished with peanuts, cucumber, boiled eggs, fried anchovies and a spicy tomato sauce.

Egg & Anise *See Anise & Egg, page 180.*

Egg & Asparagus Cooked asparagus spears are lovely dipped in soft-boiled eggs, especially if you follow Hugh Fearnley-Whittingstall's tip of slicing the tops off and then adding a little butter and a few drops of cider vinegar to each yolk for a hollandaise-like effect. Asparagus is also added to baked eggs (see Asparagus & Mushroom, page 129) and cooked in a frittata. I have an irrational dislike of frittata—or at least of its ubiquity. Don't throw it away: make a frittata. Can't think of anything? Make a frittata. If you've got eggs, you've always got a frittata! Frittatas are cozy, accommodating, practical and slovenly: the culinary equivalent of the tracksuit.

Egg & Bacon *See Bacon & Egg, page 166.*

Egg & Banana In Japan, omelettes called *tamago yaki* are made with soy and sugar and served in sushi bars as a savory dish. The sweetened omelette has classic status in French cuisine, where it might be filled with jam, fruit compote or maybe a scattering of pine nuts and served for dessert. What a great idea it is. This version is so simple you can throw it together for breakfast while you're still rubbing your eyes. Make a 3-egg folded omelette as you would ordinarily, except add 1 tbsp superfine sugar and a pinch of salt as you beat the eggs. Fry in butter and then fill with a small banana, mashed or sliced as thin as pennies, before folding.

Egg & Basil *See Basil & Egg, page 210.*
Egg & Beef *See Beef & Egg, page 48.*

Egg & Beet The distinguishing ingredients in a "kiwiburger." The McDonald's version was launched in New Zealand in 1991, and when it was later discontinued there was such widespread dismay that a campaign was launched to bring it back. At the UK's Gourmet Burger Kitchen chain, where the Kiwi chef Peter Gordon acts as consultant, the kiwiburger comes with a slice of pineapple. The beet and pineapple hang around in the background like a fun couple who didn't realize it wasn't a fancy-dress party. The egg adds richness, all right, but isn't that the French fries' job?

Egg & Bell Pepper *See Bell Pepper & Egg, page 202.*

Egg & Black Pudding *See Black Pudding & Egg, page 42.*

Egg & Cabbage *Okonomi-yaki* is often described as Japanese pizza, which is about as useful an analogy as calling salami a meaty cucumber. Roughly translated, *okonomi-yaki* means "as you like it," and is pizza-like insofar as the base is round and flat and then customized with your choice of toppings and seasonings. But in texture and flavor it's a world away. It's made by mixing chopped cabbage, grated Japanese yam and finely chopped scallions with egg, flour and water. The resulting batter is then poured onto a hot plate, shaped and, when cooked through, topped with pork, bacon, squid or *kimchi* (see Chili & Cabbage, page 204), or all of them and more besides. To finish, it's decorated with a Pollockesque drip painting of mayonnaise, spicy brown sauce, *katsuobushi* (gossamer wisps of smoked tuna) and seaweed flakes.

Egg & Caviar *See Caviar & Egg, page 151.*

Egg & Celery Celery salt and hard-boiled eggs are a classic combination. There's something in the citrusy, pine tang of celery that really lifts egg's sulfurous low notes. Quail, duck and hen's eggs are perfectly delicious served like this, although gull's eggs are the ones that make it onto the menus of fancy restaurants. Gull's eggs cost about ten times as much as hen's, not only because of the shortness of the laying season but because just a limited number of collector's licenses are granted. Celery salt is, of course, widely available to buy, but you can make your own by lightly toasting celery seeds and crushing them with sea salt. Start with a 1:6 mix and adjust to taste. Fergus Henderson, by contrast, makes celery salt by baking grated celeriac with salt, and puts a wintry spin on the combination by making a series of dents in a pile of hot, buttery mashed celeriac, placing an egg in each and baking until the white is firm and the yolk just runny.

Egg & Chicken A stranger combination than you might suppose; you'd be hard pushed to think of a Western recipe that features them both. Chicken omelette, maybe. But do *you* fancy one? There is perhaps some deeply accul-turated discomfort at serving animal and offspring together, although this is absent from Asian culinary traditions. In China, eggs are cracked into chicken soup, chicken fried rice and a chicken congee. In Japan a rice-bowl dish called *oyakodon*, or "mother and child," consists of chicken, egg and scallions simmered in a mixture of soy sauce, dashi (dried tuna stock) and mirin (rice wine), then served on rice. Apparently Paul Simon took the title of his song "Mother and Child Reunion" from the menu of a Chinese restau-rant in New York.

Egg & Chili *See Chili & Egg, page 205.*

Egg & Coconut *Kaya*, a sort of coconut jam (or curd), is made of coconut milk, eggs and sugar, and is spread on buttered toast for breakfast all over

Southeast Asia. The same ingredients make a simple and delicious coconut custard pudding. Whisk 4 eggs with ½ cup sugar, then gradually mix in 1 cup coconut milk. Pour into 4 ramekins and place in a baking dish. Carefully pour enough boiling water into the dish to come two-thirds of the way up the sides of the ramekins. Bake at 300°F for 40 minutes, until just set. See also Coconut & Cinnamon, page 281.

Egg & Cumin Cumin is a warmer, earthier, but no less delicious seed than celery, for toasting, grinding and mixing with salt for your eggs. Try caraway too, for a trio of flavored salts to serve with a handsome pyramid of unshelled soft-boiled quail's eggs and let your guests work through them, trying each salt as they go.

Egg & Dill The clean, sharp flavor of dill is a contrast to egg's sulfurous hominess. The combination is at its best when the egg's flavor has been maximized by boiling and can stand up to the insistent, citrusy greenness of dill. A simple egg and dill sandwich is hard to beat.

Egg & Ginger *See Ginger & Egg, page 303.*

Egg & Lemon *Avgolémono* ("egg lemon") is popular in Greece in both sauce and soup form. The writer Alan Davidson noted how it demonstrates the two ingredients' affinity—as do some lemony versions of mayonnaise. To make *avgolémono* soup, bring 1 quart of chicken broth to the boil and add rice or orzo pasta. Toward the end of the rice or pasta's cooking time, whisk 2 eggs with the juice of 1 lemon. Whisk a ladleful of the hot broth into the eggs, then, off the heat, whisk the lemon mixture into the chicken stock little by little. See also Celery & Lamb, page 96.

Egg & Mushroom *See Mushroom & Egg, page 79.*

Egg & Nutmeg Eggnog is the perfect restorative after an afternoon's Christmas shopping. Pour 3 tbsp rum, brandy or marsala into a cocktail shaker with 4 tbsp milk, 1 egg yolk and some ice. Shake well, then strain into a glass and give it a good freckling of fragrant grated nutmeg. It's like coming in from the cold. The initial warm shock of nutmeg gives way to soothing, thickened milk and the afterglow of rum. And this eggnog tart is so delicious it's threatening to usurp Christmas pudding. Scald ⅔ cup heavy cream. Whisk 3 egg yolks with ⅓ cup superfine sugar, a pinch of salt and a generous grating of nutmeg. Slowly whisk the cream into the eggs. Pour into a clean pan and cook gently, stirring constantly, until the mixture is thick enough to coat the back of the spoon. Remove from the heat and set aside. Thoroughly dissolve 2 tsp gelatin powder in 4 tsp hot water and stir into the custard. Add 3 tbsp rum, 1 tbsp brandy and ½ tsp vanilla extract, stir again, then strain into a glass or ceramic bowl. Grate in a little more nutmeg and put in the fridge. Before the custard sets, whisk the 3 egg whites to soft peaks. Fold them into the custard and pour into a baked, deep sweet pastry shell, 9 inches in diameter. Chill until set. Serve at room temperature, with plenty more freshly grated nutmeg on top. This also works on a crushed cookie base.

Egg & Oily Fish *See Oily Fish & Egg, page 154.*

Egg & Onion In Jewish cooking, boiled egg and scallions are chopped very finely, mixed with schmaltz (chicken fat), seasoned, chilled and served as an appetizer with challah or rye bread. Mollie Katzen's more piquant version adds parsley and watercress in a sour cream, horseradish and black pepper dressing. See also Potato & Egg, page 91.

Egg & Oyster *See Oyster & Egg, page 149.*
Egg & Parsley *See Parsley & Egg, page 189.*

Egg & Pea Make it worth cooking too much rice. In China, egg fried rice is considered a meal in its own right, not just an accompaniment, and so it should be. It's quite delicious even reduced to three basic ingredients—rice, egg and peas—but add a few more and you have the mandate of heaven. Finely chop a couple of slices of smoked bacon and a small onion and fry them in 2 tbsp peanut oil. Add about 4 handfuls of cold, cooked white rice and give it a good stir to break up the grains. Scatter over a handful of peas and, once all is well heated, drizzle the mixture with 2 beaten eggs and a shake of soy sauce. Allow the egg to cook, giving it a turn now and then. The flavor can only improve if you invest in a carbon-steel or cast-iron wok and season it properly. To the Chinese, the wok is an ingredient in itself, and skilled chefs can impart a quality to dishes that the Cantonese describe as "wok hay," or "wok breath"—a combination of flavor, heat and smokiness that comes of knowing how to get the best out of your pan.

Egg & Pork *See Pork & Egg, page 38.*
Egg & Potato *See Potato & Egg, page 91.*

Egg & Prosciutto A dish of scrambled egg yolks is the *bon vivant*'s answer to the egg white omelette. It's superbly rich, ridiculously buttery and has the simultaneously light and dense texture of clotted cream. Use 3 yolks plus 1 whole egg and a little butter. Serve with Parma ham for a luxurious take on bacon and eggs.

Egg & Sage *See Sage & Egg, page 313.*

Egg & Shellfish Dressed crab is *the* recipe for the man who likes to spend weekends taking the car apart and putting it back together again. You have to boil and shell the crab and eggs, clean out the crab shell, finely chop the egg whites, sieve the yolks, finely chop the brown meat and mix it with mayonnaise, and mix lemon juice into the white meat. The brown meat is then layered in the bottom of the crab shell with the white meat and chopped egg arranged on top in defined bands, as in a flag. It's dressed with parsley and served with a little extra mayo on the side. You eat the white meat first, appreciating its pale delicacy before moving on to the hard stuff. The brown meat is part of the crab's digestive system and contains all the flavor that implies. If all this sounds more trouble than it's worth, crab Louis is a less fussy recipe that originated on the West Coast of the United States in the early twentieth century. It's a white crabmeat and egg salad with lettuce and a "Louis" dressing that's very similar to Thousand Island.

Egg & Smoked Fish *See Smoked Fish & Egg, page 163.*

Egg & Tomato *Uova al purgatorio*, or eggs in purgatory, is a Neapolitan dish of eggs served in a thick, warm, pepper-spiked tomato sauce. North African *shakshuka* and Latin American *huevos rancheros* operate on similar principles. Some people like to scramble their egg into the tomato sauce; others prefer to keep the eggs whole by sliding them into the sauce, putting a lid on the pan and leaving it to simmer until the eggs are cooked through. The third way is to allow the white to set, tipping and rotating the pan as necessary to ensure all the translucent liquid meets the heat, and once it has, scrambling the yolk into the sauce. If this all sounds a bit too rustic, you might try the very Jamesian lunch to which Strether treats Madame de Vionnet in *The Ambassadors*. Over intensely white table linen at a little place he knows on the Left Bank of the Seine, they eat *omelette aux tomates* with a bottle of straw-colored Chablis. See also Bell Pepper & Egg, page 202.

Egg & Truffle *See Truffle & Egg, page 116.*

Egg & Vanilla Vanilla spirits away the eggy flavor that can be particularly unwelcome in pastries and desserts. Variously paired with cream, milk, sugar and flour, egg and vanilla make *oeufs à la neige* (uncooked meringues, shaped like eggs, floating in pale custard), crème caramel, crème brûlée, vanilla soufflé, vanilla ice cream and *crème anglaise*, otherwise known as custard.

Egg & Watercress *See Watercress & Egg, page 100.*

MARINE

Shellfish

White Fish

Oyster

Caviar

Oily Fish

Shellfish

This section covers bivalves and crustaceans, but oysters have their own chapter (see page 148). Mussels and clams have a saltier, stronger flavor than sweet shrimp, lobster and scallop. Crab lies somewhere in between, depending on how much of the brown meat, which has a deeper marine flavor, you mix in with the white. The food writer Alan Davidson noted that the white claw and leg meat of crab is similar in flavor and texture to lobster. Lobster might be held in higher regard, particularly by those for whom the price is immaterial, but plenty of people consider crab to be comparable, if not superior. Hugh Fearnley-Whittingstall takes the view that while lobster costs five times as much as crab, it isn't five times more delicious.

Shellfish & Almond The aroma of boiled shrimp is often described as nutty. Roasted, it becomes more specifically almond-like. The almond note in shrimp accounts for being paired with ground almonds in aromatic Indian curries, with almond-based sauces in Spain, and stir-fried with whole or chopped nuts in the Chinese-American dish shrimp and almond ding. The flavors work beautifully with rice, especially basmati.

Shellfish & Anise Anise heightens and freshens the sweetness of shellfish. Tarragon butter with lobster, and mussels cooked with fennel, are justifiably classic combinations. A dash of Pernod might replace brandy in a bisque. The simple recipe for cooking chicken breast with tarragon is easily adapted for shrimp—see Anise & Chicken, page 180.

Shellfish & Apple See Apple & Shellfish, page 267.

Shellfish & Asparagus Like peas and sweetcorn, asparagus should be eaten as soon as possible after harvesting. Once picked, it starts to consume its own sugar—more voraciously than any other common vegetable—flattening out the flavor. At its tender best, asparagus is green heaven with shrimp or sweet, fresh crab. If you're ever in England in May or June, pack up a camping stove with some plates, cutlery, a pan and some salt and head for the Suffolk coast. Keep your eyes peeled for signs advertising pick-your-own asparagus. Harvest a couple of bundles and strike on for the sea at Southwold or Aldeburgh. While one of you cusses over the stove, the other can buy fresh crab from one of the tar-black fishermen's huts. You'll also need some good bread, salty butter, a couple of lemons and a cold bottle of Fumé Blanc. And some early strawberries. Locate each other on the beach. Get some water on to boil and add the asparagus to it. Butter the bread. Open the wine. Quarter the lemons. When the asparagus is just cooked, dot it with butter and serve with plenty of crabmeat. Stare, shivering, toward Holland, as the tide fizzes on the pebbles and the gulls announce the onset of British summertime.

Shellfish & Avocado See Avocado & Shellfish, page 197.

Shellfish & Bacon *See Bacon & Shellfish, page 167.*

Shellfish & Basil The citrus and anise notes in basil make it a gracious match for shellfish. Basil contains a compound called citral, which is partly responsible for the flavor of lemon and lemongrass. A salad of lobster and mango with basil is one of Alain Senderens' signature dishes, perhaps inspired by the classic Vietnamese combination of shrimp, papaya and lemongrass.

Shellfish & Beef The term surf 'n' turf is thought to have originated in the United States in the 1960s, although its catchiness, and the cachet conferred by the payday price ticket, are probably more responsible for its enduring popularity than any particular flavor affinity between lobster and beef. They're fine—beef's flavor tends to be intensified by marine partners such as anchovy and oyster—but I find the combination of lean, dense flesh can add up to a chewy chore. And I'm not convinced that the beef does much for the lobster in return. The basic idea has, of course, been reinterpreted by various chefs as scallops with foie gras, octopus with bone marrow and monkfish tail with oxtail. And in the late 1980s the term was applied by my hungrier male colleagues to the blow-out combination of a Big Mac and a Filet-O-Fish. Alligator meat, incidentally, is said to taste like a hybrid of shellfish and veal. See also Oyster & Beef, page 148.

Shellfish & Bell Pepper Strips of red pepper turn up in shellfish sauces for pasta, in stir-fries and sticky, Spanish paella-style dishes, and in sweet and sour shrimp. Their bittersweet vibrancy perfectly offsets the rich nuttiness of shellfish. But do we ever notice them? They're like Cyrano, selflessly eloquent in the background.

Shellfish & Black Pudding *See Black Pudding & Shellfish, page 44.*
Shellfish & Butternut Squash *See Butternut Squash & Shellfish, page 228.*
Shellfish & Cabbage *See Cabbage & Shellfish, page 120.*
Shellfish & Caper *See Caper & Shellfish, page 103.*
Shellfish & Cauliflower *See Cauliflower & Shellfish, page 123.*
Shellfish & Celery *See Celery & Shellfish, page 97.*

Shellfish & Chicken *Mar y muntanya* ("sea and mountain") is a Catalan dish of meat and shellfish (often chicken and shrimp) combined in a thick, pounded nut, tomato and garlic sauce similar to a *picada*—see Hazelnut & Garlic, page 235. Chicken is also paired with shrimp or mussels in Spanish paella and Louisiana gumbo and jambalaya. The Rhône-Alpes specialty *poulet aux écrevisses* takes advantage of two local ingredients: chicken from Bresse and crayfish from the Alpine streams near the border with Switzerland. Chicken Marengo, supposedly made for Napoleon after his victory over the Austrians at the eponymous battle, consists of chicken in a tomato, garlic and wine sauce with a garnish of fried crayfish. In Brazil, chicken with a pungent mass of dried shrimp makes *xim-xim de galinha*, and in Southeast Asia, of course, untold numbers of curries and stir-fries combine chicken with shellfish both dried and fresh. Bivalves work too: clams are paired with chicken in Chinese and Korean dishes, while in Brittany chicken is sautéed with clams and samphire.

Shellfish & Chili Chili crab is huge in Singapore. The crab is fried with fresh chilies, ginger and garlic, then swathed in a sticky mixture of chili sauce, ketchup, soy, sugar and sesame oil, often with an egg stirred in at the end of cooking. It's served with rice, or a bread called *man tou*, useful for mopping up every last drop of sauce. In Malaysia and Singapore, *sambal belascan* is toasted shrimp paste and fresh chili pounded together and loosened with lime juice; a similar concoction in Thailand is known as *nam phrik kapi*.

Shellfish & Cilantro As shellfish go, mussels are particularly rich—they have a deep, briny meatiness sometimes graced with notes of butter or caramel. This makes them especially suited to the sort of ingredients used in Thai cooking to balance out the similarly deep flavor of *nam pla*, or fish sauce. Cilantro, lemongrass, lime, chili (and combinations thereof) are all great matches. Dress mussels with finely chopped cilantro and a squeeze of lime after barbecuing them on the grill in a foil tray. Should take around 8 minutes, but you'll know they're done when they yawn open.

Shellfish & Coconut Remind me of being on vacation. A shaggy coating of flaked coconut turns deep-fried shrimp into a trashy bar snack that demands cold bottles of beer and a sheaf of paper napkins. In Bahia, Brazil, a popular dish called *vatapá* combines shellfish and fish in a coconut milk stew thickened with bread. And in Thailand, shrimp bathe in a coconut milk broth called *tom yam kha kai*; the sweetness of the two main ingredients distracts you, if only temporarily, from its fiery heat.

Shellfish & Cucumber *See Cucumber & Shellfish, page 185.*

Shellfish & Cumin A cumin sauce for shellfish is given in Apicius, the oldest known cookbook in existence. The cumin is mixed with pepper, lovage, parsley, dried mint, honey, vinegar and broth. Today cumin is often paired with shellfish in India, Mexico and the American Southwest. Muddle garlic, ground cumin, ground chili and oil to make a marinade.

Shellfish & Dill *See Dill & Shellfish, page 188.*
Shellfish & Egg *See Egg & Shellfish, page 134.*

Shellfish & Garlic The richly sulfurous flavor of garlic has a multiplier effect on the flavor of all shellfish but hits a high with stewed clams. One evening on vacation, we stopped in Porto Ercole, on a mountainous peninsula connected to the Tuscan mainland by three dams like strands of mozzarella. After a swim in the warm sea, we drifted toward the restaurant—similarly adrift on a precariously tethered pontoon—that was playing the best music. No one else was around. The Italians would only just be digesting their lunch, but we'd driven all the way up from Rome, and the swim and the first glass of Soave had given us an appetite. We were debating how long we might be able to hold off eating when an enormous man with long, sun-ruined hair and a moustache to match sat down at our table. "Hi," he said, in slightly Americanized Italo-English that somehow sounded Swiss. "What would you like to eat? I'll cook you anything." This was a little like asking the young Hannibal if he fancied conquering any bits of the Roman Empire. I generally want *everything*. But we were by the sea, in Italy, and that can mean only one thing: *spaghetti alle vongole*. In short order, we were sitting in front of golden, tangled masses of garlicky pasta caught up with creamy, chewy clams, like Venus's hair before she'd had the chance to comb it through. Sprigs of flat-leaf parsley lifted the oily richness. I have a fairy godfather, and he's a chef the size of a grizzly bear who lives on a floating restaurant.

Shellfish & Globe Artichoke *See Globe Artichoke & Shellfish, page 128.*
Shellfish & Grapefruit *See Grapefruit & Shellfish, page 292.*

Shellfish & Hard Cheese Controversial, but less so than fish and cheese. In lobster Thermidor, lobster meat is served in a cream sauce laced with Gruyère and mustard. Parmesan and crab meet in a boldly flavored tart, in a soufflé, or in a mixture of spider crab, cheese and chili sauce piled back into the shell and given a quick blast under the grill. Shrimp is often paired with feta in a salad, or the feta is scattered over shrimp in a tomato-rich Provençal sauce.

Shellfish & Lamb One of the less famous meat and shellfish combinations. Nonetheless, lamb was once often cooked with cockles, and an early-nineteenth-century cookbook by John Farley contains a recipe for a roast boned leg of mutton with shredded crab or lobster meat, seasoned with lemon zest and nutmeg. Marco Pierre White includes lamb with cockles and thyme on the menu at Marco, one of his London restaurants.

Shellfish & Lemon The beach fishery at Cadgwith Cove in Cornwall is one of the last left in England. You can still see the boats huddled together on the shore in the evening. It was a summer-vacation ritual of my childhood to help my father pick out the best live shellfish from the fishery's tank after eating crab sandwiches in a converted sardine cellar nearby. The brown crab caught on the Cornish coast has a deep, seaweedy flavor, and in its shell looks, rather fittingly, like a pasty with legs. When it's fresh, it needs nothing but a squeeze of lemon juice backcombed through the meat to enhance its sweetness. The same goes for all shellfish, really, whether raw on a stand of *fruits de mer* or grilled on a platter of octopus and langoustines. A lemon wedge is all you need. Well, that and a suitably sharp Chablis.

Shellfish & Lime Deep-fried conch is delicious served with a seafood-cocktail sauce laced with lime. Conch's flavor is a cross between scallop and clam. Its texture is a cross between scallop and gym mat. You can give it the ceviche treatment (see Lime & White Fish, page 296) by pounding the flesh to make it chewable, at least by humans, and mixing it with sliced onion and lots of lime juice (which will also go some way toward tenderizing it). Overfishing is a problem, and farmed conch is more tender to the teeth, if less flavorful, than the wild kind.

Shellfish & Mango *See Mango & Shellfish, page 285.*
Shellfish & Mushroom *See Mushroom & Shellfish, page 81.*
Shellfish & Nutmeg *See Nutmeg & Shellfish, page 219.*
Shellfish & Oily Fish *See Oily Fish & Shellfish, page 156.*
Shellfish & Olive *See Olive & Shellfish, page 174.*

Shellfish & Parsley Judged by the appearance of their shells on the beach, razor clams are well named, but once caught and stacked on fishmonger's ice, flopped out of the end of their shells, the edible part looks about as treacherous as a puppy's tongue on a hot day. They taste pretty much like other clams, but there's a good deal more meat on them. Serve them cooked with white wine, garlic and parsley, or do as the Italians do and slurp them direct from the shell.

Shellfish & Parsnip *See Parsnip & Shellfish, page 221.*

Shellfish & Pea Scallops on puréed peas, or pea risotto, is something of a modern classic, but the combination of shellfish and pea is nothing new. The following recipe is from a book on medieval cooking by Hieatt, Hosington and Butler. Soak 2 cups split green peas (if necessary) and then simmer in 1 quart water until soft. Set aside to cool. In a food processor, blend 2 oz ground almonds and 1 tbsp breadcrumbs as finely as possible, then, while the machine is running, gradually add the peas with ½ tsp salt, 4 tsp white wine vinegar, ¼ tsp ground ginger, ¼ tsp ground cardamom and a pinch each of ground cinnamon and clove. Season with salt and pepper. Put the pea mixture into a pan and slowly reheat while you briefly sauté about 1 lb shelled shrimp. Arrange the shrimp on top of the purée in a bowl.

Shellfish & Peanut *See Peanut & Shellfish, page 28.*

Shellfish & Pineapple The sweet-saltiness of shellfish combined with pineapple's sweet-sourness is common to authentic Asian dishes and to their lurid, sweet-and-sour semblances in Chinese restaurants. In Indian Parsee cooking, shellfish and pineapple are paired in a sweet and sour curry with tamarind, called *kolmino patio*. And in Southeast Asia they're found in countless curries and soups. Useful to bear in mind when you're saddled with an underripe pineapple—they don't get any sweeter once they're picked.

Shellfish & Pork In Portugal it's hard to miss *porco à alentejana*, or pork stewed with clams, peppers and onions. The magic of the dish is not just the chewy little nuggets of clam but the precious juice inside their shells, which, as is the case with oysters, contributes an essential part of the flavor. One doubtful theory on how this combination arose is that Portugal's pigs, left to wander its long coastline, had a lot of seafood in their diet, with the consequence that their meat had a fishy taste that the locals covered up by cooking it with clams. But it's just as likely to have come about because shellfish and pork were abundantly available and the combination is delicious—so much so that it's traveled to Macau, a former Portuguese colony where several variations on the dish are eaten.

Shellfish & Potato *See Potato & Shellfish, page 93.*
Shellfish & Saffron *See Saffron & Shellfish, page 178.*

Shellfish & Smoked Fish Essential to fish pie, where the smoked fish provides a pungent flavor contrast to the white, and the shrimp, besides looking cute, all pink and curled in their soft bedding of mashed potato, provide a textural contrast to them both. But you don't want too much of either: the white fish should prevail. The same principle applies to *choucroute de la mer*—see Smoked Fish & Cabbage, page 162.

Shellfish & Thyme *See Thyme & Shellfish, page 319.*
Shellfish & Tomato *See Tomato & Shellfish, page 256.*
Shellfish & Truffle *See Truffle & Shellfish, page 117.*
Shellfish & Vanilla *See Vanilla & Shellfish, page 341.*
Shellfish & Walnut *See Walnut & Shellfish, page 233.*
Shellfish & Watercress *See Watercress & Shellfish, page 101.*

Shellfish & White Fish Scampi may refer to a type of shrimp in some countries, but in the UK it's a shrimp entombed in breadcrumbs or batter, usually served with fries. Monkfish used to be cut into shrimp shapes, coated in crumbs and passed off as scampi. Then someone realized that if it could be passed off as shellfish it might be rather good in its own right, and in the 1990s it increased in price and popularity to the extent that it is now on the endangered list. Jane Grigson notes that monkfish is often compared to lobster, which isn't, in her opinion, fair to either side, even if monkfish was her favorite fish to cook and eat after "the four greats"—sole, turbot, eel and lobster. Lobster and monkfish certainly have a textural similarity (they're both pleasantly chewy), and taste tests reveal that the flavor of monkfish does bear comparison with shellfish, especially in the dark, gelatinous flesh nearer the skin. Monkfish aside, John Dory and turbot are classically served with shellfish sauce, and a few mussels or clams cooked with peas in butter and white wine make a beautiful brothy sauce for any white fish.

White Fish

The flavor of saltwater fish is a result of its own delicate balancing act. On average, the salinity of seawater ranges from 3 to 3.5 percent by weight. As Harold McGee points out, animals need to keep the total level of dissolved minerals in their cells closer to 1 percent, and so sea fish offset the saltiness of their environment by filling their cells with other compounds, namely amino acids and amines, which have their own taste and flavor implications. Glycine, an amino acid, lends sweetness, whereas the glutamic acid present in shellfish, tuna and anchovies "is savory and mouthfilling." Many fin fish, however, offset the salt water flowing through their bodies with the relatively flavorless amine, trimethylamine oxide, or TMAO, which is why, in contrast to strong-flavored fish such as anchovy, most white fish are characterized by sleep-inducing flavor descriptors like "mild," "sweet" and "delicate." White fish are so often entombed in batter or slathered in sauce that it's worth trying a plate of three or four different kinds, simply fried or steamed. Once your powers of perception have acclimatized, you'll find the subtly different flavors begin to show up against the white like landmarks in a snowscape. Look out for gamy, musty, earthy and seaweedy notes. Cod can have a slightly sour, salty, cooked-potato quality, which is perhaps why it's such a hit with French fries. Monkfish can be sweet and buttery, a little shellfishy, without the full buttered-popcorn flavor of shrimp. Good sea bass is meaty, Brazil-nutty, with a faint metallic twang to match its fuse-wire gray veins. Fish covered in this chapter include cod, skate, sole, flounder, plaice, monkfish and turbot.

White Fish & Anchovy Both Hannah Glasse in the eighteenth century and Eliza Acton in the nineteenth make mention of anchovy sauce served with boiled white fish (especially flat fish) or with fried breadcrumbed cod. You can make it by adding anchovy essence to melted butter with lemon juice.

White Fish & Anise See *Anise & White Fish, page 183.*

White Fish & Asparagus When the Italians began cultivating asparagus in the seventeenth century, they usually served it on its own, whereas in France and England it was often used as an adjunct ingredient, especially to fowl or to fish like turbot, which has a sweet, buttery flavor that's an ideal match for sweet, butter-loving asparagus. Orange-flavored *sauce maltaise* is the classic accompaniment—see Orange & Asparagus, page 288.

White Fish & Bacon In *The Adventures of Tom Sawyer,* Tom and Huck cook their just-landed bass with bacon and are astonished at its deliciousness. The benefits of cooking fish with bacon fat are not to be underestimated. A little pot of rendered bacon fat is worth having in the fridge. It's good with fish, cabbage and for baking cornbread. I think of it as a cross between cooking fat and stock because it has such an enriching flavor. Buy 1 lb very fatty bacon and chop it up into lardoons or slices (there'll be bacon meat to use up once you're finished, so cut lardoons if you're tossing it with blue cheese in a salad, slices if you're planning a breakfast or club sandwich). Cook the fat over a low heat in a frying pan with ¼ cup water. Be patient: it will take at least 30 minutes. When it looks as if all the water has evaporated and all the fat melted, cool a little, then strain into a ceramic or glass storage jar and keep in the fridge.

White Fish & Caper *See Caper & White Fish, page 103.*

White Fish & Celery A rare pairing. The combination of string and bones can feel less like eating and more like some sort of therapeutic craft project. This is where celeriac—the softer face of celery—comes in. Sweeter and easier to work with, celeriac can make a great partner for all sorts of seafood. Chef Tom Aikens pairs them in dishes such as John Dory with cabbage, celeriac and horseradish, sea bass with small parsley gnocchi and cara-melized celeriac, or poached turbot with poached chicken wings, crushed celeriac, truffle gnocchi and wild sorrel.

White Fish & Cilantro *See Cilantro & White Fish, page 195.*
White Fish & Coconut *See Coconut & White Fish, page 282.*
White Fish & Cucumber *See Cucumber & White Fish, page 186.*

White Fish & Dill Although rare in other parts of Asia, dill features promi-nently in the cuisines of Laos and some parts of Thailand. The eponymous fish dish at Cha Ca La Vong, a restaurant in Hanoi, Vietnam, is the only thing they serve: white fish (a kind of freshwater catfish) yellowed with turmeric and stir-fried with dill fronds and scallions. You can re-create something like it at home with any white fish that holds its shape during stir-frying, such as John Dory, sea bass or tilapia. Just don't be tempted to skimp on the dill. Cut about 1 lb fish fillets into large bite-sized pieces and marinate for 30 minutes in a mixture of ½ tsp turmeric, 1 tsp sugar, a 1-in thumb of galangal (or ginger), finely chopped, 2 tbsp fish sauce, 1 tsp rice wine vinegar and 1 tbsp water. Fry the fish in peanut oil for about 4 minutes or until just cooked through, add sprigs of dill and shredded scallion, and fry for a minute longer. Serve in bowls on cold rice vermicelli and season to taste with roasted chopped peanuts, chili, more dill, mint, cilantro and *nuoc cham* (see Lime & Anchovy, page 293).

White Fish & Garlic *See Garlic & White Fish, page 114.*
White Fish & Ginger *See Ginger & White Fish, page 305.*

White Fish & Grape To me, sole Véronique is the classic convalescence dish: Dover sole, lightly cooked in stock, cream, vermouth and lemon juice with halved green grapes. It fulfills the vital *cucina bianca* criterion of kindliness to the stom-ach, and is so pale and delicate as to make you feel robust by comparison (plus it's a way to use up all those hospital visitors' grapes). Sole is not the *sine qua non*; any firm white fish will do. To give it its due, however, Dover sole does have a unique flavor, which develops in the muscle tissue two or three days after it's caught; unlike most fish, it's not particularly tasty straight off the boat. Other flat fish such as turbot and halibut are subject to the same effect, but not as noticeably.

White Fish & Hard Cheese A highly contentious food pairing, which is anathema to Italian restaurateurs in particular; at least one Italian restaurant in New York warns customers, "No cheese served on seafood at any time." The objection is essentially to the cheese overpowering the subtle flavor of fish, although it seems to have come about in relatively recent times— research has unearthed many older Italian cheese and fish combinations, dating back to a Sicilian fish recipe from around 400 BC. Fish and cheese are, of course, paired in French dishes, like skate with a Gruyère sauce, or bouillabaisse with grated Gruyère sunk into its depths. The British grate Cheddar on mashed-potato-topped fish pie to give it a flavor boost, while the Americans pair tuna with cheese in their classic toasted tuna-melt sandwich. I don't much like the *idea* of seafood and cheese, but do like it in practice, especially in an old-fashioned Mornay sauce with cod or haddock—which I suppose makes me as undecided as McDonald's, which uses only half a slice of cheese in its Filet-O-Fish.

White Fish & Hazelnut See Hazelnut & White Fish, page 236.
White Fish & Horseradish See Horseradish & White Fish, page 105.

White Fish & Lemon White fish is so frequently accompanied by a wedge of lemon that it can look a little lonely without it. Clearly, the lemon provides a sour counterpoint to the sweetness of fish flesh, can deodorize any stronger fish flavors and cuts through the fattiness of fried fish. But the highest compliment you can pay a piece of good fresh fish is to serve it completely unadorned, like a young girl so pretty she doesn't need a trace of makeup.

White Fish & Lime See Lime & White Fish, page 296.

White Fish & Mango Briny, nutty fresh red snapper tastes great with a mango salsa. Mango is also good paired with any white fish that will hold its shape in an aromatic coconut milk curry. In Cambodia, julienned green mangoes are cooked with mudfish (although tilapia will do just as well) in a combination of fish sauce, garlic and ginger, the delicious sourness of the fruit teasing out the sweetness of the fish. The food writer Alan Davidson describes an unusual Dutch dish of haddock with a sauce made of mango chutney, sliced tart apples, ginger and lemon juice.

White Fish & Mushroom See Mushroom & White Fish, page 82.

White Fish & Olive You might think a rule applied to fish with olives— green with delicate white fish, black for richer, oily types. It doesn't. You may find, however, that the salty oiliness of olives gives subtle white fish a welcome touch of oily-fish richness. A good-quality, extra virgin olive oil, with a clean bitter streak and the flavor of expensive green vegetables, may be all the sauce a delicate piece of fish needs.

White Fish & Orange *See Orange & White Fish, page 291.*
White Fish & Parsley *See Parsley & White Fish, page 191.*

White Fish & Parsnip "A good cook will never send salt-fish, and but a few salt meats, to table, without parsnips," wrote the horticulturalist Henry Phillips in 1822. The partnership dates back to medieval times but has recently been revived by Hugh Fearnley-Whittingstall in the form of saltfish and parsnip rösti fishcakes.

White Fish & Pea *See Pea & White Fish, page 201.*

White Fish & Potato Some pointers for anyone seeking an authentic fish-and-chip experience:
1. "Fish" and "chips" is not necessarily *fish and chips*. To qualify as the real thing—i.e., an essentially unitary quantity, *fishandchips*, wherein the chips (thick potato fries) are inseparable from the fish as opposed to just served with it—it has to come from an outlet known as a chip shop or chippy. This is not only for sentimental reasons. To achieve the essential crispness of the chip, and to cook enough pieces of battered fish at one time to stop them losing heat or going soggy, you need the sort of hard-core double-frying equipment rarely found outside the dedicated chip shop.
2. A chippy serves fish and chips, mushy peas, fishcakes, saveloys, battered sausages and pickles. At a stretch, meat pies. Avoid opportunistic "fish bars" that offer burgers, fried chicken, kebabs or pizzas.
3. Salt and malt vinegar are essential. Other types of vinegar are not acceptable.
4. The fish and chips must be served wrapped in proper chip-shop paper, which to the uninitiated looks, feels and most importantly smells like newsless newspaper, and is the secret ingredient in proper chip-shop fish and chips. It seasons them. If they're served in Styrofoam boxes, eat elsewhere.
5. Fish and chips must be eaten in appropriate surroundings. To be truly authentic you should be eating your fish and chips by the sea, or at the bus stop, or perched on the wall outside a gas station, breathing steam into your hand because you couldn't wait for the chips to cool down.

White Fish & Prosciutto *See Prosciutto & White Fish, page 171.*

White Fish & Saffron Italians turn their noses up at this combination, according to Elizabeth David in *Italian Food*, as they think the flavor of saffron overpowers the fish. She doesn't agree, and neither, surely, would anyone who's eaten a bouillabaisse. One evening in Antibes, a pharmacist slipped us the name of *the* place to get it, somewhere off the coast road between Saint Tropez and Cabasson. He didn't know the address—you just had to keep your eyes peeled. Chez Joe, it was called, or so the pharmacist's scrawl on the rumpled napkin seemed to say. We arrived at a patch of featureless headland and climbed down a rusty ladder to a narrow beach where a dozen folding tables had been pushed into the sand. "Joe," whose *chez* was a cave, attended to a large pot over a fire of driftwood, muttering to himself under a roof of dripping rock. Another man brought us a bottle of rosé from a cooler and a couple of tumblers. We drank for a while, looking out to sea, before the soup was presented along with some garlic-rubbed bread. It tasted of the view: turbulent, sea-beddy, on the vaguely sinister side of complicated. When the soup was finished, the waiter brought the fish that had been cooked in it, along with more bread and more wine. A couple of diamond-shaped almond delicacies, called *calissons d'Aix*, in place of dessert, and a glass of brandy to make you forget where you'd been and preserve the secret.

White Fish & Shellfish *See Shellfish & White Fish, page 142.*

White Fish & Thyme "I have also removed from my cooking of fish," wrote the French chef Marie-Antoine Carême, "those quantities of aromatics and spices that our forebears were wont to use for seasoning, for it is a strange delusion to believe that fish should taste of thyme, bay, mace, clove or pepper, whereas we have irrefutable evidence every day that fish cooked in salt water alone is excellent." He does go on to say that a thyme-infused court-bouillon (the stock you might use for a seafood chowder) is permissible as long as the herb isn't intrusive. Fresh thyme is a safer bet than dried, which can be too strong for fish. Lemon thyme obviously makes an excellent match.

White Fish & Tomato It's hard to beat baked bream with roasted tomatoes, or fish fingers and ketchup, but in the early 1990s researchers set out to improve on the tomato with an implanted gene from the Arctic flounder. Arctic flounders are, unsurprisingly, genetically programmed not to freeze (and rupture) in cold waters—properties the scientists thought might help tomatoes survive cultivation in harsh conditions, and the bumpy trip to market. It didn't work. Splice them as nature intended by frying your flounder in butter, with a crumb coating if preferred, and serving with roast tomatoes and boiled new potatoes, or squished in a bun with ketchup.

Oyster

An oyster's flavor is more than usually expressive of the environment that supported it. The cleanliness, mineral content and temperature of the water are important factors, and account for the common practice of identifying oysters by provenance rather than species. For example, *Ostrea edulis*, the "European flat" or "native" oyster, is variously referred to as the English Colchester, the Irish Galway and the prized and protected Belon from France. They have also been successfully maricultured in California and Maine. Order a mixed platter of oysters and hold off on the lemon while you try to detect the flavor notes connoisseurs have identified: sweet, creamy, meaty, buttery, nutty, melon, melon rind, cucumber, mineral, metallic, coppery and, of course, briny or oceanic. To experience its sweeter, more complex flavors beyond the initial shock of seawater, you'll need to give your oyster a chew before it slips down. Sour flavors like lemon and vinegar dampen an oyster's saltiness and highlight its sweetness. Horseradish, or the chili-spiked tomato ketchup known as cocktail sauce, makes for a hot-sweet-salt combination wherein the oyster also contributes its inherent umami. Oyster sauce is a thick, dark, salty condiment from China; brands vary in the quantity of oyster extractives, sugar, sodium and flavor enhancers they contain.

Oyster & Anise *See Anise & Oyster, page 181.*

Oyster & Bacon Oysters individually wrapped in half a (pre-stretched) slice of streaky bacon, impaled on a toothpick and grilled. I could never remember: angels on horseback? Devils on horseback? Pigs at the beach? Swine before pearls? Then someone explained: oysters are whiteish, like angels. Prunes are dark, and thus evil, especially when pitted and filled with mango chutney before they're enfolded in bacon. As ever, the devil has the best tunes: the extreme salty-sweet contrast of prune and bacon wins hands down for me.

Oyster & Beef When Kit buys little Jacob his first dish of oysters in *The Old Curiosity Shop* by Charles Dickens, the child takes to them "as if he had been born and bred to the business—sprinkled the pepper and the vinegar with a discretion beyond his years—and afterwards built a grotto on the table with the shells." Before stocks collapsed, oysters were so plentiful and cheap that a servant like Kit could buy three dozen of "the largest oysters ever seen" without fear of the debtor's prison. But it wasn't only to bulk out the more expensive ingredients that oysters were added to steak and kidney pudding. The salty oyster bolsters the flavor of beef. The same principle applies to the oyster-stuffed Australian carpetbag steak and the use of oyster sauce in stir-fried beef dishes.

Oyster & Caviar *See Caviar & Oyster, page 152.*

Oyster & Celery Celery makes a fresh, almost citric partner for oysters: a rather genteel combination, and never more so than in this recipe for "Oysters and Celery from Billy the Oysterman," from the charming *Esquire Handbook for Hosts*, first published in 1954. "The man," it advises, "who can ask his friends in after the theatre or other late functions . . . is a man who is going . . . to the top of his friends' hit parade." Sauté 3 chopped celery stalks in butter until tender, add 24 shucked oysters and their liquor and simmer until the oysters curl at their edges. Splash in some sweetish white wine and season gently. Remove to a dish to keep warm, cut some toast points, tap your cigarette on its initialed silver case and bask in the adoration of your chums.

Oyster & Chicken Fantastically luxurious to stuff a chicken with oysters before roasting it. Shuck them and stuff them in: it's as simple as that. If your cavity is bigger than your catch, you can always mix the oysters and their juices with buttery breadcrumbs. Or stuff a quail instead—they hold only about 3 oysters each.

Oyster & Chili My first oyster was pressed on me by a waiter in a grotty trattoria near London's Paddington station. It was the kind of place where you'd think twice about eating the breadsticks, let alone take your chances with the seafood. It didn't kill me, nor did it convert me. I ate my second in a bar in the French Quarter of New Orleans, a narrow, funereal room, cooled by an apathetic ceiling fan, neon signs fighting a losing battle against the gloom. A dive brasserie: exactly what I'd been hoping for. With his eyes half on the muted football game, a barman in sleeve clips shoveled oysters from an ice-filled trough in the bar and shunted them croupier-style toward the punters. The scene had the paradoxical exoticism of another culture's routine: everyone was bored except me. I raised a finger. The barman pushed an oyster my way. It was unlike any oyster I'd seen: a dark gray, shiny oval, monstrous as a whale's eye. It stared me down. I gave it a squirt of hot sauce and, trying to blank my mind, picked it up by its deep, craggy socket and swallowed it whole. I felt my scalp crackle. High on the oyster's dose of zinc, or the fiery sauce, or simply because I'd looked the beast in the eye and defeated it, I ordered a celebratory beer. And as it cooled my raging throat, the barman pushed something else my way. Bigger, grayer and shinier, and this time there were two of them.

Oyster & Egg Make a "Hangtown fry," invented in Hangtown in the Sierra Nevada during the Gold Rush. Some say it was made for a prospector who'd struck gold, others that it was the last meal of a convict on death row. Both myths rely on the rarity value of oysters in the desert. The likely truth is that it was devised by one of the many Chinese immigrants who found work as a cook in mid-nineteenth century California. Hangtown fry is basically fried oysters with scrambled eggs. Some recipes include bacon but the one in M. F. K. Fisher's book *Consider the Oyster* doesn't. She suggests you serve it with sausages and shoestring fries.

Oyster & Globe Artichoke *See Globe Artichoke & Oyster, page 127.*

Oyster & Horseradish *See Horseradish & Oyster, page 105.*

Oyster & Lemon Oyster purists turn their noses up at shallot vinegar or Tabasco—a raw oyster should be eaten with nothing at all, not even lemon. But I'm not a purist. For me, eating oysters is like throwing off all your clothes and jumping off the end of a jetty. Lemon juice is the mouthwatering run-up, giving way to the bracing splash of the oyster's minerality. Lemon juice also refreshes oysters deep-fried, Deep South–style, in cornmeal.

Oyster & Mushroom Some people believe that oyster mushrooms not only resemble oysters but have a faint taste of them too, like the rubbery frills on a lady's bathing cap, fresh out of the sea. Like oysters, oyster mushrooms can be quickly fried in butter and served with a squeeze of lemon.

Oyster & Nutmeg *See Nutmeg & Oyster, page 218.*

Oyster & Onion Finely chop shallots and mix with red wine vinegar for the classic French accompaniment to oysters, called *mignonette*.

Oyster & Parsley Oysters Rockefeller was devised by chef Jules Alciatore, of Antoine's in New Orleans, and has been on the menu there since 1899. Oysters on the half shell are baked with a green sauce and topped with crumbs. Alciatore took the recipe to his grave, in written form at least, and his successors at Antoine's are as tight-lipped as an unshuckable oyster. Other restaurants serve imitation dishes involving spinach but, according to Antoine's, no one has ever cracked the secret of the green sauce. Research has found its primary ingredients to be parsley, capers and olive oil, but the secret ingredient no lab test can identify is secrecy itself.

Oyster & Pork Oysters are traditionally mixed with pork and formed into sausages in the American South. They're also paired in a New England–style stuffing for poultry, and in Louisiana gumbo. But to my mind they're most electrically combined in the form of chilled, raw oysters and hot, spicy sausages. It's a perennial favorite of chefs and food writers everywhere. In Bordeaux they use *loukenkas* sausages, but mini chorizos are good too if you can find them. There's nothing like the hot, fiery snap of sausage undercut by the cool minerality of oyster. A glass of chilled white Graves and the world might fall around you like a film set.

Oyster & Watermelon *See Watermelon & Oyster, page 246.*

Caviar

Oscietra is the caviar of choice for the connoisseur: fine, nutty, complex and (some say) herbal in flavor. As with any natural product, the flavor can vary wildly, especially given the Oscietra sturgeon's omnivorous nature and its habit of evading predators by diving to the seabed, which together make its diet particularly wide. Carnivorous Beluga, whose protein-rich diet may account for the greater size of its eggs, produces a creamier, less fishy-tasting caviar. Sevruga is the commonest—or least rare—and is as black as model-village tarmac. It has a less complex, more strongly sea-salty flavor. Some estimate that stocks of Caspian Sea sturgeon may be entirely depleted by 2012. Some farmed caviars from America and Europe are nonetheless starting to receive outstanding reviews for their flavor.

Caviar & Banana If you ever feel a bit decadent tucking into bacon and eggs at breakfast, it may help to know that in pre-revolutionary Russia the tsar's children started the day with a dish of mashed banana and caviar.

Caviar & Cauliflower Cruciferous vegetables are terrific paired with salty ingredients. If you like sprouts with bacon, you'll love the classic Joël Robuchon partnership of cauliflower and caviar. He makes a gelée of caviar to pair with cauliflower cream; a couscous of Oscietra caviar, cauliflower cream and asparagus gelée; and cauliflower cream with caviar and puréed potato. See also Anchovy & Cauliflower, page 159.

Caviar & Chicken *See Chicken & Caviar, page 31.*

Caviar & Egg Frank Sinatra used to make scrambled eggs and caviar for Ava Gardner. Wolfgang Puck combines them in a buttery puff-pastry shell, which must be heaven to bite into. But asking your loved one to make puff pastry first thing in the morning might be pushing your luck, even if you look like Ava Gardner. If you're all out of caviar, note that the food writer David Rosengarten believes the best caviar eggs taste of egg yolks and butter. So you could always top buttered buckwheat blini with chopped, just-cooked egg yolks and eat them with your eyes closed.

Caviar & Hazelnut *See Hazelnut & Caviar, page 235.*

Caviar & Lemon Make an excellent sauce for scallops. Simply cook the scallops in butter, adding a little lemon juice and caviar at the end to heat through. Although lemon is often served with caviar, it's not ideal to go squeezing it on your best and most delicately flavored eggs. I once overheard an affronted hostess tell a guest, "If it's taramasalata you want, darling, I'll see if I have some."

Caviar & Oyster If you're serving caviar and oysters, it's conceivable you're not interested in the finer points of the flavor combination. You'll have more important things on your mind, like the sable trim for your speedboat. Harold McGee reckons Oscietra caviar tastes like oysters. They certainly share a decaying-frigate palette of gray, green, rust and black, and one might put any proximity in flavor down to their similar bottom-feeding habits. One of Thomas Keller's signature dishes is oysters topped with caviar served on tapioca pearls in a vermouth sabayon. At Noma in Copenhagen, chef René Redzepi, who worked under Keller at The French Laundry, makes a caviar of gelled oyster juice and serves it on a tapioca pudding.

Caviar & Potato *See Potato & Caviar, page 90.*
Caviar & Smoked Fish *See Smoked Fish & Caviar, page 162.*

Caviar & Soft Cheese Silver and stainless steel affect the flavor of caviar, which is why spoons made of mother-of-pearl or plastic are used instead. As to accompaniments, water crackers or buckwheat blini are acceptable. Less fastidious fans might forgive a side dish of chopped boiled egg. White foods certainly look fantastic with caviar, drawing attention to its iridescence by bouncing light onto each taut egg, just as white fur complements high cheekbones. Try these mini caviar cheesecakes. Using a small pastry cutter, press 1-in circles from a ½-in-thick slice of dark rye bread. Leave the cutter in place while you spoon in a ½-in depth of cream cheese and a dollop of caviar. Ease off the cutter, taking care to keep the cheesecake in shape. You'll need to wash the cutter after each one.

Caviar & White Chocolate Investigating the ability of salt to bring out the flavor of sweet foods, Heston Blumenthal paired white chocolate with ham, anchovies and cured duck before alighting on the particularly pleasing pairing of caviar. Flavor expert François Benzi noted that the two ingredients share flavor compounds that might explain their compatibility. This led Blumenthal to research the harmoniousness of pairings with compounds in common, and to conclude that while shared organo-chemical ground can often account for the compatibility of existing pairings, and, conversely, suggest unlikely pairings that turn out to work—coffee and garlic, for instance, or parsley and banana—there's no replacement for imagination and intuition. To me, the combination of caviar and white chocolate is like a heightened version of Alain Senderens' pairing of lobster and vanilla. The flavor of white chocolate is predominantly vanilla, with a slightly buttery, cream-cheesy quality that would make it all the more harmonious with the fish eggs.

Oily Fish

Oily fish are richer and more strongly flavored than white fish. Classic flavor affinities, such as gooseberry or horseradish with mackerel, cucumber with salmon, and watercress with trout, seek to cut through the fattiness. Lemon serves this function too, as well as deodorizing some of the stronger fishy aromas. Cilantro and/or lime juice are used to the same effect in Asian dishes. Briny capers, olives, prosciutto and shellfish bring out the sweeter side in sea-salty oily fish such as mackerel, herring, sardines and red mullet. This chapter also covers earthier-tasting freshwater fish and diadromous fish (i.e., species that live in both salt- and freshwater), like salmon, trout, grayling and eel, plus meatier-flavored swordfish and tuna. See also the introduction to White Fish, page 143.

Oily Fish & Almond *See Almond & Oily Fish, page 240.*

Oily Fish & Anise There are countless ways to combine fish with anise-scented herbs, spices and liqueurs, many of them quick and simple—a poached salmon steak, for instance, with a fresh, heady tarragon mayonnaise. If, however, your tastes tend more to the theatrical, you might try the recipe in Elizabeth David's *French Provincial Cooking* for fennel-stuffed grilled red mullet served on dried fennel with a measure of flaming brandy poured over it. The fennel catches and emits a puff of anise-scented smoke. Great, if you have a dine-in kitchen, where everyone can enjoy the aroma while you whip the fish off the singed twigs onto a warm serving plate. Equally sensational, if more fetishistic than rustic, is Heston Blumenthal's salmon in a tight, shiny, black licorice gel—a recipe he developed when exploring the combination of asparagus and licorice.

Oily Fish & Asparagus Marsh samphire is sometimes called "sea" or "poor man's" asparagus. You can just imagine the resentful fenman, up to his knees in brackish marsh water, muttering about the fancy folk inland, with their doilies and their salad forks and their "actual" asparagus. It's certainly served the same way, with butter or hollandaise sauce, often as an accompaniment to seafood, especially salmon. As I write this, both are in season and the samphire is four times the price of the asparagus.

Oily Fish & Avocado According to David Kamp, the California roll was first conceived in the 1960s. As the availability of raw tuna was seasonally limited in Los Angeles, Ichiro Mashita and Teruo Imaizumi, sushi chefs at Tokyo Kaikan, tried combining king crab, cucumber and ginger with avocado, on the basis that it has a buttery texture and dense fattiness somewhat reminiscent of fresh tuna. Avocados grow year-round in California, which was just as well, as it turned out Mashita and Imaizumi had invented a dish that was not only hugely popular in its own right but became a sort of training sushi; once they'd got used to the seaweed and cold, compacted rice, diners went on to try the more authentic stuff.

Oily Fish & Beef *See Beef & Oily Fish, page 49.*
Oily Fish & Beet *See Beet & Oily Fish, page 88.*
Oily Fish & Caper *See Caper & Oily Fish, page 102.*

Oily Fish & Chili Paired in countless sauces in India and Southeast Asia. Chili is, of course, more sparingly used in Western European cuisines, but the Michelin-starred Parisian restaurant Taillevent has been known to serve fresh tuna belly with Espelette peppers, lemon, capers and Serrano ham. At the other end of the price scale, Les Mouettes d'Arvor pairs sardines and chilies in handsome cans. Choose between the variant with Espelette chilies, a mild but tasty red pepper grown in the Basque Country, and the hotter bird's eye chili version. The French, to their credit, still afford the canned sardine the respect it deserves, marking the best with a *millésime*, or vintage, and a recommendation as to how long to lay them down. Chancerelle's 2001 vintage, for example, was ready to eat in 2007.

Oily Fish & Cucumber Smelts, which are small, salmon-like fish, have a smell of cucumber when freshly caught, although some compare their aroma to violets or freshly cut grass. Elizabeth David said they are one of the nicest small fish for frying. The eighteenth-century cookery writer Hannah Glasse suggested frying them in breadcrumbs and serving with fried parsley and melted butter. Along with sweet, sour, metallic, buttery notes, salmon also has a cucumber aroma when fresh, although this recedes with cooking and is replaced with a boiled-potato character. Interesting to note that the traditional accompaniments to poached salmon, boiled new potatoes and a cucumber garnish, have a deeper affinity for the fish than you might think.

Oily Fish & Cumin *See Cumin & Oily Fish, page 86.*

Oily Fish & Dill In Scandinavian and Baltic countries, there are many recipes for combining the strong, clean flavor of dill with oily fish. Gravlax, probably the most famous, is salmon cured with sugar and salt, flavored with dill and mustard, then served with more dill and mustard in the form of a sauce. The Swedish mustard used is sweeter, creamier and less pungent than Dijon or English varieties. Alan Davidson gives recipes for gravad mackerel (not as delicious as the salmon, he notes, but still pretty good), and for *makrillsopa*, a simple preparation of mackerel, dill, water, milk, peppercorns and possibly some potato. In the United States, dill is used in tuna salad sandwiches, for which freeze-dried dill works just fine. My favorite dill and fish idea comes from Nigel Slater, and isn't much more work than making a sandwich. Put 2 salmon steaks in a small baking dish and cover with a mixture of 1 very, very finely chopped onion, 1 tbsp lemon juice, 1 tbsp chopped dill and ½ cup sour cream or crème fraîche. Bake at 425°F for 15–18 minutes (just long enough to boil some new potatoes) and scatter with more chopped dill before serving.

Oily Fish & Egg How my heart sinks when salade niçoise is served with fresh tuna. Chunks of seared tuna may share their canned counterpart's affinity for egg, capers, green beans and potatoes but they lack the flaky, messy texture that is essential if the fish is to meld with the dressing and loose crumbs of egg yolk and infiltrate the entire salad, as opposed to sitting haughtily on top of it. If you're looking to impress, then Tre Torri's canned Ventresca brand, which uses tuna belly, has a buttery, creamy character that makes it a world-class treat, according to the food writer David Rosengarten. Mind you, he rates it so highly as to recommend eating it by itself, using a less exalted variety for salads—understandable when you consider that Ventresca costs roughly 20 times as much as a standard supermarket brand.

Oily Fish & Garlic This recipe from Patricia Wells is something like the fish equivalent of chicken with 40 cloves of garlic (see Garlic & Chicken, page 112). It calls for tuna, but swordfish would work well too. In fact, research has identified a "fried chicken" character in swordfish, which suggests a compatibility with garlic. Brush two 8 oz fresh tuna steaks (around 1 in thick) with oil and season them with pepper. Grill for about 5 minutes on each side, until they turn opaque but are still pink in the middle. Meanwhile, heat 3 tbsp peanut oil in a pan until hot but not smoking, add 20 thickly sliced large garlic cloves and sauté until golden. Add 1 tbsp red wine vinegar and stir to deglaze. Season the tuna, pour the sauce over it, and serve with *pipérade*—tomato, onion and green bell pepper stewed in olive oil. *Pipérade*, incidentally, is sometimes cooked with eggs too—see Bell Pepper & Egg, page 202.

Oily Fish & Ginger *See Ginger & Oily Fish, page 304.*
Oily Fish & Horseradish *See Horseradish & Oily Fish, page 104.*

Oily Fish & Lemon A juicy slice of lemon is all you need for a bowl of hot, crisp, deep-fried whitebait, a dish that's popular in various forms the world over. Whitebait are served as fritters in New Zealand and tossed in chili and turmeric in India. In the days when fish were plentiful, whitebait—a generic term for a variety of juvenile fish, or sprats, including herring—were caught in the Thames and served on summer days in pubs, moreishly salted and spritzed with lemon to help the drinks go down. See also Parsley & Potato, page 190, and Oily Fish & Dill, page 154.

Oily Fish & Lime Lemon is excellent at cutting through the fattiness of oily fish, but lime has a spicy quality that gives it the upper hand on the richest varieties. Just a quick squeeze will tease out the sweetness of fried mackerel or barbecued sardines. For an especially intense flavor, halve your limes and place them flesh-side down on the grill to caramelize a little first. Or dress a salsa of tomatoes or mangoes with lime juice to serve as a side dish. In the Arabian Gulf, fillets of oily fish are seared and served with a sauce of onions, ginger, garlic, spices and musky dried limes in a dish called *samak quwarmah*. See also Cumin & Oily Fish, page 86.

Oily Fish & Liver *See Liver & Oily Fish, page 45.*

Oily Fish & Mint In her eighteenth-century cookbook, Hannah Glasse made a special point of stipulating finely chopped mint, parsley and fennel to stuff mackerel before grilling. In Sicilian cookery the clean, powerful flavor of mint freshens oily fish dishes such as grilled swordfish, or pasta with sardines. In Thai cuisine chopped fresh mint is used to garnish salads such as *laab pla*, a spicy, tangy mix of fish, chili, mint, ground roasted rice and lime juice. The use of mint as a garnish is even more widespread in Vietnam, where a smooth, heart-shaped herb called fish mint is also prevalent. It has a slightly sour, fishy flavor and is often served with beef or grilled meats.

Oily Fish & Mushroom *See Mushroom & Oily Fish, page 80.*

Oily Fish & Onion Hawaiian *poke* (pronounced pok-ay) is like sashimi going through a rebellious phase. In contrast to the strict rules and traditions applied to the preparation of its Japanese cousin, in *poke* raw tuna is cut into chickpea-sized pieces, marinated with various ingredients, then served piled on a plate. The traditional seasonings were seaweed and roasted candlenuts, a large, oily nut with a bitter taste; now lots of other flavors, especially onion or scallion, are popular. In Japan, *toro* tuna is blanched and mixed with scallions in a delicate dashi-based soup. The meatiness of tuna sees it minced and mixed with onions in Italian *polpette* (meatballs) and in burgers. And less exotically, if no less deliciously, canned tuna is paired with onion on pizza, in tuna mayonnaise sandwiches and in the store-cupboard classic, tuna, cannellini bean and onion salad.

Oily Fish & Parsley *See Parsley & Oily Fish, page 190.*
Oily Fish & Pea *See Pea & Oily Fish, page 200.*
Oily Fish & Pork *See Pork & Oily Fish, page 39.*

Oily Fish & Potato French fries go with everything in the UK, except salmon and trout—although the environmental historian Peter Coates speculates that by 2050 declining stocks will mean most of the fish sold in fish-and-chip shops could be salmon, a prospect I find quite unappealing. There's something about the bland saltiness of sea fish that works best with the crisp saltiness of chips, while the earthiness of freshwater fish is better served by the matching earthy quality of boiled or new potatoes. Sardines and mackerel will go with both. See also Parsley & Potato, page 190.

Oily Fish & Rhubarb *See Rhubarb & Oily Fish, page 251.*
Oily Fish & Rosemary *See Rosemary & Oily Fish, page 310.*

Oily Fish & Shellfish Red mullet is prized for its exquisite balance of saltiness, sweetness and the meatiness you expect in oily fish, somewhere between the delicate flavor of salmon or trout and the rough sexiness of mackerel. Pliny thought red mullet tasted of oysters; comparisons with shellfish are common, and the combination is terrific. Try seared or grilled red mullet with a shellfish sauce, or served simply with langoustines and peas, which complement both ingredients perfectly.

Oily Fish & Thyme Grayling, *Thymallus thymallus*, is an oily freshwater fish that, when just caught and pressed right up to your nose, has an aroma of thyme. As to the taste, it's like trout except not as nice. In *Good Things in England*, first published in 1932, Florence White advised that grayling is at its best grilled and sprinkled with powdered dried thyme.

Oily Fish & Watercress *See Watercress & Oily Fish, page 100.*

BRINE & SALT

Anchovy

Smoked Fish

Bacon

Prosciutto

Olive

Anchovy

Anchovies are, of course, available fresh or marinated in vinegar (as in the Spanish tapas, *boquerones*), but this chapter largely restricts itself to anchovies preserved in oil or salt and the fish sauces of Southeast Asia, made with anchovy-like fish. The richly fishy flavor of preserved anchovies can take a bit of getting used to, but when they are cooked, the slightly rancid note lifts, leaving a rich, seared, meaty savoriness that really does heighten the flavor of other fish, meats and vegetables. Anchovy-flavored products such as anchovy essence and anchovy butter (aka Gentleman's Relish) are perhaps less popular than they used to be, but are still (fairly) widely available. Worcestershire sauce is famously made with anchovy, tamarind, vinegar, sugar and various seasonings.

Anchovy & Beef Garum is a sauce made of sun-dried anchovies or mackerel innards and brine. It's thought to have been a Greek invention, but the Romans were mad for garum, using it with all meats and fishes in much the same way that fish and oyster sauces are used in Asian cooking. A few salted anchovy fillets in a beef stew, or pushed into slits in a roasting joint, will perform a similar function. Don't worry if you're so-so about the flavor of anchovy: cooked like this, the fishiness disappears in favor of an intense savoriness that makes the beef taste meatier and somehow juicier. Cooks talk of anchovy adding an extra dimension, say in a spicy beef broth for a Vietnamese *pho bo*. Long cooking isn't essential; an anchovy and garlic butter will enhance the flavor of a steak. Alternatively, you could make a Thai weeping tiger salad by pouring a fish-sauce–based dressing, like the one in Lime & Anchovy, page 293, over slices of seared beef steak.

Anchovy & Beet *See Beet & Anchovy, page 87.*

Anchovy & Broccoli Anchovy can be a model of discretion—for example, when it bolsters the flavor of a dish and then disappears like a trusted manservant. But it's indispensable to this popular Italian pasta dish. The rich saltiness of the anchovies contrasts with the bittersweet broccoli to luxurious effect—especially if you use the purple sprouting kind, which retains the sauce in its head and its frilly leaves. In a frying pan, dissolve 6 anchovy fillets in 2 tbsp warm olive oil (not hot—you don't want to fry them). Add some chopped garlic, chili flakes and a bridesmaid's posy of cooked broccoli florets. Push the broccoli around the pan to coat it in the sauce and then serve tossed with pasta. If you top it all off with grated Parmesan, you'll have a dish fit for a peasant king.

Anchovy & Caper *See Caper & Anchovy, page 101.*

Anchovy & Cauliflower If you like the idea of Joël Robuchon's pairing of cauliflower and caviar (see page 151) but your conscience, or wallet, balks at Beluga or Sevruga, make do with the less handsome, but extremely delicious, pairing of cauliflower and anchovy. It's a popular combination in Italy, where cooked cauliflower is tossed with fried breadcrumbs, anchovy, garlic, chili flakes and parsley for a pasta sauce. They also make a salad of blanched, cooled florets in an anchovy dressing made with olive oil, red wine vinegar, mustard, garlic, onion, anchovy, lemon and capers.

Anchovy & Chili *See Chili & Anchovy, page 203.*
Anchovy & Coconut *See Coconut & Anchovy, page 279.*
Anchovy & Egg *See Egg & Anchovy, page 130.*

Anchovy & Garlic Two strong characters. Burton and Taylor in *Who's Afraid of Virginia Woolf?* Anchovy's saltiness and garlic's pungent sweetness combine in a slanging match where neither partner wins. Delicious nonetheless in anchoïade, a chilled dip from Provence in which garlic is pounded with anchovies and loosened with olive oil; and in *bagna cauda*, an Italian dish from Piedmont somewhere between a fondue and a dip. To make it, blend 1 stick butter, ¾ cup olive oil, 12 anchovy fillets and 6 garlic cloves until smooth. Transfer to a heavy-bottomed pan and heat slowly for 15 minutes, stirring occasionally. Place the pan over a burner on the table, or pour into a fondue dish (if you don't have one of these in the back of the kitchen cupboard, have a heart and give a good home to one of the hundreds of forlorn fondue sets languishing on eBay). Keep the mixture warm while you dip in raw morsels of cauliflower, fennel, celery, bread and anything else you think could do with a tasty "hot bath."

Anchovy & Hard Cheese There's an old-fashioned English recipe for "mock crab" in which grated hard cheese is mixed with a few anchovies, some anchovy essence, mustard, a little cold white sauce and a pinch of cayenne, then spread into sandwiches. Perfectly delicious if you banish the strangely offputting name from your mind. Anchovies were also used to flavor cheese straws, or laid on slim soldiers of bread, covered with grated Parmesan, chopped parsley and melted butter, and baked in the oven. Anchovy is a key ingredient of Worcestershire sauce, and, if you need conclusive proof that anchovy and hard cheese is a marriage made in heaven, try eating cheese on toast *without* a Friesian mottling of Worcestershire after you've tried it once. Or make a Caesar salad, a distant cousin of cheese on toast, by rubbing bread with oil and garlic, tearing it into croutons and toasting. Toss with torn Cos leaves, anchovy fillets, grated Parmesan, olive oil, lemon juice and either a raw egg or an egg boiled for 1 minute and no more. Make sure the ingredients are well combined before serving.

Anchovy & Lamb *See Lamb & Anchovy, page 52.*

Anchovy & Lemon Pound a couple of anchovies into a paste with a little lemon juice, then add olive oil and season to taste. This makes an excellent dressing for a salad of bitter leaves such as arugula or chicory. See also Caper & Anchovy, page 101.

Anchovy & Lime *See Lime & Anchovy, page 293.*

Anchovy & Olive Like a couple of shady characters knocking around the port in Nice. Loud and salty, they take a sweet, simple pizza margherita and rough it up a bit. They're used to make the dotted lattice pattern on pissaladière, a simple Niçois snack sold from kiosks in the old town, where its rich, oniony fug hangs around the narrow streets on summer nights. At its best, it consists of a thick rectangle of bouncy bread, a little like focaccia, spread with a rich tomato sauce, then topped with a good depth of soft, sweet onion and a sparse tic-tac-toe pattern of anchovy and olive—you don't appreciate them if they overpower every mouthful. The ideal is a detonation of brininess every few bites. See also Olive & Tomato, page 175.

Anchovy & Onion *See Onion & Anchovy, page 107.*
Anchovy & Pineapple *See Pineapple & Anchovy, page 260.*

Anchovy & Potato Great in Jansson's temptation, a Swedish variation on potato dauphinoise. Chef Beatrice Ojakangas recommends using Swedish anchovies for this dish, as they're sweeter and less salty than the more common Spanish variety. If they're not available, she suggests smoked salmon as a possible substitute. Scatter sliced onions and anchovy fillets over the bottom of a buttered shallow baking dish and cover with fat matchsticks of potato. Pour in just enough cream to cover the potatoes, then top with breadcrumbs and dot with butter. Cover with foil and bake at 400°F for 25 minutes, then remove the foil and cook for another 20 minutes. I imagine Jansson as a melancholic Stockholm detective, whose sole comfort, like real ale for Morse, is a steaming plate of salty, creamy potatoes, served by a silent blonde waitress in a dark tavern by the sea.

Anchovy & Rosemary Mix 1 tsp very, very finely chopped rosemary and a few anchovy fillets into mayonnaise for a roast beef sandwich or to serve with grilled mackerel. See also Lamb & Anchovy, page 52.

Anchovy & Sage Italians call this combination *il tartufo di pescatore*—the fisherman's truffle. Or they should. Individually they lend a meaty character to dishes. Together they're out of this world and don't need fancy ingredients to prove it. Hannah Glasse combined them with beef suet, breadcrumbs and parsley to make a stuffing for pig's ears.

Anchovy & Soft Cheese *See Soft Cheese & Anchovy, page 72.*

Anchovy & Tomato If you've ever wondered what umami is exactly, a mouthful of tomato and anchovy cooked together should settle the matter. They're paired in pizza, pissaladière and spaghetti puttanesca, but if you're lucky enough to have a glut of good tomatoes, cut them in half, put them flat-side up on a baking sheet, lay an inch of oily anchovy fillet on each, along with a grinding of black pepper, and cook for about 2 or 3 hours at 250°F. Let them cool a little, then serve mixed with cannellini beans and a few shredded slices of salami. If you have any left over, blend them to make a rich tomato sauce for a simple pasta dish, or the backbone of a sturdy bolognese sauce. See also Olive & Tomato, page 175.

Anchovy & Watercress Of course this works. It's an elegant variation on salt and pepper. Spread anchovy butter thinner than a silk stocking on two slices of white bread. Load up with watercress and press down till the stalks snap. Enjoy by the river under a broad-brimmed straw hat. And *do* cut the crusts off.

Anchovy & White Fish *See White Fish & Anchovy, page 143.*

Smoked Fish

Smoked flavor is imparted by compounds including guaiacol, which has an aromatic, sweet, smoked-sausage taste, and eugenol, the main flavor component in clove. (Incidentally, both guaiacol and eugenol are found in barrel-aged wines, which is why fish pie and oaked Chardonnay are such natural companions.) You might also detect leathery, medicinal, fruity, whisky, cinnamon, caramel and vanilla notes in smoked foods. Exactly what flavor characteristics are promoted depends on how the raw ingredients were prepared. For example, ungutted fish such as bloaters will end up tasting gamier than gutted fish. The nature of the smoking process itself is another factor, including the type of wood used and the length of the smoke. Not that your herring has necessarily seen any more wood than the pencil tucked behind the delivery boy's ear. A flavoring called liquid smoke is often used in lieu of the time-consuming, expensive smoking process, not only for seafood but for meats and tofu too. In the United States you can buy it off the supermarket shelf as a poke in the ribs for your barbecue beans. This chapter covers smoked salmon, trout, mackerel, haddock, eel and herring.

Smoked Fish & Cabbage Smoked fish is to *choucroute de la mer* as smoked ham is to *choucroute garnie*: the pervasive flavor. Aside from the familiar shredded cabbage and potatoes, *choucroute de la mer* commonly includes an assortment of shellfish, white fish and some smoked fish.

Smoked Fish & Caper *See Caper & Smoked Fish, page 103.*

Smoked Fish & Caviar Smoked salmon and caviar is a relatively common, if luxurious, pairing, but at Caviar Kaspia in Paris you can order any combination of the nine caviars offered with smoked trout, eel or sturgeon. All good flavor pairings, but nothing looks quite as striking as the pinstriped satin flesh of salmon next to the glossy dark eggs, especially on blini with a white pillow of crème fraîche or sour cream. At his restaurant in Los Angeles, Wolfgang Puck combines smoked salmon and caviar on a signature pizza. Re-create something similar by brushing a pizza base with garlic and chili oil, scattering it with onions and baking in the oven. Meanwhile, stir finely chopped shallots, dill and lemon juice into sour cream. When the base is crisp, remove, allow to cool, then cover with the sour cream mixture and a fairly comprehensive draping of smoked salmon. Scatter over some chives and some spoonfuls of caviar.

Smoked Fish & Cherry *See Cherry & Smoked Fish, page 244.*

Smoked Fish & Coconut In Thailand, *pla grop* are small fish that are slowly smoked over coconut husks, tasting of smoked bacon according to the writer David Thompson, who recommends hot-smoked trout as a convenient Western approximation of the original. In *Thai Food*, he uses *pla grop* in a smoked fish and coconut soup, in a salad with Asian citron zest and in a stir-fry with holy basil and chili.

Smoked Fish & Dill Chopped fronds of dill are used to garnish smoked salmon and sour cream buckwheat blini and are also tossed in a creamy pasta sauce with smoked salmon. Food writers Julee Rosso and Sheila Lukins pair them in a quiche, while Sybil Kapoor dresses beet tagliatelle with a smoked trout and dill sauce. See also Oily Fish & Dill, page 154.

Smoked Fish & Egg You could eat this combination all day. Kedgeree for breakfast—flaked smoked haddock mixed into spicy, turmeric-yellow rice with sweet peas and boiled eggs. For lunch, a two-hand sandwich of egg mayonnaise and the thinnest slivers of London-cure smoked salmon from H. Forman & Son. The salmon is lightly smoked and tastes of the fish rather than the smoker chimney. For high tea, creamy mashed potato topped with a fillet of smoked white fish, balancing a poached egg like a circus performer. Finally, head to the Savoy for a late supper of omelette Arnold Bennett, a smoked haddock dish of their own devising, doubly enriched with hollandaise and béchamel sauce. See also Egg & Cabbage, page 132.

Smoked Fish & Horseradish Horseradish's popularity as an accompaniment to beef has meant its affinity for other flavors, especially oily fish, has become rather overlooked—at least in the UK. Not so in Eastern Europe, the Nordic countries, and wherever Jewish cuisine is eaten. Mix some horseradish into smoked mackerel pâté and spread thickly on grainy brown toast.

Smoked Fish & Lemon Eels get a "don't eat" rating according to the Marine Conservation Society. In *The River Cottage Fish Book*, Hugh Fearnley-Whittingstall and Nick Fisher describe them as one of the best smoked fish, on account of their rich, earthy flavor, but add, "You must wrestle with your conscience if you want to eat smoked eel." One way to work up an appetite for them, I suppose. But better, perhaps, to save the lemon, pepper and horseradish cream for other earthy smoked delicacies, such as trout. A wedge of lemon is the standard accompaniment to smoked salmon. The citrus juice cuts through the fattiness of the fish and freshens up the smoky flavors.

Smoked Fish & Onion *See Onion & Smoked Fish, page 110.*

Smoked Fish & Parsley Parsley tastes of rocks, rain and lush vegetation. What could be a more fitting partner for salty Scottish smoked fish? It's the essence of the landscape in a sprig of sprightly green. Chop it very finely and sprinkle over a craggy mountain of scrambled eggs with smoked salmon, Arbroath smokies or kippers.

Smoked Fish & Pea Pea loves both fish and smoked ham, so why make it choose? Sweet garden peas are a lovely, enlivening addition to kedgeree or smoked haddock risotto.

Smoked Fish & Potato Potato with smoked fish is a staple of northern European cooking. Potato bulks out the meal, of course, but also calms the rude flavor, which would have made it particularly welcome in pre-refrigeration days, when stronger cures were used to make the fish last longer. It's a testament to the fishcake that it's outlived these practicalities and made it onto the menus of some of the world's most fashionable restaurants, as has smoked fish chowder, or the similar Scottish soup known as cullen skink.

Smoked Fish & Shellfish *See Shellfish & Smoked Fish, page 142.*
Smoked Fish & Soft Cheese *See Soft Cheese & Smoked Fish, page 74.*
Smoked Fish & Watercress *See Watercress & Smoked Fish, page 101.*

Bacon

One of the world's best seasonings. Some people believe that there isn't a flavor that bacon doesn't enhance. Smoked bacon has a stronger, saltier, spicier flavor than unsmoked. Streaky is more fatty, and the fat is both more flavorsome and sweeter than the lean. For these reasons, just one slice of smoked streaky bacon in a soup or stew can give a far more delicious result than a stock cube could ever hope to. Salty bacon brings out the sweetness of other ingredients and takes the edge off their bitterness. The aroma and flavor of cooked bacon is famously provocative to vegetarians, but synthetic vegetarian/kosher bacon flavor is available in the form of crisp pieces, bacon salt and a bacon-flavor mayonnaise. This chapter also covers cooked ham. For more about smoked flavors, see the introduction to Smoked Fish on page 162.

Bacon & Anise I predict an international breakout for *lop yuk*. Otherwise known as Chinese bacon, it's cured with soy sauce, sugar, rice wine and star anise. Among other things, it's mixed with Chinese turnip, dried mushrooms and shrimp in a savory cake served at Chinese New Year. It can also be sliced and cooked with broccoli, rice wine, soy and garlic in a dish that recalls the typical Italian pairing of fennel sausage with broccoli. Its strong anise accents make it a natural partner for seafood.

Bacon & Apple Fidget, or fitchett, pie is an old English recipe in which layers of sliced apple, bacon, onions and (sometimes) potato are mixed with water, then seasoned with salt and pepper and served under a shortcrust pastry lid. As with pork, the simple combination of bacon and apple is delicious enough not to need much embellishment, although fancier versions add cider and nutmeg, and at Bubby's in TriBeCa, Manhattan, they crumble Roquefort into a combination of cooked apple, bacon, honey and thyme and bake it in a double-crust pie.

Bacon & Avocado *See Avocado & Bacon, page 195.*
Bacon & Banana *See Banana & Bacon, page 271.*
Bacon & Beef *See Beef & Bacon, page 46.*
Bacon & Bell Pepper *See Bell Pepper & Bacon, page 201.*
Bacon & Black Pudding *See Black Pudding & Bacon, page 42.*

Bacon & Blue Cheese The Rogue Creamery of Central Point, Oregon, smokes its Oregon Blue cheese over hazelnut shells for 16 hours. The finished product has notes of caramel and hazelnut, while its gentle smokiness recalls the classic combination of bacon and blue cheese. Make a salad of this boldly flavored pair by tossing them through some chilled bitter leaves—chicory or radicchio, straight out of the fridge. It's the perfect recipe for a hot day, when you feel the need for something both refreshing and salty. See also Bacon & Apple, above, and Washed-rind Cheese & Bacon, page 61.

Bacon & Broccoli *See Broccoli & Bacon, page 124.*
Bacon & Butternut Squash *See Butternut Squash & Bacon, page 226.*
Bacon & Cabbage *See Cabbage & Bacon, page 118.*
Bacon & Cardamom *See Cardamom & Bacon, page 306.*

Bacon & Chicken A proper club sandwich should have one, not two, or three, or seven tiers—James Beard was very clear that two slices of toasted bread is the absolute maximum. Fill them with cooled grilled bacon, slices of chicken or turkey, sliced tomato, iceberg lettuce and mayonnaise. And don't stint on the mayonnaise. Serve with chips—it's said that both the club sandwich and the potato chip were invented at the Saratoga Club House in upstate New York—and a glass of champagne.

Bacon & Chili *See Chili & Bacon, page 204.*
Bacon & Chocolate *See Chocolate & Bacon, page 17.*

Bacon & Clove Before refrigeration became widespread, bacon and ham would have been far more strongly flavored. To preserve meat meant smoking it intensively, lending it a rasping pungency unfamiliar to the modern palate. Degrees of smokiness are, however, still one of the main distinguishing features of bacons and hams. Eugenol, which gives cloves their characteristic flavor, is one of the compounds imparted to food by the smoking process, and I was curious to see how differently a dish (in this case, Hugh Fearnley-Whittingstall's barbecue beans) might turn out made with smoked bacon and with unsmoked bacon and cloves. I soaked 1 lb dried beans overnight, then slowly simmered half of them with ½ lb smoked bacon chopped into bite-sized pieces, 2 small onions, quartered, 1 tbsp molasses, and 1 heaped tsp English mustard. To the other ½ pound of beans, I added ½ lb unsmoked bacon, and the same quantities of the other ingredients as for smoked, with the addition of two cloves. In the unsmoked-plus-clove version, all the ingredients kept their separate identities, the clove flavor trailing just behind the others much as the toddler spaceships follow the mother ship in *Close Encounters of the Third Kind*. In the smoked version, by contrast, the smoke flavor got into everything, even under the beans' skin. It was sharper, meatier and quite the better of the two.

Bacon & Egg The quintessentially English combination common to the full Irish breakfast, Italian spaghetti carbonara, French quiche lorraine and the classic *salade frisée aux lardons et oeuf poché*—tendrils of curly endive tangled with lardoons and topped with a soft poached egg. And American bacon and egg McMuffins and eggs Benedict, a dish I felt pretty ho-hum about until I tried the version made by the chef Denis Leary at a narrow slice of a diner called The Canteen in San Francisco. Airy muffins, soft ham, two very fragile but perfectly poached eggs, plus a hollandaise so light it was as if the chef and his commis had simultaneously leaned over and whispered "butter" and "lemon" as I put the first forkful in my mouth.

Bacon & Globe Artichoke *See Globe Artichoke & Bacon, page 126.*

Bacon & Hard Cheese Cheese making and pig rearing work in pleasing symbiosis. Ham from pigs part-fed on the by-products of dairy farming has a particularly full flavor: Parma ham pigs, for instance, are fed on whey left over from the production of Parmigiano-Reggiano. The results are so harmonious they need only be paired in a sandwich to be appreciated. In fact the availability of a fine ham and cheese sandwich is usually a reliable indicator of the quality of a country's food: in Spain a crusty roll crammed with salty Manchego and chewy, spicy *jamón ibérico*; in France a baguette of cooked ham and Emmental with a *demi* of cold Kronenbourg; and in the departure lounge of Italian airports the sort of country-prosciutto and pecorino panino that's almost worth missing your flight for.

Bacon & Horseradish *See Horseradish & Bacon, page 104.*
Bacon & Liver *See Liver & Bacon, page 44.*
Bacon & Mushroom *See Mushroom & Bacon, page 77.*

Prosciutto & Juniper During production, Tuscan prosciutto is often rubbed with a mixture of salt, rosemary, juniper, pepper and garlic to give it a more savory flavor. Prosciutto made in this way is known as *salato*, as opposed to the better-known *dolce* varieties from Parma and San Daniele. Speck, from the culturally (and linguistically) Italo-German region of Alto Adige, is a type of prosciutto soaked in brine, juniper, sugar and garlic, then dried and wood-smoked for a few weeks before aging.

Prosciutto & Melon Harold McGee writes that during its aging process, unsaturated fats in prosciutto break down and form an abundance of volatile compounds, some of which have an aroma characteristic of melon. Prosciutto and melon are, of course, a classically harmonious flavor pairing, and so simple to prepare. Orange cantaloupe is the most common choice but a less heady Galia can be as good, if not better. Whichever cultivar you use, don't overwhelm the ham with huge briquettes of melon, or use underripe melons—the lovely, buttery texture of a ripe melon is half the point of the pairing. But then it shouldn't be so ripe that its aroma catches like hair spray at the back of your throat. Make sure that neither the melon nor the ham is too cold. But then they shouldn't be warm either. And don't plate too far in advance of serving. The salt in the prosciutto is liable to draw out the melon juice, leaving you with spoiled, soggy meat. Like I said, *so* simple.

Prosciutto & Olive Tiny Arbequina olives can be hard to find but they make a very fitting flavor partner for prosciutto. What Arbequinas lack in good looks they make up for in flavor—they're nutty and buttery, with delicious background notes of tomato and melon.

Prosciutto & Pea Pellegrino Artusi (1820–1911) wrote enthusiastically about the pea dishes served in restaurants in Rome. They were the best he'd ever eaten, which he attributed to the chefs' use of smoked prosciutto. He suggests stewing them together to serve as a side dish, or using them to make a pea risotto or a pea and ham soup. Other than oversalting, it's quite hard to mess this combination up; the most difficult thing will be sourcing good, affordable prosciutto. My local deli sells rough offcuts of Serrano ham. They're hard and chewy and great for soup—that's if I manage to get them home. It's all too easy to eat them one by one out of the bag as if they were dark, salty toffees.

Prosciutto & Peach If you're lucky enough to come across good white nectarines, try one with some Parma ham. Zuni Café owner Judy Rodgers writes that the fruit has a little bitterness, a floral note and a length of flavor that brings out qualities in the ham obscured by sweeter fruit.

Prosciutto & Pear See Pear & Prosciutto, page 269.

Prosciutto & Pineapple If the definition of an intellectual is somebody who doesn't think of the Lone Ranger when they hear the "William Tell

Bacon & Olive See Olive & Bacon, page 172.

Bacon & Onion Ham and leek is a more elegantly flavored variation on bacon and onion, and makes a steamed suet pudding or pie to rival steak and kidney. The combination is especially popular in Pennsylvania and the surrounding region, where signs for "Ham and Leek Dinners" are posted outside community halls and fire stations in the spring. Wild leeks, or "ramps" as they're known locally, have a pungent flavor somewhere between onion and garlic. See also Onion & Celery, page 108.

Bacon & Orange See Orange & Bacon, page 288.
Bacon & Oyster See Oyster & Bacon, page 148.
Bacon & Parsley See Parsley & Bacon, page 189.

Bacon & Parsnip Pairing parsnips with salty bacon serves to emphasize their sweet, spicy, assertive flavor. Simmer them in a soup or make bacon and parsnip mash to serve with liver or scallops. Paired with pancetta rather than bacon, parsnips work well in a pasta or risotto. Like bacon, pancetta is cured, but it's not usually smoked (unless it's called *affumicata*). Cooked, its complex flavor is redolent of the floral and fruit notes in prosciutto, for which it can sometimes very effectively be substituted—say, wrapped around asparagus or seafood. Buy it sliced super-thin from a good deli and you can serve it *crudo* on your charcuterie board too.

Bacon & Pea When I was a kid, my parents knew a couple who without variation ate a set meal for each day of the week: lamb chops on Wednesday, spaghetti bolognese on Thursday, fish and chips on Friday—always the same, every week, year in, year out. "Even on their birthdays, Mum?" I'd ask. "Even on their birthdays." In Swedish schools, officers' messes, work canteens, the royal household and prisons, a thick pea and ham soup called *ärtsoppa* is served for lunch or dinner on Thursdays, followed by a dessert of pancakes, sour cream and lingonberry jam. Traditionally the soup was always eaten in the evening, to keep the tillers of the fields going through Friday's fast. To this day it often comes accompanied by a warm shot of *punsch*, a sweet liqueur bolstered by the rum-like Indonesian spirit arrack.

Bacon & Pineapple How good is Hawaiian pizza? Better than the surfing in Naples. But only just.

Bacon & Pork See Pork & Bacon, page 36.
Bacon & Potato See Potato & Bacon, page 89.
Bacon & Sage See Sage & Bacon, page 313.

Bacon & Shellfish Delicious in "clams casino"—clams baked on the half shell with a bacon and breadcrumb topping. Fry a mixture of finely chopped onion, red bell pepper, garlic and bacon in olive oil, spoon it onto the uncooked, open clams, sprinkle with breadcrumbs and bake in the oven.

If your clams are quite small, you'll need to chop the bacon finely so you can tell one chewy, salty protein from the other. Keep an eye on the quantities—you want the bacon to underpin the shellfish flavor, not overpower it. Bear the same thing in mind while making mussel and bacon soup, a lobster club sandwich or when fastening your scallops into a girdle of bacon.

Bacon & Thyme Bacon and thyme make an agreeable savory seasoning. Try them with Puy lentils, partridge or Brussels sprouts. Pungent thyme is used to flavor mouthwashes, toothpaste and cough medicine, and is said to have antiseptic properties. Smoked bacon also has a whiff of the first-aid box about it, as the smoking process can impart medicinal and iodine characters. Might sound unappetizing on paper, but can be very delicious in practice. After all, single-malt whiskies like Laphroaig and Lagavulin are often described as having notes of iodine, seaweed, tar and Band-Aids.

Bacon & Tomato *See Tomato & Bacon, page 253.*
Bacon & Truffle *See Truffle & Bacon, page 115.*
Bacon & Washed-rind Cheese *See Washed-rind Cheese & Bacon, page 61.*
Bacon & White Fish *See White Fish & Bacon, page 143.*

Prosciutto

Although the Italian word prosciutto *can* apply to cooked ham, in this book the term is used more generically to refer to raw, cured hams primarily from Italy and Spain. Spanish Serrano ham is considered such a delicacy that some say the only thing worthy to share a plate with it is more Serrano ham. The practice of serving fruit with dry-cured ham is better suited to gentler Parma or San Daniele ham, as sweet and softly salty as a kiss from Botticelli's Venus. Serrano is more like a snog from Titian's Bacchus on the way back from getting legless in the woods. The difference is in part down to variations in the production process. During aging, mold growth is encouraged on Serrano hams, creating mushroomy, foresty flavors, while in Parma ham mold growth is curtailed. Fat is rubbed on Parma ham to help it retain its moisture, whereas the moisture in Serrano is allowed to evaporate, concentrating the flavors. Harold McGee notes that the nitrites used in Spanish ham production are absent in Parma and San Daniele, stimulating the development of more fruity esters than are found in Serrano.

Prosciutto & Asparagus Wrap soft, salty prosciutto around asparagus while it's still hot and the fat in the meat will start to soften and release its flavor, providing a heavenly contrast to asparagus's sulfurous sweetness. The only trouble is it can get to be a habit. You start by twirling a little prosciutto around asparagus spears, move on to monkfish fillets and you're hooked, partly on the unique pleasure of feeling the thin slices of ham cling and mold

themselves to the thing-to-be-wrapped. I have a vision of whole Ch of charcuterie, the Reichstag wrapped in juniper-smoked Westphalian or the Pont Neuf in *jambon cru* from Bayonne. See also Prosciutto & V Fish, page 171.

Prosciutto & Celery I love the Motorail train that runs from the So France to Calais during the summer months. You drop your car off i quiet station—rather eerily frequented only by cars—and are shuttled ba to Avignon, or whichever city you're traveling from, to while away the fe hours before the passenger train leaves, unburdened by suitcases, car key cases of wine, glass jars of duck confit or strings of flour-dusted saucissons So unburdened, in fact, that I have without fail had to run, shoes in hand for the train before it leaves at 11 p.m. I'm never so disorganized, however, that I don't buy bread, Bayonne ham, a bottle of Chateauneuf-du-Pape and a tub of creamy, crunchy, mustardy celeriac remoulade. A last-night-of-the-vacation atmosphere invariably obtains in the Motorail bar, and a late, shared snack is always welcome before you retire to your couchette at 3 a.m. to dream that you're sleeping on a train. See also Carrot & Celery, page 223.

Prosciutto & Chestnut Parma ham pigs are often raised on chestnuts, which suggests a pleasing late-autumn combination when the usual fruit partners for ham are out of season. Nick your chestnuts at the top and roast them at 400°F for 8–10 minutes. Serve in a warm bowl and let everyone peel their own.

Prosciutto & Egg *See Egg & Prosciutto, page 134.*

Prosciutto & Eggplant The American chef Judy Rodgers likes to serve a suave, slightly smoky eggplant dip with a garnish of grated bottarga (dried tuna or mullet roe) or with thin ribbons of smoked prosciutto. "Both have a pungent feral saltiness that is perfect with the fleshy-earthy mash."

Prosciutto & Fig In *Italian Food*, Elizabeth David praises the brilliance of whoever had the idea of serving prosciutto with melon or, even better, fresh figs. I think the gentler, fruity-floral flavor of fig is less overpowering than melon, and the soft crunch of its seeds against the smooth ham is particularly gratifying. When neither fruit is in season, you might try the traditional practice, cited by David, of pinching thin slices of butter between folds of prosciutto. She thought it might be even better than figs.

Prosciutto & Globe Artichoke Arriving late one night at an unpromising-looking hotel in Ancona, I entered my room to find that someone had set out a plate of prosciutto, a basket of bread, a decanter of wine and some home-marinated artichokes, green as dollar bills but of considerably more value to a half-starved, flight-stupefied business traveler. The translucent slivers of ham were like salty-sweet silk against the artichokes, chewy, sulfurous and tinged with a delicious bitterness.

Overture," the definition of a foodie is someone who doesn't think *Hawaii!* when they see this combination. A foodie may smugly inform you that the Consorzio del Prosciutto di Parma recommends pineapple as a complementary flavor for Parma ham. And it's true that the caramel note in pineapple chimes particularly nicely with the nutty caramel flavors created during the process of curing prosciutto—flavors that are normally found only in cooked meat. Still, no foodie worth his or her *fleur de sel* would dream of threading pineapple chunks and S-bends of ham on a toothpick. But I might.

Prosciutto & Sage *See Sage & Prosciutto, page 314.*

Prosciutto & Tomato Split a panino and drizzle a little fruity olive oil on either side. Fill with silk-thin slices of prosciutto, mozzarella and seasoned tomatoes, press flat and grill. Eat while conducting an unbelievably petty argument with your boyfriend on the back of his scooter at the same time as yelling into your cell phone. If you don't have the energy for that, lay a little Serrano on a hunk of bread rubbed with garlic and tomato—see Tomato & Garlic, page 254.

Prosciutto & White Fish One of the factors that saw monkfish become an endangered species so quickly was the mania for wrapping whole tails in prosciutto. Sure, it tastes good, what with the ham's donation of fat, flavor and saltiness to the lean, light flavor of white fish, but there are more plentiful fish that take to the treatment just as well, even if they don't slice into neat rounds like monkfish. A similar flavor, incidentally, can be achieved with some fish without using prosciutto. Scale and clean a skin-on fillet of sea bass or black bream, then run the back of a knife along it, like a window cleaner's squeegee, to remove the moisture. Salt and leave at room temperature for 15 minutes, then give it another scrape and pat dry. Place the fillet skin-side down in hot oil, cook for a few minutes, then flip for a quick final cook or finish in the oven. Properly crisped, the skin is as salty, flavorful and texturally pleasing as bacon and a delicious contrast to the soft, creamy flesh. Works on oily fish such as salmon and trout, too.

Olive

Black olives are ripe green olives and are thus sweeter. Green olives tend to be sour. Both are naturally bitter, which is why to be palatable they need to be cured. Proper curing is a slow process. The fruit is packed in salt, or sundried, before being steeped in brine for months. By contrast, the industrial process, in which the olives are preserved in lye solution, takes less than a day. Both the traditional and industrial processes draw out the olive's bitter glucosides, but in the latter too much of the flavor is lost too. Find a good olive supplier and you can appreciate the variety of cultivars available, each with their own flavor and textural characteristics: Gaeta, a small, wrinkled, dry-cured olive from Italy that has a slight dried-fruit/plum flavor; Nyons, also dry-cured, then aged in brine, with a pleasingly leathery, slightly nutty flavor; tart, crunchy Picholine, a green olive that puts up a satisfyingly crisp resistance to the teeth and will cut through fattier meats in much the same way as cornichons; and Lucques, a gently flavored, buttery green olive from the South of France, touched with almond and avocado flavors. Olive oil encompasses a similarly broad range of flavors, from the mild and sweet to the dark, peppery and throat-catching. As a rule, pressings of early fruit produce greener, more peppery oils; the later, riper olives, softer flavors. Cook with the cheaper stuff and save expensive oil for salads, or for use as a sauce over warm vegetables. Flavor notes in olive oil include floral, melon, apple, butter, pepper, artichoke, herbal, tomato leaf, green leaves, green banana, avocado and grass, which leads us neatly on to the next section, Green & Grassy (see page 176).

Olive & Almond Aged West Country Cheddar and Balsamic Marinated Onions? Wild Mushroom with Garlic and Parsley? Buffalo Mozzarella and Basil? Have potato chip manufacturers gone out of their minds? It's a source of genuine puzzlement to me that the chip should have become the focal point for the rhetorical excesses of late-stage global capitalism. Something to discuss over green olives and almonds. If you miss the descriptive frills and furbelows, you can always call them Tree-grown Valencian Manzanilla Olives and Pan-roasted Marcona Hand-salted Almonds.

Olive & Anchovy *See Anchovy & Olive, page 161.*

Olive & Anise A natural combination. The olive sellers Belazu say that black, wrinkly olives have a strong licorice flavor. Deli owner and food writer Ari Weinzweig notes that green Picholine olives have an anise undertone. And *Cook's Illustrated* magazine found that the licorice-flavored liqueur, sambuca, made a harmonious marinade for olives. Extend the idea by making an anise martini, substituting Pernod for vermouth and garnishing with an olive.

Olive & Bacon A speciality of New Orleans, the muffaletta crams ham, salami, provolone cheese and a special olive mix into a sandwich so expansive it takes the jaws of a bayou alligator to bite into it. The secret is in the olive mix: chopped green and black olives with pickled vegetables, seasoned with garlic, anchovy, capers, fresh oregano and parsley. The saltiness of the olives is almost canceled out by the other briny ingredients, exposing the flavors of warm hay and butter lettuce in the green olives, overripe plum and leather in the black. Like the famous Niçois sandwich, *pan bagnat*, a mufaletta is margin-ally less delicious freshly made. Better to give the oil and aromatics time to seep into the bread. Put it at the bottom of your bag under a copy of this book, or between your thigh and your neighbor's on the St. Charles streetcar head-ing for Audubon Park. By the time you arrive, your sandwich will be not only perfectly enriched but flat enough to fit in your mouth.

Olive & Beef A good handful of stoned black olives, salty and shiny as a pirate's boots, can lend a dark, briny depth to a landlubber's beef daube. In the Camargue they call this *gardiane,* and use little black Nyons olives, which have a leathery, nutty flavor. Follow a standard beef bourguignon recipe but use a spicier wine (a Syrah/Shiraz would be ideal) and add the olives about ten minutes before serving with white rice, pasta or potatoes.

Olive & Bell Pepper *See Bell Pepper & Olive, page 203.*

Olive & Caper The first time I tried tapenade I wasn't sure whether to swallow it or smear it on my face and declare war on the hostess. It was henceforth filed in my flavor memory as a dank paste mulched in the oily seepage of discarded farm machinery. More than two tapenade-free decades later, I stayed at a *chambre d'hôte* in Rouen, where Madame la Patronne welcomed us with a tinkling decanter of chilled Muscadet and a plate of her homemade tapenade on toast. It was heavenly. We devoured it as if it were caviar. The trick is not to pulverize the ingredients, which tends to make them bitter and tarry, but to pulse them carefully and stop while the mixture is still relatively coarse. Try 1 tbsp drained capers to 4 oz unpitted olives (or 3 oz pitted), a pinch of dried thyme and roughly 1 tbsp olive oil, depending how loose you want it. Add an anchovy or two and/or some raw garlic, if you like.

Olive & Carrot *See Carrot & Olive, page 224.*
Olive & Chili *See Chili & Olive, page 206.*

Olive & Coriander Seed Olives are often split or "cracked" before steeping, so the marinade can fully penetrate the flesh. It's common practice in Greece and Cyprus, where green olives are marinated in crushed coriander seed and lemon. Rinse the brine from 8 oz unpitted green olives and pit them. In a glass or earthenware jar or bowl, mix them with ½ unwaxed lemon, cut into 8 and the pips removed, and 1 heaped tsp crushed coriander seeds. Cover with olive oil, squeeze over the other half of the lemon, add a grinding of pepper and stir. Cover and leave in the fridge for at least 12 hours. This typically Cypriot marinade might also include garlic and/or dried oregano.

Olive & Garlic Pity the green olive that's had an entire garlic clove shoved into it. It looks so uncomfortable. The best are hand-stuffed, probably by the kind of women who can, through sheer force of will, get their feet into shoes the next size down when their own isn't available. And is it worth it? You can't really taste the garlic; it just lends a slightly sinister texture, like a chemically softened bone. Better to marinate the olives in garlic and oil.

Olive & Goat Cheese A salty, tangy feta cheese meets its match in a Greek salad with meaty, inky purple kalamata olives. Kalamata, from the

Peloponnese city of the same name, are cured in red wine vinegar and brine, and have a similar strength of character to the cheese, but a juicy, vinous flavor that cuts through it too. Chop the feta, and some peeled, seeded cucumber and tomato, into 1-in chunks. Slice red onion into thin separated rings. Keep the olives whole. Dress with grassy Greek olive oil, red wine vinegar and a shake of dried oregano. Like bacon, even average feta is pleasingly fatty and salty, and the temptation can be to make do. But—like bacon—it also keeps well, and so it's worth stocking up when you can get your hands on the good stuff. By good stuff, I mean barrel-aged feta made with sheep's milk, or a combination of sheep's and goat's milk. Feta made with cow's milk is bleached, to mask the carotene that makes butter yellow, and lacks the natural piquancy and pepperiness of cheese made from the milk of hardier, hairier animals. Barrel-aged feta develops a stronger, spicier flavor than its modern, tin-aged counterpart, and can be kept in brine for up to a year. It's worth remembering that not all good feta-style cheeses come from Greece—look out for Bulgarian and Romanian brands.

Olive & Juniper *See Juniper & Olive, page 316.*
Olive & Lemon *See Lemon & Olive, page 299.*
Olive & Orange *See Orange & Olive, page 290.*

Olive & Potato Stew them together in olive oil with tomatoes, onions and garlic to serve with fish, or chop some olives into a new potato salad dressed with vinaigrette. The food writer Mary Contini uses potato and fennel sautéed in garlicky olive oil and a scattering of chopped black olives to stuff poussins. And in Heston Blumenthal's opinion, olive is the best oil for roasting potatoes, as it imparts the most flavor and reaches a higher temperature than fashionable goose fat, for example. And a higher temperature means crisper spuds.

Olive & Prosciutto *See Prosciutto & Olive, page 170.*

Olive & Rosemary A hardy combination to put you in mind of Italy. Scattered with olive and rosemary, focaccia becomes an edible postcard of the Maremma, the irrigated flatlands that span southern Tuscany and northern Lazio. A bite of olive gives the salty tang of the sea breezes that sweep in from the west, rosemary a hint of the *maquis*, so thick in places that there are local vineyard owners who claim you can taste it in the wine.

Olive & Shellfish Just the thought of the combination of salty-sweet shellfish and salty-bitter olives is mouthwatering. Especially as described by Elizabeth David: "At Nénette's they bring you quantities of little prawns, freshly boiled, with just the right amount of salt, and a most stimulating smell of the sea into the bargain, heaped up in a big yellow bowl; another bowl filled with green olives; good salty bread; and a positive monolith of butter, towering up from a wooden board. These things are put upon the table, and you help yourself, shelling your prawns, biting into your olives, enjoying the first draught of your wine."

Olive & Tomato You're late home from work. You call for a pizza. It arrives. It's horrible. Lukewarm and flabby with cheap mozzarella. Trash it, along with the pizza-delivery flyer you keep in your letter basket. Here's a quick alternative that's sweet and mouth-filling and doesn't oblige you to wrestle an uncollapsible box into a trash bag the next morning. Call it a lazy puttanesca. To serve one, put 3–4 oz pasta on to cook. Chop a garlic clove and put it in a small frying pan with 1–2 tbsp olive oil, 3 sliced sun-dried tomatoes, 6 roughly chopped, pitted black olives, 2 anchovy fillets, 1 tsp capers and some red chili flakes. Warm through. Once the pasta is cooked, drain it and toss with the sauce. Use angel hair pasta and you can be curled on the sofa with something quite delicious in 5 minutes.

Olive & Thyme *See Thyme & Olive, page 318.*

Olive & White Chocolate The chocolatiers Vosges make a white chocolate bar with bits of dried kalamata olive in it. Not so unusual when you consider other chocolate/salty combinations, such as Domori's Latte Sal (a milk chocolate with flecks of salt). See also Caviar & White Chocolate, page 152.

Olive & White Fish *See White Fish & Olive, page 145.*

GREEN & GRASSY

Saffron

Anise

Cucumber

Dill

Parsley

Cilantro

Avocado

Pea

Bell Pepper

Chili

Saffron

Saffron is inimitable. Turmeric, safflower and annatto are often used in its stead but can only ever hope to impart an approximation of its color, and maybe a little saffron-ish bitterness. Saffron combines the flavors of sea air, sweet dried grass and a hint of rusting metal—it's the spice equivalent of Derek Jarman's garden on the bleak rocky beach at Dungeness, defiantly strange and beautiful. This rarefied, and accordingly expensive, spice is most often paired with sweet ingredients, especially those pale enough on the eye and palate to show off its color and complex flavor—rice, bread, fish, potatoes, cauliflower and white beans. It also combines well with other bitter flavors, like almonds or citrus zest, and is especially harmonious with other bittersweet florals like rose.

Saffron & Almond The sweetness of almond, boosted by sugar, offsets saffron's sometimes militant bitterness, just as cigarette manufacturers attenuate harsh tobacco with sugar, chocolate and honey. (Sweet tobacco, in fact, is the chief aroma I get opening a box of dried saffron stamens.) I like to call the following recipe my Saffron Induction Cake. Heat the oven to 350°F. Place 4 or 5 saffron threads in a heatproof dish and leave in the hot oven for a minute or two. Remove and crumble the threads into a tablespoon of warm milk sweetened with a pinch of sugar. Leave to infuse while you get on with the cake mixture. Cream 1 stick butter with ¾ cup superfine sugar. Your saffron milk should by now be the color of a desert sunrise. Add it to the butter-sugar mixture and beat well. Beat in 3 eggs and a scant ¾ cup ground almonds, adding them 1 egg and ¼ cup almonds at a time. Fold in 3 tbsp plain flour. Put the mixture in a greased and lined 7-in springform cake pan and bake at 350°F for 45 minutes. Leave to cool, then cut into 8 slices and eat one every day for a week, leaving one spare to induct a friend into the cult of which you will hopefully now be a devoted member.

Saffron & Anise The food writer Anne Willan advises against combining saffron with strongly flavored herbs or spices. It's best unchaperoned, although fennel, she concedes, is a good match in fish soups and stews. For an idea of just how good, put a single thread of saffron on your tongue; when the medicinal flavor passes, you should detect clear notes of licorice. You'll also notice, next time you look in the mirror, that your teeth are the color of a leering Dickensian villain's.

Saffron & Cardamom *See Cardamom & Saffron, page 307.*

Saffron & Cauliflower In search of a flavor reminiscent of white truffle, the American food writer David Rosengarten added cauliflower to a risotto Milanese (traditionally flavored with saffron) and found the end result superb.

Saffron & Chicken Classically paired in chicken biryani and some versions of paella, where the saffron enriches the overall dish while providing a bitter contrast to the sweet chicken and rice. I also notice a distinct almond flavor emerges when saffron and chicken are cooked together, which suggests a very harmonious trio. See also Lamb & Saffron, page 55.

Saffron & Lamb *See Lamb & Saffron, page 55.*
Saffron & Lemon *See Lemon & Saffron, page 300.*

Saffron & Nutmeg These spice the traditional Cornish saffron cake, which is really more of a bread and was traditionally eaten on Good Friday, spread with clotted cream. Saffron was once grown in Cornwall, which accounts for its prevalence in Cornish cooking long after the rest of the country had given up on this expensive, high-maintenance spice. These days, of course, it's hard to find a Cornish saffron cake that uses real imported saffron, let alone the local stuff, and most bakers just dye their wares yellow: cakes with a fake tan. Which is a pity, as the astringent background flavor of saffron provides a beautiful contrast to the sweetness of the dried fruit. In the old days, bakers would leave the saffron stamens in the dough, which according to Elizabeth David suggests they understood the superior flavor of the whole stamen over powdered saffron, and that the stamens acted as a visual sign of quality.

Saffron & Orange *See Orange & Saffron, page 291.*

Saffron & Potato The liquid gold leaked by saffron makes potatoes look particularly appetizing. Maybe a primitive part of my brain thinks they've been mixed with reckless quantities of butter from pasture-grazed cows. A little saffron may be added to a Spanish tortilla or Italian potato gnocchi. In Peru, *causa* is a mashed-potato salad whose many variations boil down to what's spread between its layers: it might be a piquant mix of olives, capers, garlic and herbs, or of tuna, egg, avocado and onion in mayonnaise. The *causa* should be made a day in advance, to allow the flavors to develop and the salad to set into shape, ready to be turned out onto a serving dish. Some versions are striped with different colors of potato—blue, white and yellow, the last from either the natural color of the potato or the addition of saffron.

Saffron & Rhubarb *See Rhubarb & Saffron, page 251.*
Saffron & Rose *See Rose & Saffron, page 335.*

Saffron & Shellfish The American food writer Gary Allen points out that saffron is especially good with shellfish, as it has a slightly iodine-like, oceanic scent, redolent of just-caught seafood. Next time you order a bouillabaisse, or a saffron-scented shellfish dish, take a deep inhalation before you eat. I guarantee this will bolster your appetite just as well as a stroll by the harbor in your espadrilles.

Saffron & White Chocolate *See White Chocolate & Saffron, page 344.*
Saffron & White Fish *See White Fish & Saffron, page 146.*

Anise

This chapter covers anise seeds, licorice, fennel, tarragon, star anise and anise-flavored drinks like pastis. Anise seeds and fennel seeds share a primary flavor compound, anethol. They can be adequately substituted for one another, but what difference obtains between them is most noticeable when you nibble them straight from the jar—where anise seeds are sweet enough to please a licorice eater, fennel seeds are less sweet and greener, somehow more rustic. A better substitute for anise seeds is star anise, which is also anethol-dominated, and is the closest in flavor terms. A different, if chemically similar, compound, estragol, is primarily responsible for the anise flavor of tarragon and chervil (and is, incidentally, present in basil, too). Anise is a popular flavoring for alcoholic drinks: pastis, absinthe, ouzo, raki, arrack, sambuca and Galliano all share its licorice sweetness. It's a very combinable flavor, equally successful in sweet or savory dishes, and gets along famously with seafood and sharp fruit.

Anise & Almond Anise can do for almond what a half-decent heckle does for a tired comedy routine. They're found together in biscotti, those Italian delicacies that are somewhere between a sweet treat and hard labor. The following crumbly cookies, which wouldn't seem out of place in the bakeries of Sicily either, deliver the pleasures of this combination without the jaw ache. Cream 6 tbsp soft butter with 2 oz superfine sugar. Add 1 egg yolk, 3 oz plain flour, 2 oz ground almonds, 1 tsp anise seeds and ½ tsp almond extract. Mix well, shape the dough into walnut-sized balls and bake at 325°F for 25 minutes. When cool, sift over some confectioners' sugar. Good with coffee but at their best with mint tea.

Anise & Apple In recent years star anise has been worthy of the name, eclipsing plain old anise despite the fact that their flavor-containing oils are almost identical. The compound most responsible for their flavor is anethol, which is also sometimes used to bolster the aniseed flavor of licorice candies. Star anise is obtained from the fruit of *Illicium verum*, an evergreen tree native to China, where the spice is commonly used in pork and duck dishes. Its autumnal, cinnamon-like sweetness makes it very good with fruit, especially apples, and it works well in strudels, mulled apple juice, or spiced applesauce to serve with pork.

Anise & Asparagus *See Asparagus & Anise, page 129.*
Anise & Bacon *See Bacon & Anise, page 165.*
Anise & Banana *See Banana & Anise, page 271.*
Anise & Basil *See Basil & Anise, page 209.*

Anise & Beef Béarnaise sauce is basically hollandaise flavored with tarragon, shallots and sometimes chervil. It's most commonly served with steak, but its grassy, subtly anise-scented flavor abets most grilled meats, fish and egg dishes. See also Cinnamon & Anise, page 212, and Tomato & Anise, page 252.

Anise & Black Currant *See Black Currant & Anise, page 324.*
Anise & Carrot *See Carrot & Anise, page 223.*

Anise & Chicken The reason the chicken crossed the road. Roasted with tarragon, it's a classic of French cuisine. Served cold, the pair works well in a salad—see also Anise & Grape, page 181. If you can't get fresh tarragon, freeze-dried is an acceptable second-best in cooked dishes, and will turn a dull chicken breast into as luxurious a meal as you can make in 15 minutes. Cut 4 skinless, boneless chicken breasts into bite-sized pieces and brown in a mixture of butter and peanut oil. Cover and leave over a fairly low heat until cooked through. Remove the chicken from the pan and keep hot. Soften a couple of finely chopped shallots in the pan, then deglaze it with ¾ cup dry white vermouth. Allow to reduce a little, then return the chicken to the pan with 1 tbsp chopped tarragon leaves (or 2 tbsp freeze-dried tarragon) and 1¼ cups crème fraîche. Heat through and check the seasoning. Serve with white rice. If you have a little more time on your hands, massage roughly 2 tbsp soft butter into a whole oven-ready chicken, and scatter a little chopped tarragon over the surface and in the cavity before seasoning and roasting as normal. The Chinese might improve the flavor of chicken by placing a couple of star anise in the cavity before roasting it.

Anise & Chili *See Chili & Anise, page 204.*

Anise & Chocolate Anise seeds were just one of the spices used by the Spanish to flavor chocolate in the sixteenth century. It's an unusual combination now, but La Maison du Chocolat makes a dark chocolate ganache infused with fennel, named Garrigue after the kind of fragrant scrubland that lacerates walkers' ankles in the south of France.

Anise & Cinnamon *See Cinnamon & Anise, page 212.*

Anise & Coconut Like soy sauce, the syrupy Indonesian condiment *kecap manis* is made with fermented soy beans, but with the addition of coconut sugar, fennel and star anise. This gives it a sweet spiciness beyond the typical salty, umami notes it shares with soy. If you love the anise flavor of Thai basil leaves in a green curry, you might have just found your new favorite ketchup.

Anise & Cucumber *See Cucumber & Anise, page 183.*

Anise & Egg Tarragon joins chervil, parsley and chives in the classic French mixture *fines herbes*. Fresh, grassy and potent, they make an omelette fit for fine dining. See also Anise & Beef, page 179.

Anise & Fig *See Fig & Anise, page 331.*

Anise & Goat Cheese Star anise is very fashionable but the real aniserati have already moved on to fennel pollen. The New York chef Mario Batali pairs it with goat cheese and orange zest in tortelloni. Much has been written about the beautiful flavor of fennel pollen; the general consensus is that it's like fennel, but more so, with a sweet, almost honeyed quality particularly suited to goat cheese. However, less rarefied sources of anise flavor work equally well. Donna Hay combines goat cheese with shaved fennel bulb, pomegranate seeds, snow pea leaves and yellow pepper in a salad dressed with pomegranate juice, balsamic vinegar and black pepper.

Anise & Grape Fennel seeds might be used instead of rosemary to season *schiacciata*, the grape-scattered bread described in Grape & Rosemary, page 249. Some recipes call for anise-flavored sambuca instead of water in the dough. Less rustically, Delia Smith dresses a salad of chicken, green grapes, scallions and leaves with a combination of mayonnaise, heavy cream and chopped fresh tarragon.

Anise & Hard Cheese *See Hard Cheese & Anise, page 66.*
Anise & Lamb *See Lamb & Anise, page 52.*

Anise & Lemon Trans-anethole, the principal flavor compound in anise, star anise and fennel, is 13 times sweeter weight for weight than table sugar, according to Harold McGee. No wonder anise seeds are used to partner bitter lemon in some biscotti. Pastis, meanwhile, makes an odd but successful match for lemon in a sorbet.

Anise & Melon *See Melon & Anise, page 273.*

Anise & Mint While the British tend to marry licorice with coconut or sugary, fruit-flavored fondant in the form of Licorice Allsorts, the Dutch and Scandinavians pair it with mint, or ammonium chloride, otherwise known as sal ammoniac, in sinisterly dark confections known as *salmiakdrop* in Holland and *lakrisal* in Sweden. See also Anise & Coconut, page 180.

Anise & Mushroom *See Mushroom & Anise, page 77.*
Anise & Oily Fish *See Oily Fish & Anise, page 153.*
Anise & Olive *See Olive & Anise, page 172.*

Anise & Orange Chefs and diners alike rave about the partnership of thinly sliced fennel and orange. But spare a thought for the poor, neglected Harvey Wallbanger, as unfashionable in its combination of anise-flavored Galliano, vodka and orange juice as the (similarly well flavor-matched) piña colada. Make a pitcher for a summer party, call it a Galliciano and see if your guests aren't bowled over by its loveliness. The anise seems to enhance the citrus juiciness of orange.

Anise & Oyster A splash of anise-flavored spirit is added to Oysters

Rockefeller (see also Oyster & Parsley, page 150). But most commonly, oysters get their anise kicks from tarragon; try them raw, sprinkled with a tarragon-enriched vinaigrette, or baked with a button of tarragon butter.

Anise & Parsnip *See Parsnip & Anise, page 219.*
Anise & Pea *See Pea & Anise, page 198.*
Anise & Pear *See Pear & Anise, page 268.*

Anise & Pineapple The Australian chef Philip Searle is famous for his checkerboard dessert of pineapple, vanilla and star anise ice creams. British chef Aiden Byrne notes that pineapple and fennel are delicious paired both in savory dishes such as roasted foie gras with fennel and caramelized pineapple and in sweet dishes like yogurt with roasted pineapple and fennel foam. Or try a dash of Pernod in pineapple juice for a refreshing long drink.

Anise & Pork *See Pork & Anise, page 36.*

Anise & Rhubarb Take a lead from Mark Miller, one of the great masters of Californian cuisine, and add anise seeds to the topping for a rhubarb crumble. Roast and crush the seeds first and stir them in with the sugar. Use 2–3 tsp for a topping made with 1¼ cups plain flour, ½ cup sugar and 1 stick butter. This works for apple and plum crumbles too.

Anise & Rutabaga *See Rutabaga & Anise, page 120.*
Anise & Saffron *See Saffron & Anise, page 177.*
Anise & Shellfish *See Shellfish & Anise, page 137.*

Anise & Strawberry Anise is beautiful with strawberries—in a sauce, jam, or simply ground with sugar and sprinkled over them. As a variation on Cherries Jubilee, cook strawberries in sambuca and serve on vanilla ice cream. If you're keen to serve strawberries and cream but are saddled with inferior ingredients, the chocolatier Jean-Pierre Wybauw gives some useful tips: (a) a little anise stirred into whipped cream gives it a farm-fresh flavor, and (b) bland strawberries benefit from ten minutes or so in water with a good dash of raspberry vinegar.

Anise & Tomato *See Tomato & Anise, page 252.*

Anise & Vanilla Galliano, the sweet yellow liqueur that comes in a baseball-bat-shaped bottle to make a bar designer weep, is flavored with anise, herbs and lots of vanilla. In Spain, the Basque liqueur Patxaran is made by steeping sloes in an anise spirit (where the Brits would use gin), with the addition of aromatics such as vanilla or coffee beans, according to which brand takes your fancy. See also Anise & Orange, page 181.

Anise & Walnut *See Walnut & Anise, page 230.*

Anise & Washed-rind Cheese *See Washed-rind Cheese & Anise, page 61.*

Anise & White Fish Anise is both fish's downfall and its redemption. Fishermen sometimes keep a bottle of anise oil in their bait box to flavor the bait, as trout in particular are attracted to the smell. And just as both pigs and pork love apples, anise is a great match for fish: see the classic recipe for fish with fennel under Oily Fish & Anise, page 153, which works just as well with white fish like sea bass. If you can't get your hands on fresh fronds of fennel, try poaching fish in a Chinese stock enriched with star anise. Put about 2 quarts of water, 1 cup Shaoxing wine (a fermented rice wine), ¾ cup soy sauce, ½ cup brown sugar, 6 star anise, 2 cinnamon sticks, a 2-in piece of fresh ginger, thickly sliced, and 5 garlic cloves in a large pan. Bring to the boil and simmer for an hour. Strain the stock through a sieve and either use straight away or freeze. Use to poach any white fish, or reduce to make a velvety sauce.

Cucumber

Russ Parsons, the *Los Angeles Times* food editor, notes that cucumber varieties are of more interest to the gardener than the cook: they all have the same distinctive green aroma and flavor that we recognize, logically enough, as cucumber-like. Even the lemon cucumber, which is yellow and lemon-sized, is named after its visual resemblance to the fruit rather than any discernible lemon flavor. Any differences that do exist between varieties generally boil down to levels of bitterness, and crispness on the teeth. The texture and refreshing cleanness of cucumber clearly lend themselves to garnishes and salads, but try pairing it with sour ingredients to knock the bitterness back a bit—goat cheese, yogurt and dill all make cucumber shine, as does vinegar. Pickled cucumbers, gherkins, cornichons and dill spears are indispensable with fatty pâtés, charcuterie and heavy sandwiches. Borage, which is also covered in this chapter, is a cucumber-flavored herb that can be used in salads, to flavor alcoholic cordials or as a garnish in drinks.

Cucumber & Anise A study in 1998 by the Smell and Taste Treatment and Research Foundation in Chicago concluded that of a range of scents, women found a combination of cucumber and a licorice-flavored candy the most arousing. Canny suitors may thus dispense with the scented candles and serve fish with a simple salad of cucumber and fennel. To find out what fragrance men liked the most, see Butternut Squash & Rosemary, page 227.

Cucumber & Avocado *See Avocado & Cucumber, page 196.*
Cucumber & Caper *See Caper & Cucumber, page 102.*

Cucumber & Carrot These make a terrific pairing in a fast pickle for serving with a piece of chargrilled chicken and sticky rice or in a spectacular sandwich.

Cut a large carrot and a quarter of a cucumber into fat matchsticks. In a sieve, sprinkle them with 1 tsp salt and leave for 5–10 minutes before rinsing, gently squeezing them dry, and mixing with 4 tbsp rice wine vinegar and 1–2 tbsp sugar. Keep in the fridge until needed. Drain before using. You have to try these in a *bành mì*, the house specialty of Nicky's Vietnamese Sandwiches in New York and an ingenious mixture of indigenous and colonial French ingredients. You can make one at home by spreading mayonnaise and a good amount of (non-herby) pork pâté into a baguette (along with some sliced cooked pork too, if you like) and heating it in the oven. Once it's nice and hot, stuff the sandwich with the pickle and a thicket of fresh cilantro. Some people paint the inside of the bread with an oil, soy and fish sauce mixture before filling. Don't stint on the pickle: as well as giving the sandwich its deeply satisfying crunch, it freshens up the heavy meat with its sweet-and-sour liveliness. See also Cinnamon & Pork, page 214.

Cucumber & Cumin *See Cumin & Cucumber, page 85.*

Cucumber & Dill When I first lived in the United States, I was surprised to find sandwiches accompanied by a stubby green pickled cucumber, shiny in its foil like a miniature zeppelin. It tasted of dill, and as I was in Minneapolis I assumed this was a Scandinavian quirk, like saying "oh yah" and enjoying ice fishing. Soon I was mournfully searching under my coleslaw if lunch arrived without its pickle. The doubly fresh combination of dill and cucumber was like taking your appetite to the Laundromat. Just when consuming the second half of the sandwich, creeping up the constraining toothpick like rising dough, seemed a complete impossibility, you took a negligent crunch on your pickle and were ready for more. Pickles and gargantuan American portions are not accidental bedfellows. Neither are dill and cucumber. Dill seed is known for its digestive properties; it's the main ingredient in gripe water. Cucumbers, conversely, are known for their indigestibility. And so— aside from the beautiful flavor combination—your digestion is held in perfect balance, at least until the pound of turkey, bacon, avocado and Monterey Jack hits your stomach and you're gestating a belch that would reinflate the *Hindenburg*. "Sour pickles" are fermented in brine, not vinegar, and "half sours," fermented for a shorter time, are also available.

Cucumber & Garlic *See Garlic & Cucumber, page 112.*

Cucumber & Goat Cheese On the rue Armand Carrel in the nineteenth arrondissement of Paris, there's an unassuming brasserie called the Napoleon III. It's the sort of place you find all over France—neither rough-and-ready nor haute cuisine but the sort of mid-range establishment that in Britain often leaves something to be desired. There I was served a starter of cucumber salad with goat cheese. Nothing more than an arrangement of cucumber slices, translucently thin, topped with four slices of cheese and a garnish of flat-leaf parsley, but the cucumber's floral perfume and alkaline nature set off the cheese's lactic acid tang to perfection.

Cucumber & Melon *See Melon & Cucumber, page 273.*

Cucumber & Mint Colder than a couple of contract killers. Add yogurt, also known for its cooling properties, and you have a form of gastronomic air-conditioning found the length and breadth of the "tsatsiki belt" that runs between India and Greece. It's called *cacik* in Turkey, *raita* in South Asia, *talatouri* in Cyprus. Each cuisine imposes its subtle variation—*cacik* often includes lime juice, *raita* onion—but the core remains the same. Dried or fresh mint may be used. The English use mint and cucumber to lend a summer-garden freshness to drinks, most famously to Pimm's and lemonade, although the combination is so good it's worth adding to other sweet, fruity concoctions on a hot day.

Cucumber & Oily Fish *See Oily Fish & Cucumber, page 154.*
Cucumber & Onion *See Onion & Cucumber, page 108.*
Cucumber & Peanut *See Peanut & Cucumber, page 28.*

Cucumber & Pork The French delicacy *rillettes* is succor to the pâté lover who's built up a resistance to richness. Goose, duck or pork meat is cooked very slowly in fat, then shredded and cooled with enough of the fat left on it to bind the meat into a mouth-coating, variegated brown-and-white paste. It's usually spread on toast, into whose crisp cavities the heated fat melts, and served with a ramekin of tiny pickled cornichons, whose vinegariness cuts through the fat and distracts you from the feeling that your arteries are furring like the crystal boughs of Magic Trees.

Cucumber & Rhubarb *See Rhubarb & Cucumber, page 250.*

Cucumber & Rose Share a summery, herbal quality. This natural affinity is picked up and extended in Hendrick's Gin, handcrafted in small batches by William Grant & Sons in Ayrshire, Scotland. Oddly for a Scottish product, the addition of Bulgarian rose and a cucumber mash was inspired in part by the quaintly English notion of eating cucumber sandwiches in a rose garden. Neither is distilled *into* the drink, like the other botanicals, but are added during the final blending process to preserve their flavor, as you might a delicate herb in a curry or a stew. The cucumber lends Hendrick's its distinctive freshness, the rose a hint of sweetness.

Cucumber & Shellfish The freshest shellfish has a soft saltiness that offsets the bracing minerality of cucumber. Exaggerate the contrast with these Chinese-inspired sesame-shrimp toasted sandwiches, served hot and crisp with a brisk cucumber garnish. Peel a 4-in length of cucumber, cut it in half lengthways and slice into thin semi-circles. Mix with 2 tsp rice wine vinegar and a few pinches each of salt and sugar. Leave in the fridge while you make the sandwiches. Dry-fry 2 slices of smoked bacon, leave to cool, then pulse in a food processor until quite finely chopped. Add about ½ lb cooked peeled shrimp, 1 tsp sesame oil, 1 tsp soy sauce and 1 tbsp sesame seeds. Pulse a couple of times until the mixture has the texture of white crabmeat. Brush vegetable oil onto 8 slices of white bread, crusts removed, and make up 4 sandwiches by piling the shrimp mixture on the *unoiled* sides. Sprinkle extra sesame seeds over the oiled top of each sandwich and toast in a sandwich maker until golden on the outside. Serve while hot, with a little of the cold cucumber on the side.

Cucumber & Strawberry *See Strawberry & Cucumber, page 258.*
Cucumber & Tomato *See Tomato & Cucumber, page 254.*
Cucumber & Watermelon *See Watermelon & Cucumber, page 245.*

Cucumber & White Fish In defense of Dr. Johnson, when he said that cucumber should be "well sliced, and dressed with vinegar and pepper, and then thrown out as good for nothing," he was only voicing the commonly held medical opinion of the day. Nonetheless, he should have tried cucumber with fish. Pickled, and in salsas, cucumber makes a crisp, cool counterpoint to fried, spicy or fatty dishes like goujons, smoked salmon pâté, fishcakes and fish kebabs. Chopped pickles, along with capers, provide the bite in tartare sauce. Some note that cooking cucumber intensifies its flavor while retaining a little of its soft crispness, and recommend cutting it into strips and stir-frying it with firm white fish.

Dill

Dill takes its name from the Norse word *dilla*, meaning to lull, as the seeds are said to have a relaxing effect on the muscles. They're anything but relaxing to the palate, however: dill seeds have a considerably stronger, sharper flavor than dill weed (i.e., the fresh herb). Dill weed (which is what these entries mean by "dill," unless otherwise specified) strikes me as having a nervy flavor in keeping with its frayed fronds. It initially seems sweet, before a sour, clean taste takes over, to the benefit of rich fish, meat and creamy dishes. It also rubs along very well with other sour ingredients like lemon and vinegar. Not a flavor, in other words, for a Sunday afternoon slumped in front of a boxed set. Dill is complex, demanding and opinionated. Think Velma in *Scooby-Doo* (basil is Daphne).

Dill & Avocado Both grassy, but in different ways. The flavor of dill is a neat, blue-green lawn tended with an assiduousness verging on mania. Avocado is the rich greensward whose springiness speeds you downhill on a moorland walk. Mix the buttery flesh of a Hass avocado (the one that looks as if it's been covered in woodchip and spray-painted purple-green) with dill clippings and a little vinaigrette and spread generously into a crayfish, tuna or chicken sandwich.

Dill & Beef The essence of a Big Mac. Odd that Big Macs are so popular, particularly among British teenagers, who aren't known for their love of dill or dill pickles. (Admittedly, judging by the pavement outside my local branch, a lot of them jettison the pickles, but there's still a heavy hit of dill flavor in the mysterious orange sauce.) My husband calls this Big Mac pie. Try it on your friends and see if they make the connection. Corned beef may strike you as an odd choice, but I've tried this using freshly ground beef and it doesn't work anything like as well. Line an 8-in tart pan with shortcrust pastry. Slice and seed 2 tomatoes and lay the slices on the bottom. Mash a 12-oz can of corned beef with 4 tbsp dill relish, 1 tbsp dried dill and 1 tbsp yellow mustard. Spread the mixture over the tomatoes, cover with a pastry lid, brush with milk or beaten egg, then scatter with sesame seeds. Bake for 35–40 minutes at 375°F. Serve hot or cold.

Dill & Beet See Beet & Dill, page 87.

Dill & Coconut A common pairing in Indian and especially Laotian fish and vegetable curries. In Laos dill is treated more like a vegetable, and the entire plant, stalk and all, is often thrown in the pot with the fish and other vegetables. Dill and coconut also crop up, rather more unexpectedly, as flavor notes in some red wines aged in American oak barrels. The coconut flavor comes from lactones in the wood, and dill is one of the herbaceous notes naturally present in oak. Worth trying to detect them in a Ridge Lytton Springs, safe in the knowledge that if all you can taste is wine it'll still be a terrific bottle of Californian Zinfandel. For more about oak flavors in wine, see Vanilla & Clove, page 340.

Dill & Cucumber See Cucumber & Dill, page 184.
Dill & Egg See Egg & Dill, page 133.

Dill & Lamb Finnish *tilliliha* ("dill meat") is a simple stew of slow-simmered lamb or beef, to which vinegar, sugar, cream and dill are added toward the end of cooking. In Greece, the offal of just-slaughtered lamb is mixed with dill in a soup called *mageiritsa*, eaten at Easter to break the Lenten fast. And in Iran, heavily dilled rice is alternately layered with skinned fava beans and cooked lamb (shank or chop meat), possibly with the addition of turmeric or saffron or both, in a dish called *baghali polo*.

Dill & Lemon See Lemon & Dill, page 298.

Dill & Mint Spearmint and dill seed contain different forms of the same naturally occurring flavor compound, carvone. In either case the carvone molecule is the same shape but the mirror image of the other, and thus perceived differently by the flavor receptors in our noses and mouths. There are hundreds of these, each evolved to receive different molecules according to their shapes, a bit like a vastly complex version of a toddler's shape-sorting puzzle. Spearmint contains the left-oriented form of the molecule (l-carvone), dill the right (d-carvone). Aromatically they are quite distinct, and thus unlikely to be effective substitutes for each other, unless, of course, you're prepared to cook the entire recipe backwards.

Dill & Mushroom *See Mushroom & Dill, page 79.*
Dill & Oily Fish *See Oily Fish & Dill, page 154.*

Dill & Pea Used to hook up with diced potatoes and carrots in a Russian salad, now as rare a summer sighting as socks with sandals, found only in hotel restaurants and old-fashioned tapas bars. But dill's brightening quality is ample reason to revive this dish. Peel and cube 1 lb new potatoes, then simmer until just tender. Do the same for ½ lb carrots. Cook ⅔ cup frozen peas. Let them all cool, then mix with a dill mayonnaise. You could add a drained, diced dill pickle for an extra bit of bite. Serve with cold cuts and crisp lettuce.

Dill & Pork Cabbage, leek and chives are often mixed with pork in Chinese dumplings, as a freshening counterpoint to its meatiness. So is dill, though far less frequently, except in dishes originating in northeastern China. Here dill weed, along with beet and parsley, was cultivated by the indigenous population for the large number of Russian immigrants who had arrived to work on the construction of the Chinese Eastern Railway. You can make Chinese dumplings in minutes if you can get your hands on the dumpling wrappers. Chinese supermarkets keep them in the cooler and they freeze very well. To make about 15, mix together ½ lb finely minced pork, 1 tbsp chopped dill, a crushed garlic clove, 1 tbsp soy sauce, 1 tbsp Shaoxing wine, a shake of sesame oil and a little salt. Place about 1 tbsp of this on each wrapper, brush the edges with water, then fold the wrapper over the filling and press the edges together to seal. Cook in boiling water (or a tasty stock) for 5–8 minutes. Serve with soy sauce.

Dill & Potato In Poland, a potato salad without dill is like a union leader without a bushy moustache. In India, potatoes are cubed and fried with garlic, turmeric and chili, then tossed with plentiful quantities of dill. Dill (called *sowa* or *shepu*) is indigenous to India, but the variety has a lighter flavor than its European relative.

Dill & Shellfish In August in Sweden, dill and crayfish are paired in the grand outdoor feast known as a *kräftskiva*, or crayfish party. The crayfish are boiled in water (and sometimes beer) flavored with crown dill, which is straightforward dill harvested after the plant has flowered, supposedly giving it a stronger flavor. The crayfish are slurped from their shells with simple accompaniments of bread, beer, aquavit and a strong hard cheese riddled with tiny holes called *Västerbotten*.

Dill & Smoked Fish *See Smoked Fish & Dill, page 163.*
Dill & White Fish *See White Fish & Dill, page 144.*

Parsley

Parsley's fresh, green, woody notes are described as "generic" by Harold McGee, which is, according to him, why the herb complements so many foods. It is at its best with briny ingredients, especially ham and all types of fish, to which it brings a welcome coolness, and a bitterness that offsets the salt-sweetness in meat. Its generic herbal flavor also makes it great for mixing with other herbs. The flat-leaf variety usually has a stronger flavor and leaves that are more tender than those of curly parsley.

Parsley & Bacon Thickly sliced ham is traditionally served hot with a roux-based parsley sauce. Some recipes call for ham stock and a little cream, but milk alone makes for a purer parsley flavor and a gentler counterpoint. In France they make a jellied terrine of ham and (lots of) parsley called *jambon persillé*; the one from Burgundy has the best reputation. Parsley's fresh, green flavor makes a cool, clean contrast to the saltiness of the meat.

Parsley & Beef *See Beef & Parsley, page 50.*

Parsley & Caper A pair of green avengers, battling the palate-numbing tedium of fried foods. Pitch them, possibly in the form of a *salsa verde*, against fried eggplant slices, battered fish and crumbed escalopes. Great too with more strongly flavored fatty foods—roughly chop and stuff into sardines before a skin-blistering turn on the barbecue.

Parsley & Carrot *See Carrot & Parsley, page 225.*
Parsley & Cilantro *See Cilantro & Parsley, page 193.*

Parsley & Egg American cookbook writer Fannie Farmer specifically recommends a garnish of parsley for poached eggs if they're cooked for an invalid. I'd recommend them if you're cooking for a geologist. Parsley has a lean, metallic crispness and there's an echo of that minerality in parsley's scientific name, *Petroselinum*, or "rock celery." To my mind, cooked egg white has the mineral edge of Perrier in cans, and so caution should be exercised when combining it with parsley—especially in an egg white omelette—unless you get your kicks from licking rocks.

Parsley & Garlic By reputation, parsley is the Hail Mary to the sin of garlic breath. Hold a little back when you chop it finely with garlic for a *persillade*, mixed with breadcrumbs for lamb. Or stir the two into sautéed potatoes just before the end of cooking, when you'll retain some of parsley's bright minerality while taking the pungent edge off the garlic. The juice and zest of a lemon added to parsley and garlic will create a *gremolata*, used as a last-minute seasoning, most famously in *osso buco*. Or hold the lemon and add olive oil for the simple *chimichurri* sauce served with roasted meats in Argentina (see Beef & Parsley, page 50, for more about that). Seek out the

Spanish *rojo* variety of garlic, which has an excellent reputation for flavor, even in its raw state.

Parsley & Lemon *See Lemon & Parsley, page 300.*

Parsley & Mint I love tabbouleh. There's a great little place called Istanbul Meze on Cleveland Street in London that serves it as it should be—overcome with parsley. You need about five parts parsley to one part mint, and much less cracked wheat than you get in supermarket versions (it should come as a pleasant surprise—oh, there's a bit—rather than the dominant force). As to the parsley, chop so much that you can hardly bear to chop more. Then chop more. Be as restrained with the tomato as you were with the wheat. Dress with plenty of lemon juice, a little oil and a touch of crushed garlic. If the result is as juicily green as a mouthful of meadow, you've got it the right way round. The herb-light, wheat-heavy stuff they sell in shops gets it back to front: it should be called *heluobbat*.

Parsley & Mushroom *See Mushroom & Parsley, page 81.*

Parsley & Oily Fish For a cold parsley sauce to serve with fried mackerel or herring, fold 5 tbsp finely chopped parsley into 1¼ cups whipped heavy cream with Tabasco to taste and some seasoning.

Parsley & Oyster *See Oyster & Parsley, page 150.*

Parsley & Potato Lest old potatoes be forgot: didn't you just love those fist-size floury spuds, peeled, quartered and simmered till soft and pale yellow, with a halo of buttery fuzz? In Britain they've all but disappeared. In Lisbon they're everywhere: in stews and chunky soups, beside garlicky pork chops and robust sausages. Any attempt to disguise their country frumpiness, other than an offhand strewing of chopped parsley, is mercifully resisted. Seek them out at Pateo 13, one of those small, family-run restaurants that materialize as casually as a cat or a line of bright washing as you explore the labyrinthine backstreets of Alfama, the oldest district of the city. Traditionally Alfama was renowned for *fado*, a genre of folk music characterized by *saudade*, which is one of those untranslatable Iberian concepts roughly corresponding to our notions of nostalgia and loss. Sit on a parched terrace drinking tart, spritzy *vinho verde* and watch the host turn sardines on a grill the size of a church organ. A stainless-steel tray of boiled potatoes patterned with parsley provides a perfect foil for the charred, meaty fish. You can't help feeling something like *saudade* for the simple boiled potato, in its prime here but unsung elsewhere, pushed out by mashed potatoes and the ubiquitous French fry.

Parsley & Shellfish *See Shellfish & Parsley, page 141.*
Parsley & Smoked Fish *See Smoked Fish & Parsley, page 164.*
Parsley & Walnut *See Walnut & Parsley, page 233.*

Parsley & White Fish There was a time when a piece of passed-out parsley was to be expected on almost any savory plate. It was gastronomic junk mail: something you were wearily accustomed to throwing away. But as a partner for white fish it's indispensable. The saltiness of the fish is offset by the herb's leafy freshness. For a parsley sauce to serve with firm white fish such as cod or haddock, add a generous amount of chopped parsley to a béchamel sauce and season with lemon zest and a little lemon juice. Or, for a lighter, faster variation, dip your fish fillets in seasoned flour, cook in olive oil and add a little white wine and fish stock (or water). Allow to reduce a little and, when the fish is cooked through, scatter with chopped parsley. See also Lemon & Parsley, page 300.

Cilantro

In common with its lookalike, flat-leaf parsley, cilantro is characterized by fresh, green and woody notes. But in place of parsley's cold-rain flavor, cilantro is more redolent of the monsoon, with hints of warm earthiness and fruity citrus peel. It has a bittersweet taste, and is often used as a garnish, partly because its flavor isn't heat-stable and therefore needs to be added at the end of cooking, and partly because it has so many useful tempering qualities. Cilantro calms saltiness, deodorizes fishiness, cuts fattiness and lends a cooling note to hot, spicy food.

Cilantro & Avocado A popular pairing in salsas and guacamole. They have harmonious, grassy flavors in common, but the freshness of cilantro cuts through avocado's fattiness, especially in the spin taken on the pairing by Ferran Adrià at El Bulli, where the avocado is deep-fried in tempura batter. Cooked avocado is not to everyone's taste—it's thought that tannins in the flesh become increasingly bitter when cooked—but I've not been able to detect anything unpleasant in gratinéed avocado, which used to be popular in trattorias, or in my own experiments with tempura. Whether or not you can handle the double fat-whammy of deep-fried avocado, even with the mediation of cilantro, is another matter.

Cilantro & Chicken *See Chicken & Cilantro, page 32.*
Cilantro & Chili *See Chili & Cilantro, page 205.*

Cilantro & Coconut Cilantro may well have sailed into our hearts on a sea of coconut milk but that shouldn't limit the way these two are used together. Sweet, juicy coconut is a natural partner for fresh, zesty cilantro in this flatbread. Put 1½ cups self-rising flour in a food processor with ½ tsp baking powder, ½ tsp salt, 1 tsp sugar and 2 tbsp soft butter and pulse until the mixture resembles breadcrumbs. With the motor running, drizzle ½ cup lukewarm milk or water through the tube and your mixture will become a dough. Remove and knead for 5 minutes. Cover with a damp cloth and leave to rest for 20 minutes. Divide the dough into 4 and roll each piece into a rectangle roughly 8 x 6 in. Mentally dividing each in half lengthwise, scatter the right-hand side with some cilantro and flaked coconut, then fold the left side over and gently roll out until it's as thin as it will go. Set the grill to high and line the grill pan with greased foil. Grill your flatbreads just below the heat source for 90 seconds, then turn and grill on the other side for about 60 seconds, by which time they should have developed some golden patches. Brush with butter and serve.

Cilantro & Coridander Seed *See Coriander Seed & Cilantro, page 337.*

Cilantro & Cumin Toast cumin and you bring out a bitter note in its warm earthiness. Cilantro shares these qualities, although where cumin's earthy bitterness is downbeat, smoky and autumnal, cilantro's is zingily fresh. Together with chili, these two are found in Indian curries, dhals and chutneys, and sometimes in Mexican bean and meat stews. Add toasted ground cumin and chopped cilantro to a guacamole, especially if the avocados are a little lackluster. Also try adding toasted cumin to scrambled eggs with chopped cilantro. So simple you'll have time to make your own flatbread to scoop it up with—see Cilantro & Coconut, page 191.

Cilantro & Garlic In Spain, *sofrito* is a mixture of slowly fried onions, garlic, tomato and sometimes bell pepper. In other Hispanic cuisines the term might denote a similar mixture that's added to dishes in its raw state. In Puerto Rico they add cilantro and culantro—the leaf of a related plant, *Eryngium foetidum*, which is like cilantro but more cilantro-y. *Culantro* has long, saw-toothed foliage and looks more like a salad leaf than a herb. Keep an eye out for it in Caribbean food shops. Puerto Ricans make *sofrito* in huge batches and use it in soups, stews, rice dishes, and anything that will benefit from its perfumed zinginess. Roughly chop 2 large Spanish onions, 1 green pepper, the peeled cloves of half a garlic bulb and a fistful of cilantro (or half each of cilantro and culantro). Finely chop some fresh chili in whatever amounts you can tolerate—although the authentic choice is the mild *aji dulce*, which has a smoky, fruity character rather than intense heat. Add a few tablespoons of olive oil, and maybe 1 tomato, skinned and chopped, and whiz in a food processor until finely minced. Keep it a little loose by adding water a tablespoon at a time. The mixture can be stored in the fridge for a few days, or you could freeze it in ice-cube trays.

Cilantro & Goat Cheese *See Goat Cheese & Cilantro, page 58.*

Cilantro & Lamb The Lebanese dish *yaknit zahra* translates as "cauliflower stew," although lamb and the large amount of cilantro used dominate its flavor. The Indian flatbread *keema naan* is often stuffed with lamb and cilantro, but cilantro addicts may prefer the following version of the lamb and spinach dish *saag gosht*. Cilantro has a tendency to lose its flavor when cooked but the sheer amount used in this recipe, plus the inclusion of the stalks, ensures its leafy freshness survives, mingling with the spinach, the lamb juices and yogurt to make a curry that's very luscious in the mouth but not in the least bit rich or cloying. In an ovenproof pot, fry a chopped large onion in a little butter and oil until soft. Add a crushed garlic clove, 1 tsp salt, a thumb of fresh ginger, finely chopped, 1 tbsp ground coriander, ½ tsp turmeric and chopped fresh chili to taste. Stir, and cook for a few minutes, then add 2 lb diced lamb. Brown the meat and stir in 1 package (defrosted) frozen whole-leaf spinach and most of a good bunch of cilantro—leaves roughly chopped, stems quite finely chopped. Gradually add 1¼ cups yogurt. Cover, transfer to the oven and cook at 325°F for 2 hours. Before serving, stir in the cilantro you've held back, to restore some of the top notes lost in the cooking. Serve with rice.

Cilantro & Lemon *See Lemon & Cilantro, page 298.*

Cilantro & Lime The first time I ordered Vietnamese beef and cilantro noodle soup, a wedge of lime turned up in it, I put it to one side and left it there, grinning at me as it leaked tangy beef stock into my paper napkin. Lime, I thought, is sweet. Bear in mind I was brought up on Rose's lime cordial and its matching alien-plasma marmalade: sweet and then some. The more I ate Vietnamese and Thai food, however, the more lime began to turn up in savory dishes, most often in partnership with cilantro. It wasn't long before I was ordering dishes simply because they came with lime and cilantro, which is a bit like buying a song because you like the backing vocals, and no worse a habit for that. Cilantro and lime are the *wooh woohs* in "Sympathy for the Devil"—completely and utterly indispensable.

Cilantro & Mango *See Mango & Cilantro, page 283.*
Cilantro & Mint *See Mint & Cilantro, page 321.*

Cilantro & Orange When people say cilantro adds "zest" to food, they're closer to the literal truth than they may realize. Harold McGee writes that the main component of cilantro's aroma is a fatty aldehyde called decenal, which is also responsible for the waxy note in orange peel. Decenal is highly unstable, so cilantro quickly loses its characteristic aroma when heated, and is mostly used uncooked, as a garnish. Try a salad of orange slices, sweet onion and radish, strewn with torn cilantro. Some Thai spice pastes include cilantro root, which contains no decenal, instead contributing woody, green notes not dissimilar to parsley. Cilantro is often used to pep up orange, mandarin and other citrus notes in manufactured flavorings. Cilantro's oranginess is markedly more apparent in its fruit—see Orange & Coriander Seed, page 289.

Cilantro & Parsley Poor parsley. Cilantro was once the herb of choice in South America, Asia and the Caribbean, while parsley held sway in Europe and North America. Slowly and surely cilantro is doing for parsley what the gray squirrel did for the red. It's the most consumed herb in the world. But, if it can't beat cilantro, parsley can join it. It shares a green grassiness with cilantro and so can be used to dilute it in a dish without skimping on the quantities of herb. And you needn't feel too sorry for parsley in any case. It'll always have ham.

Cilantro & Peanut Substitute cilantro and peanut for basil and pine nuts and you have a delicious Vietnamese-style take on pesto. Roughly chop a handful of cilantro and half a handful of roasted peanuts. Holding some of each back for garnish, mix them with a sheet of cooked (and still warm) egg noodles with 1 tbsp peanut oil and a dash of fish sauce, maybe with a shake of dried chili flakes and a squeeze of lime. Garnish and serve immediately.

Cilantro & Pineapple See *Pineapple & Cilantro, page 261.*
Cilantro & Pork See *Pork & Cilantro, page 38.*

Cilantro & Potato Fans and foes of cilantro often cite its earthy flavor, which I think makes it a winning partner to potato, sustaining some of the earthiness lost in boiling or mashing. *Caldo de papa* is a potato soup eaten in many Latin countries, especially Colombia. Made with generous amounts of cilantro and with either beef or beef stock, it's renowned as a hangover cure. I don't know about that, but it would be easy enough to rustle up *with* one. Homemade beef or veal stock would be heaven, but I've made it with bought concentrate and, as the potato starts to disintegrate, the soup thickens, and deepens in flavor. So: cook a chopped large Spanish onion in oil, then add 1 lb floury potatoes, peeled and cut into small-roast-potato-sized chunks. Cover with 2 cups beef stock and simmer until the potatoes are on the point of falling apart. Stir in 2 generous handfuls of chopped cilantro and some slices of fresh green chili.

Cilantro & Shellfish See *Shellfish & Cilantro, page 139.*

Cilantro & Tomato Dancing partners in a salsa. So popular that sales of prepared salsa have overtaken ketchup in North America—and that's not counting the huge amounts made from scratch. Basil needs to watch its back . . . Somewhere in the warehouse district in Minneapolis, a short gust of freezing wind from the Mississippi, stands the Monte Carlo, a supper club dating from 1906, when the lumber trade was at its height. The place isn't retro: you're just late. It has a tin ceiling and a copper bar, and serves the kind of martinis to make you see the Prohibitionists' point. Once your eyes have uncrossed, you'll find yourself with the appetite of a timber baron, easily satisfied with one of the Monte Carlo's trademark flatbread pizzas, topped with tomato, havarti cheese and cilantro pesto. Every time I ate one I thought, they should make more pizzas with cilantro. And that, basil, is the way the world ends. Not with a bang, but a pizza.

Cilantro & Watermelon See *Watermelon & Cilantro, page 245.*

Cilantro & White Fish Cilantro's citric quality makes it a good match for fish. Like lemon, it complements the delicate flavor while countering its "fishiness." In Southeast Asian cuisine, with its heavy use of fish sauce, shrimp paste and dried fish, cilantro is an essential balancing element. See also Lemon & Cilantro, page 298.

Avocado

No wonder it's hard to stop grazing on avocados: they taste like grass and have the texture of butter. Delicate avocado goes well with other subtly flavored ingredients, such as mozzarella and crustaceans; the latter love the light anise note in avocado flesh. Often buried under stronger flavors, like lime and garlic, the flavor of avocado arguably runs second to its lovely, unctuous texture, and the cooling, fatty quality it brings to sandwiches, salads and salsas. Avocado oil is even more delicately flavored than the flesh, and lacks the grassy note, for which you'd be advised to seek out an olive oil instead.

Avocado & Bacon As rich as Jonathan and Jennifer Hart and just as gorgeous. When these two meet it's not exactly murder, but you might want to have your cardiologist on speed dial. The green flavor of avocado makes a fresh counterpoint to bacon's heavy, salty meatiness, but it's at least as fatty. Get the max out of these in a salad made with baby spinach or in a hefty whole-wheat sandwich with mayonnaise.

Avocado & Blue Cheese *See Blue Cheese & Avocado, page 63.*
Avocado & Chicken *See Chicken & Avocado, page 31.*

Avocado & Chili The Gwen avocado is a variety derived from the Hass cultivar in 1982 and is worth seeking out for the hint of smokiness that *Saveur* magazine notes is redolent of chipotle chilies. If Gwens aren't available, you can always make an avocado and chipotle chili soup with any variety of ripe avocado you can get your hands on. For two servings, blend the flesh of 2 avocados with the juice of 1 lime, ⅓ cup yogurt or sour cream, ½ cup water, 1 tsp chipotle chili paste and some salt. Check for taste and texture (add a little more water if it needs thinning) and chill for a short while before serving garnished with a minimal pinch of cayenne pepper.

Avocado & Chocolate Adherents of the raw-food movement make a silky chocolate mousse from these two, using cocoa powder for the chocolate flavor, sometimes supplemented by banana or date. Avocado takes care of the texture. The aim of the rawist is to eat as much of their food uncooked as possible. If this sounds quite hardcore, rawism's a cakewalk compared to being an instincto. Instinctos eat only one, raw ingredient per sitting, selected according to what smells right. This might be fruit, vegetable, egg, fish, badger or grub, as long as it's uncooked, unseasoned and neither juiced nor ground. The method is straightforward: the instincto sniffs at a range of foods until they find one that smells instinctively right (instinctos are easy to spot at the Pizza Hut salad bar). In time they reach a state of such instinctive receptiveness that they only have to pass, say, a freshly road-killed rabbit to stop and with lizard-like relish snatch it from its final resting place in the hedge. Then they feast on it until they can eat no more—or "reach their stop" (in this regard they're less easy to spot at the Pizza Hut salad bar). A typical instincto meal might thus consist of 52 egg yolks at a single sitting, or 210 passion fruit in a day.

Avocado & Cilantro *See Cilantro & Avocado, page 191.*

Avocado & Coffee An unlikely mix to Western tastes, maybe, but the avocado is a fruit, and is treated as such in Vietnam, Indonesia and the Philippines. It's puréed into a shake with milk and sugar (or condensed milk) and sometimes flavored with a coffee or chocolate syrup. In Mexico, avocados are sprinkled with sugar or rum to make a simple dessert.

Avocado & Cucumber I make a cold soup with these two. Blend the flesh of an avocado with a peeled, seeded cucumber and a squeeze of lemon juice, then season to taste. Avocado dominates, but the fresh, green notes of cucumber aren't far behind. Thick, silky, but refreshing. Like a guacamole with its clothes off.

Avocado & Dill *See Dill & Avocado, page 186.*
Avocado & Grape *See Grape & Avocado, page 247.*
Avocado & Grapefruit *See Grapefruit & Avocado, page 292.*

Avocado & Hazelnut According to the Mexican food expert Diana Kennedy, avocados can contain notes of hazelnut and anise. The Guatemalan-Mexican hybrid Fuerte—"the strong one"—is particularly prized for its hazelnut flavor. Avocado leaves are dried and used for seasoning stews, soups and bean dishes in Mexico. They too have an anise-like flavor, and food writer Rick Bayless recommends a combination of bay leaf and anise seeds as a substitute.

Avocado & Lime Lee Hazlewood and Nancy Sinatra singing "Some Velvet Morning," Lee the velvety baritone note of avocado, Nancy the high-pitched lime that cuts through the smoothness just when you're getting too comfortable. At once beautiful together and distinctly separate. The association may largely be down to my having played the song over and over again, driving down Highway 1 in California on my honeymoon. While others toast their future together over lightly grilled fish and flutes of champagne, served on a lapping pontoon in the Indian Ocean, we stopped at striplit *taquerías* and ate burritos the size of shotputters' forearms. Tender strips of grilled steak come rolled in a floury, slightly crisp tortilla packed with rice, beans, sour cream, mouthwatering, lime-laced guacamole and salsa so fiery you have to grip the sides of the little plastic serving basket till your eyes stop streaming tears. On the subject of guacamole, some say that leaving the avocado stone in will prevent discoloration. My view is that if the guacamole's around long enough to find out, you're not making it right.

Avocado & Mango Delicious together but you have to be accurate with your timing. Avocados don't ripen on the tree—their leaves supply a hormone to the fruit that inhibits the production of ethylene, the ripening chemical. That's why you have to buy them unripe and pinch them on the nose every day until there's give. Although mangoes *do* ripen on the tree, most of the mangoes for sale in supermarkets have about as much give as the stone inside them, which means leaving them at room temperature until they soften and release their complex, spicy perfume. If you haven't managed to synchronize your avocado to your mango, put the ripe one in the fridge, where the cold will arrest the ripening process, and nip out for some fresh crab to serve with them as a reward for your patience. Mix the crab with mayonnaise and press into a ramekin, followed by a layer of chopped mango and another of avocado, mashed or chopped and mixed with lime juice. Turn out, like a sandcastle, onto a plate with some tender watercress leaves. Or simply make an avocado and mango salsa and serve with freshly fried crab cakes.

Avocado & Mint For a salad dressing, blend an avocado, 4 tbsp yogurt, 1 tbsp olive oil, a handful of mint leaves and a pinch of salt.

Avocado & Nutmeg *See Nutmeg & Avocado, page 217.*
Avocado & Oily Fish *See Oily Fish & Avocado, page 153.*
Avocado & Pineapple *See Pineapple & Avocado, page 260.*

Avocado & Shellfish If God hadn't meant us to fill avocado cavities with seafood in Marie Rose sauce, he wouldn't have made the stone so big, would he? Simon Hopkinson and Lindsey Bareham suggest mixing home-made mayonnaise with store-bought ketchup and hot sauce. For about ½ lb peeled shrimp, stir 4–5 tbsp mayo with a few drops of Tabasco, 1 tbsp tomato ketchup, a little lemon juice and 1 tsp cognac. See also Grapefruit & Avocado, page 292.

Avocado & Soft Cheese Mozzarella is the classic cheese pairing for avocado. If the cheese is made without gums and fillers, the two should share a mild, milky sourness that combines with the soft springiness of mozzarella and the velvet depth of avocado to create comfort food suitable for the summertime. Add slices of tomato for an Italian *tricolore*.

Avocado & Strawberry *See Strawberry & Avocado, page 257.*

Avocado & Tomato Out of mozzarella? Make do with a *bicolore*. Or pair them in a sandwich with a jalapeño mayonnaise. See also Walnut & Chili, page 231, and Avocado & Soft Cheese, above.

Pea

A tiny pouch of sweet and savory: under its simple grassy sweetness the garden pea is a rich source of umami. Peas make harmonious matches with salty seafood, bacon and Parmesan, and the herbal flavors of mint and tarragon. The field pea, most commonly eaten dried in the form of split green peas, is the garden pea's country cousin. It's less fresh tasting, in part because of the farinaceous texture of its flesh, which swells and bursts from the skin when cooked. Marrowfat peas are also a type of field pea. Peas and green bell peppers share dominant green flavor compounds.

Pea & Anise The sweet anise flavor of tarragon has a slightly bitter edge that suits pea, which is sweet enough itself, very well. Try these light pea fritters with a sharp tarragon cream sauce. Defrost 1 lb frozen peas and pat them dry, then roughly process them with 1 egg, 1 tbsp olive oil, 4 tbsp plain flour and some seasoning. Using a pair of dessert spoons, shape the pea mixture into 12–15 quenelles. Shallow-fry them in vegetable oil for a few minutes, turning them twice so they cook on all 3 sides. Serve 3 to a plate with a dollop of sauce made from finely chopped fresh tarragon leaves stirred into seasoned crème fraîche. See also Pea & Chicken, page 199, and Pea & Hard Cheese, page 199.

Pea & Asparagus Raw asparagus tastes like freshly podded peas, but when cooked it takes on a nuttier, more savory flavor. Cooked peas, by contrast, retain their sweetness, as well as their own strong savory character. The combination of asparagus and pea might sound light but it's remarkably flavorful, and very good in risottos or egg-based dishes. The "asparagus pea," incidentally, is claimed by some to re-create the delicious subtleties of asparagus. Sadly, it doesn't.

Pea & Bacon See Bacon & Pea, page 167.

Pea & Beef From *The Horticulturist and Journal of Rural Art and Rural Taste* (1851): "Everybody knows how to cook peas, or at least everybody thinks so—and everybody boils them. That is excellent, but by no means the only way to taste this vegetable in perfection; an Old Digger may not be supposed to know much about cooking, but in fact no place lies so close to the kitchen as the kitchen-garden, and it must be a dull digger who does not know something of what the cook does with his 'truck'. So I will tell you that the neatest little dishes that any cook ever sends to the table are very small joints of lamb or veal, or perhaps a pair of spring chickens, stewed in a close pot or stew pan very gently, over a slow fire, for two or three hours, till quite done, with peas—butter, pepper, and salt being added, of course. The juices of the meat penetrate the peas, and the flavor of the peas is given to the whole dish, so that I doubt that there was more savory dishes among the flesh-pots of Egypt, than one of these stews."

Pea & Chicken Liking pea and chicken is about as interesting as liking warm sunshine. But the combination needn't be as boring as it sounds. *Poulet à la clamart* is a dish to please adventurous and conservative palates alike. Clamart, a suburb about five miles to the southwest of central Paris, has long been famous for the quality of its peas, and lends its name to any dish that makes extensive use of them. In *poulet à la clamart*, pea and chicken's stale marriage comes under the refreshing influence of light, anise-scented tarragon. Brown 8 chicken pieces in a mixture of butter and olive oil. Add about 2 oz (two strips) smoked streaky bacon, cut into matchsticks, a dozen trimmed and chopped scallions, 2 tbsp chopped tarragon leaves and half a glass of white wine, then shake the pan. Season, cover with a lid and cook in the oven at 325°F for 30 minutes. Add 1 lb peas and 2 tsp sugar, then return to the oven for another 30 minutes. Transfer the casserole to the burner and the chicken to a warmed plate. Add 1 shredded Cos lettuce to the sauce, give it a good stir and cook, covered, over a low heat for 10 minutes. Finally, stir in about 1¼ cups heavy cream and heat gently. Taste, adjust the seasoning and serve with a garnish of parsley.

Pea & Dill See Dill & Pea, page 188.
Pea & Egg See Egg & Pea, page 134.

Pea & Globe Artichoke In his 1891 cookbook, Pellegrino Artusi admits his recipe for artichoke and pea pie is strange but will nonetheless appeal to many. Me included. Parboil 12 fresh globe artichoke bottoms and 1 cup peas, drain them, then cut the artichokes into eighths. Slowly stew both vegetables with 4 tbsp butter and some seasoning until cooked. In a separate pan make a sauce with 1 tbsp butter, 1 tbsp plain flour and ½ cup meat stock and stir into the vegetables. Place this mixture in a heatproof dish in layers, sprinkling grated cheese (Parmesan or pecorino) between them, and cover with a shortcrust pastry lid. Brush with a beaten egg yolk and bake at 425°F for 10 minutes, then 350°F for 20 minutes or until the crust has browned. If you fancy something lighter than pie, another classic Italian recipe is to stew chopped artichoke hearts with fava beans, sliced onion and pancetta for about 15 minutes, then add peas and cook for an additional 10–15 minutes, keeping the mixture moist with a few tablespoons of white wine and/or water as necessary.

Pea & Hard Cheese On St. Mark's Day, April 25, the Doge of Venice would be presented with a dish of *risi e bisi*, made with the pick of the new vegetables arriving at the Rialto market. Edward Lear would have appreciated it, too: it's essentially a soup that you eat with a fork. The classic version adds a stock made of the pea pods to peas and onions cooked with a (very) short grain rice called *vialone nano* and finished with plenty of grated Parmesan. Venetian cuisine is full of sweet-sour combinations, furnished in this case by the contrast of sugary pea and sharp, sour cheese. Bacon, parsley and (less frequently) fennel are common additions to the dish.

Pea & Horseradish Is there a deeper cultural chasm anywhere than the uses to which the English and Japanese put the marrowfat pea? The English boil them to a green sludge and ladle them into a Styrofoam cup, fragments of pea skin protruding like the eyelids of crocodiles; the Japanese turn them into neat, brittle, deep-fried spheres in pale wasabi crash helmets, vacuum-packed in foil bags that look like pretentious design magazines costing $30.

Pea & Lamb *See Lamb & Pea, page 55.*
Pea & Mint *See Mint & Pea, page 323.*

Pea & Oily Fish According to New England tradition, gardeners make sure to plant their peas by Patriots' Day (April 19), in the hope that they'll be ready for the traditional Independence Day feast of poached salmon, fresh green peas and new potatoes. Strawberry shortcake is served for dessert.

Pea & Onion Petits pois are a small breed of pea with a super-sweet flavor. Canned petits pois are sometimes just small peas untimely ripped from the pod, and will have neither the sweetness nor the tenderness of the variety they're passing themselves off as. For *petits pois à la française* you should really seek out the real thing, fresh or frozen, to give the lettuce in the dish its due as a bitter counterpoint to the sweetness. The small onions often stipulated in French recipe books are hard to find outside France but the white socks of scallions, shorn of their tendrilly toes, will do just as well, as will the sliced white bits of leek. Slice a bunch of scallions (or the white parts of 3 leeks), soften them in 2 rounded tbsp butter, then add a few handfuls of shredded butter lettuce (or a couple of Little Gems, cut into wedges), ⅓ cup stock or water, salt and ½ tsp sugar. When the lettuce has wilted, add 1 lb frozen petits pois and simmer for about 10 minutes, stirring infrequently so as not to damage the peas. Taste for seasoning and add a little more butter before serving.

Pea & Parsnip *See Parsnip & Pea, page 220.*

Pea & Pork As fitting a pair as legs in breeches, according to Ibsen's *Peer Gynt*. Deemed worthy of inclusion in the 1850 *Oxford English Dictionary*, the most famous pork and pea dish in Britain (particularly in Wales and the Black Country) is faggots and peas. Faggots are essentially pig's offal mixed with pork belly, onion, sage and mace, then rolled into balls, either wrapped in a caul or breadcrumbed, and baked. Opinion is divided over whether dried or fresh peas make the best accompaniment, even if tradition would have favored dried, as pigs were slaughtered long after the pea harvest was over. *Crépinettes* are the French equivalent and, wrapped as they are in caul, look similar to faggots but are usually made with a more standard sausage meat mixture of lean and fatty pork, seasoned with sweet spice and sage, thyme or parsley. Additions of chestnut, truffle or pistachio are common.

Pea & Potato *See Potato & Pea, page 93.*
Pea & Prosciutto *See Prosciutto & Pea, page 170.*

Pea & Rosemary The field pea, which is dried and split and used primarily these days for soup, is thought to be a close relation of the far more cultivated, vibrant garden pea. Where the garden pea is sweet, with a grassy-fresh flavor attributable to the compound that gives green peppers their characteristic aroma, field peas are more akin to another member of the legume family, the lentil. They have a meatier flavor, and are most commonly cooked with ham or bacon, but rosemary's woody eucalyptus notes suit them too. The following is a recipe for soup, but can be cooked with less water to make a side dish for lamb or pork. Soften a chopped onion in olive oil, then add ½ lb dried green split peas, a sprig of rosemary and about 6 cups of stock. Bring to the boil, turn down the heat and simmer for 1–1½ hours, until the peas are soft, seasoning toward the end of the cooking if necessary. Add boiling water if the pan is running dry. You can reduce the cooking time by pre-soaking the peas for a few hours. You might alternatively use yellow split peas, but note that they tend to be milder and sweeter than the green ones and have a less earthy taste.

Pea & Shellfish *See Shellfish & Pea, page 141.*
Pea & Smoked Fish *See Smoked Fish & Pea, page 164.*

Pea & White Fish Surf 'n' turf is all very well but fish goes best with ingredients that truly taste of turf, not just graze on it—and there's nothing much more grassy than the garden pea. (Other grassy flavors like parsley, fennel and tarragon also make classic pairings.) The natural sweetness of pea emphasizes the savoriness of the fish, whether in a humble Styrofoam cup of mushy peas with fish and chips or the turbot served with scallops, pork belly and pea purée at Michael Caines's Michelin-starred Gidleigh Park. And as pea loves salty ingredients, it pairs exceptionally well with *bacalao*, or salt cod.

Bell Pepper

Green peppers are immature red peppers. Green and red differ substantially in flavor but for the purposes of concision share one chapter here. Yellow and orange have more in common with the flavor of red than with green. As you might imagine from an unripe fruit, green peppers are more bitter and have a fresher, grassier flavor than the heat-ripened red, which is sweeter and fruitier. They can sometimes, but not always, be substituted for one another, although you wouldn't stuff a bitter olive with green pepper, and red peppers don't have quite the freshening quality that makes green so good in a stir-fry.

Bell Pepper & Bacon The combination of salty bacon and oily red pepper lends dishes a sort of smoky, sweet charriness without the hassle of setting up the barbecue. It might remind you of chorizo too. A little goes a long way: even a couple of slices and a wrinkled, punctured old pepper will season a hearty rice or chickpea dish.

Bell Pepper & Beef Face it, you don't go to a barbecue to eat well. Even good cooks produce dog chews of cindered meat and trays of chicken à la salmonella. In my experience, only one outdoor-cooked meal ever came within a smoke wisp of the campfire meatfeasts that made my stomach rumble during cowboy films, and it was in the middle of the New Forest in the 1970s. It was August, baking hot; I must have been eight or nine. We were with my parents' best friends—a chef, originally from Italy, and his wife. Having found a shady spot by a stream, Piero built a fire while my father made a rope swing. I busied myself making star-shapes on the swing until a whiff of searing meat lured me off the rope, over to where the grown-ups were holding sticks—real sticks from the forest floor!—threaded with beef and green peppers over the fire. It tasted like nothing I've ever eaten since. Maybe because the beef was aged rump steak, cut into chunks large enough to stay tender and juicy on the inside while crisp, salty and charred on the outside. Maybe because the peppers were quartered like cups so they caught and held the juices of the meat. And maybe because real and highly illegal wood fires will always lend more flavor to food than barbecue briquettes.

Bell Pepper & Chicken *See Chicken & Bell Pepper, page 31.*
Bell Pepper & Chili *See Chili & Bell Pepper, page 204.*

Bell Pepper & Egg I refer anyone bemused as to why there should be an album of music featured in *The Sopranos* called *Peppers and Eggs*, when said album doesn't contain a song of that name, to a paper published by Dr. Maria-Grazia Inventato, chair of Cultural Studies at the University of Eau Claire, Wisconsin: "Peppers and Eggs: Red-Blooded Males and Mother-Worship in Italian-American Crime Culture." As she says, "For many mafiosi, the capsicum, with its erect stalk, blood-red juice, and consanguinity with the violent fieriness of the chili pepper, stands as a potent symbol of Latin masculinity, at once contradicted and confirmed by the strong maternal attachment represented by the egg. However, it should be noted that the recipe requires both the *softening* of the pepper, and the *scrambling* of the mother-symbol, a double-trauma whose significance will not be lost on anyone familiar with mother-son dynamics even in second- or third-generation immigrant communities." Fry a couple of sliced red (or green) peppers until soft and juicy, scramble in 4 large eggs, then season and pile into 4 crusty ciabatta or white rolls. The recipe was traditionally served during Lent, when meat was forbidden, but it tastes more feast than fast to me. The Basque dish of *pipérade* similarly combines chopped peppers, onion and tomato with scrambled eggs. See also Chili & Egg, page 205.

Bell Pepper & Eggplant *See Eggplant & Bell Pepper, page 82.*

Bell Pepper & Olive A green olive stuffed with a sliver of red pepper is the classic gin martini garnish, although anchovy and Gorgonzola make tasty, if less attractive, alternatives. Some experts say the meatier quality of an unpitted olive is a better match for gin's juniper-dominated flavor. They might have a point—think how well juniper goes with game—but the stuffed olive definitely pips it when it comes to sparing drinkers the inconvenience of hacking up seeds like a parrot while sitting at the bar trying to radiate seductiveness or discourse on American foreign policy. And besides, the red bull's-eye, pierced through the middle with the toothpick, serves as a handy visual metaphor for the effect a properly made martini has on the brain.

Bell Pepper & Onion *See Onion & Bell Pepper, page 107.*
Bell Pepper & Shellfish *See Shellfish & Bell Pepper, page 138.*
Bell Pepper & Soft Cheese *See Soft Cheese & Bell Pepper, page 72.*
Bell Pepper & Tomato *See Tomato & Bell Pepper, page 253.*

Chili

The heat of chili can make it difficult to appreciate its various flavor characteristics. As with their close relatives, bell peppers, green chilies are unripe and have a fresh green-bean/pea flavor. Red chilies are sweeter and less easy to characterize. Dried chilies are sweeter still, and tend to the same rich, fruity, sometimes smoky, leathery flavors as sun-dried tomatoes and olives. Chili fans might like to familiarize themselves with the flavors and heat intensities of a range of cultivars, and with chili products such as oils, sauces, vodkas and pastes.

Chili & Almond Spicy, nutty romesco sauce comes from Tarragona, 70 miles southwest along the coast from Barcelona. It's made with pulverized almonds (and/or hazelnuts), tomato, garlic and a dried sweet chili pepper from which the sauce takes its name, otherwise known as the ñora. Ñoras are always dried, are red-brown in color and have a smoky flavor. They're one of the dried red peppers that are ground to make Spain's ubiquitous pimentón, or Spanish paprika. If you can't find ñoras, try anchos. Romesco sauce is classically served with chargrilled *calçots*, a leek-like vegetable, and seafood. But why deprive lamb chops or barbecued chicken of its company?

Chili & Anchovy *Phrík náam pla* is a simple mixture of sliced fresh chilies and fish sauce that you'll find on most tables in Thailand. It's a less unfamiliar idea than it might seem at first, operating on the same principles as salt and pepper but turned up a few notches. Sea salt is rarely used as a seasoning in Thailand—it's considered a rather crude ingredient. Peppercorns, although still used in Thai sauces and pastes, lost ground to chili peppers as the major contributor of heat when the latter arrived from the New World in the sixteenth century.

Chili & Anise When spicy Italian sausages are called for but you can't get any, try a few pinches of fennel seeds and crushed dried chili with a good-quality, herb-free pork sausage. See also Pork & Broccoli, page 37.

Chili & Avocado *See Avocado & Chili, page 195.*

Chili & Bacon Sweet, crumbly cornbread tends to be cakey until you lend it the savoriness of bacon. This recipe is quick and easy enough to rustle up warm for breakfast. Whisk 2 eggs in a large bowl, then add 1⅔ cups plain yogurt and 4 tbsp melted butter and whisk again. Sift together ½ cup plain flour, 2 tsp salt and 1 tsp bicarbonate of soda, gradually stirring it all into the egg mixture. Then add 1¾ cups fine cornmeal and fold in 4 crumbled, well-cooked bacon slices and 1 tsp dried chili flakes. These will lend some welcome smokiness, as well as a flicker of heat, but you could also use an equivalent amount of finely chopped fresh green jalapeños. Grease a muffin tray (use the fat from frying the bacon), fill to about halfway with the mixture and bake at 400°F for about 20 minutes, until golden and risen. Alternatively, pour the whole lot into a 9-in square pan and cook for 25–30 minutes before cutting into squares. Whatever the shape, it's excellent spread with cold cream cheese and a piquant chili jelly.

Chili & Beef *See Beef & Chili, page 47.*

Chili & Bell Pepper Although they're both members of the capsicum family, bell peppers lack the capsaicin that gives chili peppers their fire. Chili heat is measured on the Scoville scale, with habanero and Scotch bonnet chilies scoring 80,000–150,000 and jalapeños 2,500–8,000. Bell peppers score a big fat zero. But remove the seeds and ribs from a green jalapeño and, with its heat reduced, a nibble will reveal a very similar flavor to its mild-mannered relative. They share a dominant flavor compound, 2-methoxy-3-isobutylpyrazine, which is also present in Cabernet Sauvignon and Sauvignon Blanc wines. Similarly, red chilies taste like red peppers, although unlike green they lack a shared single character-impact compound (i.e., a compound that can summon up an ingredient in one sniff). Nonetheless, green or red, big or small, all are highly combinable. Try an Indian chili-laced potato mixture stuffed into green peppers: mix roughly mashed potato with onion fried with an Indian spice blend and chili, stuff into halved, seeded peppers, put a little knob of butter on each and bake until the peppers are cooked through—about 30–40 minutes at 350°F. You can speed up the process by blanching the peppers first.

Chili & Broccoli *See Broccoli & Chili, page 125.*
Chili & Butternut Squash *See Butternut Squash & Chili, page 226.*

Chili & Cabbage *Kimchi* is a Korean side dish of pickled vegetables, often made by fermenting cabbage and chilies in brine. Before the arrival of chilies from the New World, ginger and garlic provided the kick; now very hot, sweet red peppers called *koch'u* have taken their place, in both fresh and coarsely ground form. The cabbage is usually the sweet Chinese variety, kept whole in the very best *kimchi*, but often shredded to keep the jar size manageable. Plenty of other vegetables, including cucumber and eggplant, are given the *kimchi* treatment; several different types of *kimchi* might be offered at the same meal. Buy some from an Asian grocery if you feel the need to try it before making your own and burying it in the garden to mature.

Chili & Cauliflower *See Cauliflower & Chili, page 122.*
Chili & Chicken *See Chicken & Chili, page 32.*
Chili & Chocolate *See Chocolate & Chili, page 18.*

Chili & Cilantro Frequently used together. Green chilies have a particularly pleasing affinity for cilantro, lending a sharper freshness to its lush, just-after-the-rain grassiness. Blend them with a little lemon juice and sugar to serve as a condiment.

Chili & Coconut Coconut milk enfolds Thai ingredients in a sweet, forgiving embrace. It knocks the sharp edges off lime, shushes foul-mouthed fish sauce and soothes the heat of chili, whose active component, capsaicin, is soluble in fat but not in water. The medium-hot *phrik chii faa*, or "sky-pointing chili" (known as the Japanese chili elsewhere), is often used in Thai curries, most often in dried form but sometimes fresh.

Chili & Egg Experience and innocence; as uncomfortable an incongruity as a baby with its ears pierced. Until the plate of *pad thai* arrives, that is, sweetly fragrant of scrambled egg and garnished with rounds of fresh red chili. Or *huevos Mexicanos*—the patriotic breakfast that scrambles egg with the colors of the country's flag; green chili, red tomato and white onion. Literalists, or committed carnivores, might want to add some chopped eagle and serpent.

Chili & Eggplant *See Eggplant & Chili, page 83.*

Chili & Garlic Finding out that *not all food in Italy is good* was as big a disappointment as discovering that Santa was a fraud. Within the first few days of our vacation in the Italian lakes, my sister and I had exhausted our restaurant options. On day four we hatched Plan B: eat nothing but *aglio, olio e peperoncino*. Spaghetti with garlic, oil and chili. It's not always on the menu, but rest assured that the kitchen will have the ingredients, and be sufficiently aroused by the mark-up not to turn down the request. *Aglio olio* is the sort of dish Italians throw together late at night for friends, and if you try it yourself you'll see why—it's mysteriously more delicious, and satisfying, than the sum of its parts. Certainly good enough to eat four nights in a row. While your spaghetti is boiling, warm enough olive oil in a pan to coat the pasta without drenching it. Finely slice some garlic (a clove per serving, as a rough guide), slide it into the oil and fry until it just turns golden. Don't burn it. You might alternatively fry the garlic whole or crushed, depending on how garlicky you want it—sliced is the *via media*. Remove the garlic, or don't (depending, again, on levels of garlic tolerance), shake in a teaspoon of dried chili flakes and toss your perfectly cooked spaghetti in the flavored oil. Serve with chopped parsley and grated Parmesan.

Chili & Ginger *See Ginger & Chili, page 302.*
Chili & Goat Cheese *See Goat Cheese & Chili, page 57.*
Chili & Hard Cheese *See Hard Cheese & Chili, page 68.*

Chili & Lemon The heat of a chili can disguise its fruity qualities. The habanero, for example, is so fiery you'll almost certainly have missed its lovely citrus notes, which pair so nicely with lemon in both savory and sweet sauces. As the name suggests, Aji Lemon Drop is another variety of citrusy chili pepper, but it also scores highly on the Scoville scale of chili heat. In the United States, Toad Sweat, a range of chilified dessert sauces, includes a lemon-vanilla sauce with a hint of habanero, good with ice cream and fruit salads.

Chili & Lime *See Lime & Chili, page 294.*
Chili & Liver *See Liver & Chili, page 44.*
Chili & Mango *See Mango & Chili, page 283.*

Chili & Mint *Vada pav,* a popular fast-food snack indigenous to Bombay, is essentially a patty of mashed potato combined with onion and spices, deep-fried in chickpea batter and spread with a vibrant chutney made of mint leaves ground with fresh green chilies, lemon juice and salt. It comes in a soft white roll. Chili-mint jelly is delicious, and easy to make at home: simply add a couple of finely chopped jalapeños to a standard mint jelly recipe.

Chili & Oily Fish *See Oily Fish & Chili, page 154.*

Chili & Olive The pleasing counterpoint of sweet flecks of chili in olives is lent scientific weight by research conducted at the University of California, Davis. Capsaicin, the compound responsible for chili heat, also suppresses our ability to detect bitterness, thus moderating the harshness of the olives.

Chili & Orange *See Orange & Chili, page 288.*
Chili & Oyster *See Oyster & Chili, page 149.*

Chili & Peanut Is it the peanut's mission to neutralize chili? Capsaicin, the sinisterly flavorless, odorless, heat-giving compound in chili, is fat- but not water-soluble, and thus foiled (some say) by oily peanut. Peanut also comes armed with tryptophan, which induces sleep, and undermines the endorphins produced by the body to counter the pain of chili heat (and which, it's argued, constitute the otherwise counterintuitive pleasure of removing the top of your scalp with a shrimp vindaloo). Or are they working together? Chili and peanut combine to make a complex garnish for soups and noodle dishes, are ground into a paste for satays, and in Mexico take center stage in *mole de cacahuate,* a smoky sauce made with lots of dried chilies and peanuts.

Chili & Pineapple *See Pineapple & Chili, page 261.*

Chili & Pork Sweet dried red chili meets pork in chorizo sausages, which were exotic a decade ago, then suddenly hit the big time and have been painting the world's menus red ever since. According to the strength of pimentón used to flavor it, chorizo comes *dulce* (sweet/mild) and *picante*, although the spice is also available in *agridulce*, or bittersweet, form. Pimentón is made from a variety of chilies, including ñora and choricero, and also varies regionally and according to production technique. In La Vera valley in Extremadura, the ripe peppers are hung in smokeries in the fields and left for up to two weeks to develop the characteristic deep, rich quality that's substantially different from the flavor of chilies left to dry in the sun. There's a lot to be said for trying out the full range of pimentón varieties to understand quite how different one is from the other, and (note to self) not just because they come in such pretty cans. Paprika from Hungary is similarly prepared with dried red chilies and is available in a similar range of styles, but is said to be fruitier than pimentón. It's used in roughly the same way pimentón is used in Spain—i.e., with everything.

Chili & Potato *See Potato & Chili, page 90.*
Chili & Shellfish *See Shellfish & Chili, page 139.*

Chili & Tomato In the early 1990s, a tapas bar opened near where I worked in London. A pokey little place, with room for maybe 20 or 30 customers, but the food was great. Well, I say the food, but what I really mean is the *patatas bravas*. "Fierce potatoes," as any tapas lover will know, are fried potato wedges or cubes, draped in a tomato sauce spiked with fiery pimentón. Here they came with *aioli* too, Barcelona-style. Sure, we tried the manzanilla clams, the spicy *albondigas*, the tortilla and the *gambas al pil-pil*. But after a few visits all we ever ordered was the *bravas*. Bravas, bravas, bravas. In time, people would make the pilgrimage to Soho, as they do to the famous Las Bravas in Madrid. Or they might have done had the place not abruptly shut down. I wonder in hindsight if the owners, dreaming of converting numb British palates to the glories of Spanish cuisine, had grown tired of running a glorified chip shop, packed with increasingly ample women glugging over-oaked Rioja and stubbing out Marlboro Lights in smears of garlic mayonnaise.

Chili & Walnut *See Walnut & Chili, page 231.*
Chili & Watermelon *See Watermelon & Chili, page 245.*

SPICY

Basil
Cinnamon
Clove
Nutmeg
Parsnip

Basil

Sweet basil is the warmest, most fragrant, beautiful, fresh and irritatingly likeable of herbs. It has strong notes of spice—clove, cinnamon, anise and tarragon—combined with a minty grassiness that's particularly noticeable when it is pulverized in quantity to make pesto. Basil comes in many "flavor variants," such as lemon, lime, cinnamon and the licorice-flavored Thai basil, all of which share a rich spiciness and an enlivening freshness. Dried basil is no substitute for fresh, but it's not to be sniffed at. Sure, with drying, the beautiful, perfumed top note disappears, but carefully dried basil develops a spicy, licorice-like, mint character that works well in fish stews or baked lamb dishes.

Basil & Anise If you love basil and star anise, you'll walk over hot coals for Thai basil. Identifiable by its purple stem, Thai basil has a licorice flavor with a hint of cinnamon. Unless you live near a Thai or Chinese supermarket, it's probably easier to grow your own than find it to buy, and well worth it if you cook a lot of Thai curries or stir-fries.

Basil & Chicken A sort of wish-fulfillment combination: less charming together than you'd hope they'd be. Basil mayonnaise is good in a chicken baguette, and sweet basil leaves are an adequate substitute for holy or Thai basil in a Thai-style stir-fry, but please: spare us the chicken with pesto on pasta.

Basil & Clove Some of sweet basil's spiciness comes from eugenol, which gives cloves their distinctive flavor. I've heard of Italian housewives putting a pinch of ground cloves in their pesto if the basil isn't pulling its weight. The flavor of holy basil, used in Thai cooking, is dominated by eugenol and, when fresh, shares some of clove's anesthetic effect on the mouth. Which is maybe why the herb is often paired with chili in stir-fries such as *bai kaprow*: the basil takes the edge off the chili's heat. Pound 3 cloves of garlic, fresh red chili to taste and a little salt into a paste. Heat 2 tbsp peanut oil over a high heat and cook the paste for 30 seconds. Add 2 minced or thinly sliced chicken breasts and stir-fry until cooked through. Pour in a sauce mix of 3 tbsp water, 2 tbsp soy sauce, 1 tsp sugar, ¼ tsp white pepper and a handful of holy basil leaves. Serve with plain rice. Season to taste with the sauce in Lime & Anchovy, page 293.

Basil & Coconut Freshly torn basil leaves make an ideal counterpoint to rich, fatty coconut. Strewing standard sweet basil over a creamy Thai curry works wonders, not just on the flavor but also by giving it the floating-market prettiness that makes Thai soups so appetizing to the eye. Even better, if you can get it, Thai basil has a similar flavor to its sweet relative, but with stronger aniseed notes.

Basil & Egg

When in the mood to make eggs green
(To clarify just what I mean,
Not "eco" green, like Prius cars
But *colored* green, like men from Mars),
In your mental pan you should
Consider only what tastes good.
Rule out at once the croquet lawn,
The waistcoat of a leprechaun.
Green ink will only make you ill,
As would a mashed-up dollar bill.
Avocado'd be a waste:
In scrambled eggs it's hard to taste.
Green tea—a very current fad—
Would be preposterously bad.
Leeks and lettuces lack the might:
When cooked they'll be less *green*, more *white*.
Peas and peppers taste persuasive,
But their color's not pervasive.
Broccoli all but makes the mark,
Except its green's a trifle dark;
Celery, okra, spinach, kale
In hue and flavor mostly fail.
Of the green things that we tested
Basil simply can't be bested.
To scrambled eggs you add pesto,
Stir it well, and then, *hey presto*—
Green eggs. Season. Serve with ham.
(Do not prepare for folk named Sam.)

Basil & Garlic *See Garlic & Basil, page 112.*

Basil & Goat Cheese In tomato and mozzarella salad, basil's all ring-a-ding sweetness and light, but in rougher company it can turn quite raunchy, as in pesto. Pair it with a lusty goat cheese—preferably something piquant, well chilled, and soft enough to be spread on these pesto scones. Place 2½ cups plain flour in a food processor with a pinch of salt and 1 tbsp baking powder and pulse to mix. Add 6 tbsp diced cold butter and resume pulsing until the mixture has the texture of breadcrumbs. Mix 4 tbsp pesto with ½ cup cold milk and add through the feed tube with the machine running. Mix until you have a dough, adding a little more milk or flour respectively if it's too dry or too wet. Remove the dough, knead briefly on a floured surface until smooth, and roll out to ½ in thick. Cut out rounds with a 2½-in biscuit cutter, place on a greased baking sheet and bake at 425°F for 12–15 minutes.

Basil & Hard Cheese *See Hard Cheese & Basil, page 67.*

Basil & Lemon *See Lemon & Basil, page 297.*
Basil & Lime *See Lime & Basil, page 294.*

Basil & Mint A mint character can often be detected in basil, especially when dried, although this is not always welcome. Ligurian basil is the preferred type for making pesto because of its lack of mintiness. Make the most of both herbs' abundance in summer with this zucchini pasta. Cut 4 zucchini into 1-in-thick coins and place on a well-oiled baking tray. Drizzle with more oil and roast at 350°F for about 30 minutes. Meanwhile, cook ½ lb pasta (conchiglie or farfalle) so that it's *al dente* by the time the zucchini is well browned on both sides and a bit collapsed. Drain the pasta and mix it with the zucchini and the oil in the tray. Stir in a couple of handfuls of grated Parmesan and some finely chopped basil and mint. Garlic cloves, roasted with the zucchini and peeled, are a good addition.

Basil & Raspberry *See Raspberry & Basil, page 330.*
Basil & Shellfish *See Shellfish & Basil, page 138.*

Basil & Soft Cheese Basil could be considered the herbal equivalent of the allspice berry, combining the flavors of clove, cinnamon, mint and anise. And where spices can be awkward to include in a sandwich, basil can just be smoothed onto the bread like a transfer, preferably with some sliced mozzarella, prosciutto and tomato. Aside from the beautiful flavor released by its essential oils, it will contribute a satisfying snap on the teeth—the plant world's answer to the pleasures of biting into a sausage.

Basil & Tomato Tomato's sweet and sour flesh loves the bitter, spicy flavor of fresh basil. With both at hand, even the least experienced cook can make something delicious—consider them culinary stabilizers. Combine them in a thick sauce for pasta or pizza (see Garlic & Basil, page 112), a soup, a risotto, an omelette or a tart. If that's too much trouble on a hot summer's day, in less than ten minutes you could chop and pile them onto toasted rustic bread for bruschetta, or made a salad with mozzarella, or a panzanella salad with yesterday's bread. Or try tossing chopped tomato and torn-up basil leaves through just-cooked angel hair pasta with a little olive oil and seasoning; so rewarding for such little effort. Jars of ready-made tomato and basil sauce should blush for shame. See also Tomato & Clove, page 254.

Basil & Walnut Walnut and basil mingle pleasingly with cooked green beans. Toss the just-drained beans in warm walnut oil and scatter with torn basil, maybe with some chopped walnuts too. If you don't like the resinous flavor of pine nuts in pesto you'll find that walnuts make a fine, and far less opinionated, substitute.

Cinnamon

Homey *and* exotic. Sweet, warm, slightly bitter cinnamon flavors apple pies, Christmassy confections, Moroccan tagines and salads, Mexican *moles* and Indian dhals. Cassia (*Cinnamomum cassia*), cinnamon's rougher cousin, comes in shards of bark in contrast to true cinnamon's neat scrolls. It's considerably cheaper and so is sometimes sold ground with the real stuff by spice merchants. Sniff it and you'll detect something of a star-anise quality—it's sometimes called Chinese cinnamon—although for me it's so redolent of cola you can almost hear it burp.

Cinnamon & Almond *See Almond & Cinnamon, page 238.*

Cinnamon & Anise Chinese five-spice powder is a combination of ground cinnamon, star anise, fennel seed, clove and Sichuan pepper. If you're a few spices short, even a stick of cinnamon and a carpel of star anise can give bought beef stock a rich, Oriental depth. Throw in shreds of leftover roast beef, rice noodles, scallion, mint and chili for a phony (but delicious) Vietnamese *pho*.

Cinnamon & Apple *See Apple & Cinnamon, page 265.*

Cinnamon & Apricot The old English name for cinnamon was "pudding stick." It's like a wand to transform all manner of sweets. Poach apricots with a little water, a cinnamon stick and sugar to taste.

Cinnamon & Banana *See Banana & Cinnamon, page 272.*
Cinnamon & Beef *See Beef & Cinnamon, page 47.*
Cinnamon & Blueberry *See Blueberry & Cinnamon, page 336.*

Cinnamon & Butternut Squash Sweeter than a basket of kittens sharing a lollipop. Delicious roasted together or paired in a soup. See also Ginger & Butternut Squash, page 301.

Cinnamon & Cardamom *See Cardamom & Cinnamon, page 306.*

Cinnamon & Carrot Carrot has a woody character with hints of pine. Cinnamon is the dried inner bark of a tropical tree. Bring them together in a classic carrot cake and plant an arboretum in your flavor memory.

Cinnamon & Cherry Whirl like folk dancers in Hungarian cherry soup. Pit 1 lb sour (e.g., morello) cherries. Put the stones in a freezer bag and bash with a rolling pin until some of them are cracked. Empty into a pan with 2 cups wine (Riesling, Bordeaux and fruity rosés are all commonly used in this type of soup), a 3-in cinnamon stick, ¼ cup sugar, a strip of lemon zest, the juice of ½ lemon and a pinch of salt. Bring these to a simmer, cover and cook for 15 minutes. Strain the mixture into a clean pan over a medium heat. Hold back a few tablespoons and mix them with 1 tsp cornstarch to make a paste. Add this to the warming, boozy cherry juice and stir until it thickens a little. Add the cherries and simmer for about 10 minutes, until soft. Leave to cool a little, then stir in ⅔ cup sour cream and liquefy in a blender. Serve chilled as an appetizer or a dessert, perhaps using more sugar in the latter case. If you can only find sweet cherries, use just 1–2 tbsp sugar to begin with, add a little more lemon juice, then adjust to taste accordingly.

Cinnamon & Chocolate A popular combination in Mexico, where they go nuts for it in both drinking chocolate and chocolate bars. It's also found in their famous *mole* sauce for meat—see Chocolate & Chili, page 18. Elsewhere in the world, vanilla has replaced cinnamon as the aromatic of choice for flavoring chocolate—although Nestlé in Canada recently launched a limited-edition cinnamon Kit Kat.

Cinnamon & Clove Kicks the holly and the ivy into touch. A simmering pan of mulled wine. Plum pudding. Mince pies. Such a gorgeous combination shouldn't be limited to Christmas or sweet stuff. Drop a 2-in cinnamon stick and 3 cloves into a pan of basmati rice while it's cooking to make a nonspecifically festive accompaniment to curry. Add ½ tsp turmeric if you like it yellow.

Cinnamon & Coconut *See Coconut & Cinnamon, page 281.*
Cinnamon & Coffee *See Coffee & Cinnamon, page 24.*

Cinnamon & Fig There's a kind of sherry called Pedro Ximénez that's as syrupy and dark as old-fashioned cough mixture and as sweet as figgy pudding. Simmer 15 chopped dried figs in ½ cup warm water, ½ cup Pedro Ximénez and ½ teaspoon ground cinnamon until the liquid is thick and sticky. Cool slightly, and serve with vanilla ice cream.

Cinnamon & Ginger *See Ginger & Cinnamon, page 303.*

Cinnamon & Grapefruit A half grapefruit sprinkled with cinnamon is a fine partnership. So how about this grapefruit cheesecake with a cookie base enriched with cinnamon and honey? Melt 6 tbsp butter and stir in 10 crushed crackers or digestive biscuits, 2 tsp ground cinnamon and 2 tbsp honey. Press into a 8-in loose-bottomed flan tin. Stir 3 tbsp confectioners' sugar into 1 lb mascarpone cheese, then carefully fold in the juice and grated zest of 2 grapefruits plus the finely chopped flesh of another. Spoon onto the cookie base and refrigerate for a few hours before serving. You could alternatively set the topping with gelatin but I like the loose, unctuous creaminess the simpler recipe gives.

Cinnamon & Lamb As night falls on the Djmaa el Fna in Marrakech, the air fills with the aroma of spiced meat cooking on charcoal grills, drawing in locals and tourists, weaving like charmed snakes between neat stacks of sheep's heads, to sit at smoky stalls lit by gas lanterns. Harassed parents wishing their kids were as eager to arrive at the dinner table might try the same trick—simply grill lamb chops with a sprinkle of cinnamon. It's great in a slow-cooked lamb tagine, too, or in minced lamb kebabs made with finely chopped onion. Try roughly 1 tsp cinnamon per pound of meat.

Cinnamon & Lime *See Lime & Cinnamon, page 294.*

Cinnamon & Mint A hot yet herbaceous mix that works well with red meat. Or add a few cinnamon sticks to the mint syrup in Mint & Watermelon on page 323 to make a cinna-mint variant. Dilute with sparkling water for a long, refreshing drink.

Cinnamon & Orange After a few seconds in the mouth, a piece of cinnamon should take on a complex character as its essential oils rehydrate, according to the spice expert Tony Hill. Essences of orange and cedar come first, followed by a heat reminiscent of clove or mild pepper. Cinnamon not only combines beautifully with orange but its sweetness offsets the fruit's sharpness. Orange slices sprinkled with a few pinches of ground cinnamon and a drop or two of orange-flower water are a classic combination. Or try cinnamon added to the cake in Orange & Almond, page 287.

Cinnamon & Peanut *See Peanut & Cinnamon, page 27.*
Cinnamon & Pear *See Pear & Cinnamon, page 268.*
Cinnamon & Pineapple *See Pineapple & Cinnamon, page 261.*

Cinnamon & Pork In Vietnam this pairing is found in a popular kind of charcuterie called *cha que*. Blitz 1 lb lean pork leg in a food processor with 3 tbsp fish sauce, 3 tbsp water, 2 tbsp peanut oil, 1 tbsp cornstarch, 1 tsp ground cinnamon, 1 tsp baking powder, ½ tsp salt and a little white pepper. Place on a baking sheet covered with foil and fashion into a rough cylinder about 3 in thick. Wet the palm of your hand and smooth the surface of the cylinder before pricking it all over with a toothpick, pushing the pick all the way through to the foil underneath. Bake at 375°F for 25–30 minutes, until the cylinder is light brown on top and a skewer comes out clean. When cool, cut into ¼-in-thick slices. This makes an authentic addition (or alternative) to pâté in *bành mì* sandwiches—see Cucumber & Carrot, page 183.

Cinnamon & Soft Cheese Chalk might seem a better match for cheese, but these two are a common pairing in America. Mind you, everything goes with cinnamon in America. Sure, Europeans like cinnamon in cakes and pastries, but not necessarily in chewing gum, candy, breakfast cereal, cola, coffee, tea and booze. In every airport and mall in the United States, the Cinnabon baker slides a fresh tray of rolls out of the oven every 30 minutes and gives them a rich cream-cheese frosting. Cinnabon rolls are basically butch Danish pastries. They're deeper, stickier and deliver a far stronger

Nutmeg & Chocolate *See Chocolate & Nutmeg, page 20.*
Nutmeg & Egg *See Egg & Nutmeg, page 133.*
Nutmeg & Eggplant *See Eggplant & Nutmeg, page 84.*

Nutmeg & Hard Cheese Nutmeg has a crispness that lends itself to fatty foods. Macaroni and cheese is a case in point. The flavor of nutmeg is a complex mixture of pine, flower, citrus and warm peppery notes from a compound called myristicin, which is said to give nutmeg its hallucinogenic qualities. When nutmeg is added to bland foods, like cheesy pasta, custards and potato, the complexity is especially notable—even more so if you add enough to begin hallucinating about it.

Nutmeg & Lamb Long before I learned to cook, I ate a "spaghetti bolognese" in a taverna in Corfu. It was delicious, but unfamiliar for a reason that dawned on me only recently, cooking a Greek meal in which the minced lamb was flavored with sweet, peppery nutmeg: they were using their moussaka sauce on the pasta. To make *spaghetti corfiote*, chop ½ small onion and soften it in 3 tbsp butter or lard. Add 1 lb minced lamb and cook until well browned. Stir in 1 cup tomato passata, 1 cup white wine, 1 cup water, 2 tbsp chopped parsley, ¼ nutmeg, freshly grated, and a good pinch of sugar. Season, bring to the boil, and cook gently for about 1 hour, stirring intermittently, until most of the liquid has been absorbed. Add an extra grating of nutmeg, check the seasoning and serve on pasta. The taverna served this on bucatini, a long, thin pasta similar to spaghetti but hollow, like a tiny water main. Note that some moussaka recipes substitute cinnamon for nutmeg, in a similar vein to the Greek dish *pastitsio*—see Beef & Cinnamon, page 47.

Nutmeg & Onion *See Onion & Nutmeg, page 110.*

Nutmeg & Oyster Nutmeg is a classic companion for oysters in cooked dishes, especially with cream. It's the seasoning that's used for oyster loaves, which are said to have originated in Massachusetts in the late eighteenth century. Cut lids from the tops of some crusty bread rolls, scoop out their soft, bready center and paint the cavity with melted butter. Put the hollowed rolls and lids in a 400°F oven for no more than 10 minutes while you cook the filling. Take enough oysters to fill the rolls and stew them along with their liquor, and the torn-up excavated bread, in some butter for 5 minutes. Add a little cream and a grating of nutmeg and cook for another minute. Fill the hot rolls with the oyster mixture, replace their lids and serve.

Nutmeg & Parsnip Some chefs have noted that the flavor of parsnips is partly reminiscent of nutmeg. That's because parsnip contains myristicin, a key component of nutmeg's flavor, which is particularly noticeable when the vegetable is roasted. I'd steer clear of baking pears with nutmeg for pudding: the combination is too parsnip-like for comfort.

cinnamon hit—maybe because in the U.S. (in common with China, Vietnam and Indonesia) cinnamon flavor comes from the cinnamon-like cassia plant, *Cinnamomum aromaticum*. Cassia grows in China and Indonesia and is harder and darker than the true cinnamon from Sri Lanka that we get in the UK. It's also stronger, hotter, simpler and generally more intense in the mouth compared to true cinnamon's sweeter, more complex flavor. Basically cassia's the hard stuff, which may explain why America is addicted.

Cinnamon & Strawberry *See Strawberry & Cinnamon, page 257.*

Cinnamon & Thyme The recipe for Colonel Sanders' unique blend of 11 herbs and spices is kept in a vault in KFC's headquarters in Louisville, Kentucky. I, on the other hand, make no secret of this finger-lickin' crumb crust for Fitzrovia baked chicken. Thoroughly mix 4 tbsp seasoned flour with 1 tsp cinnamon and 1 tsp dried thyme. Sprinkle it over 6 skinless chicken thighs, dip the thighs in beaten egg and coat in breadcrumbs. Bake on a rack at 400°F for 30 minutes, turning halfway through. Add a pinch or two of cayenne to the flour if you want some heat. Serve in a bucket, although a washing-up bowl works just as nicely.

Cinnamon & Tomato Cinnamon adds a warm bass note to shrill tomato. Try a pinch instead of sugar to sweeten canned tomatoes. A subtly spiced tomato sauce does wonders for meatballs, lamb shanks, shrimp and eggplants. See also Beef & Cinnamon, page 47.

Cinnamon & Walnut *See Walnut & Cinnamon, page 232.*
Cinnamon & Watermelon *See Watermelon & Cinnamon, page 245.*

Clove

The word clove is derived from *clavus*, Latin for nail, which, aside from the visual likeness, suits the spice's hard, direct, hotheaded flavor. The nail association becomes literal in the case of *clouté*, wherein a clove is used to fasten a bay leaf to half an onion in order to season soups, stews and sauces. The singular flavor of clove often sees it paired with other flavors to modify or round it out, although British clove-flavored boiled sweets are one exception; another is the cordial made with pink cloves from Zanzibar, once popular in the south of England and still made today. In Indonesia, the majority of cigarettes are flavored with clove and, when fresh, Thai holy basil has a similar flavor and numbing effect on the lips.

Clove & Apple *See Apple & Clove, page 265.*
Clove & Bacon *See Bacon & Clove, page 166.*
Clove & Basil *See Basil & Clove, page 209.*

Clove & Beef Use a clove or two in beef stew (as most French recipes do for beef *pot au feu* or stock) or a pinch of ground cloves in gravy. The flavor of clove will be scarcely distinguishable but its bittersweet warmth enhances the beef in a way that makes it seem darker and more concentrated. Just don't forget to remove whole cloves before serving.

Clove & Cinnamon *See Cinnamon & Clove, page 213.*

Clove & Coffee Like cardamom and coriander seeds, clove is sometimes ground with coffee, specifically in Ethiopia. Quite a double hit of bitterness, this one, so if you prefer a softer, more latte-style beverage, heat up a mugful of milk with a couple of cloves in it, then discard them and add to the coffee. See also Cardamom & Cinnamon, page 306.

Clove & Ginger Clove needs company. It's rare to find it in anything other than an ensemble recipe, perhaps because most of its flavor comes from just one compound (eugenol), and needs complementary flavors to round it out and make it subtle. The spice blend *quatre-épices* combines clove, ginger (or cinnamon), nutmeg and white pepper, and has a hot, sweet, fruity character that gives depth to meat dishes, especially pork. See also Cabbage & Pork, page 119. Five-spice, incidentally, isn't just *quatre-épices* plus one—see Cinnamon & Anise, page 212.

Clove & Hard Cheese *See Hard Cheese & Clove, page 68.*
Clove & Onion *See Onion & Clove, page 108.*
Clove & Orange *See Orange & Clove, page 289.*

Clove & Peach A fair few fruits contain clove's identifying flavor compound, eugenol, and peach is one of them. Perhaps that's what made pickled peaches such a popular recipe in the American South. You peel the peaches before you stick the cloves in. Although who would have the heart to pierce that delicate, downy skin?

Clove & Pork Clove pairs very well with pork, not just on the grounds of flavor but because it acts as a natural preservative. The eugenol and gallic acid in clove have a strong anti-oxygenic effect at fairly low concentrations, so as food companies respond to growing consumer unease at synthetic additives, we may see a lot more of this partnership. Which would be no bad thing.

Clove & Tomato *See Tomato & Clove, page 254.*
Clove & Vanilla *See Vanilla & Clove, page 340.*

Nutmeg

The botanical name for nutmeg, *Myristica fragrens*, makes it Bond girl—appropriately enough for such an exotic, beautiful of a spice, apt equally to make sweet, creamy dishes less cloying ous vegetables less bitter. Mace, nutmeg's outer coating, is com same flavor compounds but in different proportions, and con greater quantities of essential oil. They can be used intercha always opt for fresh rather than the pre-ground form of either easier to grate than lacy pieces of mace, and is a little cheaper to

Nutmeg & Apple I love using nutmeg, partly because of the Also because it's simultaneously rich and fresh, and wonderful up warm apple purée. Grate a light dusting over it and serve wit cream. Like falling in love in the autumn.

Nutmeg & Avocado Avocado mixed with nutmeg is said aphrodisiac effect on men. Avocado contains vitamin B6, whi edly stimulates the production of testosterone. Nutmeg turns Together they transform the mildest-mannered mama's boy Bardem. Grind a little nutmeg over cold avocado soup or avocc with shellfish, then stand back.

Nutmeg & Butternut Squash Butternut squash, lovely as it is the palate to sleep. That's where nutmeg comes in—to keep th esting. Italians use nutmeg in a pumpkin and ricotta filling for America, it forms part of the spice mix in pumpkin pies and baked custards. But it doesn't need heavy dairy to come to life: try so grated on butternut squash that's been roasted with a mixture of 2 oil to 1 part balsamic vinegar.

Nutmeg & Cabbage Nutmeg can have an enlivening effect o greens. Dot cooked cabbage or sprouts with butter and speckle t bare winter trees decked in fairy lights, with freshly grated nutmeg.

Nutmeg & Cauliflower *See Cauliflower & Nutmeg, page 123.*

Nutmeg & Celery Many nineteenth-century recipes for celer are seasoned with nutmeg, or its conjoined twin, mace. In *The* Sportsman (1855), Elisha Jarrett Lewis demurs. Either she found it th of vulgarity to flavor one's celery or she was suffering from an attack cision: "If fond of spices, put in a little mace and a clove or two; w however, recommend it. A shallot or so, a bay leaf, lemon juice, or parsley, might also be advocated by some of our friends." See also C Chicken, page 96.

Nutmeg & Potato *See Potato & Nutmeg, page 92.*
Nutmeg & Rutabaga *See Rutabaga & Nutmeg, page 121.*
Nutmeg & Saffron *See Saffron & Nutmeg, page 178.*

Nutmeg & Shellfish Nutmeg, or a combination of nutmeg and mace, is traditionally used to season potted shrimp. Mace is the lacy outer shell of the nutmeg fruit. Both have a clear, citrusy, fresh-pepper quality that makes them equally natural partners for creamy or buttery shrimp and lobster dishes.

Nutmeg & Tomato In Bologna they use nutmeg to tamp down the harsh acidity tomato can bring to a ragù. For an even silkier sauce, try adding milk to the meat once it's browned, in place of some of the stock or tomato you might otherwise have used, and your bolognese will be as round and smooth as a Brancusi. See also Nutmeg & Lamb, page 218.

Nutmeg & Vanilla *See Vanilla & Nutmeg, page 340.*

Nutmeg & Walnut A few pinches of grated nutmeg can enhance the flavor of walnut cakes and cookies in the same way that sweetmeats made with hazelnut are heightened by a teaspoon of cocoa powder.

Parsnip

Cooked parsnip has the sweet, slightly earthy flavor typically associated with root vegetables but with its own strong hints of nutmeg and parsley. Raw, its flavor is rather dispiriting, especially in combination with its fibrous, woody texture, for which you might as well substitute a macramé plant holder. Roasted or mashed, using plenty of fat and salt, parsnip takes on a gorgeous sweet spiciness that can recall coconut or banana.

Parsnip & Anise There's a trace of herbal, anise flavor in parsnips. Nibble one raw and you should get it. It's subtle—don't expect pastis on the *pétanque* pitch at St. Paul de Vence—but it's there. A welcome change from the usual curry treatment, then, is to pair parsnip with tarragon in a soup.

Parsnip & Bacon *See Bacon & Parsnip, page 167.*

Parsnip & Banana When bananas weren't available in Britain during the Second World War, mock bananas were made from parsnips. Aside from their ivory flesh, parsnip and banana share a sweet spiciness. Try a piece of roasted parsnip, close your eyes and think of banana—more convincing than carrots posing as apricots, another wartime substitution. Add a little ground cloves and rum to them and it's even more of a match. Now that bananas are no longer threatened by U-boats, a mock parsnip made of banana might lend a Caribbean lilt to your Sunday roast, but better to pair them in a moist cake—parsnips were once as popular a cake ingredient as carrots are now.

Parsnip & Beef Carrot and parsnip are similarly sweet, but parsnip's greater complexity of flavor makes it a more interesting partner for beef. Serve roast beef with parsnips roasted in the dripping, a basic beef stew with parsnips cut into chunks, or oxtail cooked in red wine with a parsnip and potato mash. On a less homey note, Charlie Trotter makes a terrine of beef cheek, salsify and endive with a parsnip purée. And as parsnip has a passion for salty flavors, I pair slices of the Italian air-dried salted beef, bresaola, with slices of parsnip and Parmesan bread made to Delia Smith's recipe.

Parsnip & Chicken *See Chicken & Parsnip, page 33.*
Parsnip & Hard Cheese *See Hard Cheese & Parsnip, page 69.*
Parsnip & Nutmeg *See Nutmeg & Parsnip, page 218.*

Parsnip & Pea It's generally thought that the compounds most responsible for parsnip's characteristic flavor are terpinolene (also found in some essential oils derived from pine trees), myristicin (present in nutmeg and, to a lesser extent, parsley and dill) and 3-sec-butyl-2-methoxypyrazine, which has a slightly musty, green-pea quality and is present at higher levels in parsnip than in any other vegetable, conceivably accounting for the unique affinity parsnips have for peas. Try them together in a winter pea soup. The parsnip acts as a thickener and also brings a lightly spicy flavor; it's like a sunny winter's day. Peel and chop 3 parsnips, then cook in olive oil with a sliced onion until softened. Add 3 cups vegetable or chicken stock, bring to the boil and simmer for 10 minutes. Then add about 1 lb frozen peas and season. Bring back to the boil, simmer for 5 minutes, then leave to cool a little. Liquefy, reheat and serve.

Parsnip & Pork The parsnip has recently gained "exotic" status among French gourmands, who historically considered it fit only for feeding livestock. Parsnips are not just excellent for fattening the herd: they're known to have an enriching effect on the flavor of pig meat, and are often fed to prosciutto pigs. Reciprocally enough, in the nineteenth century the flavor of parsnips was often improved by boiling them with salt pork, then mashing and frying them in butter—which sounds great, but would probably also improve the flavor of cotton balls.

Parsnip & Potato The introduction of the potato to Europe in the sixteenth century was responsible for the decline in parsnip's popularity, and entirely did in the skirret, which has a flavor somewhere between the two. Salsify, another pale root, is sometimes called the oyster plant, as it's said to taste like oysters, although I'd say a delicate, more savory parsnip is probably closer. Scorzonera, sometimes known as black salsify, looks and tastes similar to regular salsify but has a dark skin, as its name suggests.

Parsnip & Shellfish Scallops will perch on almost anything, as long as it's puréed. Parsnip is no exception. Nestle the scallops on top and enjoy the complementary sweetness and nuttiness.

Parsnip & Walnut *See Walnut & Parsnip, page 233.*

Parsnip & Watercress Watercress and roasted parsnip make a good basis for a winter salad. Both have a bitter spiciness, offset in watercress's case by its freshwater cleanliness and in parsnip's by its sweetness. Toss cubes of warm roasted parsnip with watercress leaves, croutons and crumbled blue cheese, then dress with a piquant French dressing. If you grow parsnips, the White Gem variety, whose leaves are edible when very young and have a recognizable parsnip flavor with a satisfyingly bitter aftertaste, would make an interesting addition to this dish.

Parsnip & White Fish *See White Fish & Parsnip, page 146.*

WOODLAND

Carrot

Butternut Squash

Chestnut

Walnut

Hazelnut

Almond

Carrot

Carrots vary both in their levels of sugariness and the pine/parsley flavor notes that can tend to a certain woodiness. They have a strong flavor affinity for other umbellifers, and their sweetness, and hint of woodiness, make for a harmonious match with the same quality in nuts. Nantes carrots have a particularly good reputation for flavor.

Carrot & Anise Guy Savoy offers a carrot and star anise soup as an *amuse-bouche* at his restaurant, and uses the same flavors to complement lightly grilled lobster. The combination is perfect: star anise flatters carrot's fresh, woody quality, which is all to the good, as carrot flourishes under a surfeit of praise. John Tovey makes a homey carrot and anise side dish by mashing 1 lb cooked carrots with 1 tbsp butter and 1 tbsp Pernod. Try with confit of duck.

Carrot & Apple Children tend to like this sweet and sour salad. Grate both and dress with a little neutral-tasting oil and lemon juice. Strew with a potluck handful of mixed seeds, nuts and raisins for a more exciting texture.

Carrot & Beef *See Beef & Carrot, page 46.*

Carrot & Cabbage Both crisp, clean and spicy-sweet when raw. Carrot brings a little fruitiness to the mix. A dependable basis for a coleslaw with apple, celery, walnuts, caraway or blue cheese (or all of them). I'm more than happy with just cabbage, carrot and a little onion. Unless it's to be eaten straight away, you can salt the shredded cabbage first, leave for about an hour, rinse, drain, then dry. This will draw out some of the water and prevent the coleslaw from turning soupy. It also gives the dish a more complex, savory quality in place of the sweet, innocent fruitiness.

Carrot & Cardamom Combined in a popular Indian dessert *gajar halwa*, which is served at wedding feasts and during the autumn festival of Diwali. There are many ways of making *gajar halwa*, but essentially it's finely grated carrot, slowly simmered in milk until most of the moisture has been absorbed, then enriched with sugar, ghee (or unsalted butter/flavorless oil) and crushed cardamom seeds. Garnishes vary, but almonds, pistachios and dried fruit are typical. Eat warm or at room temperature. Some say an accompaniment of vanilla ice cream is good, but I think its sweetness calls for the contrasting sourness of crème fraîche or thick yogurt.

Carrot & Celery When it comes to diet foods, which do you prefer, the carrot or the stick? That's right: neither. Or both, soused in dressing. In a game of word association, my response to "France" would not be "Paris" or "terroir" but "celeriac remoulade." A pot of that, the same of *carottes* *râpées*, some slivers of Bayonne ham, a boxed ripe Camembert, a baguette, a bottle of Côtes du Rhône. The perfect picnic. Inevitably I'll forget to bring cutlery and have to make do scooping up heaps of vegetable with the bread. To make *carottes râpées*, finely grate the carrot (preferably in a food processor, which makes for silkier strands) and toss with a dressing made with a light-flavored olive oil, lemon juice, a touch of Dijon mustard and pinches of sugar, salt and pepper. For celeriac remoulade, grate your celeriac (not too fine, or it'll taste like coconut), toss with lemon juice to prevent it from browning, then dress with a mayonnaise mixed with a little grain mustard and season to taste.

Carrot & Cinnamon *See Cinnamon & Carrot, page 212.*

Carrot & Coconut One of the signature dishes of Wylie Dufresne, chefowner of wd~50 in New York, is Carrot-Coconut Sunny-Side-Up—a perfect replica of a fried egg. The (fully pierceable) yolk has flavors of carrot and maple, while the white tastes of coconut and cardamom. Those of us without the necessary industrial gum pastes (or know-how) to whip this up in our own kitchens might have to limit ourselves to a coconut frosting on a carrot cake.

Carrot & Cucumber *See Cucumber & Carrot, page 183.*

Carrot & Cumin It's rare to sit down at a Moroccan feast without being offered a plate of sweet carrots in a robust, cumin-flavored dressing. Cut into crinkly discs, like the metal from which Ingersoll keys are cut, they remind me, cumin's efforts notwithstanding, of the sorry rounds of orange matter you get in cans. Homegrown carrots, on the other hand, tugged from the ground when they're long, thin and pointed, are perfect for this combination. Toss them, washed but unpeeled, in olive oil and sprinkle with cumin before roasting. This will intensify both their earthiness and their sweetness, coaxing the sugars to the surface, where they caramelize and mingle deliciously with the spice.

Carrot & Hazelnut *See Hazelnut & Carrot, page 234.*

Carrot & Olive I love those inexpensive restaurants—often Turkish or Greek—where they bring you a complimentary dish of wrinkly black olives and sticks of raw carrot to nibble on while you read the menu. It shows a little care. And the combination of oily, salty olives with fresh, sweet carrot stimulates your taste buds while you consider your options.

Carrot & Onion To make carrots *à la nivernaise*, first blanch peeled whole young (or chopped older) carrots for a few minutes and drain. Stew them slowly in butter and just enough stock to cover, until all that's left of the liquid is a small amount of thick syrup. At the same time, cook peeled small onions via the same method, bar the blanching. Combine the carrots and onions with seasoning and a little sugar. Just a little: remember how candied both carrots and onions can get anyway. And if you're scrumping them from Mr. McGregor's garden, watch out for the cat.

Carrot & Orange Over eighty years ago, the French polymath Paul Reboux wrote, "I am not interested in a compote of spiders, a salmis of bats or a gratin of blindworms. All I want to do is break the traditional association of certain ingredients in cookery. My mission is to ally the unexpected with the delectable." Having thus set out his stall, he goes on to give a recipe for a salad of lettuce leaves and sliced cooked potato in a creamy dressing, garnished with orange zest and carrots, both cut into one-inch lengths and "as thin as pine needles." This, claims Reboux, is guaranteed to capture the attention of the gourmand. "Orange? Carrot? How is it that the orange tastes of carrot and the carrot of orange?" A game for the palate to play.

Carrot & Parsley Carrot flavor can be accentuated by adding one of its umbelliferous relatives, such as parsley, parsnip, cumin, coriander or dill. A fine scattering of parsley is often used to finish both carrot soup and Vichy carrots (new carrots cooked over a low heat in a combination of butter, sugar, salt and the non-chalky water of the Vichy region of France).

Carrot & Peanut *See Peanut & Carrot, page 26.*
Carrot & Rutabaga *See Rutabaga & Carrot, page 120.*

Carrot & Walnut Carrot juice has a sweet, fruity flavor and can taste a little like a thin milkshake. Walnuts, when fresh, also have a milky sweetness. The last of the summer's carrots with the first of the year's walnuts will make a lovely salad to accompany good bread and cheese. As they age, and thus become woodier, they might need a little sweetening up, say, in a carrot cake.

Butternut Squash

Butternut squash and pumpkins share a chapter as they have similar flavor affinities and are often used interchangeably. Bear in mind, nonetheless, that butternut squash is naturally sweeter and will require less sugar in sweet recipes. There's also a difference in texture: the dense flesh of butternut squash is fine-grained and silky, whereas pumpkin can be rather fibrous. The pronounced sweetness of butternut squash works well with salty ingredients; cut through with something spikily sour; or, in combination with the density of its flesh, as a background for potent herbs such as rosemary and sage.

Butternut Squash & Almond In northern Italy they fill pillows of pasta called tortelli with amaretti cookies and pumpkin. The bittersweet almond cookie lifts the sweetness of the pumpkin. Nadia Santini, chef at Dal Pescatore in Mantua, makes legendary pumpkin and amaretti tortelli, which also contain mostarda di Cremona (fruits preserved in a thick mustard syrup), nutmeg, cinnamon, clove and Parmesan.

Butternut Squash & Apple One autumn lunchtime at the Beacon restaurant on West 56th Street in New York, I ordered the pumpkin and apple soup. The waiter brought it in a pitcher, poured it into the bowl and put a puff of spiced spun sugar on top. It fainted instantaneously into the soup, leaving a trace of cinnamon sweetness and the gentlest lick of heat. Back in London a week later, trying to re-create the experience in my own kitchen, I left the soup chilling in the fridge as I riffled through eBay for a cotton candy machine. I came to my senses. Who needs another large device, solely in order to make a soup taste *slightly* better, when you already have to stack tea towels on top of sieves on top of an ice-cream machine? And in the meantime, as it happened, the flavor of apple had been intensifying in the fridge, to the extent that by the time I got around to eating the soup it didn't need a spicy puff of spun sugar at all.

Butternut Squash & Bacon The sweet-saltiness of this combination reminds me of crabmeat, which was the inspiration behind this landlubber's version of crab cakes—butternut squash and bacon cakes with lime mayonnaise. Grate ½ lb butternut squash. Combine with 4 cooked, crumbled bacon slices, a generous handful of breadcrumbs and a pinch of salt. Bind together with about 1 tbsp mayonnaise. Shape into cakes about 2 in in diameter and ½ in deep, then leave to firm up in the fridge for about 30 minutes before deep-frying in vegetable oil. Top each cooked cake with a little mayonnaise mixed with lime juice and zest, with an extra sprinkle of zest to finish.

Butternut Squash & Blue Cheese *See Blue Cheese & Butternut Squash, page 63.*

Butternut Squash & Chestnut Sounds like a couple of fat Shetland ponies, but a good winter combination nonetheless. Try them paired in a wild rice pilaf. Or detect both flavors in a Potimarron, a French heirloom squash variety whose name is derived from *potiron* (pumpkin) and *marron* (chestnut). Bright orange and the shape of an outsize fig, the Potimarron has a wonderfully aromatic chestnut flavor that's made it widely popular, and fairly widely available. In Japan it's known as *hokkaido* and is often cooked in miso.

Butternut Squash & Chili Squashes make great fillings for pies and pasta because they're dry(ish), and their sweet, uncomplicated flavor lets other ingredients have their say. A quesadilla provides the perfect opportunity to pitch butternut squash against chili. Take a tortilla that'll fit in your frying pan. Top with mashed butternut squash to a depth of ½ in–1 in, scatter over a sour-hot mixture of pickled jalapeños and chopped fresh chilies, then a layer of grated mild Cheddar. Top with another tortilla and cook over a medium heat, on both sides, until heated through and the cheese has melted. Cut into quarters and serve.

Butternut Squash & Cinnamon *See Cinnamon & Butternut Squash, page 212.*

Butternut Squash & Ginger *See Ginger & Butternut Squash, page 301.*

Butternut Squash & Goat Cheese Sweet beets are a classic match for salty goat cheese, but butternut squash is even better. Roast cubes of squash and the edges will caramelize and take on a honey flavor—and we all know how gruff old goat cheese takes to a slick of golden honey. Some pair squash and goat cheese in a gratin but I find that a little heavy; better, in my opinion, to toss them in a sharp dressing with baby spinach leaves, or in a couscous with chickpeas, toasted pine nuts, finely chopped red onion and lots of chopped parsley and mint.

Butternut Squash & Lime *See Lime & Butternut Squash, page 294.*

Butternut Squash & Mushroom *See Mushroom & Butternut Squash, page 78.*

Butternut Squash & Nutmeg *See Nutmeg & Butternut Squash, page 217.*

Butternut Squash & Pork Take kindly to all sorts of stews. Can be given a Western slant in cider with apples, bacon and onions, or sent East in chicken stock with soy, Shaoxing rice wine, five-spice and brown sugar. To help thicken the sauce, flour the pork—shoulder, ideally, cut into 1-in cubes— before browning. Add the butternut squash pieces 20–30 minutes before the end, as they'll disintegrate if cooked for too long.

Butternut Squash & Rosemary To roast these together is to remember how intensely sweet and fragrant they both are—so much so that they can carry a whole dish, like this rib-sticking butternut squash, cannellini bean and rosemary stew. Chop an onion and soften in olive oil, adding 3 finely chopped cloves of garlic toward the end. Add 1 lb peeled, seeded butternut squash cut into 1-in cubes, 2 cans of cannellini beans, drained, 1 can of cherry tomatoes, the very finely chopped spikes of 1 sprig of rosemary, 1 cup dry white wine, 1 cup water and some seasoning. Stir, bring to the boil, then cover and simmer for about 20 minutes. Remove the lid and continue cooking until the butternut squash is tender and most of the liquid has evaporated. Serve this as it is, or top with breadcrumbs mixed with a little Parmesan and grilled until lightly browned. If you feel more than usually excited by the smell of this dish, you may be interested to note that butternut squash and rosemary are close relatives respectively of pumpkin and lavender. In a study of a range of scents by the

Dr. Alan Hirsch's Smell and Taste Treatment and Research Foundation in Chicago, men found the fragrance of pumpkin pie with lavender the most arousing; for women it came second—see also Cucumber & Anise, page 183.

Butternut Squash & Sage *See Sage & Butternut Squash, page 313.*

Butternut Squash & Shellfish You had us at the butter, say lobsters, who love butter more than anything: you could just poach them in it. Squash is a lovely match, providing a harmonious sweet nuttiness as well as an appropriately expensive texture, variously described as like velvet, satin or silk.

Chestnut

Like pine nuts and cashews, one of the sweeter nuts, but it's chestnut's low fat content that makes it unique. Nonetheless, as with its oily cousins, roasting—over an open fire, naturally—brings out its full flavor and sweetness. Its earthy quality makes it a good match for other autumnal flavors, such as game, mushrooms and apples. Chestnuts can be ground into a flour for rather rustic-flavored bread, pasta and cakes, or stuffed to the cell walls with sugar to make purées and *marrons glacés*. A chestnut liqueur called *crème de châtaigne* is also available.

Chestnut & Butternut Squash *See Butternut Squash & Chestnut, page 226.*

Chestnut & Cabbage Having cut crosses in your Brussels sprouts and nicks in your chestnuts, you'll have the knife skills to whittle everyone their Christmas presents. Cutting a vent in chestnuts is essential to stop them exploding in the oven. You really needn't bother with sprouts: it'll only risk overcooking them. And you'll need all your energy to peel the chestnuts, and suppress the yodel of agony when a shard of superheated shell insinuates itself behind your thumbnail. If you have gone to the trouble of gathering them yourself, don't squander it by eating them too soon—sweet chestnuts need a few days to dry out, giving the starch a chance to convert to sugar, thereby intensifying the flavor. The sprout and chestnut pairing is delicious, of course, but why not try braising chestnuts with apple and red cabbage, or, if your hands have recovered, stuff a cabbage, adapting the recipe in Cabbage & Pork on page 119, using chestnut and bacon in place of the pork and beef.

Chestnut & Celery A *brunoise*, or fine dice, of celery is recommended as a pairing for chestnuts in soup, not only for the flavor compatibility but because celery and chestnut have similar cooking times, according to the French food writer Madame E. Saint-Ange.

Chestnut & Chicken Edward A. Bunyard writes that "the chestnut finds its best end within a bird of some sort—preferably a dead bird. Here its soft pastiness makes an admirable ground on which more pungent flavours may display themselves, a more attractive background than the (nearly) all conquering potato." Bunyard was most likely thinking of how chestnuts pair with fattier, boldly flavored game birds such as goose or duck, although the pairing is just as common in stuffings for chicken and turkey.

Chestnut & Chocolate *See Chocolate & Chestnut, page 18.*

Chestnut & Lamb For those who find apricots and prunes too sweet, chestnuts make an excellent sweetish substitute in a North African lamb stew or tagine. Add vacuum-packed chestnuts to the simmering pot about 10 minutes before the end of cooking and they'll hold their shape.

Chestnut & Mushroom *See Mushroom & Chestnut, page 78.*

Chestnut & Pear Hold back some of the chestnuts you bought for the stuffing at Christmas and serve them the following day in a salad of chopped pear, the best bits of dark turkey meat and some dark green leaves.

Chestnut & Pork Alexandre Dumas opines that chestnuts are very good with all meats. In *Le Grand Dictionnaire de Cuisine*, he gives a recipe for a chestnut purée to serve with pork sausages. Peel roasted chestnuts, removing the thin inner skin too. Soften them in butter, white wine and a little stock, then purée. Mix the cooking juices from the sausages into the chestnut mixture before serving. Chestnuts have an unusually high carbohydrate content for a nut, which accounts for their traditional use as a potato substitute; it also allows them to be ground into flour for baking. Its farinaceous quality earned the sweet chestnut the nickname "bread tree" in Corsica. The chestnut is also used to brew Pietra, Corsica's first native beer, which has a roasted, smoky quality.

Chestnut & Prosciutto *See Prosciutto & Chestnut, page 169.*

Chestnut & Rosemary *Castagnaccio* is a Tuscan cake originally made with nothing but chestnut flour, olive oil, water and salt. Through more prosperous times, it has accrued a variety of rosemary, pine nuts, dried fruit, walnuts and candied peel. Chestnut flour has a curious aroma, somewhere between cocoa and silage, which makes for a rustic flavor experience. And I mean really rustic: it tastes like the floor of a shepherd's refuge. Not to everyone's tastes, then, but if you want to try it for yourself, mix 2 cups sifted chestnut flour and a pinch of salt to a smooth batter with 1½ cups water. Stir in some finely snipped rosemary leaves and pour into a buttered 9-in flan tin—it's a shallow cake. Scatter with pine nuts, brush generously with olive oil and bake at 375°F for 30–45 minutes, until the top is dark brown and fretted with cracks like old varnish. You'll notice the recipe contains no raising agent, and pretty much the only sweetness comes from what's naturally present in the chestnuts, so if you have guests you might want to recalibrate any expectations the word "cake" may arouse.

Chestnut & Vanilla Most commonly combined in *marrons glacés*, a sweetmeat so sugary it warrants a wanted poster in every dental surgery. They are made with particularly sweet, large, single-kernel nuts, infused over days with a vanilla-flavored syrup before a final glazing with an even more concentrated syrup. In France Clément Faugier makes a *marron glacé* spread that tastes like a subtly nutty toffee, with the pleasantly powdery quality you'd expect of a chestnut. It's served simply mixed with whipped cream or spread onto thin crêpes.

Walnut

After almond, the second most popular nut in the world. Walnuts are easy to take for granted in cakes and candy, where their wrinkly lobes are often more interesting than their mild flavor. Toasting brings out their full character—stronger nutty notes with a hint of nicotine bitterness. Walnuts work particularly well with other "brown" flavors, such as cinnamon, nutmeg, maple syrup, honey and pears. They can be bought pickled, steeped in syrup, as a liqueur (*nocino* or *vin de noix*) and as walnut oil, which makes a great dressing for vegetables and salads.

Walnut & Anise I urge you to try ½ tsp crushed anise or fennel seeds in a sticky walnut or pecan pie (pecans and walnuts are nearly always interchangeable, although pecans are sweeter, less bitter and easier to slide under doors). Simply stir the spice into the mixture before decanting it into the pastry shell. Brings a fresh note to what can be a rather monotonous sweetness. If sweet's not your thing, try a dressing made with walnut oil and tarragon vinegar.

Walnut & Apple See *Apple & Walnut, page 267.*

Walnut & Banana Here's a novel idea for bananas. Novel because it calls for ones that aren't overripe. Peel a pristine banana and slice it into coins. Top each one with a walnut half, using a neat dollop of dulce de leche as cement. The art here is to get the dulce de leche onto the banana without it clinging to the spoon and trailing messily. It's worth the effort, though. Twice as satisfying, even, as a carpaccio of chilled Milky Way. If the bananas *are* overripe, the traditional combination is banana bread mixed through with walnuts that show in each slice like stranded jigsaw pieces.

Walnut & Basil *See Basil & Walnut, page 211.*
Walnut & Beef *See Beef & Walnut, page 51.*

Walnut & Beet As wrinkly and red-faced as a couple of weather-beaten peasant farmers. Give them a fresh start by mixing broken walnut pieces and chunks of roasted beet and sweet potato into cooled, cooked red quinoa (which is crunchy, and has something of a roast-walnut flavor itself). Dress with a maple syrup vinaigrette—just add maple syrup where you might ordinarily use honey.

Walnut & Blue Cheese *See Blue Cheese & Walnut, page 65.*

Walnut & Broccoli Combine them in a pasta dish or a stir-fry: broccoli makes an incongruously healthy addition to the heart-furring dish described at Walnut & Shellfish, page 233.

Walnut & Carrot *See Carrot & Walnut, page 225.*
Walnut & Cauliflower *See Cauliflower & Walnut, page 123.*

Walnut & Celery Celery and walnut share distinctive aroma compounds called phthalides, which you can also detect in lovage. Chew a walnut and you will clearly taste the connection. This flavor overlap ensures they combine beautifully in stuffing and in Waldorf salad. As celery also has a magical effect on chicken stock (see Celery & Chicken, page 96), try this recipe, which brings all three together in a soup. In a little butter or oil, gently fry a chopped onion, a large potato, peeled and cubed, and 4 or 5 chopped celery stalks. When the onion has softened, add 3 cups chicken stock, bring to the boil, then turn down the heat and simmer until the potatoes and celery are cooked. Finely grind 2 oz walnuts and set aside. Liquefy the soup, return it to the pan and add the nuts. Stir until thickened. Serve with some suitably coarse brown bread.

Walnut & Cherry *See Cherry & Walnut, page 244.*
Walnut & Chicken *See Chicken & Walnut, page 35.*

Walnut & Chili *Chiles en nogada* are stuffed green chilies served with a ground walnut sauce and scattered with pomegranate seeds. A signature dish of Puebla in south-central Mexico, *chiles en nogada* was devised to celebrate Mexican independence, as the green chilies, white walnut sauce and red pomegranate seeds match the colors of the national flag (see also Chili & Egg, page 205, and Avocado & Tomato, page 197). In the book *Like Water for Chocolate*, the dish is served at the climactic wedding of Tita's niece, Esperanza, to Dr. Brown's son, Alex, where it makes everyone so randy that Pedro (spoiler alert) expires after a bout of post-prandial lovemaking (even Mrs. Beeton knew you should always wait 20 minutes to let the meal go down).

Walnut & Chocolate *See Chocolate & Walnut, page 22.*

Walnut & Cinnamon Walnuts have a kinship with sticky-sweet autumnal flavors like cinnamon, toffee, maple syrup and apple. There's even a walnut cultivar called Poe that's said to taste like butterscotch. The common Persian walnut is sometimes called the English walnut in the United States, because it was imported from (if not necessarily grown in) the UK, and had to be distinguished from the native black walnut. The British used to call it the Madeira nut, because that's where *they* imported it from, although I like to think it might have something to do with the way the nut's rich sweetness is offset by the astringency of its skin, just as golden Madeira wines (full-flavored with notes of caramel, nuts, dried fruit, marmalade and toffee) are balanced by their pronounced acidity.

Walnut & Coffee See *Coffee & Walnut, page 25.*

Walnut & Eggplant Frequently paired in Russian and Turkish dishes and in Georgia, where baby eggplants are halved, fried and stuffed with a combination of ground walnuts, garlic and cilantro mixed with chopped onion, celery, tarragon vinegar and paprika. They're served at room temperature, scattered with pomegranate seeds. In Lebanon, baby eggplants are almost split lengthways, stuffed with walnut and garlic and preserved in oil. Italian chef Giorgio Locatelli stuffs pasta parcels with a mixture of walnut, eggplant, ricotta, nutmeg, egg and Parmesan. He recommends buying unshelled walnuts while they're in season (December to February), for their freshness and intensity of flavor.

Walnut & Fig See *Fig & Walnut, page 333.*

Walnut & Garlic In the early eighteenth century, *aillade* sauce from the Languedoc was described by a M. Duchat as "a mess which the poorer sort make with garlic and walnuts pounded together in a mortar, and which prepares the stomach for the reception of certain meats of an undigestive and disagreeable nature. As for the *aillade* itself, it is so much admired by some persons of distinction, even in Italy, that the historian, Platina, could not forbear telling the world, that a brother of his would often put himself in a sweat by the pains he took in preparing this ragoo." Poorer sort or not, you can make this mess in a food processor, although the flavor will be better if you pound it in a mortar. Pulse about 1 cup walnuts and 4 garlic cloves until crushed. Season, then slowly add ⅔ cup olive oil (or a mixture of olive and walnut oils) until you reach sauce consistency. For a Turkish variation known as *tarator*, add 3 crustless slices of white bread in chunks, 2 tbsp lemon juice, 1 tsp red wine vinegar and ⅓ cup stock to the garlic and walnuts before pouring in the oil. Traditionally this sauce is served with roast meat, but works well with a spinach, red pepper and chickpea paella.

Walnut & Goat Cheese See *Goat Cheese & Walnut, page 60.*
Walnut & Grape See *Grape & Walnut, page 249.*

Walnut & Hard Cheese Cheese expert Patricia Michelson writes that wet walnuts enhance strong-flavored cheeses like Parmesan and pecorino. Wet walnuts have a fresh-milk flavor similar to the note lost in cheese during the aging process. That's not to say that dried walnuts aren't good partners too. They're particularly well matched with aged Gouda, with its deep butterscotch flavor and color.

Walnut & Mushroom *See Mushroom & Walnut, page 82.*
Walnut & Nutmeg *See Nutmeg & Walnut, page 219.*
Walnut & Orange *See Orange & Walnut, page 291.*

Walnut & Parsley Parsley adds welcome freshness to walnut's woody astringency. They're often paired in a sauce related to the more famous pesto. Heat 3 tbsp olive oil in a pan and add 2 chopped garlic cloves, 3 oz finely chopped walnuts and a handful of roughly chopped flat-leaf parsley. Season and allow to warm for a few minutes. You might toss this sauce with pasta, or use it over gnocchi, beets or a cheese soufflé.

Walnut & Parsnip Like gnawing on Pinocchio's leg. Counter excess woodiness by removing the tough cores from older parsnips.

Walnut & Pear *See Pear & Walnut, page 269.*

Walnut & Shellfish There's a dish they serve in American Chinese restaurants that tends to divide humanity fairly straight down the middle: deep-fried shrimp tossed in a honey and lemon mayonnaise and mixed with candied walnuts. Tribe A sees the potential in chewy, salty shellfish mixed with sugar-crisp, slightly spicy nuts. Tribe B suppresses instant emesis at the thought of the calories and sheer inauthenticity. I'm with Tribe A.

Walnut & Soft Cheese *See Soft Cheese & Walnut, page 75.*

Walnut & Vanilla Black walnut is a boisterously flavored alternative to the common Persian variety, with a fruitier, musty quality. It lacks the paint-like notes that in Persian walnuts recall the smell that hits you on reopening an old can of enamel. Persians also tend to be woodier and more astringent. Native to northeastern America, the black walnut is prized for its wood, which is used in furniture making and for the stocks of upscale shotguns. They're notoriously hard shelled, and have to be whacked with a hammer, driven over by a station wagon or taken to the local shelling depot. Once cracked, they're often paired with vanilla to make legendary cakes, ice cream and a sweetmeat called divinity candy, made with sugar, egg whites and corn syrup. The result is like a cross between a meringue and a nougat.

Walnut & Washed-rind Cheese *See Washed-rind Cheese & Walnut, page 62.*
Walnut & Watercress *See Watercress & Walnut, page 101.*

Hazelnut

Hazelnuts have a sweet, buttery flavor, with hints of cocoa that make them a sensational match for chocolate. Hazelnuts grown in the Piedmont region of Italy are particularly renowned for their quality. Less glorious specimens can taste woody, with a metallic streak, rather like pencil shavings. Like all nuts, hazelnuts become more flavorful when heated: research has shown that the key hazelnut flavor compound increases tenfold when the nuts are roasted. The bittersweet richness of hazelnuts is also good ground into a coating for delicate, buttery seafoods or as the basis for a luxurious sauce. Hazelnut oil is useful for baking and is a treat in dressings. Frangelico and *crème de noisette* are hazelnut-flavored liqueurs. And health-food shops sell a hazelnut butter that tastes like the inside of a delicious praline truffle on a spoon.

Hazelnut & Almond Emergency Cake. It's one of life's paradoxes that the need for cake often arises from conditions that make you incapable of either making it or going out to buy it. Ideally, the following ingredients would be kept under glass with a little hammer hung underneath, but it was desperate improvisation that initially led me to this combination. A ciabatta roll was excavated from the bottom of the freezer, defrosted in the toaster, given a drizzle of hazelnut oil, sprinkled with a few almonds and sultanas picked from my husband's muesli, squashed together and eaten while still warm. It tasted amazing, somewhere between fruitcake and pain au chocolat, without the sickliness of either. Happily the bread's air holes turned out to be perfect pockets for the fruit and nuts, deepening its similarity to fruitcake.

Hazelnut & Apple See *Apple & Hazelnut, page 265.*
Hazelnut & Avocado See *Avocado & Hazelnut, page 196.*

Hazelnut & Banana Hazelnut butter, not to be confused with *beurre noisette*—see Hazelnut & White Fish, page 236—is a posh version of peanut butter. It tastes like a Ferrero Rocher denuded of its chocolate shell but without the sugar, although still possessed of a mellow sweetness that brings out the fruity side of banana. Spread it onto a loaf of fine white bread *before* you slice it (to prevent the bread from ripping) and make sandwiches with slivers of banana. Or thin a little out with milk and maple syrup to make a sauce for banana pancakes, scattered with roasted nuts.

Hazelnut & Carrot Recipes for carrot cake often call for vegetable oil rather than butter or margarine, so you might add some depth of flavor by using hazelnut oil. It's not cheap, but neither does the oil last very long, so if you have a bottle nearing its use-by date this recipe will amply consecrate its memory. If you find the hazelnut flavor too strong, dilute it with sunflower oil, which has a hint of hazelnut flavor itself. Ground hazelnuts can be used to flavor your baking too—you'll need to substitute them for a third to a quarter of the flour called for in cake recipes. By all means grind them yourself in a food processor, but roast them first to activate all the flavor compounds, then remove the bitter skins—unless you're making a coarser cake, in which case leaving the skins on can be rather good.

Hazelnut & Caviar Could hazelnut be the finest flavor in the world? It is detectable in aged white Burgundies, champagne, oysters, *jabugo* ham, Sauternes, Beaufort and Comté cheese, French farmhouse butter, lamb's lettuce, toasted sesame seeds, wild rice, many of the most revered potato cultivars and Oscietra caviar.

Hazelnut & Cherry *See Cherry & Hazelnut, page 244.*

Hazelnut & Chicken Filbertone is a key characteristic of hazelnut flavor. It has nutty, cocoa flavors with some meaty, earthy notes, which might suggest a Mexican *mole* sauce (see Chocolate & Chili, page 18). Hazelnut flavor is intensified by roasting or dry-frying the nuts; a good pairing for chicken is the *picada* sauce given in Hazelnut & Garlic, below. Or use them to thicken an aromatic stew like the one in Almond & Chicken, page 238, as they do in Turkey, where so many of the world's hazelnuts are grown. Or simply combine cold roast chicken, toasted hazelnuts and arugula with fresh figs in an autumnal salad.

Hazelnut & Chocolate *See Chocolate & Hazelnut, page 20.*
Hazelnut & Coffee *See Coffee & Hazelnut, page 24.*

Hazelnut & Fig Frangipane, the lovely, soft nut paste used to fill French fruit tarts, is usually made with almonds, but it needn't be. Hazelnut is a great alternative, particularly for a tart made with velvety, dark-skinned figs. Beat 4 oz skinned, toasted ground hazelnuts with 1 stick soft unsalted butter, ½ cup superfine sugar, scant 3 tablespoons plain flour, 2 eggs and ½ tsp vanilla extract. Pour this into a 9-in blind-baked sweet pastry shell, place your figs on top, slightly pressing them in and bake for 25 minutes at 350°F. Some like to position the figs so their luxuriant red interiors face up, but I prefer to see them rump-up, protruding from the golden paste like baby elephant seals softly snoring on the beach.

Hazelnut & Garlic Spanish *picada* is, at its most basic, a sauce made with pounded nuts, garlic, bread and oil. Sometimes hazelnuts are used, sometimes almonds and sometimes a combination. There are endless variations that might include saffron, tomato, parsley or pine nuts. It's not unlike the French *aillade* and Turkish *tarator* described in Walnut & Garlic, page 232. Use it to thicken stews (see Almond & Chicken, page 238), to mix into meatballs before shaping, or to serve as a sauce with meat or seafood. Take 15 hazelnuts, 15 almonds (both roasted and skinned) and a slice of good white, crustless bread that's been fried in olive oil, and pound them together in a mortar with 2 garlic cloves. When the mixture has become a paste, add 1 tbsp olive oil and seasonings as desired.

Hazelnut & Pear Although its flavor dissipates somewhat on cooking, hazelnut oil is very flavorful drizzled on steamed vegetables or mixed with vinegar in salad dressings. It combines particularly well with raspberry vinegar for a salad of goat cheese, pear and lamb's lettuce—a dark green leaf (also known as mâche) with its own trace of hazelnut flavor.

Hazelnut & Raspberry *See Raspberry & Hazelnut, page 330.*

Hazelnut & Rosemary Heston Blumenthal once gave a recipe for hazelnut and rosemary couscous in which he toasted the couscous in peanut oil in a pan before adding the water. But not just any water. He heated it with fresh rosemary stems to infuse it with flavor, before discarding them and reheating the water. After adding it to the couscous, he raked skinned, roasted chopped hazelnuts, finely snipped rosemary, butter and plenty of seasoning through the cooked couscous to finish the dish. You could also combine hazelnut and rosemary in a sweet cookie or ice cream, providing the nuts are roasted to boost their sweet, chocolatey character.

Hazelnut & Strawberry *See Strawberry & Hazelnut, page 258.*

Hazelnut & Vanilla The hazelnut liqueur Frangelico, from the Piedmont region of Italy, is made with toasted wild hazelnuts, vanilla, cocoa and a number of the usual secret ingredients. It has a smooth, buttery, hazelnut flavor with a noticeable vanilla finish, and can be whisked to great effect into whipping cream—say 2 tbsp Frangelico and 2 tbsp confectioners' sugar to 1 cup cream. Serve with any number of chocolate, pear or raspberry desserts.

Hazelnut & White Fish *Beurre noisette* is most often served with fish and vegetables, but it's good on poultry too. Butter is heated in a pan and, when the milk proteins and sugar in the whey caramelize, it turns brown, taking on a nutty color and flavor. If allowed to go darker, it becomes *beurre noir*, which is traditionally served with skate, brains or eggs. Sometimes a few chopped hazelnuts are added to a *beurre noisette*, making it *beurre de noisette*.

Almond

There are two distinct kinds of almond flavor, bitter and sweet. Bitter almond is the pronounced marzipan flavor that is found in almond extract, almond essence and Amaretto. The extract is derived from bitter almond kernels and from other stone fruits, such as apricot and peach; it must be treated to remove its cyanide content before it is suitable for human consumption. The primary compound responsible for bitter almond flavor is benzaldehyde, which was first synthesized in 1870 and is the second most used flavoring in the world. Aside from bitter almonds, it is also naturally found in a number of foods, including the prince mushroom and the cinnamon relative, cassia. Bitter almond goes particularly well with fellow members of the rose family—stone fruits, berries, apples, pears and rose itself. It's used to boost the flavor of other nuts; you may detect a hint of bitter almond in some dyed green pistachio ice creams. The sweet almond is the world's most popular nut. It has hints of bitter almond flavor but is mild, milky and slightly grassy when raw, and richer when roasted, with a slight toffee-popcorn flavor. Sweet almond's soft, rounded flavor makes it highly compatible. Almond milk and almond butter are available from health-food shops.

Almond & Anise *See Anise & Almond, page 179.*
Almond & Apple *See Apple & Almond, page 263.*

Almond & Apricot Charles Arrowby in Iris Murdoch's *The Sea, the Sea* rhapsodizes on how well anything made with almonds goes with apricot, thumbing his nose at the peach along the way. At one point he prepares a meal of lentil soup, chipolatas with boiled onions and apples stewed in tea, accompanied by a light Beaujolais and followed by almond shortcake with dried apricots. Clearly he's no stranger to cans and packets, but he's right about the apricot and almond. For a plain almond cookie to serve with poached dried apricots or an apricot fool, follow the recipe in Anise & Almond on page 179 but leave out the anise.

Almond & Asparagus The flavor of asparagus is often described as nutty. It's not surprising, then, that it has such an affinity for almonds, particularly toasted almonds, whose buttery quality complements the sulfurous bass notes of asparagus. Cook a handful of slivered almonds in a knob of butter over a low heat for 6–7 minutes, until golden. Remove from the heat, add 1 tsp lemon juice and ½ tsp salt, and pour over your cooked spears.

Almond & Banana *See Banana & Almond, page 271.*

Almond & Blackberry There's no such thing as a free crumble. The price you pay for the astonishing late-summer abundance of inkily delicious blackberries is to walk away with snags in your sweater and stains on your skirt. Almond's sweetness calms some of blackberry's wild spiciness in this blackberry and almond crumble. Butter a deep 8-in dish and fill with blackberries to a depth of about 2 in. Sprinkle with 2 tbsp sugar. In another bowl, rub 1½ cups plain flour and 6 tbsp butter together until they have a breadcrumb-like consistency. Stir in ⅓ cup golden superfine sugar and a small handful of chopped, toasted sliced almonds. Sprinkle this on top of the blackberries and bake at 400°F for about 30 minutes.

Almond & Black Currant *See Black Currant & Almond, page 324.*

Almond & Blueberry Fine in your muesli, but you wouldn't want to meet them at a party, would you? Almond is pale, characterless and shy unless toasted. Blueberry is usually just plain insipid. The Spartan, Ivanhoe and Chandler cultivars of blueberry are worth looking out for, but beware the faint-praiseworthy Duke, described variously as mild and moderately flavored. Things improve considerably if you add some almond extract or, even better, Amaretto, to whipping cream and whisk until the cream holds its shape. Fold in a handful of blueberries and some toasted almond slivers, holding back a few of the almonds to sprinkle over the top.

Almond & Butternut Squash *See Butternut Squash & Almond, page 226.*
Almond & Cardamom *See Cardamom & Almond, page 306.*
Almond & Cauliflower *See Cauliflower & Almond, page 121.*
Almond & Cherry *See Cherry & Almond, page 243.*

Almond & Chicken In Spain, the Moorish influence is easily discernible in almond-thickened recipes such as *gallina en pepitoria*, a favorite dish at fiestas. Fry 3 lb chicken joints in olive oil until golden, then set aside. Soften a finely chopped large onion in a little oil with 2 crushed garlic cloves and a bay leaf. Put the chicken back into the pot and pour over ⅔ cup fino sherry and just enough chicken stock or water to cover. Bring to the boil, put a lid on and leave to simmer while you make the *picada* in Hazelnut & Garlic (page 235), except use 30 almonds and no hazelnuts and add a pinch of saffron, a few pinches of ground cloves and 1 tbsp parsley. When you have a paste, add it to the chicken mixture and continue to simmer until the chicken is completely cooked (45–60 minutes from beginning to simmer). Just before serving, you may need to remove the chicken and keep it warm while you boil the sauce to reduce and thicken it. Many recipes suggest adding a couple of finely chopped boiled egg yolks at the end for the same reason. Serve with boiled rice.

Almond & Chili *See Chili & Almond, page 203.*
Almond & Chocolate *See Chocolate & Almond, page 17.*

Almond & Cinnamon These work well together in cakes, pastries and cookies. They also meet in one of the strata of Morocco's legendary *bastilla* pie. Spiced poached pigeon meat, very lightly scrambled eggs and a mixture of ground almonds, ground cinnamon and sugar are layered between superfine leaves of *warqa* pastry. Be warned, *bastilla* is fiddly and time-consuming to make. It's faster to fly to Fez and back than attempt it at home—meaning, of course, it's utterly exquisite. As is *keneffa*, the lesser-known sweet version that's often served at weddings. This intersperses the same delicate *warqa* pastry with cinnamon and almonds. My anglicized take on this treat is to make puff pastry horns and fill them with a cinnamon- and almond-flavored *crème pâtissière*.

Almond & Coconut When vanilla extract is called for in coconut cakes, cookies and puddings, try substituting almond extract for half of it. You'll get the same flavor-bolstering, rounded quality but with a nuttier, more sympathetic edge.

Almond & Coffee *See Coffee & Almond, page 22.*
Almond & Fig *See Fig & Almond, page 331.*
Almond & Garlic *See Garlic & Almond, page 111.*

Almond & Ginger Gingerbread and marzipan on cold night air is the smell of a Christmas market. Add the warm notes of cinnamon, clove and lemon basking in simmering red wine, with chestnuts and bratwursts scorching over coals, and you can enjoy your own miniature *Weihnachtsmarkt* in the comfort of your own home. It's traditional to cut the gingerbread into the shape of buildings and people, the marzipan into animals. In 1993, Roland Mesnier, the Clintons' pastry chef, made a gingerbread White House and no fewer than 21 marzipan likenesses of Socks the cat.

Almond & Grape *See Grape & Almond, page 247.*
Almond & Hard Cheese *See Hard Cheese & Almond, page 66.*
Almond & Hazelnut *See Hazelnut & Almond, page 234.*
Almond & Lamb *See Lamb & Almond, page 52.*

Almond & Lemon Ground almond soothes lemon's sharpness in cakes and tarts. In Kew, southwest London, deep little tarts called maids of honor are baked and sold at The Original Maids of Honour Shop. They're said to date back to the time of Henry VIII—in Hilary Mantel's novel *Wolf Hall*, Thomas Cromwell sends flat baskets of them to console Anne Boleyn's ladies-in-waiting. The recipe is a secret, but broadly speaking they're puff pastry tarts with a cheesecake-like lemon and almond filling. Farther north, Lancaster lemon tart is a variation on Bakewell tart (see Raspberry & Almond, page 329) that omits jam in favor of a thick spread of lemon curd under the cooked almond and egg mixture. In Italy, *torta della nonna* is usually made with lemons and pine nuts, or occasionally almonds—perhaps when *la nonna* has squandered her pension on sweet wine. And in both Italy and Spain, ground almonds are combined with lemon in a moist cake the Spanish call *torta de almendras de Santiago*. You could try using the whole fruit (zest, juice, pith) as described in Orange & Almond on page 287, increasing the amount of sugar a little to compensate for the lemon's extra sourness, but expect quite a mouth-drying result. Better to skip the pith in lemon's case.

Almond & Melon Grapes may be the standard garnish for the cold Spanish soup *afo blanco* (see Garlic & Almond, page 111) but melon is sometimes used instead. You might be able to make perfect spheres with your melon baller but they don't burst in the mouth quite as satisfyingly as an ice-cold grape.

Almond & Oily Fish According to Elizabeth David, the French writer Jean Giono dismissed trout and almonds as packaging rather than cooking (she didn't have much time for it herself). Alice Waters was more positive. In fact, her biographer recalls that the chef once visited a restaurant in Brittany where a meal of cured ham and melon, trout with almonds, and a raspberry tart constituted her idea of good food.

Almond & Olive *See Olive & Almond, page 172.*
Almond & Orange *See Orange & Almond, page 287.*

Almond & Peach There's a pleasing symmetry to the classic Italian dessert of baked peaches with amaretti. The cookies, which share an almond flavor with peach stones, are crushed and stuffed in the cavities left where the stones have been removed. Like the stones, the leaves of the peach tree have a peachy-almond flavor, and can be steeped in water and sugar to make a delicious syrup used in drinks, sorbets and fruit salads.

Almond & Pear *See Pear & Almond, page 268.*
Almond & Raspberry *See Raspberry & Almond, page 329.*
Almond & Rhubarb *See Rhubarb & Almond, page 250.*

Almond & Rose In France you'll find an almond and rose syrup called Orgeat that tastes like Amaretto getting ready for a date. It's mixed with water to make a refreshing summer cordial, with rum, lime, orange Curaçao and mint to make a Mai Tai, and with pastis to make a Moresque. In Iran, slivers of almond coated with rosewater-flavored sugar are served on festive occasions. If you like *ajo blanco*, the soup from Andalusia (see Garlic & Almond, page 111), you might try a sweet version in which rosewater and a little honey replace the garlic. A similar principle operates in the *sharbat* of milk, almonds and rose petals offered by the bride's family to the groom's at Muslim weddings in Bangalore.

Almond & Rosemary Almonds tossed in oil and then lightly salted have a surprising smoky-bacon flavor enhanced by a sprinkling of chopped rosemary. Serve with chilled fino sherry before dinner. Alternatively, fry the almonds in olive oil for an extra-toasty flavor—use ½ lb almonds to 1 tbsp very finely snipped rosemary.

Almond & Saffron *See Saffron & Almond, page 177.*
Almond & Shellfish *See Shellfish & Almond, page 137.*
Almond & Strawberry *See Strawberry & Almond, page 257.*

Almond & Vanilla Partners in crime. Pass themselves off as pistachio in green-dyed ice cream. The color is okay, but the flavor is a flagrant con. Neither vanilla, almond, nor a combination of the two could ever be mistaken for the buttery-egg-yolk/toasted-cereal peculiarity of roasted pistachio. Even if you *don't* need convincing, it's worth putting this to the test by making a Swedish *toscatårta*, a vanilla-scented sponge cake topped with shiny, sticky caramelized almonds, reminiscent of the famous almond tart made by Lindsey Shere, co-owner and original pastry chef at San Francisco's Chez Panisse. Incidentally, Shere includes a splash of both vanilla and almond extract in the pastry, a tip to bear in mind if you're making any sort of nut or stone-fruit pie.

Almond & White Chocolate *See White Chocolate & Almond, page 342.*

FRESH FRUITY

Cherry

Watermelon

Grape

Rhubarb

Tomato

Strawberry

Pineapple

Apple

Pear

Cherry

In common with the apple, a fellow member of the *Rosaceae* family, cherries come in sweet and sour form, with something of a gray area in between. And like apples, they have a predominantly fresh, green, fruity character and an almond-scented seed, although almond flavor is far more strongly present in cherry stones than in apple seeds—so much so, in fact, that the bitter-almond compound, benzaldehyde, forms the basis of synthesized cherry flavoring. In real cherries, almond flavor is most easily detectable when the fruit is cooked with the stones in. Cherries also have floral and spicy notes, plus a tannic quality that is particularly noticeable in the dried form of the fruit. Cherries are especially lovely with spices such as vanilla and cinnamon, and are paired with them to make cherry liqueurs. Kirsch and maraschino are clear, distilled spirits with a clean cherry flavor that goes very well in fruit salads or with flambéed fruit.

Cherry & Almond Cherry stones and bitter almonds contain a volatile oil that can be used to produce a compound called benzaldehyde. Benzaldehyde is the second most popular flavor molecule in the U.S. flavor-and-fragrance industry, after vanillin, and is used to synthesize both almond and cherry flavors. Sip a cherry cola and think of Amaretto and you'll taste the proximity. It's the cherry stone that tastes most like almond, not the flesh. Unless it's a maraschino cherry. Brined while unripe and green, maraschino cherries were originally steeped in a liqueur made from the juice, stones and leaves of cherry, which gave them a lovely, drunken marzipan flavor. These days most maraschinos are steeped in sugar syrup and given an almond flavoring, which doesn't taste half as nice—so if you want the best garnish for your Manhattan you'll have to steep your own, or seek out an artisan producer. Make a Manhattan by stirring aboutr 2¾ tbsp good rye or bourbon, 2 tbsp sweet vermouth and a dash of Angostura bitters with ice and then straining over a maraschino cherry.

Cherry & Banana *See Banana & Cherry, page 271.*

Cherry & Chocolate A winning combination. Even if you turn your nose up at Black Forest gâteau, or Ben and Jerry's Cherry Garcia ice cream, you can't deny that cherry, with its fruit *and* nut flavor, is a natural partner for chocolate. Sour morello cherries, grown in the orchards of the Black Forest in southwestern Germany, are the usual choice for combining with choco-late. Delia Smith makes a roulade of them, while Nigella Lawson makes chocolate muffins with morello cherry jam mixed into the batter. If you're not in the mood for baking, a good chocolate counter should furnish a half pound of cherries slouching kirsch-drunk in their chocolate shells.

Cherry & Cinnamon *See Cinnamon & Cherry, page 212.*
Cherry & Coconut *See Coconut & Cherry, page 280.*
Cherry & Coffee *See Coffee & Cherry, page 24.*

Cherry & Goat Cheese Sweet cherries work well with young goat cheese, particularly when fresh, grassy examples of both are available. Keep your eye out for them in high summer and, if you love long walks punctuated by picnics but can't bear to lug around a hamper or a cooler, this less-is-more lunch is ideal. Wrap fresh cherries and a few rounds of goat cheese in a paper bag with only as many slices of crumbly, nutty bread as you plan to eat, then head off to your picnic spot. I like to eat some of the cheese piled on the bread first and then relish the rest of it with the cherries, ending the meal with the remaining cherries on their own. Edward Bunyard writes that to enjoy cherries at their best you should walk through an orchard just after they've been harvested and pick the ripe remainders, so maybe you should plan your route accordingly.

Cherry & Hazelnut Added to your muesli, a handful of hazelnuts and some dried sour cherries recall a mug of sweet, milky tea and a cigarette. Dried cherries have a tangy sweet-sourness with strong hints of tobacco and a clear tannic tea flavor, which marry well with hazelnuts' milky sweetness.

Cherry & Lamb See *Lamb & Cherry, page 53.*
Cherry & Peach See *Peach & Cherry, page 278.*

Cherry & Smoked Fish Hugh Fearnley-Whittingstall suggests a seasonal combination of hot-smoked sea trout and a tart morello cherry compote made with fresh stoned cherries simmered with a little brown sugar. Serve both warm with a watercress salad and slices of walnut bread.

Cherry & Vanilla The anise-like quality of pure Tahitian vanilla is particularly complementary to cherry flavor, which will no doubt put you in mind of Cherries Jubilee. Created by Escoffier for Queen Victoria's Golden Jubilee in 1887, the dish originally consisted of sweet cherries cooked in a thick, sugary sauce to which kirsch or brandy was added before being set alight. In later versions, Escoffier ladled the cherries over vanilla ice cream, which is almost certainly how you'd be served it today. Cooks who value their eyebrows might settle for cherry clafoutis, a French dessert of cherries cooked in a sweet batter, sometimes flavored with almond, although vanilla is more common. The cherries are left unpitted, as the stones bring a balancing bitterness to the dish. Put 1 lb sweet cherries in a buttered 9-in round ovenproof dish. Split a vanilla pod and scrape the seeds out into a saucepan containing 1 cup milk. Throw in the pod while you're at it, then scald the milk and set aside to cool. Whisk 2 eggs, 1 egg yolk and ⅔ cup superfine sugar until well combined. Add 6 tbsp melted butter and whisk some more. Sift in ⅓ cup plain flour and whisk until smooth. Remove the vanilla pod and whisk the milk into the batter. Pour this over the cherries and bake for about 30 minutes at 400°F, until set and golden on top.

Cherry & Walnut The sort of partnership Alan Bennett might get wistful about. Poor walnut is in danger of being squeezed out by more newly fashionable nuts—macadamia, pecan, pine—and glacé cherries by overqualified, under-sweet fruits such as blueberry and cranberry. Make a cherry and walnut cake, just for old times' sake.

Watermelon

A bite of cold watermelon: like cherryade and cucumber juice with a hint of grass, chilled into a soft granita. Mark Twain called the watermelon "chief of this world's luxuries," and if you get your hands on a ripe one you may be inclined to agree. Look for a watermelon that's as heavy as a medicine ball and sings a low B-flat when you thump it. As a rule, the seeded varieties taste better than the seedless ones. Watermelon juice is the color of rubies and has a lovely sweet flavor with a vegetable undertone, a little like carrot juice. The compatibility of watermelon flesh with other ingredients is limited by its tendency to waterlog them, but herbal and sour flavors make fine pairings.

Watermelon & Chili A favorite combination in Mexican candy. Watermelon lollies come with a chili sherbet for dipping, watermelon gummy sweets with a sugar and chili coating, and watermelon hard candy with chili powder in the center.

Watermelon & Chocolate The Sicilian dish *gelo di melone* is a watermelon soup thickened with cornstarch, sweetened with sugar, spiced with cinnamon and flavored with either crushed pistachio, grated chocolate or candied peel, or a combination of all three.

Watermelon & Cilantro Sniff a slice of watermelon and it might remind you of a hot day at the beach—the watery, salty tang, with the candied wafts of a distant fairground. Combine it in a salsa with cilantro, a little red onion and some fresh green chili, and it's more dripping jungle after a rainstorm. (I need to get out more.)

Watermelon & Cinnamon The Baptiste mango of Haiti is said to have the fragrance of pine and lime, and a flavor reminiscent of sweet watermelon with a mild cinnamon aftertaste. Chances are readers outside of Port-au-Prince won't be finding one in their local greengrocer's. But you can make the watermelon sorbet from Jane Grigson's *Fruit Book*, in which the sugar syrup is enriched with Haiti-conjuring cinnamon. Try the cinnamon-flavored sugar syrup in Lime & Cinnamon on page 294, except with unflavored sugar. When it's cool, add to strained watermelon juice with a squeeze of lemon until it tastes right. Freeze.

Watermelon & Cucumber Watermelon is related to cucumber and shares many of its flavor characteristics. Not surprising, then, that watermelon goes well with cucumber's classic partners, especially feta cheese and mint. It's also worth considering it as a cucumber replacement in, say, a gazpacho, where its fruity take on cucumber's wateriness lends an essentially savory dish a lovely melting-Popsicle quality.

Watermelon & Goat Cheese *See Goat Cheese & Watermelon, page 60.*

Watermelon & Lime Serve watermelon pieces with lime zest, juice and a sprinkle of corrective sugar if needed. Or combine them in a long, refreshing drink, like they do in Mexico. The crisp flesh of watermelon holds its chill particularly well, and the noise it makes when cut into chunks for liquidizing is like the creak of boots on snow.

Watermelon & Melon Although part of the same family as cantaloupes, ogdens, honeydews and Galias, and combinable with all of them, watermelon is not of the same genus. It lacks the fruity-smelling esters that are characteristic of its cousins. That's not to say that cantaloupes and the other sweet, petite melons don't have some of the vegetal funk that is common to the squash family. Their seedy bellies have a composty low note that often comes as an unpleasant surprise, like hearing an otherwise delicate woman speak in a booming baritone.

Watermelon & Mint *See Mint & Watermelon, page 323.*

Watermelon & Oyster The cucumber notes in watermelon work well with oysters, which in turn bring out the juicy sweetness of the fruit. And far from being swamped by the wave of wateriness, oyster positively basks in it. At O Ya in Boston, chef Tim Cushman combines *kumamoto* oysters with watermelon pearls and a cucumber mignonette. Along similar lines, a 2006 menu at El Bulli in Spain included a selection of seaweeds arranged like a garland around a piece of watermelon covered with an intensely sea-flavored foam. You ate the seaweeds in order of increasing saltiness, and the watermelon provided sweet solace at the end.

Watermelon & Pork In 2006 The Fatty Crab, Zakary Pelaccio's Malaysian street-food restaurant in New York's West Village, won *Time Out New York*'s award for Most Deliciously Unhealthy Salad with its watermelon pickle and crispy pork appetizer. Pork belly is marinated in *kecap manis* (a kind of soy-dark, Satanic ketchup), rice vinegar, fish sauce and lime juice, roasted, then cubed and deep-fried. The cubed pork is mixed with cubes of sweet watermelon dressed with seasoned sugar, lime and vinegar, and little pieces of pickled white watermelon rind. Finally it's topped with a thatch of scallion, cilantro and basil, and a shake of white sesame seeds.

Watermelon & Rosemary *See Rosemary & Watermelon, page 312.*

Watermelon & Tomato Delicious paired in salads and salsas. Shake Shack in New York combines them in its frozen custard—a sort of posh soft-serve ice cream. It's one of their special flavors, as are coffee and doughnuts, cucumber and mint, and raspberry jalapeño.

Grape

Impossible to write about grapes without writing about wine; the sheer variety of wine grapes, and of the flavors coaxed from them by vinification, is simply astounding. Regrettably, the same can't be said for the sorts of table grape available in shops. Broadly speaking, the difference between table and wine grapes is that the former are bred to be thin-skinned and seedless, and thus easy to eat. The skin of a grape is where much of its flavor resides, and it also gives red and rosé wines their color. The table/wine divide is not absolute, however: some varieties fulfill both functions. Muscat wine, for example, has the unusual quality of tasting like the grapes it was made from (consider how often wine tastes of apples, grapefruit, gooseberries, apricots, melon and black currants, and how rarely of grapes). With their honeyed, floral quality and notes of rose and coriander seed, muscat grapes are also delicious plucked off the bunch. Other white table grapes have similar flavors, while red grapes can have hints of strawberry or black currant, at least when given time to develop on the vine. This, sadly, is rare, and all too often we're saddled with slightly unripe, seedless varieties whose flavor could at best be described as generically fruity. The unobtrusive, sweet-sour fruitiness of grape works well with other fruits, or as a refreshing counterpoint to meat. You might want to savor the good ones by themselves.

Grape & Almond Likeable, bland grape and almond must be sent to extremes to arrest your attention. Freezing the grapes concentrates the flavors, and the resulting texture is somewhere between a sorbet and a Gummi Bear. You could bring out the flavor of almonds by toasting them for almond brittle; serve the brittle and the frozen grapes with the cheese course in separate glass or silver bowls, the brittle broken into sticky fragments. Another take on this pair is almond-studded cantuccini cookies dipped into rustic, raisiny Vin Santo, the (usually) sweet Tuscan wine that's made with dried grapes. See also Garlic & Almond, page 111.

Grape & Anise *See Anise & Grape, page 181.*

Grape & Avocado The California Table Grape Commission helpfully suggests that grapes can in most cases be used in place of tomatoes. Remember that next time tomatoes are out of season and you need something to serve with your mozzarella and avocado. Not sure it works both ways, even in the case of the small tomato cultivar called Green Grape, which like its namesake has translucent skin and a zesty flavor.

Grape & Blue Cheese It goes without saying that port with Stilton is one of the great cheese and wine pairings. They share a richness, but it's the contrast between the sweet wine and the salty cheese that makes it so satisfying. Sauternes with Roquefort works on the same principle. Similarly, a black grape that floods your mouth with deep, sweet, spicy juice will make a good partner for Stilton, not just on the cheeseboard but in a salad of lamb's lettuce with a few nuts.

Grape & Chicken Grapes complement light meats such as chicken or, more famously, delicately flavored quail. Quail with grapes, stalwart of the high-gloss coffee-table cookbook, often calls for vine leaves and wine too; in the bad old days you might have been required to blanch, peel *and* seed your grapes before adding them. These days the practice of seeding green grapes is next to obsolete, thanks to the market dominance of the seedless Thompson variety, developed in mid-nineteenth-century California by the English viticulturist William Thompson. Thompsons have medium-thick skin, so you'll still need to peel them for your quail or chicken dish. As with tomatoes, blanching them beforehand will make the job that much easier if they're not quite ripe.

Grape & Hard Cheese See *Hard Cheese & Grape, page 69.*
Grape & Melon See *Melon & Grape, page 274.*
Grape & Peach See *Peach & Grape, page 278.*

Grape & Peanut Grape is the jelly of choice for a classic American peanut butter and jelly sandwich (see also Black Currant & Peanut, page 325). Grape jelly is made with Concord grapes, a flavor experienced in the UK only in the form of imported juice, or occasionally in "grape-flavored" sweets that, until you try the fruit itself, invite you to ponder the nature of subjectivity ("Grape? Are you *kidding*?"). I finally sampled a Concord grape from a stall in New York's Union Square Market. Crikey, I thought. It tasted simultaneously cheap and very expensive, like a penny candy heavily laced with jasmine. Back home, I get my Concord fix from Welch's grape juice, which makes a lovely perfumed sorbet. Mix 2 cups unsweetened juice with 1 cup sugar syrup, chill, then freeze in the usual way. Serve with a scoop each of peanut-butter and brown-bread ice cream for a peanut butter and jelly sandwich sundae.

Grape & Pineapple On a hot summer's afternoon, seedless grapes and chunks of super-ripe pineapple in a bowl of iced water will keep everyone chilled, at least until the ice-cream van arrives.

Grape & Pork Pellegrino Artusi suggests serving sausages with grapes, on the assumption that the sweet and sour fruit will partner pleasingly with the pork. Simply cook the sausages (Italian pork, if you can get them), and when they're nearly done, add whole grapes to the pan and cook until they collapse. See also Chili & Anise, page 204.

Grape & Rosemary Never mix the grape and the grain, unless you're making *schiacciata con l'uva*, a kind of bread that Tuscans cover with luscious crushed grapes at harvest time. The bread itself is not dissimilar to focaccia but is baked longer and is crisper for it. The grapes used are wine cultivars, either semi-dried or fresh from the vine, with the seeds left in. Rosemary or fennel might be sprinkled over the bread before it's baked. Eat warm from the oven while watching Éric Rohmer's *Autumn Tale*, which makes viniculture look a deal more romantic than it undoubtedly is.

Grape & Soft Cheese *See Soft Cheese & Grape, page 74.*

Grape & Strawberry The grape known in Italy as *fragola uva*, the strawberry grape, is an American import known in its country of origin as the Isabella. Its juice has a pronounced strawberry flavor and can be mixed with Prosecco, in the style of a Bellini, to make a Tiziano (I don't know much about art, but I know what I like to drink). Isabella and a wild strain of grape native to North America were crossed to create the more commonly grown Concord (see Grape & Peanut, page 248), and they share distinct candy notes. *Fragola uva* can be grown in the UK: there's a vine on the roof garden of the Coq d'Argent restaurant in the City of London.

Grape & Walnut Jeremy Round wrote about the excellent autumn pairing of mild, creamy, fresh walnuts and muscat-flavored grapes. He recommended the richly sweet Italia variety, a milder muscat grape that's most often used for wine-making.

Grape & White Fish *See White Fish & Grape, page 144.*

Rhubarb

Rhubarb is a vegetable native to Siberia. The leaves are poisonous; it's the pink petioles, or stalks, that we eat. Once its intense sourness has been countered with sufficient quantities of sugar, the flavor becomes fascinating—a combination of aromatic, candied strawberry notes with a cooking-apple fruitiness, plus a strong, thick note redolent of a greenhouse full of ripening tomatoes. The fruity notes are able to withstand cooking, and retain their freshness even after sugar has been added. Rhubarb is best paired with overtly sweet ingredients, such as maple syrup, honey or anise, then smothered with even more sweetness in the form of vanilla, almonds, cream or butter. Some take advantage of its gooseberry/cooking-apple sourness to pair rhubarb with fatty meat and oily fish.

Rhubarb & Almond The Scottish dessert cranachan originally consisted of cream, cream cheese, honey and toasted oatmeal—the sort of recipe that was invented before it was compulsory to worry about your heart. More recently it's likely to include whisky and raspberries. As the only raspberries you can buy in January are those tasteless grenades of pure sourness they sell in supermarkets, one Burns Night I devised this spin-off using rhubarb instead. Rhubarb has a similarly pleasing sharpness to raspberry, and I used Amaretto in place of the whisky—partly because almond is a natural partner for rhubarb, partly because its sweetness and lower alcohol content make it less of a hot shock than Scotch. Toast ½ cup oatmeal until golden brown and set aside to cool. Cut 6 stalks of rhubarb into 1-in pieces, place in a buttered ovenproof dish, sprinkle over ¾ cup sugar, cover with foil and bake at 350°F for about 30 minutes. Remove and cool. Whip ¾ cup cream until it holds its shape, then fold in four-fifths of the toasted oatmeal, the cooked rhubarb, 2 tbsp Amaretto and 2 tbsp honey. Divide between 4 dishes and top each with the remaining oatmeal and a Highland fling of toasted almond slices.

Rhubarb & Anise *See Anise & Rhubarb, page 182.*
Rhubarb & Black Pudding *See Black Pudding & Rhubarb, page 43.*

Rhubarb & Cucumber I used to love dipping raw rhubarb stalks in sugar and biting off sections that always threatened to be face-tighteningly sour. The secret was to overdo the sugar, which is surely what rhubarb's celery-like, natural scoop shape encourages you to do. The effect was like a marginally healthier version of the sour gummies you find in candy stores—the mouthwatering citric shock at once tempered and intensified by the intense, crunchy sweetness. Until I read Paula Wolfert, I'd never thought of dipping rhubarb in salt, which a friend of hers reported people doing in Turkey. Further investigations yielded an Iranian recipe for thinly sliced cucumber and rhubarb tossed and left to stand for a while in salt, then mixed with arugula, lemon juice and a little mint, which Wolfert suggests as an accompaniment to poached salmon.

Rhubarb & Ginger *See Ginger & Rhubarb, page 305.*

Rhubarb & Juniper At Alinea, in Chicago, Grant Achatz offers a dish of rhubarb cooked in seven ways—pairing it with goat milk jelly, for example, with green tea foam and with gin. Rather than mask it with sugar, Achatz likes to highlight the inherent sourness of rhuburb, which he plays off against bold flavors such as lavender and bay leaf. I pair it with juniper in a gin and rhubarb sorbet; like vodka, the gin lends the sorbet a satisfying smoothness.

Rhubarb & Lamb You can see how rhubarb might work in the sweetly spiced, fatty tagines of North Africa, or in this Iranian *khoresh*. Soften a large onion in a mixture of peanut oil and butter, then add 1 lb cubed lamb and brown. Stir in a pinch of saffron and 1 tsp pomegranate syrup, then pour in enough water to cover the meat. Simmer, covered, for 1½ hours.

Half an hour into the cooking, fry a finely chopped large bunch of both parsley and mint in butter and add to the *khoresh*. About 5–10 minutes before the stew is ready, add 3 stalks of rhubarb, cut into 1-in pieces, stir once, cover and leave until the rhubarb is cooked but still holding its shape. Fried herbs are often added to *khoresh*; although fresh herbs generally lose their flavor when cooked, in this instance they're present in sufficient quantities for some flavor to survive, and their bulk serves the secondary purpose of thickening the sauce. Serve with basmati rice.

Rhubarb & Mango Chef Richard Corrigan rhapsodizes over the combination of rhubarb and Alphonso mango, whose seasons, he notes, overlap. He poaches the rhubarb in a rosemary-flavored sugar syrup with a splash of grenadine, leaves it to cool, then serves it in a bowl with slices of Alphonso mango, nutmeg-freckled vanilla ice cream and stem ginger shortbread.

Rhubarb & Oily Fish Sour gooseberry might be the better-known companion for mackerel but rhubarb can be equally ruthless in tackling the fish's oiliness, and the combination is just as delicious. Soften 2–3 finely chopped shallots in vegetable oil, add 3 chopped rhubarb stalks and cook, covered, over a medium heat until the rhubarb is well broken down. Stir in 2–3 tsp red wine vinegar and 2 tsp brown sugar, and cook for a few minutes until most of the vinegar has evaporated. Taste, season, and serve with crackling-hot mackerel.

Rhubarb & Orange *See Orange & Rhubarb, page 290.*
Rhubarb & Pork *See Pork & Rhubarb, page 40.*
Rhubarb & Rosemary *See Rosemary & Rhubarb, page 312.*

Rhubarb & Saffron Rhubarb invites experimentation. The *New York Times* food writer Mark Bittman described his trials with a number of flavor pairings. He rejected tarragon, mint, cumin and coriander before settling on saffron, in which he detected an almost smoky "elusive depth of flavor" that works perfectly with the rhubarb. He recommends serving them with simply cooked fish. See also Rhubarb & Juniper, page 250.

Rhubarb & Strawberry A standard pairing in the United States, where it's hard to find a rhubarb pie or tart that doesn't include strawberry. Ripe rhubarb contains a juicy, floral strawberry flavor, and the two share pronounced fresh, green notes. A tip from Edmund and Ellen Dixon, writing in 1868, is to add a couple of tablespoons of strawberry jam to your rhubarb tart, which, they claim, will result in a pineapple flavor.

Rhubarb & Vanilla Rhubarb is easy to grow, and by the late nineteenth century most people in England had a crown or two in their garden. In *Plenty and Want*, John Burnett cites a week's worth of menus for a typical English family in 1901. Rhubarb makes it into no fewer than six meals. Such ubiquity and ease of cultivation was a blessing during the two world wars but by the second half of the twentieth century the nation had grown heartily sick of rhubarb, boiled to a pulp and burbling in lumpy custard, especially as so many new foods became available. Rhubarb might have disappeared altogether had it not been for the diplomatic skill of vanilla. Which is just as well, as rhubarb is one of vanilla's best friends: the contrast of luxurious, floral creaminess with rhubarb's mouthwatering sour-fruity flavor—somewhat akin to passion fruit but without the heavy muskiness—is completely heavenly. And so rhubarb clung on, skirting obsolescence on the back of rhubarb crumble with vanilla ice cream and rhubarb tart with proper custard (which is out of this world, by the way, if you cook the rhubarb and custard together in the tart). And those shocking pink and yellow boiled sweets. Now, of course, rhubarb is popular again, thanks to renewed interest in seasonal food and growing your own. And how does it repay vanilla for sticking by it through the hard times? By looking for newer, more exotic partners, of course. See Rhubarb & Saffron, page 251, and Rhubarb & Mango, page 251.

Tomato

Buying decent canned tomatoes is a good deal easier than buying tasty fresh ones. Look for Italian canned plum tomatoes—if it says San Marzano D.O.P. on the label, they'll be good. The original San Marzano cultivar was more or less wiped out in the 1970s by the cucumber mosaic virus, but the kind available today is an approved replacement, with a fleshy texture that's hard to beat for cooking. There are also some excellent organic brands from California. Canned tomatoes have been cooked a little (common for canned products) and will have developed a more sulfurous, jammy, spicy flavor than their raw counterpart. Raw tomatoes taste sour, sweet and salty, and have leafy, fruity and floral flavors. They're at their best when they've been left to ripen properly on the vine, developing their optimal bold sweetness and acidity. If you can't grow your own, store-bought cherry tomatoes are often the best bet for flavor. Or you could invest the money you might otherwise have wasted on premium-priced, but stingily flavored, varieties in a bottle of high quality balsamic vinegar, which will furnish the sweetness and acidity the tomatoes lack. (Incidentally, the flavor of strawberry is also intensified by balsamic vinegar; see Tomato & Strawberry, page 256, for their alleged interchangeability.) Tomato has an umami taste too. Heston Blumenthal, acting on a hunch that tomato seeds contain more flavor than the flesh, worked with scientists from Reading University to confirm that they are in fact richer in glutamic acids, which not only makes them particularly tasty but also boosts the flavor of other ingredients.

Tomato & Anchovy *See Anchovy & Tomato, page 161.*

Tomato & Anise Add a few tablespoons of tomato purée to tarragony Béarnaise sauce and it becomes the rosy-pink *sauce Choron*, named after

Alexandre Étienne Choron, the great French chef. When the Prussian siege of Paris in 1870 cut off supplies to his restaurant, he turned to the zoo in the Bois de Boulogne and devised a menu that included elephant consommé, bear shanks in a roasted pepper sauce and cat with rat. Assuming supply lines are open as you read this, you might serve *sauce Choron* with salmon, fishcakes or red meat. As to the purée, you want the good stuff. Giorgio Locatelli writes that a quality tomato purée should taste good straight out of the tube, with a flavor reminiscent of tomatoes dried naturally in the sun.

Tomato & Avocado *See Avocado & Tomato, page 197.*

Tomato & Bacon What a contrast! Salty bacon and sweet-sour tomato. Famously good with crisp, bittersweet lettuce in a BLT. Not so good in the first dinner I cooked my then husband-to-be, pasta all'amatriciana. A safe bet, you'd think. I fried a dozen chopped slices of pancetta in olive oil until crisp, removed them from the pan and set them aside. Having softened a sliced onion in the bacon fat, I emptied in a can of whole plum tomatoes, broke them up with my wooden spoon and added 1 tsp each of chili flakes and sugar before seasoning. While that slowly simmered, I slipped into the bedroom and crammed all the discarded shoes, recipe books, home-gym equipment and candy wrappers into the wardrobe before nonchalantly returning to the kitchen to give my sauce a stir. A little more sugar, I thought, to balance the sharpness of the tomato—a sharpness that lingered on my palate, worryingly, as I withdrew to the bathroom. So much so that once I'd applied my eyeliner I had to pop back to the kitchen and add an extra teaspoon. And so on, teaspoon after teaspoon, until the sauce had taken on a candied glaze and might as well have been called pasta alla diabetica. I added the bacon to the sauce and reheated it. Cooked the pasta and drained it. Mixed it with the sauce and served it with pecorino romano. He ate it, but I'm sure I saw him wince. Nervous hosts should note that stress and anxiety can dull the ability to taste. Research carried out on people with seasonal affective disorder has shown a marked difference in the sufferer's taste thresholds during the winter months.

Tomato & Basil *See Basil & Tomato, page 211.*
Tomato & Beef *See Beef & Tomato, page 51.*

Tomato & Bell Pepper Not long after the second summer of love came the first summer of Delia. The ungainly chili con carne pots of our youth were cleared away in favor of the roasted vegetable couscous salad with harissa-style dressing, oven-roasted ratatouille and chicken basque in *Delia Smith's Summer Collection*. And, of course, the Piedmont roasted peppers, which Smith had eaten in London at Bibendum, whose chef, Simon Hopkinson, had in turn eaten them at Franco Taruschio's Walnut Tree Inn in Wales. If 1993 had a flavor, it was halved red peppers stuffed with half a skinned tomato, an anchovy fillet, a little garlic and some olive oil, roasted and then garnished with a scattering of torn basil. See also Bell Pepper & Egg, page 202.

Tomato & Caper The volcanic soil on the Greek island of Santorini is said to give the capers grown there a unique intensity of flavor. A quantity of them are sun-dried until pale and very hard, then cooked in a tomato sauce to accompany the island's prized yellow split peas. See also Potato & Tomato, page 94, and Olive & Tomato, page 175.

Tomato & Chicken *See Chicken & Tomato, page 35.*
Tomato & Chili *See Chili & Tomato, page 207.*
Tomato & Chocolate *See Chocolate & Tomato, page 22.*
Tomato & Cilantro *See Cilantro & Tomato, page 194.*
Tomato & Cinnamon *See Cinnamon & Tomato, page 215.*

Tomato & Clove Eugenol is the predominant flavor compound in dark, spicy cloves. It's responsible for some of basil's flavor too, and occurs naturally in tomatoes. Clove is often added, along with other spices, as a flavor enhancer in cooked tomato dishes and sauces, such as ketchup. See also Basil & Clove, page 209.

Tomato & Cucumber Although tomato and cucumber may seem to be the lifeblood of gazpacho, the dish pre-dates their arrival from the New World. Gazpacho was originally made with bread, garlic, olive oil, vinegar and water—a less fancy version of *ajo blanco*. The first gazpacho I made—or had a hand in making—was in Portugal, on a day I'd cheated death twice (landslide, drowning). It took 12 of us to prepare it, variously assigned to the peeling, grating, chopping, blending, sieving and so on: a kitchen orchestra conducted by a brusquely efficient hostess. Even with a dozen of us, it took an hour. But it was the best gazpacho I've ever tasted, tangy and seemingly containing every fresh salad flavor under the Cascais sun.

Tomato & Egg *See Egg & Tomato, page 135.*

Tomato & Eggplant Both members of the nightshade family and, used incautiously, can be deadly for the cook. You have to manage tomato's acidity and the bitterness of eggplant, while not undercooking the eggplant or using so much oil that it oozes from its pores. No wonder so many *imam bayildi* can be as delicious as a plate of boiled tennis shoes. Handled sensitively, however, this Turkish dish of eggplants stuffed with garlic and tomatoes, among other things, can be sublime, as can *melanzane parmigiana* (baked eggplants in a cheese-enriched tomato sauce), *pasta alla norma* (the classic Sicilian pasta dish with eggplants, tomato, basil and ricotta) and caponata (a cold, sweet and sour eggplant and tomato salad, also from Sicily).

Tomato & Garlic *Pa amb tomàquet*: the only way to start the day in Catalonia. Not a bad way to finish it either. Served with most meals, it's a simple dish of bread, toasted, or a day old so it's hard enough not to tear when you rub on the garlic, then the tomato, pieces of whose flesh, along with the juice and the seeds, get caught in the air pockets and enrich the flavor. To finish, it's drizzled with olive oil and sprinkled with salt. It's usually eaten on its own, though sometimes accompanied by anchovies or *jamón jabugo*. In his *Teoria i pràctica del pa amb tomàquet*, which runs to more than 200 pages, Leopold Pomés tells of a famous musician who liked to take alternate bites of tomato bread and chocolate. See also Chocolate & Tomato, page 22.

Tomato & Ginger *See Ginger & Tomato, page 305.*
Tomato & Hard Cheese *See Hard Cheese & Tomato, page 71.*

Tomato & Horseradish Mix tomato ketchup with horseradish to make a simple sauce for raw oysters or cooked shrimp. A little dash of chili sauce will pep it up if the horseradish alone is too tame. See also Celery & Horseradish, page 96.

Tomato & Lamb Like coconut, tomato can add instant summer to a dish: it even lightens roast lamb. In Greece they cook lamb joints with tomatoes and add orzo, the rice-shaped pasta, toward the end. The pasta soaks up the lamb and tomato flavors. All it needs is a simple spinach salad on the side. Noting how long it will take to roast the meat to your taste, make slits in your lamb joint, poke slivers of garlic into them and season. Transfer the joint to a rack in a roasting pan and place on the middle shelf of an oven preheated to 375°F. Roughly chop the contents of two 14-oz cans of tomatoes, then stir in some oregano, a bay leaf, ½ tsp sugar and some seasoning. When the meat has an hour left to cook, working quickly, take the roasting pan out of the oven, set the lamb aside and drain the fat from the pan. Pour the tomatoes into the pan, rest the lamb on top of them and return to the oven. When the lamb's cooking time is up, remove it and wrap loosely in foil to rest. Stir 1 cup orzo and ½ cup boiling water into the tomatoes and return to the oven. Once the meat has rested for 15 minutes, carve it, by which time the orzo should be cooked. Double-check it is, then serve slices of lamb on the pasta. See also Lamb & Anise, page 52.

Tomato & Lemon A light squeeze of lemon will remove the metallic taste from a can of tomatoes. Worth knowing, especially if you've just retrieved a can from the depths of the cupboard, as the canned flavor develops with age.

Tomato & Lime *See Lime & Tomato, page 296.*
Tomato & Mushroom *See Mushroom & Tomato, page 81.*
Tomato & Nutmeg *See Nutmeg & Tomato, page 219.*
Tomato & Olive *See Olive & Tomato, page 175.*

Tomato & Onion Tomatoes open up perhaps the widest gap between ideal and actual flavor. Elizabeth David writes about the tomato and onion salad she ate every day one summer in Spain, noting that both ingredients were so delicious they could only have been spoiled by the addition of cucumbers, olives or lettuce. She goes on to lament that such a pleasure was becoming increasingly rare in England. It's almost nonexistent now. The only tomatoes that have ever really lived up to the ideal for me were the ones I ate at a beach restaurant in El Puerto de Santa María, between Jerez and Cádiz on the Costa de la Luz. It had garish plastic tablecloths covered by a layer of rough paper, held in place by fluorescent green table-clips. The tomatoes were fringed with green, the onions, just, well, onions. But both were exquisite: the tomato deep with chlorophyllic greenhouse flavor, the onions so sweet you could have blindfolded someone and convinced them they were slices of fruit.

Tomato & Peanut *See Peanut & Tomato, page 28.*
Tomato & Pork *See Pork & Tomato, page 41.*
Tomato & Potato *See Potato & Tomato, page 94.*
Tomato & Prosciutto *See Prosciutto & Tomato, page 171.*
Tomato & Sage *See Sage & Tomato, page 314.*

Tomato & Shellfish *Linguine alle vongole* is served either *rosso* or *bianco*—with or without tomatoes. I'm firmly in the *bianco* camp, but if you prefer *rosso* you may be more comfortable than I was with the concept of Clamato, a sort of liquefied *vongole rosso* (minus the pasta) in much the same sense that gazpacho is liquefied salad. A mix of tomato juice, clam broth and spices inspired, in fact, by Manhattan clam chowder, Clamato is the sort of novelty item you bring back to your vacation rental, and, as I did, stick in the fridge, taking the bottle out now and then to send the air bubble on a futile mission to the bottom and back, searching for signs of lurking clam meat. Steeling myself, largely by reasoning that I liked tomato-shellfish concoctions like shrimp Provençal, lobster américaine, the Catalan fish stew *zarzuela*, and the Italian-influenced San Franciscan seafood stew cioppino, I popped the lid. And fell in love. Clamato is pure liquid umami, nothing more challenging to the palate than a very savory, salty tomato juice, perfect for Bloody Marys. In Mexico they mix it with beer to make a *michelada*.

Tomato & Soft Cheese *See Soft Cheese & Tomato, page 74.*

Tomato & Strawberry These are interchangeable, according to some scientifically minded chefs, as the two share many flavor compounds. In the mid-1990s, Ron G. Buttery and his team discovered that tomatoes contain what's known as the strawberry furanone, also found in raspberry, pineapple, beef, roasted hazelnuts and popcorn. Later research discovered the highest concentrations were found in homegrown tomatoes in high summer. Try substituting one for the other in your favorite strawberry and tomato dishes. Strawberry, avocado and mozzarella salad is a no-brainer. How about strawberries in your burger or tomatoes on your fruit tarts? Wimbledon may never be the same again.

Tomato & Thyme *See Thyme & Tomato, page 320.*
Tomato & Vanilla *See Vanilla & Tomato, page 342.*
Tomato & Watermelon *See Watermelon & Tomato, page 247.*
Tomato & White Fish *See White Fish & Tomato, page 147.*

Strawberry

The most popular berry in the world, and one of the sweetest tasting. When ripe, fresh strawberries contain a combination of fruity, caramel, spice and green notes. Some cultivars have strong pineapple flavors. Wild strawberries have flavor notes in common with wild grapes, and can have a distinct, spicy clove character. The strawberry makes harmonious matches with warm, sweet spices and with other fruits, and its natural candy quality comes to life in the company of sugar or dairy (famously cream, but also yogurt, fresh cheeses and buttery pastry).

Strawberry & Almond Blend ½ lb ripe strawberries with 2 tbsp confectioners' sugar and 3 tbsp each of Amaretto and water for a sauce to go with ice cream or Madeira cake. Sweet and sharp, like the descant to your favorite hymn.

Strawberry & Anise *See Anise & Strawberry, page 182.*

Strawberry & Avocado Prue Leith acknowledges that the combination of strawberry dressing and avocado might seem bizarre but points out that mixing oil and strawberries makes a sort of vinaigrette, with the strawberries taking the place of the usual wine vinegar. Purée ½ lb strawberries with ⅓ cup olive oil and ⅓ cup sunflower oil mixed together, adding the oils in increments until the balance seems right. Season with pinches of salt, pepper and sugar to taste. This should be enough for 3 avocados, peeled, stoned, halved and neatly sliced (i.e., 6 servings). Top with toasted almond slivers.

Strawberry & Chocolate *See Chocolate & Strawberry, page 21.*

Strawberry & Cinnamon Strawberries have a hint of cotton candy about them. Cinnamon loves sugar and fruit. Warmed together, the pair gives off a seductively seedy fug of the fairground. For an irresistible sweet snack, dig out the sandwich toaster, butter 2 slices of white bread, spread 1 slice (on its unbuttered side) generously with strawberry jam, and the other with more butter and a good shake of ground cinnamon. Sandwich together, with the just-butter sides facing out, and press in the toaster till the bread is crisp, golden and, essentially having been fried rather than toasted, more like a doughnut than plain old jam on toast. Do wait until the lava-hot jam has cooled a little before biting or you won't be able to tell anyone how good this is. You could also try the combination in a sorbet, milkshake or a layered sponge cake filled with strawberry jam and fresh whipped cream, topped with a shake of cinnamon sugar.

Strawberry & Coconut The French chef Michel Bras halves and slices fresh strawberries, then sets them into a coconut cream to make an elegant terrine. Against the pure white background, the strawberry slices look as delicate as Japanese printed fans. It's a restrained take on a partnership that might otherwise scream cupcakes, iced cookies and experiments with jelly.

Strawberry & Cucumber At wedding breakfasts in provincial France, newlyweds were traditionally served a soup made with strawberry, borage (a cucumber-flavored herb), thinned sour cream and sugar. Borage is often planted next to strawberries in the garden, as they are believed to have a stimulating effect on each other's growth, which is perhaps where the pertinence to weddings comes from. Like all enduring partnerships, they improve each other's flavor too. Slice some strawberries as thin as a bride's nightie and layer them with borage leaves in delicate tea sandwiches enriched with cream cheese.

Strawberry & Grape See Grape & Strawberry, page 249.

Strawberry & Hazelnut The flavor of roasted hazelnuts is deep without being overpowering. It's great with strawberries, allowing the fruit's flavor to shine in a way that chocolate rarely can. It's even better when strawberry's sweetness is further emphasized by sugar—say, in a hazelnut meringue roulade filled with strawberries and whipped cream. For the hazelnut meringue, whisk 4 large egg whites to soft peaks. Continue to whisk while gradually adding 1 generous cup superfine sugar. Fold in 4 oz toasted ground hazelnuts, then spread the mixture onto a lined 10 x 14-in Swiss roll pan. Bake for 20 minutes at 375°F. Remove from the oven, cool, then turn out, peeling off the paper lining if still attached to the meringue. Cover with whipped cream and chopped strawberries. While the meringue is still pliable, roll it up with care. It's not supposed to look perfect. And if you really make a mess of it, you can always go nuts, break it into pieces and call it Eton mess—a classic English dessert of broken meringue, berries and heavy cream. Hazelnuts share strawberry's love of toffee-nosed flavors, so a swirl of caramel sauce is an excellent addition.

Strawberry & Melon See Melon & Strawberry, page 275.

Strawberry & Mint Freshly torn mint, a squeeze of lemon juice and a few pinches of sugar can really amplify strawberry's sweetness. Heston Blumenthal describes mint as a classic partner for strawberry, and one that he initially considered, along with coconut, black pepper, olive oil and wine, when developing his dish of macerated strawberries with black olive and leather purée and pistachio scrambled egg.

Strawberry & Orange See Orange & Strawberry, page 291.
Strawberry & Peach See Peach & Strawberry, page 278.
Strawberry & Pineapple See Pineapple & Strawberry, page 262.

Strawberry & Raspberry Like wearing black and navy blue: they *can* work together. Raspberry is the black: classic, classy, sophisticated. Works well with other classics such as chocolate and vanilla. Strawberry is the navy blue. Seems straightforward, safe even, but is really quite a tricky one to match. You need to be very sure, combining raspberry and strawberry, that you wouldn't be better off just plumping for one or the other. Note that when they are combined, it's almost always for decorative purposes—on tarts, cheesecakes or pavlovas—but that as soon as the berry is blended or juiced beyond recognition they tend to go it alone, in jams, ices and drinks, which has to tell you something about the flavor combination.

Strawberry & Rhubarb *See Rhubarb & Strawberry, page 251.*

Strawberry & Soft Cheese Strawberries have both buttery and creamy notes, which is why they go so well with cream, clotted or otherwise. There's also a distinctive cheesy note in there, which might explain the near-mystical aptness of a strawberry topping on cheesecake. So, as a variation on grapes, why not add some to your summer cheeseboard? Ideal with a young Brie or a Brillat-Savarin—a triple-cream cheese that's glorious with strawberries.

Strawberry & Tomato *See Tomato & Strawberry, page 256.*

Strawberry & Vanilla A fine partnership. They are paired in a millefeuille, the vanilla in the form of a pastry cream, or in the tidy glazed tarts in the windows of French pâtisseries. But—hear me out on this one—isn't strawberry better with unadulterated dairy? With whipped, pale-yellow cream in an Eton mess (see Strawberry & Hazelnut, page 258) or a pavlova, sitting pretty on a dense cheesecake, or, best of all, with clotted cream on a scone? For me, the combination of strawberry, with its strong candy flavor, and perfumed vanilla is too much—it almost tastes synthetic. On a slightly more positive note, of the three permutations available to the consumer of "classic" Neapolitan ice cream, strawberry and vanilla is the least bad. Isn't it time for a new take on Neapolitan? It was originally made from a combination of all sorts of flavors, including pistachio, raspberry and coffee. And then it became that dull greatest-hits compilation of chocolate, vanilla and strawberry: a threesome of convenience. Pistachio in place of the chocolate would be an improvement. Rhubarb, strawberry and vanilla, a delight.

Strawberry & White Chocolate In fancy chocolate shops, I sometimes see slabs of white chocolate spattered with clots of freeze-dried strawberry, like stucco after a shoot-out. White chocolate makes for a better combination with strawberry than milk or dark because, like strawberry, and gangland comparisons, it's a little cheesy.

Pineapple

A cocktail all on its own. When properly ripe, pineapples combine an array of juicy fruit flavors with a spicy, boozy, candy quality redolent of the fruit's classic partners—vanilla, rum, coconut and caramel. As they stop ripening once picked, commercially available pineapples are often seriously under-ripe, having failed to reach their full flavor and ideal balance of sourness and sweetness. To check the ripeness of a pineapple, smell its base—the fruitlets there are the oldest, and therefore the sweetest and most fragrant.

Pineapple & Anchovy *Nuoc nam* is a Vietnamese dipping sauce made of fish sauce (in Southeast Asia, usually made with fermented anchovies), lime juice, chili and sugar. A similar but even more pungent dipping sauce for beef or fried fish is made by mixing a thicker, unstrained fish sauce, *mam nem*, with pulverized pineapple, chili, sugar, garlic and lime juice.

Pineapple & Anise *See Anise & Pineapple, page 182.*

Pineapple & Apple According to the *Oxford English Dictionary*, the first instance of the word "pineapple" in English dates back to 1398, when it was used to describe the fruit of the pine tree, i.e., the pinecone. The first recorded use of "pineapple" as the common name for the tropical fruit *Ananas comosus* is attributed to the diarist and botanist John Evelyn in 1664; it's thought the term was coined by European explorers struck by the similarity of the fruit to what we now refer to as pinecones. "Apple" had long been used to describe not only fruits that resembled apples themselves but all manner of fruit and vegetables; the Anglo-Saxon poet, Aelfric, uses *eorþaeppla* (earth apples) as a synonym for *cucumeres* (cucumbers). As it happens, a fresh, sweet, green-apple character is one of the fruit flavors you can detect in pineapple; correspondingly there are apple cultivars that taste profoundly of pineapple, like the Allington Pippin, the Claygate Permain and the Pitmaston Pine Apple.

Pineapple & Avocado In 1557 a Brazilian priest wrote that pineapple was so "immensely blessed by God" that it "should be cut only by the holy hands of Goddess Venus." In the same century, the Spanish historian Fernández de Oviedo wrote that pineapple was similar to peach, quince and very fine melons, and tasted "so appetizing and sweet that in this case words fail me properly to praise the object itself." With typically English restraint, the adventurer Edward Terry, who traveled to India in 1616, described the flavor of pineapple as a "pleasing compound made of strawberries, claret-wine, rosewater and sugar." In the nineteenth century, the German philosopher and poet Heinrich Heine spoke of pineapple in the same breath as fresh caviar and Burgundian truffles. With reviews like that, no wonder it's such a diva. Pineapple won't allow gelatin to set, curdles cream and reduces other foods to a pulp if you're not careful. It's least problematically served *à la Garbo*, i.e., alone. Jane Grigson feels much the same way about avocado, with the exception of a handful of "magnificent partnerships"—including, incidentally, pineapple. Dice them for a salsa or slice them thinly and layer into a fried fish sandwich.

Pineapple & Bacon *See Bacon & Pineapple, page 167.*

Pineapple & Banana The cherimoya, or custard apple, looks like a cross between a Granny Smith and an armadillo, and tastes like a cross between pineapple, banana and strawberry.

Pineapple & Blue Cheese *See Blue Cheese & Pineapple, page 65.*

Pineapple & Chili Buying a ripe pineapple is a bit of a lottery. Leaves that come away on tugging as an indicator of ripeness is an old wives' tale. Sniffing its bottom isn't: the juice in this part of the fruit tends to be sweeter, and when ripe its perfume will penetrate the armored peel. If it smells as if it's been in the pub all afternoon, avoid it. I've found that medium-sized fruits tend to be sweeter and more flavorful than the larger varieties; and you're generally safer buying them in winter and spring. If you do end up with a sour pineapple, try dipping pieces in chili and salt, as they do with green mangoes in Southeast Asia and Mexico. Sweet pineapple is good mixed with fresh red chilies too, especially in a salsa to accompany fish, or chopped up very finely and served with a mango sorbet.

Pineapple & Chocolate The Québécois chef and Japanese-food enthusiast David Biron serves a chocolate club sandwich with "fries" made of pineapple. Strawberry and basil take the place of tomato and lettuce.

Pineapple & Cilantro Author Leanne Kitchen notes pineapple's affinity for Asian dishes and ingredients such as curries and cilantro. Pineapple and cilantro are a common pairing in Mexico, where they're both grown. See Pineapple & Pork on page 262, or try this striking black bean soup when you have some ham stock. Soak 1 cup black beans in water overnight, drain and rinse, then bring them to the boil in a quart of the stock. Cover and simmer for 45 minutes–1 hour, by which time the beans should be soft. Remove about a quarter of the contents, blend and add back to the soup to thicken it. Before serving, stir in some cilantro and shredded fresh pineapple (cut a peeled, cored pineapple lengthways into 8 spears and then chop along the grain for short shreds). You can add some shredded ham if you fancy that, and if you don't have good ham stock, you can always try softening an onion and some smoked bacon in the soup pot before adding the beans and a quart of water.

Pineapple & Cinnamon Pineapple, like strawberry, combines with sugar and cinnamon to create a flavor not unlike a natural version of spun sugar. A similarly lovely caramelized effect can be achieved in a cinnamon and pineapple tarte Tatin.

Pineapple & Coconut *See Coconut & Pineapple, page 281.*
Pineapple & Grape *See Grape & Pineapple, page 248.*

Pineapple & Grapefruit Take a sip of pineapple and grapefruit and you're more likely to be transported to a hammock in the tropics than to the moon. But in 1969 the crew of Apollo 11 washed down their first lunar meal—bacon squares, peaches, sugar-cookie cubes—with a pineapple and grapefruit drink. I assumed this would have been Lilt, the pineapple-and-grapefruit soda with "the totally tropical taste" made by Coca-Cola and drunk by the teenage me in vast quantities, until I found out it was neither sold in the States nor launched until 1975, and wouldn't have been appropriate anyway. Drinks taken on space missions had to be ones you could rehydrate, and, since carbon dioxide bubbles lack buoyancy in weightless conditions, even if an astronaut did manage to slurp down the frothy mass they wouldn't be able to burp, leaving them with a gassy back-up they'd have to wait till re-entry to relieve. Zero gravity is also believed to have an adverse effect on flavor perception, as the particles by which we detect aroma are less likely to make it to our olfactory bulbs. Chances are the big flavors of pineapple and grapefruit were very welcome. For more about the particular potency of grapefruit flavor, see Grapefruit & Shellfish, page 292.

Pineapple & Hard Cheese *See Hard Cheese & Pineapple, page 70.*
Pineapple & Mango *See Mango & Pineapple, page 285.*
Pineapple & Orange *See Orange & Pineapple, page 290.*

Pineapple & Pork Worth traveling to Mexico just for *tacos al pastor,* a street snack containing spiced pork topped with fresh pineapple and spit-roasted. As it cooks, the pineapple juice flows over the meat, simultaneously creating a dark-brown, caramelized crust and tenderizing the meat by means of a protease enzyme called bromelain, which breaks down its collagen. The pork is served sliced in warm corn tacos, garnished with sweet onion, pineapple, lime juice and lots of cilantro.

Pineapple & Prosciutto *See Prosciutto & Pineapple, page 170.*
Pineapple & Raspberry *See Raspberry & Pineapple, page 330.*

Pineapple & Sage Pineapple sage (*Salvia elegans*) is used to flavor drinks or fruit salads. Dolf de Rovira Sr. writes that it tastes like pineapple or piña colada. Imagine dancing to Wham's "Club Tropicana" in white stilettos with a bunch of pineapple sage in one hand. George would have loved it; Andrew would have thought you were weird.

Pineapple & Shellfish *See Shellfish & Pineapple, page 141.*

Pineapple & Strawberry On the basis of its acidity, strawberry is one of the fruits that the food writer Richard Olney singles out as particularly good with *lots* of apple in it. You might think it's not going to work, but have faith. In a large bowl, mix 1 cup roasted (or blanched), skinned and ground hazelnuts, 1 cup plain flour, 1 stick very soft butter, ½ cup sugar, 2 tsp baking powder, 1 egg, 1 tbsp hazelnut oil and 1 tsp cocoa powder. Peel, core and quarter 3 medium cooking apples, then halve each quarter and cut each piece into 4. Fold them into the mixture and transfer to a greased and lined 8-in springform cake pan. Bake at 350°F for about 45 minutes. You can serve this as a rustic dessert on its own, with caramel sauce or with ice cream. The hazelnut flavor will be more pronounced once the cake has cooled. Bear in mind that with so much apple it's very moist, and so is best eaten within two days.

Apple & Horseradish Grate a tart green apple and mix it with 1½ tbsp grated fresh horseradish. Add 3 tbsp sour cream, a pinch each of salt and cayenne and ¼ tsp each of lemon juice and brandy. Serve with cold meats, especially beef and duck.

Apple & Liver *See Liver & Apple, page 44.*
Apple & Mango *See Mango & Apple, page 283.*
Apple & Nutmeg *See Nutmeg & Apple, page 217.*

Apple & Orange Incomparable maybe, but not incompatible. Nigel Slater suggests cooking a couple of peeled dessert apples, segmented into eight, in 4 tbsp butter for 6–7 minutes. Transfer to a warm dessert dish. Add 2 tbsp brown sugar to the pan and stir for 2–3 minutes while the appley butter caramelizes. Pour in the zest and juice of one big orange followed by ⅔ cup heavy cream. As soon as the mixture begins to bubble and thicken, pour it over the apple and divide between two plates.

Apple & Peanut *See Peanut & Apple, page 26.*

Apple & Pear Sara Paston-Williams makes the point that this combination must be very old, as pear and apple were the first two fruits grown in Britain. She gives a recipe for pears in nightshirts, in which whole pears are poached in cider, set on a bed of spiced apple purée, covered in meringue and baked until crisp and golden. You could replace the apple with a purée of quince, which, like apple and pear, is a pome fruit. Quince is famous for its heavy, sensual perfume, which at autumn farmers' markets hangs as thickly in the air as Dior's Poison on the platforms of London Underground stations circa 1987. Quince's aroma is a combination of apple, pear, rose and honey, with a musky, tropical depth. Jane Grigson believed that it couldn't be beaten as a flavoring for apple or pear tarts. Grate or finely chop a quince and mix it into a pie or tarte Tatin. Be sure to include the skin, where most of the flavor compounds are concentrated.

Apple & Pineapple *See Pineapple & Apple, page 260.*

pineapple, along with raspberries and orange juice. There can be an overlap between the flavors, and some believe the best strawberries have a hint of pineapple about them. If you want to grow your own pineappley strawberries, look out for Cleveland and Burr's New Pine, described in Edward James Hooper's *Western Fruit Book* (1857).

Pineapple & Vanilla Dole, the pineapple growers, published the recipe for pineapple upside-down cake as part of a marketing campaign in the 1920s. It was an instant success, and rightly so—the pairing of toffeeish, slightly caramelized pineapple on creamy vanilla sponge cake is delicious. Other upside-down cakes have come in its wake (cranberry and peach, maple and pear, orange and cardamom), but none matches the magical fragrance of the original.

Pineapple & White Chocolate *See White Chocolate & Pineapple, page 344.*

Apple

Inseparable from the refreshing, fruity, green flavor of apples is their crucial balance of sour and sweet. Imagine a continuum running from sweeter varieties such as Fuji and Gala, through middling Golden Delicious, tart Braeburn, Pink Lady and Granny Smith to the Bramley cooking apple at the sour extreme. The cultivar also dictates the presence of idiosyncratic flavor characteristics, such as floral (rose) or fruit notes like damson, pear, pineapple, strawberry and rhubarb. Apples might contain spicy notes too, such as nutmeg and anise, dairy notes like butter, cream and cheese, nutty notes—especially near the core, as the seeds have an almond flavor—or hints of honey, wine and bubblegum. Apples are the most cultivated fruit in temperate climates, not only on account of their superb flavor but because they are so versatile. Apart, of course, from their consumption raw, they're excellent cooked in cakes, pies and puddings or made into jelly, piquant sauces, juice, cider and brandy.

Apple & Almond Butterfly an almond croissant, spread thickly with cream cheese on one side and apple purée on the other, and close its wings. More delicious than strudel.

Apple & Anise *See Anise & Apple, page 179.*
Apple & Bacon *See Bacon & Apple, page 165.*

Apple & Beet I adore beets for their warm earthiness, which reminds me of the smell of garden centers. If, however, this is not your bag, a sharp apple will temper its richness. Try one part diced Braeburn to two parts diced cooked beet. Oh, and keep the apple skin on for the texture contrast. In time, the beets stain the mix a uniform red, which is pleasingly confusing to the brain when, expecting the soft resistance of beet, you crunch into a sweet morsel of apple. Both flavors work well with horseradish, so maybe mix a little into some mayonnaise for a dressing. Or walnut—another great match. Apple and beet with a walnut-oil-based dressing. Delicious with grilled oily fish.

Apple & Black Pudding I first ate this simple combination in a brasserie called Aux Charpentiers in the St. Germain district of Paris. Once through the heavy door, you have to part a thick, faded velvet curtain to enter the dining room. This made me feel like Edmund stepping through the wardrobe into Narnia, except instead of arriving in a magical land of snow and ice and being seduced by Turkish Delight, I arrived in a 1930s Parisian workingmen's café and was seduced by black pudding. Make this on a chilly autumn night when you yearn for something satisfying but snappy. Choose an eating apple that holds its shape, preferably a tangy one to offset the rich spiciness of black pudding. Peel and core the apple and cut into 8 wedges. Cook in a tablespoon each of butter and peanut oil until softened and lightly browned. Heat the black pudding in the same pan when the apples are nearly done: it's already cooked, so it takes only a few minutes to heat through. Serve with nothing but a glass of cold, peachy Viognier.

Apple & Blackberry Like Simon and Garfunkel: perfectly respectable solo careers, can go octuple platinum together. Apple is Simon, by the way, the dominant partner. Blackberry does the high notes. Blackberries have a spicy character, although not a specific spice. Pick a bagful of blackberries. Mix these in a pan with 4 peeled, cored and chopped cooking apples and about ¼ cup sugar. Cook on a low heat for about 20 minutes, until the apples are soft and the blackberries have stained the mixture a shiny crimson. Check for sweetness and add more sugar if necessary. We used to call this stewed fruit, but supermarkets now insist on calling it compote, which is more elegant on the ear but misses the sweet rustic rowdiness. Just a little cream will sweeten it further.

Apple & Blueberry No doubt who does all the work in this partnership. A little apple will give a fruity boost to the diffident blueberry. Try them in a tart or crumble—the one that you make with the blueberries you bought to eat instead of chocolate-covered peanuts but which now sit in your fridge as baggy as your good intentions.

Apple & Butternut Squash *See Butternut Squash & Apple, page 226.*

Apple & Cabbage You might pair pineapple or orange with spicy raw cabbage in a coleslaw, but apples are the only fruit that go with sulfurous slow-cooked cabbage. That said, braised red cabbage with apple and onion (and the optional addition of bacon and/or chestnuts) is one of the best side dishes for pork you could think of. Just make sure to add something acidic to the braising liquid, such as lemon juice or red wine vinegar, to prevent the cabbage from turning blue.

Apple & Carrot *See Carrot & Apple, page 223.*
Apple & Celery *See Celery & Apple, page 95.*

Apple & Cinnamon A classic. The spice graces the sharpness of apple with a sweet, slightly woody warmth. Like the sitar on a Stones track. Similarly, shouldn't be overdone.

Apple & Clove In Robert Carrier's opinion, no apple pie is complete without a hint of clove. In Elizabeth David's, no apple pie is edible *with* one. Not wishing to sit on the fence, but it might simply come down to the apples and the time of year. An apple that packs plenty of character and acidity, like many cooking varieties, will make a lovely contrast to soft, buttery pastry, crumble or sponge, with no need to call for any backup on the flavor front. Some dessert apples also make the grade; Simon Hopkinson recommends Golden Delicious for apple tarts. Most dessert apples, however, don't have enough acidity for that gorgeous fresh fruitiness to survive the cooking process, and are better eaten raw. Nonetheless, if needs must, the flavor of cooked dessert apples can be improved by a squeeze of lemon juice and a sweet spice such as clove or cinnamon. If you grow your own apples, or keep fresh apples bought from a farmer's market in cold storage, you'll notice that over time both cooking and dessert apples become less acid and more sweet: even a puckersome Bramley can develop into an eater, albeit a sharp one, by March.

Apple & Coriander Seed *See Coriander Seed & Apple, page 337.*

Apple & Hard Cheese Glorious. A friend once told me rather sniffily that the plowman's lunch was a marketing confection. But I'm hardly going to turn my back on a centuries-old combination of cheese, apple and bread simply because somebody gave it a slightly uncool name. A wedge of tangy mature Cheddar or Stilton, like a road sign warning of a steep hill ahead, with a whole apple, half a loaf of crumbly brown bread and some home-pickled onions and chutney. The sharpness of the apple cuts through the salty creaminess of the cheese, making it just the thing with good beer or cider. You could also try a sharp hard cheese with apple pie, as is the custom in Yorkshire and parts of America. Some take the cheese on the side, others bake it under the pastry crust. In Wisconsin a law was once passed banning the consumption of apple pie without cheese; Eugene Field (1850–95) wrote a poem to the pairing, and in *Taxi Driver* Travis Bickle orders apple pie with melted cheese in a coffee shop, some say in reference to a similar request made by the psychopath Ed Gein in exchange for a full confession.

Apple & Hazelnut The combination of hazelnut and apple can make you wish for summer to be over and done with. Stuff a pork loin with them, make a hazelnut pastry for your apple pie or try my autumn cake. This has

Apple & Pork Apple-fed pork was only one of many of the benefits that came from giving your pigs the run of the orchard. They also fertilized the ground and, in scoffing themselves to a healthy weight, cleared it of pest-attracting fallen fruit. The Old Spot pig from Gloucestershire is, in fact, also known as the orchard pig, and legend has it that its black spots are bruises caused by apples. On the plate, these two are made for each other. With a plate of proper roast pork, by which I mean one with a curly roof of crackling, your apple pulls back the curtains and throws open the window of your palate. Make more applesauce than you need. Lots more. Nobody ever had too much of it—it's so versatile. Put 2 lb of peeled, cored, chopped Bramleys, ⅓ cup sugar and 1–2 tbsp water in a saucepan. Bring to the boil over a medium heat, cover and cook for about 5 minutes, keeping an eye on it and stirring now and then, until you have the texture you like. Taste for sweetness and adjust. See Apple & Almond, page 263, and Nutmeg & Apple, page 217.

Apple & Rose *See Rose & Apple, page 334.*
Apple & Sage *See Sage & Apple, page 312.*

Apple & Shellfish Grate a cold, sharp apple into soft, sweet crab mayonnaise to freshen it up.

Apple & Soft Cheese *See Soft Cheese & Apple, page 72.*
Apple & Vanilla *See Vanilla & Apple, page 339.*

Apple & Walnut These two have lots of robust, autumn flavor matches in common. Mix them with beets, orange and watercress and the result is like New England in October.

Apple & Washed-rind Cheese *See Washed-rind Cheese & Apple, page 61.*

Pear

Pear is less acidic than its relative the apple, and less hardy, but it isn't as dainty as you might think. The characteristic flavor of pear survives the canning process and, even more impressively, distillation into brandy. The Williams pear is the variety most commonly used in canning and brandy, as well as in many pear-flavored products. The Doyenne du Comice is a highly regarded pear for eating raw. Unveiled in 1849, it has been treasured for its quality ever since. It's a butter pear, as are Bosc and Anjou, similarly prized for their rich, aromatic flavor and (as the name suggests) buttery texture. The sweet, vinous character of pear makes it a great partner for sharp cheeses or ingredients with a tannic edge, such as walnuts and red wine. The juicy crispness of the Nashi (or Asian) pear suits similar flavor combinations but has a more pear/melon character.

Pear & Almond A natural couple: classy and restrained. Save them from an excess of tastefulness by making an unctuous pear and almond croissant pudding. Like bread and butter pudding in an Armani suit, and great if you have three almond croissants to use up—which is unlikely, admittedly, so buy three more than you need. Cut them into 1-in strips, arrange in a buttered shallow 1-quart dish with a couple of pears, peeled, cored and sliced. Scald 2 cups each of milk and heavy cream together in a pan. Whisk 1 egg, 4 egg whites and 1 tsp almond extract with 3 tbsp sugar, and add the milk mixture to the eggs. Pour over the pears and leave to stand for 10 minutes, before baking in the oven at 350°F for 45 minutes.

Pear & Anise Asian, or Nashi, pears taste like pear but have an apple's crisp bite. This makes them particularly enjoyable raw in salads—finely sliced fennel bulb is a perfect, subtly perfumed partner. Or play up their Asian origins by peeling, coring and poaching them whole in a syrup flavored with star anise. Excellent with plum ice cream, but vanilla will do. See also Beef & Pear, page 50.

Pear & Apple *See Apple & Pear, page 266.*

Pear & Banana Banana-flavored candy and pear drops share a fruit ester called isoamyl acetate, which is also released by a honeybee's sting apparatus, acting as an attack pheromone to alert other bees to the presence of something, or somebody, it would be a good idea to sting a lot. The moral of which is, choose your treats wisely when strolling past beehives.

Pear & Beef *See Beef & Pear, page 50.*
Pear & Blue Cheese *See Blue Cheese & Pear, page 65.*

Pear & Cardamom Can combine in an upside-down cake or a tarte Tatin, but they're particularly good when the pear flavor retains some of its crisp fruitiness as a contrast to the lovely floral character of cardamom, say in a sorbet. Just make a cardamom-flavored sugar syrup to add to your puréed pear. A little poire brandy will pep it up.

Pear & Chestnut *See Chestnut & Pear, page 229.*
Pear & Chicken *See Chicken & Pear, page 33.*
Pear & Chocolate *See Chocolate & Pear, page 21.*

Pear & Cinnamon Cook pears in an unflavored sugar syrup and it becomes clear why they are normally poached in cinnamon and wine. Without strong flavors to bolster them, cooked pears can all too easily taste like overboiled turnips. For the same reason, cinnamon is a welcome addition to hearty hot pear puddings such as tarte Tatin, upside-down cake and clafoutis. Stuck for more traditional apricots, dates or prunes, I once used dried pear in the Moroccan cinnamon-sugared sweet couscous known as *seffa*. Toast a handful of sliced almonds until golden, and set aside. Snip 4 dried pears into pea-sized pieces. Empty 1 cup couscous into a bowl and mix in 4 tbsp brown sugar. Pour over 1 cup boiling water, add 2 tbsp butter, cover and leave for 5 minutes. Rake the couscous with a fork to separate the grains. Add most of the pear and almonds, sprinkle over 2 tsp orange-flower water and ½ tsp ground cinnamon and stir well to combine. Serve piled into small bowls with a little pear and almond on the summits. In Morocco, *seffa* is often accompanied by a glass of milk or buttermilk. A small, chilled maple-syrup lassi in Moroccan tea glasses works well for me.

Pear & Goat Cheese *See Goat Cheese & Pear, page 59.*

Pear & Hard Cheese Peter Graham quotes an old French proverb: *Oncque Dieu ne fist tel mariage / Comme de poires et de fromage.* "Never did God make a marriage / Like that of pears and cheese." The Italians are more aggressive: *Al contadino non far sapere quant'è buono il cacio con le pere*—"Don't tell the peasant how good cheese is with pears." I tried this once, striding through an Apennine valley with my nose in the air. "Come here, my good man," I said. "Do you know how well a Williams pear goes with fontina?" Whereupon he chased me off his land with a stick. The principle applies wherever you are. Try a nutty Bosc pear with mature Cheddar, or a Comice with Brie. You can't go wrong, really.

Pear & Hazelnut *See Hazelnut & Pear, page 236.*

Pear & Pork Apples are such a popular pairing with pork that it's hard for pears to get a look in. Nonetheless they marry nicely with the sweet notes in the meat. One reason apples work so well is that their acidity cuts through pork's fattiness, so pears will be better roasted with leaner cuts such as fillet. Mind you, the London bakers Konditor & Cook make a pie topped with puréed pear and slices of fatty chorizo and it's magnificent.

Pear & Prosciutto I've seen them combined on a pizza, in panini and in any number of ritzed-up salads, but really they want nothing more than to be left alone together. They're perfect.

Pear & Walnut A mellow autumnal pairing, classically enlivened by piquant blue cheese in a salad. Start by nibbling one of your pears to see if they need peeling—if it's not too tough, keep the skin on for flavor and texture. Core, quarter and slice 2 pears and drop into 2 cups acidulated water (i.e., water with 1 tbsp lemon juice added). Wash, drain, and, if necessary, chop your leaves: watercress, radicchio or chicory would work well. Crumble or cut 1 cup blue cheese. Roughly chop a generous handful of walnuts, toasting them first for a fuller flavor, or leaving them raw, which is fine. Make a dressing with 3 tbsp walnut oil, 1 tbsp olive oil, 2 tbsp sherry vinegar and some seasoning. Dress the leaves and toss with the walnuts, half the blue cheese and the drained pears. Scatter over the rest of the cheese and serve. Pear and walnut will also make a rich cake or tart.

Pear & Washed-rind Cheese *See Washed-rind Cheese & Pear, page 62.*

CREAMY FRUITY

Banana

Melon

Apricot

Peach

Coconut

Mango

Banana

Fresh banana, when it still has a streak of green on the skin, has a noticeable astringency at the heart of its bland, slightly grassy flavor. As it ripens, this develops into the familiar fresh, fruity banana flavor with a distinct note of clove. By the time the peel is mottled with brown, the fruit's flavor is reminiscent of vanilla, honey and rum, as if anticipating its conversion into banana bread or its flambéing in a pan. Banana has a great affinity for roasted flavors such as coffee, nuts and chocolate, and for heavily spiced flavors like rum.

Banana & Almond A banana split without a sprinkling of toasted almond flakes? Like a Bee Gee in a buttoned-up shirt.

Banana & Anise Giorgio Locatelli pairs tiramisu with a banana and licorice ice cream. Sounds pretty good to me. Licorice has spicy and salty notes, both of which marry well with banana. Skeptics might try chasing a banana with a licorice toffee.

Banana & Bacon Wind thin streaky bacon slices around peeled bananas, secure with a toothpick and grill for 8–10 minutes, turning frequently. Can't you just taste this in your mind's mouth? The bacon's powerful saltiness held back by the sweetness of the banana? Not in the least bit sophisticated, but fun.

Banana & Cardamom *See Cardamom & Banana, page 306.*
Banana & Caviar *See Caviar & Banana, page 151.*

Banana & Cherry Kirsch, the clear spirit distilled from stone-in cherries, has a strong note of bitter almonds that betrays the flavor's family relations (almonds are the seed of a stone fruit, or "drupe," that's closely related to plums, peaches, cherries and apricots). Splashed on fruit, kirsch triangulates the sweet-sourness with a delicious bitterness. There's an old French recipe called *bananes baronnet* that sounds far more tra-la-la than it is to make. Slice a banana per serving, then squeeze over a little lemon juice, add a sprinkling of sugar, 2 tsp kirsch and finally 1 tbsp (unwhipped) heavy cream. Mix thoroughly and serve.

Banana & Chicken Banana meets chicken in chicken Maryland, which usually consists of fried breaded chicken, fried banana and cornbread (or fritters), served with a creamy gravy. The last dish you'd associate with the Riviera set, and yet in *Tender Is the Night* we find Nicole Diver leafing through a recipe book in search of it. And it was served in the first-class restaurant of the *Titanic* on the night it sank. Jamie Oliver suggests a baked take on the combination, in which banana-stuffed chicken breasts are wrapped with bacon and baked on fresh sweetcorn kernels and cannellini beans in white wine, heavy cream and butter.

Banana & Chocolate Slash the skin of an unpeeled banana almost end to end without damaging the fruit. Make slits in the flesh about an inch apart and then push a chunk of chocolate into each. Pinch the skin back together, wrap in foil and put on the barbecue embers for 5 minutes. Open and eat the warm chocolatey, gooey banana with a spoon.

Banana & Cinnamon The presence of the clove-flavored compound eugenol increases in banana as it ripens. Think how spicy mottled bananas are. Consequently they're particularly harmonious with other spices—for example, cinnamon or vanilla in banana bread, that great redeemer of the ruined banana. If you have only one banana, peel it and halve it lengthways, then sprinkle with 1 tbsp flour and ¼ tsp ground cinnamon. Fry in 2 tsp each of butter and peanut oil and serve. It looks better on the plate with a scoop of ice cream, but it's sweet and satisfying enough by itself.

Banana & Coconut *See Coconut & Banana, page 279.*

Banana & Coffee Before Brangelina and Bennifer there was banoffee, the most cloying, and enduring, supercouple of all: banana and toffee. Banoffee pie consists of a pastry or graham cracker base topped with caramel, banana slices and coffee-flavored cream—so you could say the etymology of "offee" splits two ways. I'd choose a digestive biscuit base over pastry, as its salty, malty flavor sets off the fresh green notes in just-ripe banana better. Coffee beans contain highly volatile aldehydes and esters that lend coffee its fragrant, floral notes and hints of sweet spices such as clove. Banana also has floral and clove components, so the two flavors combine rather pleasingly. Note how the coffee and bananas aren't just lending flavor; they contribute bitter and sour tastes that strain against the pie's desire to become a sweet, cloying headache.

Banana & Egg *See Egg & Banana, page 131.*
Banana & Hard Cheese *See Hard Cheese & Banana, page 67.*
Banana & Hazelnut *See Hazelnut & Banana, page 234.*
Banana & Parsnip *See Parsnip & Banana, page 219.*

Banana & Peanut Fried peanut butter and banana on white was Elvis Presley's favorite sandwich. Some say there was bacon in there too, but my contact at the Elvis Presley Estate Archive confirms that this was not the case. The confusion arose, they told me, from another of Elvis's sandwiches that has passed into myth. Seized with nostalgic cravings one night in Graceland, he bundled some friends into his private jet and flew a thousand miles to Denver, Colorado, there to feast on a local specialty—an entire loaf hollowed out and filled to the brim with peanut butter, grape jelly and fried bacon.

Banana & Pear *See Pear & Banana, page 268.*
Banana & Pineapple *See Pineapple & Banana, page 261.*

Banana & Vanilla We stopped in Benson, an old mining town 25 miles north of Tombstone, Arizona. Excusing himself with an actual tug of the Stetson, my cowboy got out and came back with a can of oil that stops horses' hooves from cracking, a birthday cake that looked as if it were made out of shaving foam and a bouquet from a store where the flowers were cooled like corpses in refrigerated cabinets. Sweltering in the truck, I felt like an abandoned pet. Noting how hot and tired I was, he opened the truck door and, with all the unironic courtesy of the Old West, escorted me to Dairy Queen, where he bought me a cup of banana and vanilla ice cream. Both flavors have floral, spicy notes to them, and the almost citric sourness of banana is beautifully blunted by creamy, sweet vanilla. I relished each tiny, cool spoonful, kicking my heels on the side of the dusty road as an endless freight train clanked past along the railway, and my cowboy finished his unlikely chores.

Banana & Walnut *See Walnut & Banana, page 230.*

Melon

This covers cantaloupe, Galia, Charentais and honeydew, while watermelon, a relative of melon cultivars, has its own chapter (see page 245). When allowed to ripen, all varieties of melon become sweet and share a basic melon flavor note; other than that there's some variation, notably in depth of flavor and in the presence of fruity (pear and banana), floral and sulfurous characters. Cantaloupes develop a particularly floral and persistent flavor. Galias are known for their sweetness, and have a cucumber streak, sometimes with a hint of glue. Melons are easily mixed with other fruit, but their wateriness is one reason for their limited number of classic flavor affinities.

Melon & Almond *See Almond & Melon, page 240.*

Melon & Anise Stuck for things to do with the stubborn melon, chefs used to cut it in half, then get it half-cut on some sort of booze: port, cassis and champagne were the popular choices. At Petersham Nurseries, Skye Gyngell serves slices of Charentais melon with crushed toasted fennel seeds and a splash of sambuca.

Melon & Cucumber From the same family. Melons, especially the Galia variety, share cucumber's green, grassy flavor notes. Harold McGee writes that Galias also contain sulfur compounds that give them a deeper, savory dimension. Combine them in a salsa, a chilled soup or a mint-dressed salad.

Melon & Ginger A marriage of convenience. Ginger was traditionally paired with melon because its warmth and stomach-soothing properties were thought to counter melon's chilliness and resistance to easy digestion. With a tasting panel, I tried ground, preserved, fresh, crystallized and liqueur ginger with five different varieties of melon, and all combinations proved considerably worse than melon with no ginger at all. Even honeydew tasted worse with ginger, which is saying something, as this melon variety has always reminded me of those brands of attractively packaged European chewing gum that lose their flavor within one grind of the teeth. The marriage is therefore annulled. In this household at least.

Melon & Grape Melon will fare just fine in a mixed fruit salad but it enjoys a particularly happy relationship with the grape. Simply served as a twosome, grape's rather plain, crisp flavor contrasts well with melon's peculiar and slightly tropical alkaline fruitiness. The textural contrast—melon's soft granularity against the jelly pop of grape—only adds to the pleasure. Melon notes, by the way, often turn up in Chardonnay wines, including *blanc de blancs* champagne, made from 100 percent Chardonnay grapes. If you're planning on juicing a melon to make a cocktail, make sure you taste it before adding it to the fizz: it can sometimes become a little too cucumbery when liquefied.

Melon & Mint Forget ginger. Mint is melon's real best friend. In Syria they combine the two in a drink with milk, yogurt and a little sugar. Or you could try them in a soup, using a few different types of melon. Cut them into neat pieces (or balls) and float them in the slightly sweetened, liquefied juice of the remaining flesh, scattered with torn mint leaves.

Melon & Orange The Melon de Cavaillon consortium suggests flambéing its Charentais melons in orange liqueur and serving them on a nest of dark chocolate tagliatelle. You might alternatively serve them with a simple drizzle of Cointreau, but then again the tutti-frutti, honeysuckle flavor of Cavaillon melons is so heavenly you may prefer to eat them as they are. So prized have they been throughout the ages that, when asked by the mayor of Cavaillon if he would donate some of his books to the town library, Alexandre Dumas pledged his entire oeuvre in return for an annuity of 12 Cavaillon melons. To experience them at their very best, you really have to go to Provence in midsummer—although if you (or a friend) do make the trip, bear in mind that a bottle of the Cavaillon melon syrup made by the Domaine Eyguebelle will give a year-round glimpse of the deep beauty of this flavor.

Melon & Prosciutto *See Prosciutto & Melon, page 170.*

Melon & Rose For a fruity take on the classic Indian dessert *gulab jamun*: find the muskiest, most floral cantaloupe melon you can, cut it in half and seed it. With a melon baller, scoop out as many spheres as possible, then drape them in a chilled, rose-flavored syrup. Let the flavors infuse before serving. See also Rose & Cardamom, page 334.

Melon & Strawberry According to one of my trusted reference books, melon fritters with strawberry sauce is a popular combination in France. When I tried making fritters with a Charentais melon, my kitchen smelled of the most exquisite jelly doughnuts imaginable, but they disappointed in the eating. The melon flavor survived the hot oil treatment, just about, but the texture was quite unpleasant. Better, perhaps, to stick to simple, unheated pairings, like cantaloupe with wild strawberries, as suggested by Elizabeth David. Anna del Conte, by contrast, is adamant that melon and strawberry are not a match, but notes how good they are individually sprinkled with superfine sugar and balsamic vinegar. Some say kiwi fruit combine the flavors of melon and strawberry: see if you can detect them yourself.

Melon & Watermelon See *Watermelon & Melon, page 246.*

Apricot

Apricots are sour-sweet, with a creamy, floral character and a mixture of fresh and tropical fruit notes. Dried, they lose some of their perfume and take on a sweeter, cheesier character. When dried with sulfur dioxide, they are particularly tangy; without it, they tend to the toffee-fruity. Apricots have a great affinity for dairy flavors as well as other fruity florals.

Apricot & Almond See *Almond & Apricot, page 237.*
Apricot & Cardamom See *Cardamom & Apricot, page 306.*

Apricot & Chocolate Even when sweetened, the sharpness of apricots persists in bringing a fruity tang to bitter dark chocolate. That they work together is undisputed. How they should be paired is more contentious. Take the notorious chocolate-and-apricot-flavored Sachertorte. The Hotel Sacher and the Demel bakery in Vienna had a full-scale bun-fight over who owned the original recipe. The main point of contention appears to have been the correct deployment of apricot jam. The bakery spreads it only under the chocolate icing that tops the chocolate sponge gâteau. The hotel, which won the right to name its cake the original Sachertorte, also uses apricot jam to sandwich the layers together.

Apricot & Cinnamon See *Cinnamon & Apricot, page 212.*
Apricot & Cumin See *Cumin & Apricot, page 84.*

Apricot & Ginger Sweet apricot and hot ginger pair successfully in spicy chutneys, sauces for pork or a stuffing for duck. It's also worth considering their sweet applications, say in a soufflé, cake or cookie. The Parisian tea salon Ladurée offers apricot and ginger macaroons, as well as strawberry and poppy, orange and saffron, and jasmine and mango. But macaroons remind me of Brad Pitt. They're undeniably good looking and have an appealing rough edge, yet I can't muster up any desire for them.

Apricot & Goat Cheese *See Goat Cheese & Apricot, page 57.*

Apricot & Hard Cheese Tomás Graves tells how Mahón, an unpasteurized cow's milk cheese imported to Majorca from Menorca, was traditionally eaten with the different fruits that came into season as the cheese ripened, making ideal complements to each stage of its maturity. *Nísperos* or "loquats" (an orange-colored fruit comparable in flavor to apple) came first, followed by apricots, grapes and figs. Younger, milder cheeses suited the acidity of earlier fruit, while the heavier sugar content of fruits consumed later in the year, like dried apricots, prunes and raisins, paired better with more mature, fuller-flavored cheese.

Apricot & Lamb *See Lamb & Apricot, page 53.*
Apricot & Mango *See Mango & Apricot, page 283.*
Apricot & Mushroom *See Mushroom & Apricot, page 77.*

Apricot & Orange Omelette Rothschild has been on the menu at London's Le Gavroche for 35 years. It's an apricot and Cointreau soufflé, essentially: sweet, velvety and not half as rich as its name suggests.

Apricot & Peach Get along just fine, but arguably too alike to make an arresting combination. Both are creamy, floral and fruity, with a low note of almond. Peaches, however, are creamier and more complexly fruity, while apricot has the stronger floral, lavender notes.

Apricot & Pork This combination got my vote on the "sausage trail" at the yearly food festival in Ludlow, Shropshire. For a small fee, you can sample the efforts of half a dozen local butchers, marking down your score on a sheet soon spotted with mustard and grease. The town smells like scout camp, its streets filled with impromptu inspectors wearing abstracted expressions as they chew over the merits of each banger. In the end, the salty, fatty porkiness of a sausage cut through with the sharp sweetness of apricot edged it over a crumbly number made with pork from that "It-pig," the Gloucester Old Spot, and given gentle hints of eucalyptus from the sage.

Apricot & Raspberry *See Raspberry & Apricot, page 329.*

Apricot & Rose Elizabeth David recommends baking rather than stewing to get the most flavor out of dried apricots. Soak if necessary and then bake in a covered dish, with just enough water to cover, for an hour at 350°F. David says this gives them a roasted, smoky flavor. Try this with the addition of a few drops of musky rosewater, which makes for a frankly erotic combination. Eat on the banks of the Nile.

Apricot & Rosemary *See Rosemary & Apricot, page 309.*

Apricot & Vanilla Most apricots get sweeter *and* sourer when dried, which is what makes them such lovely snacks—the *bien-pensant* alternative to sour gummies. But the sharpness and fizz is down to the sulfur dioxide that's used to preserve the bright orange color. It halts the natural oxidation process in the fruit, which left to its own devices would give you a far sweeter—and browner—result, as in the famous Hunza apricot. These have to be soaked and cooked, but it's worth it for their intensely honeyed, toffee flavor. They're exquisite with vanilla, so serve with good homemade custard or posh ice cream. Can't get Hunzas? Not in the mood to make custard? Some say sharon fruit, a variety of persimmon, has a sweet apricot-vanilla flavor.

Peach

The peach's affinity for dairy products extends beyond cream to the bolder flavor of blue cheese. Peach also works well with the rich oiliness of nuts—coconut and almond are particularly harmonious. The flavor of peach is complexly fruity, combining a range of tropical and drupe fruits, including raspberry. Nectarine is not particularly different in flavor terms, but a creative flavorist can add a fuzzy character to synthesized peach to distinguish it from its smooth-skinned relative.

Peach & Almond *See Almond & Peach, page 240.*
Peach & Apricot *See Apricot & Peach, page 276.*

Peach & Blackberry When cooked, late-summer peaches are rich enough to suit the heavier, spicier flavors of autumn, like blackberry. Combine them in a cobbler—a baked fruit pudding with a scone-like topping arranged like paving stones over the fruit. A crumble with aspirations, in other words. Peel, stone and slice 4 peaches and mix them with a couple of handfuls of blackberries in a baking dish. Sprinkle over 3 tbsp sugar and dot with a little butter. Place 1½ cups plain flour, 4 tsp sugar, 1 tsp baking powder and 5 tbsp butter in a food processor and pulse until you have a mixture that resembles breadcrumbs. Add 4 tbsp milk and a lightly beaten egg through the feed tube. Pulse until the mixture becomes a dough. Remove it and knead briefly. Roll it out to a thickness of ½ in, cut out as many 1½–2-in discs as you can and lay them over the fruit. Bake at 400°F for about 30 minutes. The blackberries will bring a spiciness of their own, but including 1 tsp ground allspice (or mixed spice) in the dough will take it even deeper.

Peach & Blueberry More American than apple pie. Pair them in a cobbler (see Peach & Blackberry, above) or a tart. White-fleshed peaches tend to be very sweet and less acidic than yellow-fleshed varieties. They're also often more perfumed, with delicate hints of jasmine and tea, which is a sublime match for the floral but astringent blueberry.

Peach & Blue Cheese *See Blue Cheese & Peach, page 65.*

Peach & Cherry All that's left, other than a crusty roll, by the time the fun starts in *Le déjeuner sur l'herbe*. What exactly Manet was depicting has been variously interpreted, but my guess is that the naked woman on the left had been deep in conversation with the two men until the hungry artist interrupted to ask if there was any cheese or ham, or perhaps a little celeriac remoulade, left. This would go some way to explaining the slightly aggressive blankness of her expression—a blankness at once designed to disguise her embarrassment at having polished off the nice bits and to face down Manet for clearly having turned up late. The guy in the middle is keeping out of it. The woman in the background is looking for that bottle of rosé they put in the river to chill.

Peach & Clove *See Clove & Peach, page 216.*

Peach & Grape Occasionally you'll come across a peach or nectarine with distinctly vinous flavors. Conversely, peach is a frequent flavor note in wines, especially Chardonnays, Rieslings, Semillons and the Italian sparkling wine Prosecco. The pair are most famously combined in the Bellini, signature cocktail at Harry's Bar in Venice, but may also have inspired the French custom of poaching peaches in Sauternes. By the way, elsewhere in Venice, in a bar in Campo Santa Margherita, you might try a *spritz al bitter*, which classically combines Prosecco, mineral water and Campari. After your first couple of sips, when you're wondering if you can take any more bitterness, a dish of green olives arrives and proves that you can.

Peach & Mango *See Mango & Peach, page 284.*

Peach & Orange A little orange will make peach flavor more vivid, as in a Fuzzy Navel, which pairs orange juice with peach schnapps.

Peach & Prosciutto *See Prosciutto & Peach, page 170.*
Peach & Raspberry *See Raspberry & Peach, page 330.*

Peach & Strawberry Which is cream's best friend, peach or strawberry? Both contain dairy flavors that give them a natural affinity for cream. Either of them stirred into whipped cream will make a simple and delicious dessert. Peach and strawberry make a lovely layered vacherin—see Coffee & Black Currant, page 23, using fresh fruit in place of the ices.

Peach & Vanilla As a society lady at the turn of the twentieth century, you were nobody until you'd had a peach-based dessert named after you. For the actress Sarah Bernhardt, Escoffier created *pêches aiglon*, peaches poached in vanilla syrup, served on vanilla ice cream and topped with crystallized violets and spun sugar. The actress and singer Blanche d'Antigny, who was the model for Zola's *Nana*, got *coupe d'Antigny*, half a peach poached in vanilla syrup, on alpine strawberry ice cream, topped with rich unpasteurized cream. Princess Alexandra, wife of Edward VII, was immortalized in peeled peaches with kirsch and maraschino cherries—not unlike the Empress Eugénie, whose eponymous dessert was further garnished with wild strawberries and served with a champagne sabayon. Quite who inspired the *coupe Vénus*—a *Carry-On* confection of peach halves suggestively topped with cherries—history fails to record. See also Raspberry & Peach, page 330.

Coconut

Like other nuts, coconut has a mild, milky, fruity flavor when fresh but becomes more boldly flavored when toasted or baked, taking on a creamy, nutty sweetness. In common with sweet almond, it has a flavor notably compatible with others, both sweet and savory. Coconut milk is usually made with grated coconut flesh and water, yet the canned brands can be surprisingly variable, especially in sugar content. Chaokoh, a Thai brand, has a good reputation for flavor, a low sugar content, and is particularly suitable for savory dishes. Coconut cream, separate from the milk, is also widely available, as are coconut milk powder, desiccated or flaked coconut (both sweetened and unsweetened), coconut flour, extract, essence and water, and the coconut-flavored rum, Malibu.

Coconut & Almond See *Almond & Coconut, page 239*.

Coconut & Anchovy In Southeast Asian cooking, adding fish sauce to coconut milk is like giving your stew or curry a central nervous system. The two are also paired in one of Malaysia's favorite dishes, *nasi lemak*—creamy coconut rice served with *ikan bilis* (small dried anchovies), cucumber, peanuts, boiled egg and a spicy sauce. Like its paler-flavored cousin, kedgeree, it's most often eaten at breakfast time.

Coconut & Anise See *Anise & Coconut, page 180*.

Coconut & Banana *Kluay buat chii*, bananas stewed in coconut milk with sugar and salt, is a popular sweet treat in Thailand. The name means "bananas ordaining as nuns"—Thai nuns wear white robes and have their heads and eyebrows shaved during ordination. Dissolve ½ cup sugar in 1¾ cups coconut milk, add a pinch of salt and 4 bananas cut into bite-sized pieces, and simmer until warmed through.

Coconut & Basil See *Basil & Coconut, page 209*.
Coconut & Beef See *Beef & Coconut, page 47*.

Coconut & Beet Give the beet a much-needed vacation from its traditional northern European flavor affinities by pairing it with a hot-country ingredient like coconut. Like many root vegetables, beets can be combined with coconut milk in a soup, but an even more tantalizing idea comes from chef-restaurateur Cyrus Todiwala, who stuffs samosas with diced beets and grated fresh coconut, seasoned with mustard seeds, curry leaf, cumin and chili, with a little potato to bind.

Coconut & Cardamom Not to be trusted. In Indian rice puddings and *barfi*, a fudge-like sweet, the complex flavor of cardamom, with its hints of citrus and eucalyptus, throws a veil of sophistication over the childish sweetness of coconut. You come around, uncrossing your eyes, realizing quite how much fat, sugar and white carbohydrate this delicious combination has hoodwinked you into ingesting.

Coconut & Carrot *See Carrot & Coconut, page 224.*

Coconut & Cherry Back in the 1980s, a chocolate bar called Cabana came and went. Coconut studded with glacé cherries, topped with caramel and covered in milk chocolate. So sweet it made your teeth throb in unison, as if you'd strayed too close to the speakers at a disco. Coconut, like cherry and almond, is a drupe (or stone) fruit, and among its dominant creamy, nutty notes you can detect some fruitiness. Cherry, also fruity and nutty, is a particularly good match, and both work very well with chocolate. Judge for yourself with my cover version. I make them in a silicone mold divided into the ideal bar shapes, 3 in x 1 in x 1 in. Thoroughly beat an egg with ¼ cup sugar, then stir in 1¼ cups shredded coconut and about 15 quartered glacé cherries. Press this mixture into 10 bar shapes and bake for 15 minutes at 350°F. Leave to cool while you make a toffee by melting 2 rounded tbsp sugar, 2 rounded tbsp butter and 2 tsp corn syrup in ⅓ cup condensed milk over a low heat. Turn up the heat, bring to the boil and cook for 4–5 minutes, by which time the mixture will have turned a caramel color. Allow to cool for a minute, then spread it over the coconut bars while they're still in their molds. Once cool, cover with chocolate—milk if you want to be authentic.

Coconut & Chicken When the Canadian composer Colin McPhee went to Bali in the 1940s to write about the island's music, he kept records of daily life there, taking particular interest in the meals prepared by a woman called Madé, who cooked for him regularly. After barbecuing chicken over coals, she would shred it, then pound it with grated coconut to allow the nut's oils to mingle with the meat. To this she added another pounded mixture of onions, ginger, red pepper, spice and fish paste cooked in coconut oil, before pouring over thick coconut milk and then lime juice. Madé insisted that the meal, served with rice, should be eaten with the hands, because they contributed to the flavor of the food. Cutlery, with its chill and metallic taint, would only get in the way.

Coconut & Chili *See Chili & Coconut, page 205.*
Coconut & Chocolate *See Chocolate & Coconut, page 19.*
Coconut & Cilantro *See Cilantro & Coconut, page 191.*

Coconut & Cinnamon A popular combination in Cuba, where they're combined in a rice pudding called *arroz con coco* and in *coco quemado*, which is similar to *flan*, the baked custard pudding so popular in Spain. To make *coco quemado*, put a cinnamon stick, 2 cloves, ½ cup water and ⅔ cup heavy cream in a saucepan and scald. Set aside for 5 minutes to infuse, then add ⅔ cup coconut milk and 3 tbsp brown sugar. Stir over a low heat until the sugar has melted. Beat 2 eggs and 1 yolk together and gradually whisk the milk mixture into them. Strain through a fine sieve and divide it between 4 ramekins. Place them in a roasting pan of hot water and bake at 325°F for 45 minutes. Eat hot or cold.

Coconut & Dill *See Dill & Coconut, page 187.*
Coconut & Egg *See Egg & Coconut, page 132.*

Coconut & Lemon Coconut gets a refreshingly citric lift from lemongrass. Even if it lacks the sharp acidity of lemon, the vivid citrus-floral character of lemongrass lightens the weighty fattiness of coconut. The lemoniness comes from citral, a combination of two compounds that exists at a low level in the essential oil of lemon and dominates that of lemongrass. Lemon verbena and lemon myrtle also contain high levels of citral and are sometimes used as substitutes for lemongrass. Kaffir limes, native to Southeast Asia, have a flavor and aroma closer to lemon than ordinary limes, due to the potent citrus-herbal character of their constituent compound, citronellal. You might also detect a slight flavor of pears, and a leathery, waxy quality consonant with its famous glossy leaves. Kaffir lime zest has a similar flavor and is sometimes added to curry pastes, as is the small amount of pungent, perfumed, sour juice. Lemon balm can be used as a substitute for kaffir lime leaves, as it contains comparably high levels of citronellal. Both lemongrass and kaffir lime partner with coconut for curries, seafood broths and delicately spiced chicken. And don't overlook their sweet applications—they make great panna cotta and ice cream, or you can infuse coconut milk with lemongrass and use it for the recipe in Mango & Coconut, page 284.

Coconut & Lime *See Lime & Coconut, page 295.*
Coconut & Mango *See Mango & Coconut, page 284.*
Coconut & Peanut *See Peanut & Coconut, page 27.*

Coconut & Pineapple Don't knock the piña colada. Passé it may be, but it's based on a real flavor affinity. Split a ripe pineapple in half, put your nose to the flesh and inhale, and you'll detect strong rum and coconut notes. Piña coladas are particularly delicious made with pineapple juice, white rum, ice and the fresh water and jelly found inside a green (immature) coconut. The jelly nut's viscosity and coconut-custard-like flavor make the more commonly used coconut cream or milk redundant.

Coconut & Pork *See Pork & Coconut, page 38.*

Coconut & Raspberry The madeleine of my childhood was a sponge cake in the shape of an inverted flowerpot, coated in raspberry jam, rolled in coconut and topped with a glacé cherry. Try dissolving *that* in a cup of linden-flower tea. It's an idea worth adapting for simple vanilla cupcakes if you're fed up with cloying pompadours of frosting. Warm some seedless raspberry jam and spread it on top of the cakes with a teaspoon, stopping just short of the edges—it's quite easy to make a neat circle if your cakes aren't too risen. Sprinkle with flaked coconut.

Coconut & Shellfish *See Shellfish & Coconut, page 139.*
Coconut & Smoked Fish *See Smoked Fish & Coconut, page 163.*
Coconut & Strawberry *See Strawberry & Coconut, page 257.*

Coconut & Vanilla You're walking along the South West Coast Path in Devon, hair finned by the breeze, buoyed by the sense that when the air smells this good you could live on it, when a coconut cream pie crashes into your consciousness. You sniff around suspiciously. It's the *gorse*. However unfriendly the yellow-flowered thatch of thorns might look, it radiates deep wafts of coconut and vanilla custard. I like my coconut cream pie topped with vanilla whipped cream and toasted coconut shavings that smell like hot, sweet, butter-coated popcorn.

Coconut & White Chocolate *See White Chocolate & Coconut, page 343.*

Coconut & White Fish White fish and coconut are combined in the famous curries of Thailand and the *laksa* dishes of Malaysia. Less well-known are *amok*, one of Cambodia's national dishes, which consists of fish cooked in fragrant coconut milk and wrapped up in banana leaf, and *molee*, a Keralan Christian specialty in which fish is rubbed with turmeric and salt, then cooked in a coconut gravy with curry leaves, garlic, chili and onions that have first been fried in coconut oil.

Mango

Mangoes, the fruit of a tropical evergreen tree, generally taste sweet and have fruity, creamy and floral flavors, often with a hint of resin. There are, however, many cultivars and producing countries and, depending on its origins and level of ripeness, your mango might equally taste like a stringy canned peach soused in turpentine as a sleek fruit steeped in Gewürztraminer and vanilla cream. Mangoes make a harmonious match with other fruity, spicy and creamy ingredients, but care must be taken for the flavor not to dominate. The classic partner is fruity, spicy, bold-flavored lime: the two make a simple dessert. *Amchoor* is a mango powder prized in Indian cooking for the sourness it lends to dishes.

Mango & Apple Crisp, sour green apples remind mango of its youth. Shredded green papaya or mango provides the spicy Thai salad *som tam* with its satisfying crunch, but if you can't get your hands on either of those, use a Granny Smith apple instead. They're a little more porous than green mango, and you'll need to get some lime juice on it right away to arrest the browning, but it's a good enough substitute not to deny yourself a treat that, aside from its fresh crunchiness, holds the four main elements of Thai cuisine in perfect balance: chili heat, sweetness (from the sugar—traditionally palm), sourness (from the lime) and the saltiness brought by funky fish sauce. When I first discovered how easy *som tam* was to make, I got through a bottle of fish sauce faster than an unwatched kid gets through ketchup. Halve, blanch and cool a handful of green beans. Make the dressing in Lime & Anchovy, page 293. Crush a few tablespoons of peanuts. Core and coarsely grate a couple of Granny Smiths. Halve about 10 cherry tomatoes. Put the grated apple into a bowl and toss with lime juice to prevent browning. Add all the other ingredients, mix in the dressing and serve immediately. Ripe mango and apple, for their part, get along famously. Mango is definitely the older sibling, dominating apple if unchecked, but held in balance they make a mouthwateringly sweet-sour juice, apple's freshness enfolded by the deep, heady creaminess of mango.

Mango & Apricot These are characteristic notes in the ice wine made at Niagara-on-the-Lake in Ontario. The manner of production is incredibly romantic (though presumably less so for the grape pickers). The grapes are left on the vine through the autumn and into deep winter, when they're harvested at below-freezing temperatures, often by moonlight. The frozen grapes are pressed and, because their sugar and flavor-imparting compounds have a lower freezing point than their inherent water, the highly concentrated juice can be separated from the water and skins. The resulting wine is both sweet and pingingly acidic, with pronounced fruit. Noble-rotted Rieslings can also fetch up mango and apricot, and have their own romance, grown in vineyards steeped in mist on the banks of the Rhine.

Mango & Avocado See *Avocado & Mango, page 197.*
Mango & Cardamom See *Cardamom & Mango, page 307.*

Mango & Chili In Southeast Asian and Mexican cuisine, green mango is served dipped in a mixture of equal parts superfine sugar and salt pounded with red chili. Sometimes the fruit is dipped in lime juice before the salt mix. You'll adore this if you liked mouthwateringly sharp sherbet as a child. Try the same dip with tangy apple, pineapple or guava.

Mango & Cilantro Mango has a great affinity for cilantro, sharing pine, citrus and floral notes, and the two are frequently paired in Asian and

Mexican dishes. It's striking, given how polarizing their flavors can be, to reflect how popular mango and cilantro have become in the last ten years. Cilantro is now the bestselling herb in the UK, even if many people find its flavor "soapy" or (conversely) "dirty." Chef and writer Julia Child thought it had a dead taste; others detect a nylon, doll's-hair quality. Mango can have a resinous turpentine flavor, and in fact naturally contains trace amounts of kerosene. See also Cumin & Oily Fish, page 86.

Mango & Coconut An important part of ripe-mango flavor comes from the luscious lactones that murmur of coconut at the back of your palate. Coconut sticky rice and mango is sold all over Thailand, from market stalls and shacks by the side of the road. To make it at home, soak 1 cup glutinous rice for at least a few hours, if not overnight. Drain it, line a steamer with cheesecloth and steam the rice for 20–25 minutes, until it's cooked. Meanwhile, you can get on with dissolving sugar and salt into coconut milk over a low heat. Start out with roughly 2 tbsp sugar, a couple of pinches of salt and 1 cup coconut milk. Put the cooked rice in a bowl, let it cool a little, then gradually feed it the coconut milk until it can't hold any more. Peel and slice a mango, taking care not to reduce it to mush, and serve it on the side of the rice, finishing the dish with a scattering of black sesame seeds, if you have them.

Mango & Cumin Like mango and cilantro, a combination common to India and Mexico. In India, green mango might be stirred into a cumin-scented dhal. A Mexican salsa of black beans, red onion and mango benefits from a pinch of earthy cumin. Rasoi Vineet Bhatia, an upscale Indian restaurant in Chelsea, London, sometimes includes a mango and cumin lassi with coconut and fudge ice cream on its menu, for which it just might be worth reversing the order of your meal.

Mango & Ginger *See Ginger & Mango, page 304.*

Mango & Lime Fresh lime squeezed on ripe mango is one of the world's greatest food pairings. They're both pugnacious flavors, although lime's slightly harsh, medicinal qualities are offset by the floral notes it shares with mango. Lemon, milder than lime, with sweeter rose and fruit notes, can't take mango to the same heights.

Mango & Mint *See Mint & Mango, page 322.*

Mango & Orange Both mango and orange have citrus and floral characters, but when mixed they're subordinated to mango's complex blend of fruity resin and evergreen flavors. It's as if the orange was never there.

Mango & Peach Both turn up as flavor descriptors for Chardonnay, at least in Europe and North America. Jeannie Cho Lee, Korea's first Master of Wine, throws an interesting light on both the cross-cultural difficulties of communicating flavor and the possibilities for considering it differently.

In Asia, the reference points for Chardonnay might include pomelo, dried mango, egg custard and *wakame* seaweed, the latter used to describe more austere, mineral-style wines.

Mango & Pineapple Preside over a library of fruit flavors. Over the full range of cultivars, mangoes might have notes of peach, pineapple, tangerine, banana, watermelon, pear, black currant, guava, apricot, green apple, cherry, fig, sweet grapefruit, red grape, ripe melon, plum, lemon zest or passion fruit. The flavor of pineapple can recall strawberry, orange, peach, apple, banana, raspberry, jackfruit and pear.

Mango & Rhubarb *See Rhubarb & Mango, page 251.*

Mango & Shellfish A delightfully natural partnership. Mango imparts the freshening citrus notes that go so well with shellfish, but also has hints of coconut that harmonize with the nuttiness of shrimp and scallops. Thread some marinated shrimp on skewers and grill or barbecue them for the few minutes they take to go orangey-pink, turning once. Serve with a neat heap of shiny mango salsa. See also Avocado & Mango, page 197.

Mango & White Fish *See White Fish & Mango, page 145.*

CITRUSY

Orange

Grapefruit

Lime

Lemon

Ginger

Cardamom

Orange

All citrus fruits lead double lives, the flavor of their juice being quite different from that of the peel. In the manufacturing of juice and other orange products, once the juice has been pressed from the ripe fruit, the peel is pressed separately to extract the aromatic liquid from its oil glands. Another oil, of a different character again, is distilled as a by-product when the juice is concentrated. These oils are used in soft drinks, or blended back into orange juice to improve the flavor without recourse to synthesized additives. All freshly squeezed fruit juice deteriorates pretty rapidly, so it's always best to squeeze it on demand. Of all citrus fruits, orange is by far the most popular, especially when the term extends, as it does in this chapter, to mandarins, satsumas, blood and bitter oranges, as well as the dominant sweet orange. Its breadth of flavor characteristics guarantees that orange is highly compatible with other flavors. Sweet orange has the fruitiest flavor of all citrics, containing hints of mango and pineapple among its layers of generic citrus flavors, along with mild hints of spice and herb. Mandarins share the pleasing sweet-sourness of sweet orange, although the zest has a more noticeable herbal accent. Blood oranges usually add a berry, specifically raspberry, note to the sweetness. The zest of bitter oranges, such as Sevilles, has a stronger, waxy flavor with a hint of lavender. With the addition of plenty of sugar, their extremes of bitterness and sourness are what make marmalade so full-flavored and interesting. Bitter oranges are also used in most orange-flavored liqueurs, including Cointreau, Grand Marnier and Curaçao, and orange-flower water is made from their blossoms. Dried orange peel, for use as a flavoring, can be bought in Chinese and Middle Eastern supermarkets.

Orange & Almond Claudia Roden's legendary orange and almond cake is unusual in that it calls for pretty much every part of the orange save the seeds: the zest, with all its oil glands, the pith, the segment membranes and the vesicles—those wondrous juice-containing structures that look like tiny eyedroppers. It's this thorough exhaustion of the fruit's possibilities that gives the cake its deeply musky, spicy character, redolent of marmalade but without the sticky sweetness. The almond contributes the dense texture that makes the cake equally at home on the pudding plate as on the afternoon tea plate, especially if served with cream or a compote. In a nutshell, you boil 2 oranges in water for nearly 2 hours, until completely soft. Once cool, quarter them, discard the seeds and process the oranges to a pulp. Beat 6 eggs in a large bowl, then mix in 2¼ cups ground almonds, 1¼ cup sugar, 1 tsp baking powder and the orange pulp. Pour into a greased and lined, deep, round 9-in cake pan. Bake at 375°F for an hour. Between you and me, the oranges can be microwaved in a few minutes if you're short of time.

Orange & Anise See Anise & Orange, page 181.
Orange & Apple See Apple & Orange, page 266.
Orange & Apricot See Apricot & Orange, page 276.

Orange & Asparagus Gangly asparagus might seem an unlikely match for voluptuous orange but it works. *Sauce maltaise*, a hollandaise flavored with blood oranges, was created especially for asparagus. Boil ⅓ cup blood orange juice until reduced to about 2 tbsp, then add the zest of 1 orange and simmer for 1 minute. Stir into a hollandaise made with 4 egg yolks and serve immediately.

Orange & Bacon A marmalade-glazed ham is a thing of wonder. Use a marmalade that's made with plenty of Seville oranges; if it's too sugar-heavy, you're in danger of making ham with jam, which will please no one but Dr. Seuss. Seville orange marmalade has a deep, bitter tang that will counter the ham's saltiness. Eating this is only part of the fun. Rubbing a whole jar of marmalade into a large joint of meat is a rare sensual pleasure.

Orange & Beef A bouquet garni of pared orange zest, bay leaf, thyme and parsley is often recommended for slow-cooked beef dishes, such as Elizabeth David's beef and wine stew with black olives. In case you need *two* good reasons to try it, Fiona Beckett writes that dried orange peel used in this way enhances the richness of medium-bodied red wines.

Orange & Beet At The Fat Duck, Heston Blumenthal teases diners with an orange-colored jelly candy that has the flavor of beet and a crimson jelly candy with the flavor of orange. The beet candy is made with golden-colored beets and the orange one with dark red blood oranges. The waiters playfully suggest you start with the orange one.

Orange & Carrot *See Carrot & Orange, page 225.*

Orange & Chili Rick Bayless describes a ripe, orange habanero chili as having an aroma of passion fruit, apricot, orange blossom and herb, with a noticeable piquancy. The flavor is similar to the aroma, with added notes of sweet, tangy tangerine. You may be able to detect these notes in the teacup of tears you will have shed trying this most fearsomely fiery of peppers. Fruit flavors are often detectable in dried chilies too, and not just habanero: prune and raisin are commonly cited. If habanero's too hot, infuse olive oil over a low heat for about half an hour with a strip of orange peel and a couple of dried chilies. Strain off the peel and chilies and drizzle the oil over fish.

Orange & Chocolate Both orange zest and orange-flower water have been used to flavor chocolate since at least the seventeenth century. The combination of orange and chocolate has endured where other once-common flavorings, like black pepper and anise, have faded into obscurity. Surprising, then, that Terry's famous Chocolate Orange was originally an apple. Launched in 1926, the apple proved so popular that an orange version was launched four years later. When both went back into production after the war, orange quickly overtook apple in popularity and the latter was discontinued in 1954. The Chocolate Orange finally grew up in 1975 when the dark chocolate version was introduced, its bitter astringency complementing the sweet muskiness of orange oil far better than frumpy, facetious milk.

Orange & Cilantro *See Cilantro & Orange, page 193.*
Orange & Cinnamon *See Cinnamon & Orange, page 214.*

Orange & Clove In *The Size of Thoughts*, Nicholson Baker comments on the sensuous pleasure of writing on an eraser with a ballpoint pen. I get a comparable kick studding a firm orange with cloves. Some use these as pomanders, but I let them bob around like limpet mines in a sea of mulled wine. The fresh citrus and smoky spice can give the dullest grog a spark.

Orange & Coffee *See Coffee & Orange, page 25.*

Orange & Coriander Seed Look at a coriander seed close up. It could be a peeled mandarin from the kitchen table in a doll's house. Bite it and you'll find it has an orange flavor too; like marmalade or Seville orange peel with a cedar background. Serve orange segments in a syrup flavored with coriander seed for a layered orange flavor (and an update on the classic oranges in caramel), or explore coriander's spicy orange character by pairing it with some of the fruit's classic flavor affinities, such as cinnamon, duck, lemon or cranberry.

Orange & Fig *See Fig & Orange, page 333.*
Orange & Ginger *See Ginger & Orange, page 304.*
Orange & Grapefruit *See Grapefruit & Orange, page 292.*

Orange & Hard Cheese Cheese with all sorts of fruits—grapes, apples, pears, quince—is, of course, completely uncontroversial, as are citrus fruits on cheesecake, yet Cheddar with marmalade is apt to raise eyebrows. But think about a rich, salty, mature Cheddar and how delicious it might be cut through by the bittersweetness of marmalade—there's a real balance of flavor there. For a sandwich, try grating the cheese and using a fine-cut marmalade, as thick hyphens of orange might prove too dominant. Walnut bread would be excellent. Alternatively make "jam" tarts using marmalade mixed with grated Cheddar. Fill little cheese pastry tarts and bake at 425°F for about 15 minutes, taking special care not to burn them.

Orange & Juniper *See Juniper & Orange, page 316.*

Orange & Lemon St. Clement, the patron saint of designated drivers, lends his name to the mix of orange juice and bitter lemon. Bitter lemon's bitterness comes not only from lemon but from quinine, the colorless, odorless alkaloid that puts tonic water in such agreeable ill temper. The adult palate can generally take only so many sweet drinks before tiring of them. This is less of a problem with alcoholic drinks, which generally have a balancing bitterness to them. Orange and lemon zest (or the harder-core mixed peel) are also used to flavor, and balance the sweetness of, puddings and cakes. The Meyer lemon, big in the United States but a rare find in the UK (because it doesn't travel well), is a lemon-orange hybrid, with a pronounced floral nose and flavor. It's low in acid, so is seen as a sort of sweet lemon, as Sevilles are sour oranges—see Orange & White Fish, page 291.

Orange & Lime *See Lime & Orange, page 295.*
Orange & Mango *See Mango & Orange, page 284.*
Orange & Melon *See Melon & Orange, page 274.*
Orange & Mint *See Mint & Orange, page 322.*

Orange & Olive *La Cucina Futurista* was published in 1932 by the poet Marinetti, after he'd spent a few years traveling around Europe organizing banquets that featured wild flavor pairings and recipes with titles like Excited Pig, Elasticake, Steel Chicken and Piquant Airport. My favorite is called Aerofood. While the sound of an airplane motor and something suitable by Bach plays loudly from the kitchen, the diner is served, from the right, a plate of kumquats, black olives and fennel. Rectangles of silk, sandpaper and velvet are served from the left and the diner must eat with the right hand while simultaneously stroking these with the left. Meanwhile the waiter spritzes the nape of the diner's neck with a carnation perfume. Carnations have a rose-clove fragrance—try Santa Maria Novella's Garofano. The plane might prove trickier to source.

Orange & Onion *See Onion & Orange, page 110.*
Orange & Peach *See Peach & Orange, page 278.*

Orange & Pineapple All the *joie de vivre* of a Hawaiian shirt without the stigma of wearing one. Pineapple blinds you to orange's dark side—the bitterness, the complexity, the just-detectable whiff of sulfur. The two share tangerine, fruity, green flavors, and pineapple juice is sometimes added to orange juice to give it a more natural orange flavor.

Orange & Rhubarb Orange zest and rhubarb are often paired, especially in crumble, but not in my kitchen. I find the neediness of super-sour rhubarb and the belligerence of orange zest pull in different directions; the flavor equivalent of patting your head and rubbing your stomach. Even when rhubarb is cooked in milder orange juice, which doesn't have the pushy bitterness of zest, my palate can detect neither harmony nor pleasing contrast.

Orange & Rose Orange-flower water and rosewater are often treated as interchangeable in recipes. Unsurprisingly, floral notes are dominant in both, but there's a citrus lick to orange-flower water, which is extracted from the blossoms of the bitter (Seville) orange tree. Good flower waters are made in Iran, while in Lebanon an artisan company called Mymouné uses traditional distillation methods and no artificial ingredients. The key for both is to use them by the drop, not the teaspoon, so they bring a mysterious background note to the dish, like a shimmer from a zither, not the thudding of the drum. They're traditionally used in North African cooking in lamb and chicken tagines and all sorts of almond puddings and cakes; in France to flavor madeleines; and to add a floral note to fruit juices (especially orange) and grated carrot salads. See also Cinnamon & Orange, page 214.

Orange & Rosemary *See Rosemary & Orange, page 311.*

Orange & Saffron Citrus flavors pair well with saffron. Orange and saffron turn up in Mediterranean fish stews and North African tagines but also make a great couple in cakes and cookies. Soak a pinch of saffron in a tablespoon of warm milk and add it to a Victoria sponge cake mixture. Sandwich together with marmalade. Very Moorish.

Orange & Strawberry Strawberries Romanoff was created for Tsar Alexander I by the legendary chef Marie-Antoine Carême. It's a sparkling combination. Hull some strawberries, marinate them in a 50:50 mixture of orange juice and orange liqueur, then stir through some crème Chantilly—see Vanilla & Raspberry, page 341.

Orange & Thyme *See Thyme & Orange, page 318.*
Orange & Vanilla *See Vanilla & Orange, page 341.*

Orange & Walnut Use the walnuts and clementine in the toe of your Christmas stocking to make a relish with cranberries or a salad with bitter green leaves, or mix them with thick yogurt and a whirl of maple syrup for breakfast.

Orange & Watercress Check the sweet, sour and bitter boxes. Add something salty (olives, perhaps) for a perfect salad. Good with duck. See also Apple & Walnut, page 267.

Orange & White Fish Until the eighteenth century, oranges were used with fish much as lemons are today. At that time most oranges were sour Sevilles, as opposed to the sweet varieties that have dominated the market more recently. For an authentic eighteenth-century experience when Sevilles are hard to come by, a mixture of two sweet oranges to one lemon is an effective substitute, even if the aromatic quality isn't quite the same. In a similar vein, Mark Hix adds a little bitter-orange-flavored Curaçao to his sole Véronique. As for sweet oranges, Alan Davidson notes their great affinity with a firm, strong-flavored fish called *mérou* (grouper), and gives a recipe for a sauce made with 1½ oz each of butter and flour whisked with 1⅛ cups meat stock/bouillon, ½ cup orange juice and a pinch of salt. *Sauce maltaise*, flavored with blood orange, is often paired with firm white fish—see Orange & Asparagus, page 288.

Grapefruit

The lumbering old uncle of the citrus family. Grapefruit shares the generic citrus flavors and some of the tropical fruitiness of orange, alongside a more pronounced herbal, woody flavor, but what really sets it apart is its musky, sulfurous character. Can work well with typical citrus-family partners such as seafood, but has a particular affinity for ingredients that share its bitter streak, like blue cheese and green leaves. Ruby grapefruits tend to be sweeter than yellow ones.

Grapefruit & Avocado This combination is a modern classic in a salad with lobster, plump shrimp or fresh crab. A café in Montpellier calls this *salade fraîcheur* and serves it with a shot glass of gazpacho on the side. On a witheringly hot afternoon, it was enough to rehydrate the body and the soul. The brightness of the flavors is one thing, but there's also pleasure in feeling the soft butteriness of avocado against the grapefruit's vesicles, tautly rippled like wet sand after the tide's gone out.

Grapefruit & Blue Cheese See *Blue Cheese & Grapefruit, page 64.*
Grapefruit & Cinnamon See *Cinnamon & Grapefruit, page 213.*
Grapefruit & Juniper See *Juniper & Grapefruit, page 315.*

Grapefruit & Orange Grapefruit is a hybrid of sweet orange and pomelo, the large, yellow-green citrus fruit that looks like a bloated pear. It is, of course, far closer in flavor and appearance to the pomelo, but shares, among other things, orange's affinity for Campari. Both make refreshing (and variably bitter) sorbets paired with Campari, but the respective effects they have on the drink are more easily understood in the highball glass. Orange juice is the way in, the stabilizers on the frightening bike ride into Campari territory, helping to suppress its weird herbal notes and bitterness so stringent that your brain, at least on the first few attempts, flicks into alert mode at the influx of toxicity. Grapefruit, on the other hand, is pretty bitter itself, with its own extraordinary, herbal-fruity notes, and mixed with Campari makes a gloriously complex drink.

Grapefruit & Pineapple See *Pineapple & Grapefruit, page 262.*
Grapefruit & Pork See *Pork & Grapefruit, page 39.*

Grapefruit & Shellfish The main identifying flavor compounds in grapefruit are nootkatone, mercaptan and naringin, which sound like the sort of user names you'd find on a Web forum debating the relative merits of *Deep Space Nine* and *Star Trek: Voyager*. Appropriately enough, perhaps, as of all the citrus family grapefruit surely has the most alien flavor. Nootkatone, the most important (i.e., "grapefruity") compound, has a warm, woody character that is present but suppressed in other citrus fruits by their more dominant compounds. Grapefruit mercaptan, which is primarily responsible for the smell of grapefruit and lends a musky, tropical top note to the flavor, has one of the lowest aroma thresholds known to science. That is, it's so potent that it can be detected in extremely minute quantities—0.0001 parts per billion, to be exact. Grapefruit's odd, standoffish flavor has earned it a special place in *haute cuisine*; at Les Jumeaux, the Parisian restaurant run by the Gillaizeau twins, scallops are served with a split pea purée and pink grapefruit sauce; at Chez Jean, they're sautéed with chanterelle mushrooms, Chinese noodles and slices of pink grapefruit. More recently the vogue for grapefruit has been challenged by the yuzu, a small, wrinkled fruit big in Japan. Yuzu has a lemon-lime flavor, with hints of tangerine, grapefruit and pine, and is prized primarily for its aromatic peel, although the juice is used too. A very rough substitute can be made with equal parts lime and grapefruit juice. At Jean Georges in New York, your waiter might spray your plate of scallops with a spritz of yuzu juice. Don't hit him.

Grapefruit & Watercress Share a strong minerality and bitterness that set off full-flavored, fatty proteins. Blue cheese would be the obvious choice—especially if it's sweet, salty Roquefort, with its own underlying mineral quality. Remove any oversized stalks before tossing the leaves with skinless grapefruit segments and crumbling over the cheese. Maybe add some walnuts, too. Duck would be excellent in place of the cheese.

Lime

The hardest, sharpest member of the citrus family. Oil made from lime peel is strong and spicy, with pine, lilac and eucalyptus notes. Much of what is produced is used for flavoring cola. Lime juice is notably acidic, which accounts for its palate-cleansing qualities, and can seem quite salty squeezed on a salsa. Its sourness and bitterness work wonders set against a correspondingly determined sweetness—caramel in cola, for instance, butternut squash, or condensed milk in a key lime pie. The cultivar most widely available in the UK is the Persian lime, of which the citron and key lime are thought to be the parents. Key limes are smaller, thicker-skinned and, unlike most Persian limes, have seeds. They're said to have a distinct flavor but, having hauled a bag home from the States to make an authentic key lime pie, I couldn't detect much difference. Lime flavor is also contributed by musky dried limes (whole or ground), spicy lime pickle, lime marmalade and lime cordial. For kaffir lime, see Coconut & Lemon, page 281.

Lime & Anchovy Meet in a Thai salad dressing of lime juice and fish sauce—or in *nuoc cham*, the Vietnamese dipping sauce served with spring or summer rolls. Everything you love about fish plus citrus, cubed. Even with a lighter fish sauce, lime will always be the underdog, trying to shine a light on fish sauce's dark thoughts and failing beautifully. Proportions really are a matter of taste, and balance with the other flavors in your recipe. I pound 2 garlic cloves with 1 chili, then add 2 tbsp lime juice, 2 tbsp fish sauce and a pinch of sugar. Addictively delicious on a homemade *som tam* salad—see

Mango & Apple, page 283. Vegetarians can buy a "fish sauce" made with soy beans. See also Pineapple & Anchovy, page 260.

Lime & Avocado *See Avocado & Lime, page 196.*

Lime & Basil You can grow this combination yourself by hunting down lime basil seeds—they're fairly widely available. In Thailand, where they have basil varieties coming out of their ears, lime basil is most frequently used with fish.

Lime & Beef Lime is sour, sweet, a little bitter, used in some cultures as a salt substitute, and above all intensely flavored, with a strong hint of tropical fruit. Beef steps up to the plate, answering lime's acidity with a metallic tang of its own. The dressing in Lime & Anchovy, page 293, is often served with seared beef on a salad with lots of chili, in a dish known as "weeping tiger." In Vietnam, lime wedges are served with beef *pho*, and as part of a dipping sauce for "shaking beef," marinated in garlic and soy and served in lettuce leaves.

Lime & Butternut Squash The heavy sweetness of butternut squash welcomes some interference from sharp, spicy lime. Pair them in a chowder, roast chunks of squash in a combination of olive oil and lime juice, or try tempura pumpkin with a soy-lime-sesame dipping sauce. For a Thai take on the combination, simmer chunks of pumpkin or squash in spiced-up coconut milk. When they're cooked through, stir in a mixture of lime juice, fish sauce and palm or brown sugar. See also Butternut Squash & Bacon, page 226.

Lime & Chicken *See Chicken & Lime, page 32.*

Lime & Chili Diego Rivera and Frida Kahlo. This turbulent twosome brings fervor to a vast range of Mexican dishes, including *chapulines*—grasshoppers fried in lime juice and chili powder and eaten as a snack. For those without the inclination, or patience, to catch grasshoppers, peanuts in their skin are served the same way, as are fried plantain, corn on the cob, watermelon, barbecued shrimp and Doritos. Happily for chili-lime addicts, the seasoning can be bought pre-mixed by the jar, so you can put it on your cornflakes if you like. In India, lime and chili are paired in lime pickle (the best thing since marmalade in a cheese sandwich—see Orange & Hard Cheese, page 289). Finally, should you find the sweetness of bottled chili sauce a little cloying, it can be freshened up with a squeeze of lime to make a versatile dipping sauce.

Lime & Chocolate *See Chocolate & Lime, page 20.*
Lime & Cilantro *See Cilantro & Lime, page 193.*

Lime & Cinnamon These are cornerstones of cola flavor, in which it's fairly common for cinnamon to appear in the form of cassia, a related spice with a harsher, more strongly flavored profile and a good match for lime's own pugnacious qualities. Vanilla is another of cola's typical ingredients, alongside caramel, nutmeg, orange, lemon, coriander, and coca-leaf extract. I combine lime and cinnamon in a sorbet. Make a simple syrup by gently heating 1 cup vanilla sugar and several sticks of cassia (or cinnamon) in 1 cup water until the sugar has dissolved. Bring to a simmer, then cool. Keep in the fridge until the cinnamon has imparted a strong flavor to the syrup, then strain. Juice 3 limes, strain and add to ¾ cup of the cinnamon syrup with 2 tsp lemon juice and ½ cup water. Chill until very cold, then freeze according to your usual method. Serve in a caramel basket to extend the deconstructed cola theme.

Lime & Coconut Combined with the tropical drowsiness of coconut, lime is the niggling intimation that you should get out of the hammock and exert yourself. Lime frosting on a coconut cake needs a knife-edge intensity to set off the sleepy sweetness. A grating of fresh coconut and a squeeze of lime can make a pineapple seem sweeter and juicier—as they do with fish, cooked or raw: they're paired in a popular ceviche (see Lime & White Fish, page 296). In India, strips of fresh coconut are mixed with lime juice, crushed garlic and chilies and served with curry.

Lime & Cumin Quite a power struggle. Let them fight over barbecued meat, roasted corn or tomato salsa. Or add mint to make *jal jeera*, a cooling drink popular in India, in which pounded roasted cumin seeds are mixed with pounded mint, salt, lime juice and water.

Lime & Ginger *See Ginger & Lime, page 303.*
Lime & Lemon *See Lemon & Lime, page 299.*
Lime & Mango *See Mango & Lime, page 284.*

Lime & Mint Cubans take mint and lime and add rum to make mojitos, the ubiquitous cocktail of the last decade. What's the secret of its success? My theory is that the sour lime and refreshing mint come together with the kick of rum and the sugar rush to create something that sits at the crossroads of amphetamine and aromatherapy. Put 1 tbsp superfine sugar in a highball glass with 2–3 tbsp lime juice. Add a leafy sprig of mint and fill to about a third full with soda water. Muddle to dissolve the sugar and release the essential oils in the mint. Add about 3 tbsp Havana Club and a handful of ice cubes. Listen to their seismic creak. Stir, then garnish with more mint and a straw. If strong enough, the mojito will produce a pleasurably painful perceptual sharpening, like putting on glasses when you're already wearing contact lenses. Or try a lime sorbet served with a shot of rum and garnished with mint leaves.

Lime & Oily Fish *See Oily Fish & Lime, page 155.*

Lime & Orange These two intertwine to create a taut support for tequila in the high-wire act that is the margarita, the most popular mixed drink in the United States. A good margarita holds its extremes of sweetness, sourness, bitterness and salt in thrillingly tenuous balance. It should make you gasp and stretch your eyes. Some don't care for the salt, feeling that a good tequila has a salinity of its own, but I can't resist it. The salt emphasizes the sweet-sourness, and has the teasing effect of keeping you thirsty while you're drinking.

Lime & Peanut *See Peanut & Lime, page 28.*
Lime & Shellfish *See Shellfish & Lime, page 140.*

Lime & Tomato Sangrita, a popular drink in Mexico, is made with tomato, lime and orange juice with a dash of chili. It's sipped alternately with tequila. Some ditch the orange juice, making it a liquid form of my favorite basic salsa. Chopped tomatoes with lime juice have enough complexity of flavor to count as a salsa on their own. Try half a lime squeezed over 2 diced tomatoes and see just how salty and mouth-filling it tastes. Add a few drops of chili sauce and a little chopped onion for extra piquancy. Serve with a handful of tortilla chips.

Lime & Watermelon *See Watermelon & Lime, page 246.*

Lime & White Fish Nothing says you're on first-name terms with your fishmonger like serving a ceviche. Strips of raw fish are cured in lime (or lemon) juice and commonly mixed with chopped onion, bell pepper, chili and cilantro. Once the fish has had a chance to "cook" in the lime juice, coconut milk might be added for a creamier result. In Ecuador, ceviche is often served with roasted corn kernels or popcorn; in Peru, sweet potato is a more typical accompaniment. Ceviche is said to have come about when the Spanish brought the Arabic notion of cooking with fruit to South America. As a culinary version of Chinese whispers, one couldn't hope for a more surprising outcome.

Lemon

Lemon zest contains a compound called citral that is immediately recognizable as lemony. Aromatic notes of rose, lavender and pine are also present, as is a slight herbaceousness; all these are released when you grate lemon zest. The juice has a clean, fresh flavor, dominated by citric acid. It can be used in moderation to give an unidentifiable lift to a dish, or in greater quantities to lend a definite lemon flavor. Lemon is highly combinable and adaptable: in sweet or savory dishes, as a seasoning or star ingredient, in everything from apéritifs to petits fours. Pull it to extremes in a sweet lemon tart, sour lemon candy, bitter-lemon soft drinks and Middle Eastern salted preserved lemons. Lemon flavor is also dominant in the Italian liqueur limoncello, in lemon curd and in the exquisite lemon-sherbet hard candy that smell like lemon myrtle. For more about lemon-flavored herbs, see Coconut & Lemon, page 281.

Lemon & Almond *See Almond & Lemon, page 239.*
Lemon & Anchovy *See Anchovy & Lemon, page 160.*
Lemon & Anise *See Anise & Lemon, page 181.*

Lemon & Asparagus In their book *Urban Italian*, Andrew Carmellini and Gwen Hyman enthuse about this partnership, noting how in a risotto "the lemon cuts through the richness of the dish, opens up the asparagus flavor, and keeps things fresh." What I love about their recipe is its good housekeeping. Nothing is wasted: the woody bottom inch of the stalks is used to flavor the stock, their middle sections are cooked, puréed and stirred into the rice just as it's ready, while the tips are blanched and added whole to finish. It reminds me of *risi e bisi*—see Pea & Hard Cheese, page 199.

Lemon & Basil As suggestive of summer as a bucket and spade. Combine their mood-lifting citrus and licorice flavors in a simple pasta. A *relatively* simple pasta: I once witnessed four Italian men argue for an hour over the exact means of preparing this properly. To serve two, cook 7 oz spaghetti until *al dente*. Meanwhile, heat 2 tbsp olive oil in a small pan and soften a finely chopped shallot in it. Add 2 tbsp white wine and simmer for a few minutes before adding the juice and zest of an unwaxed lemon. Season and turn the heat down. When the pasta is ready, drain it and stir in the lemon sauce with a couple of handfuls of grated Parmesan, a small handful of torn basil leaves and 2 tsp butter.

Lemon & Beef Big in Italy, where the lemons are legendary. They squeeze them on bresaola, the salt-cured, air-dried beef fillet served in gossamer slices like Parma ham. It's silky, gamy and sometimes has a musty flavor. Raw beef, sliced a little thicker than bresaola, is called carpaccio, and gets the same lemon treatment. In Florence, it's not unusual to serve *bistecca alla fiorentina*—hefty slabs of T-bone steak seared over wood or charcoal—with a lemon wedge, which not only lifts the beef flavor but arbitrates between it and the local Chianti. Some say lemon makes an inferior Chianti taste fruitier and less, um, rough.

Lemon & Blueberry Poor blueberry. Its pretty, floral notes are masked by the blurt of sour juice it emits when bitten. Similarly, the acidity of lemon detracts from its floral perfume. But a problem shared is a problem halved: with the redemptive sweetness of sugar or honey, both flavors get the chance to shine, and make a heady, perfumed combination in cakes and puddings.

Lemon & Broccoli *See Broccoli & Lemon, page 125.*
Lemon & Caper *See Caper & Lemon, page 102.*
Lemon & Caviar *See Caviar & Lemon, page 151.*

Lemon & Chicken If the world was organized the way it might be, you could plot how lemon-chickeny a dish was on the Lemicken scale. A soft Thai chicken broth delicately scented with lemongrass—a 2. At 4, chicken rubbed with lemon, then roasted with a lemon in the cavity. A roast chicken baguette with tangy lemon mayonnaise—a 5. A thick, spicy, slow-cooked Moroccan tagine made with chicken thighs and preserved lemons— a 9. A 10 is the battered fried chicken with vivid yellow sauce that I am too ashamed to order in Western Chinese restaurants.

Lemon & Chili *See Chili & Lemon, page 206.*

Lemon & Chocolate Not an easy combination to pull off, but when it works it can be sublime. I dream of sinking a long spoon into alternate layers of aromatic, zingy lemon custard and dark chocolate ganache, striped like a bumblebee in a tall glass. Joël Robuchon's take on the pair is to serve lemon-scented madeleines with little pots of chocolate.

Lemon & Cilantro This recipe spreads faster than gossip. I know because I gave it to somebody, who gave it to somebody else, who then made it for me and asked if I'd like to have the recipe. The cheek, I thought, before I remembered *I* got it from a supermarket recipe card. It originally called for cod, but I've yet to find a white fish that's not woken up by the combination of lemon and cilantro. Combine ¾ cup breadcrumbs with the zest of a lemon, a fistful of finely chopped cilantro (use the thin parts of the stalk too), 4 tbsp melted butter, a pinch of chili flakes, salt and pepper. Pat the mixture down on 4 skinned fillets of white fish lined up on a greased baking tray. Bake for 20–25 minutes at 400°F if you're using a fish of a cod-like texture, or adjust accordingly.

Lemon & Coconut *See Coconut & Lemon, page 281.*
Lemon & Coriander Seed *See Coriander Seed & Lemon, page 338.*
Lemon & Cumin *See Cumin & Lemon, page 85.*

Lemon & Dill The citrus character of dill is created by d-limonene, a compound it shares with lemon, although it's not necessarily lemony. Some detect orange or generically citrusy notes. The knee-jerk reaction is to use lemon and dill for fish, but in Greece, where the partnership is extremely popular, lemon and dill are used far more widely: with lamb, with vegetables, and mixed into rice with onion and pine nuts to serve with feta. The combination is valued for its freshening effect and its ability to emphasize the sweetness of other ingredients.

Lemon & Egg *See Egg & Lemon, page 133.*

Lemon & Ginger Fresh ginger is commonly described as zesty or citrusy and makes a very harmonious match for lemon. Lemon sauce is an established partner for ginger puddings, lemon icing is good on ginger cakes and the two combine in a hot toddy whose deliciousness may unfortunately pass you by when you're most in need of one. Being clogged up is a good, if unwelcome, demonstration of how much your nose contributes to flavor recognition. The taste receptors in your tongue detect sweetness, saltiness, bitterness, sourness and "umami," or savoriness, but it's your olfactory nerve endings that detect the subtler differences between the lemony flavor in ginger and the lemony flavor in lemon. Try this toddy while your olfactory bulb is in full working order. Put a ¼-in piece of fresh ginger in a glass with the juice of ¼ lemon, 1–2 tsp honey and 1 tbsp whisky, rum or brandy. Top up with boiling water, stir and leave to infuse until it's cool enough to drink.

Lemon & Globe Artichoke *See Globe Artichoke & Lemon, page 127.*

Lemon & Goat Cheese A citric character is inherent in many goat cheeses, including the award-winning Cerney from Gloucestershire and the cute Innes Button from Staffordshire. Juliet Harbutt writes that Innes Button "dissolves on the palate, leaving a trail of almonds, wild honey, lemon, white wine and tangerine," which, as pushing buttons goes, pushes mine.

Lemon & Juniper *See Juniper & Lemon, page 316.*

Lemon & Lamb Succulent preserved lemons are slow-cooked with lamb in a Moroccan tagine. Sharp lemon is cooked with fatty lamb in Greece. Brown a 3 lb joint in olive oil with garlic and then braise in a tightly covered pot with the juice of a couple of lemons and some oregano. If it begins to run dry, add a little water, but not too much: the point is to serve the lamb in slices with its concentrated, lemon-spiked juices. Traditionally the lamb is served well cooked rather than rare, with roast potatoes, rice or white beans.

Lemon & Lime Cockney rhyming slang for crime (as in, *It was the lemon of the century, mate*). And all too appropriately: lime will bludgeon lemon to death given half a chance. Lemon and lime are, of course, from the same family, and their juices have a lot of flavor elements in common, but lime has an overpowering spicy pine and lilac character—meaning that lemon juice is useful for diluting lime juice, but you can't expect to taste both fruits unless you use the zest. The lemony, rosy, herbal characters are far more apparent in the peel.

Lemon & Mint *See Mint & Lemon, page 322.*
Lemon & Oily Fish *See Oily Fish & Lemon, page 155.*

Lemon & Olive In Morocco, thick, silky slices of preserved lemon and army-fatigue green olives combine to make chicken interesting. At once bitter, bold and tangy, they bring a welcome counterpoint to the sweetness of tagines or fruity couscous dishes. They might also be paired in a salsa served with goat cheese tart or in a salad to accompany oily fish.

Lemon & Orange *See Orange & Lemon, page 289.*
Lemon & Oyster *See Oyster & Lemon, page 150.*

Lemon & Parsley A simple, freshening, unobtrusive couple, always eager to lend a hand in the professional kitchen, either in the form of *beurre maître d'hotel* (a compound of butter and parsley) or in the preparation known as *à la meunière*, wherein fish, usually sole, is dipped in flour, fried in clarified butter, then served with brown butter, lemon juice and chopped parsley. See also Parsley & Garlic, page 189.

Lemon & Potato I once saw a chef berate a cooking-show contestant for pairing these two. Maybe the way they were combined wasn't terribly appetizing, but to claim that potato and lemon can't work together would baffle the Greeks, who often include potatoes in lemon-sauced dishes. In India, roughly mashed potato is mixed with lemon juice, breadcrumbs, cilantro and chili, formed into pancakes, then deep-fried and served as a snack with chutney and yogurt. New potatoes might be dressed with lemon-infused olive oil or a lemon-laced vinaigrette. Mashed potato with lemon and black pepper can be very good with fish. And Mrs. Leyel gave a recipe for lemon cream pie, made by mixing a grated potato, the zest and juice of a lemon, a cup of sugar and a cup of water, then baking it in a double crust.

Lemon & Rose *See Rose & Lemon, page 335.*

Lemon & Rosemary Lemon tart is classically served with crème fraîche, which has a slight sourness that works well with the sweetened, yet still sour-at-heart lemon curd. But the best partner I've ever tasted for a frisky lemon tart was rosemary ice cream. After much experimentation with steeping rosemary in cream or milk, which works but requires a bit of preparation in advance, I now use rosemary essence from The Hop Farm in Kent. Beat 2 egg yolks with ¼ cup superfine sugar and 2 tsp cornstarch until smooth. Scald 1¼ cups whipping cream, remove from the heat and beat in 2 tsp liquid glucose. Gradually add the cream mixture to the egg mixture and whisk together before returning all to the pan and heating, stirring constantly, to a custard thickness. Remove, pour into a clean bowl and cover the surface of the custard with cling wrap. Leave to cool, stir in about 20 drops of rosemary essence, then chill in the fridge before freezing.

Lemon & Saffron Work like a couple of vacation reps to keep paella's orgy of disparate ingredients in order—chicken, rabbit, snails, green beans, bell peppers, mussels, shrimp, butterbeans, globe artichokes, rice. The saffron suffuses the rice to sound a faint but consistent background note. With a flamenco dancer's contemptuous flick of the wrist, squeeze lemon over the entire dish before serving. This way, some mouthfuls are alive with citrus zing, others with a quieter, relieving sweetness. It's this balance of tastes that makes paella so dangerously easy to eat in vaster quantities than more monotonous meals, which must explain why Spanish cookshops sell paella pans as big as the satellite dishes at Very Large Array, New Mexico.

Lemon & Shellfish *See Shellfish & Lemon, page 140.*
Lemon & Smoked Fish *See Smoked Fish & Lemon, page 163.*

Lemon & Thyme So popular it comes in at least three strengths. For maximum impact, combine chopped thyme with the zest and juice of a lemon—excellent with fish, lamb or chicken, or as a dressing for fried artichokes. Then there's soft-leaved lemon thyme, which furnishes a full but beautifully soft lemon top-note with a herbal murmur underneath, obviously delicious anywhere that lemon and thyme might work in a gentler form. Subtlest of all is the flavor of Meyer lemons, thought to be a hybrid of lemon and mandarin, with a trace of the latter's mild thyme flavor and a sweeter, less acidic freshness than the true *citron limon*. The pairing of lemon and thyme is becoming quite common in sweet foods; I've recently seen them combined in recipes for lemon drizzle cake, a cheesecake and an ice cream. See also Orange & Lemon, page 289.

Lemon & Tomato *See Tomato & Lemon, page 255.*
Lemon & White Chocolate *See White Chocolate & Lemon, page 343.*
Lemon & White Fish *See White Fish & Lemon, page 145.*

Ginger

Native to Southeast Asia but now widely cultivated, ginger varies considerably in flavor depending on where it is grown. In general, fresh ginger is lemony, woody and earthy, with a kick of heat. Jamaican ginger grown in Jamaica is known for its fine quality. Jamaican ginger grown in Nigeria and Sierra Leone is characterized by its richness, pungency and a camphorous element that particularly distinguishes it from the more lemony varieties and betrays its relation to cardamom. Australian ginger is said to be the lemoniest, as of all varieties it contains the most lemon-flavored citral in its oil. Popular the world over as both a sweet and a savory ingredient, ginger is available fresh, dried, ground, glacé, pickled, preserved in syrup, juiced and in both soft and alcoholic drinks.

Ginger & Almond *See Almond & Ginger, page 239.*
Ginger & Apricot *See Apricot & Ginger, page 275.*

Ginger & Beef Beef loves tangy flavors, and you'll find it paired with ginger in Thai and Chinese stir-fries and in the crunchy, battered ginger beef created by two Chinese sisters in Calgary, where it's become something of a local specialty. A more unusual instance of the pairing is in crushed ginger cookie crumb for beef croquettes and schnitzels.

Ginger & Butternut Squash In most recipes butternut squash and pumpkin can be substituted for one another—say, in pumpkin pie where, once the spices and sugar have been added, few people would be able to spot the difference. Pumpkin pie spice is a blend of ginger, cinnamon, clove, nutmeg and allspice. The same ingredients make up a popular English blend now rather drily called "mixed spice." It used to be called pudding spice, which was a far better name, not only because it's more evocative but because it avoids confusion with allspice, which is a different thing altogether. As an aside, allspice is not a blend but a single spice, which looks like a black peppercorn and has a flavor dominated by eugenol, the compound that gives clove its flavor. It's called allspice because it also contains notes of cinnamon and nutmeg. See also Butternut Squash & Rosemary, page 227.

Ginger & Cabbage Stir-fried with cabbage, fresh ginger compensates for the crisp spiciness cabbage loses when cooked. Spring greens are particularly good cooked this way, as they have dense leaves that don't get so messy and flaccid when heated, and a robust, bittersweet, zesty flavor.

Ginger & Cardamom Cardamom is a member of the ginger family, a relationship that is more detectable to the nose and mouth than the eye. Both have warming properties with a distinct citric note. In India they're renowned for their beneficial effects on the digestion, and are commonly paired in a drink called *panakam*, made by dissolving jaggery (palm sugar) in hot water with dried ginger and ground cardamom, then straining and leaving to cool. I use this combination for *inji jamun*, a variation on the classic *gulab jamun*—see Rose & Cardamom, page 334—flavoring the little caky *jamuns* with cardamom as standard, but substituting my own ginger syrup for the usual rose. Make the *jamuns* by putting 1 cup powdered milk, 6 tbsp plain flour, 1 tsp baking powder, 1 tsp ground cardamom and 1 tbsp butter, cut into pieces, in a food processor, pulsing to combine, then adding water a tablespoon at a time until it becomes a silky dough. Shape into about 18 balls the size of lychees. Slowly dissolve 2 cups sugar in 2½ cups water with 4 or 5 thick slices of bruised fresh ginger. Deep-fry the *jamuns* in hot oil a few at a time, turning the heat down to medium once they're in the oil to allow them to brown slowly. Remove and drain on kitchen paper before submerging in the ginger syrup. Leave to cool and soak up the syrup, overnight if possible. Remove the ginger slices and serve chilled or at room temperature.

Ginger & Chili Chili shows ginger what heat is, which is why manufacturers sometimes add a little to ginger ale to give it that extra kick. Chili and ginger are paired in the sauce that goes with Hainanese chicken rice, originally from Hainan Province off the south coast of China but massively popular in Malaysia, Taiwan and Singapore, where the locals have pretty much adopted it as their national dish, and have dedicated several restaurant chains to it. As it's so popular, you might want to give it a whirl. The sauce is simply fresh chili pounded with fresh ginger and garlic, sometimes loosened with a little vinegar, lime juice or stock. The chicken is poached whole in salted water with ginger and scallion. The rice should be cooked with some fat from the cavity in the chicken and the water it was cooked in, now a fragrant stock. The bird is served in pieces at room temperature, drizzled with soy and sesame oil, with the ginger and chili dipping sauce on the side. Sounds easy, but the skill is in arranging the skinless chicken neatly on the serving dish, almost as if you're trying to reassemble it.

stronger in eucalyptus or floral-citrus flavors. Whichever dominates, those fresh notes are good for cutting through fattiness, especially with ingredients that let the spice's complexity of flavor shine—e.g., cream, chocolate, nuts or buttery rice.

Cardamom & Almond The Nordic countries take a disproportionate amount of the world's cardamom crop and sprinkle it liberally in their cakes, buns and pastries. Finnish *pulla* is a cardamom-spiced plaited sweet bread, Norwegian *goro* is a thin, crisp, cardamom-flavored wafer that looks like the cover of an old Bible. *Fattigman* is made with the same dough but shaped with a special cutting tool and deep-fried. On Shrove Tuesday in Sweden, they make a cardamom-scented bun called a *semla*, whose top is lopped off and its scooped-out center filled with almond paste and a blob of cream. The bun's top is then replaced and dusted with sugar. Non-bakers will find them seasonally available in Ikea.

Cardamom & Apricot Cardamom and apricot are paired in cakes such as apricot Danish pastries, in crumbles and in jams. Dried apricots, poached in syrup with cardamom, are a specialty of Kashmir. Or try this apricot tart with a luxurious cardamom *crème pâtissière*. Whisk ¼ cup superfine sugar with 3 egg yolks. Continue whisking while you add 2½ tbsp each of plain flour and cornstarch. Scald 1¼ cups milk with 3 bruised cardamoms and ½ tsp vanilla extract. Remove the pods and gradually stir the milk into the egg mixture. Transfer the lot to a pan and bring to the boil until large bubbles break on the surface. Lower the heat and cook for about 5 minutes, until very thick. Cool and use to fill a cooked 9-in sweet pastry shell. Top with about 15 poached, peeled apricot halves and glaze with apricot jam.

Cardamom & Bacon Black cardamoms are closely related to green ones. They share their warm aromatic flavor but are stronger, a little more bitter, and are roasted over a fire to dry them out, giving them a smoky flavor, like a green cardamom after a Laphroaig bender. Add a few pods to a stew or soup to impart a flavor subtly reminiscent of bacon.

Cardamom & Banana Banana and cardamom combine in a soothing raita. Some slice the banana, but I like to mash a couple of not particularly ripe ones and mix them with ¼ tsp ground cardamom, a pinch of dried chili and ⅔–1 cup yogurt, depending on the looseness demanded by the main dish (a robustly hot, tangy lamb curry makes an excellent partner).

Cardamom & Carrot *See Carrot & Cardamom, page 223.*
Cardamom & Chocolate *See Chocolate & Cardamom, page 18.*

Cardamom & Cinnamon Like Krishna and his mortal consort, Radha, legendary lovers. Together they add a sweet, aromatic note to Indian and Pakistani milk desserts. I combine them in a chai-like hot drink that gives cocoa a run for its money. Pour a mug of milk into a pan with a stick of cinnamon and 2 or 3 crushed cardamom pods and slowly bring to the boil.

Ginger & Chocolate Bite into a piece of decent chocolate-covered stem ginger and the snap of dark chocolate should give way to a sugary rasp as your teeth sink into the nap of the ginger's juicy fibers. The dark chocolate's bitter complexity and its cool, almost menthol quality provide the perfect contrast to the ginger's sweet heat. I pair them in a florentine-style cookie. It's not as pretty as the jewelry-box variegation of glacé fruits but it's far and away more flavorful. Melt 1 tbsp unsalted butter in a saucepan over a low heat. Add ¼ cup superfine sugar, 2 tsp plain flour and 2 tbsp heavy cream, and bring slowly to the boil. Simmer for 1 minute, then stir in 2 oz chopped glacé ginger and 2 oz sliced almonds, and remove from the heat. When it has cooled to room temperature, place teaspoonfuls of the mixture on lined baking sheets, pat them down and shape them into neat rounds. Make sure to keep the blobs an inch or two apart to prevent them fusing into one sticky agglomeration in the oven. Bake at 375°F for about 12 minutes. Give them a few moments to harden, then transfer to a cooling rack. Melt 2½ oz quality dark chocolate and paint it over the smooth side of the florentines, scoring wiggly lines into the warm chocolate with a fork.

Ginger & Cinnamon The heart and soul of the gingerbread man. Historically the dough was bolstered by honey and pepper; these days it's likely to be molasses and a pinch of clove. To make a gingerbread-flavored syrup for coffee, add 2 cinnamon sticks and 1 tsp vanilla extract to the syrup in Ginger & Cardamom, page 302.

Ginger & Clove *See Clove & Ginger, page 216.*
Ginger & Coffee *See Coffee & Ginger, page 24.*

Ginger & Egg In China, "century eggs" are made by plastering raw hen or duck eggs in a muddy mixture of lime, salt, pine ash and water, then storing them for weeks (or longer) in an earthenware jar or buried in the earth. When the shell is peeled off, the white will have set and turned an amber color, like cold tea jelly, and the yolk a dark gray-green. They have a sulfurous odor and a pungent ammonia flavor, and are served with slices of pickled ginger as a casual snack. The English preserved egg is boiled, peeled and pickled in vinegar and is commonly sold in chip shops and pubs. In the pub, a tradition persists of burying the pickled egg in a packet of scrunched-up potato chips (a good barman will do this for you). Salt and vinegar chips are the typical flavor used, but Worcestershire sauce, with its sweet and sour spiciness, is a superior choice.

Ginger & Eggplant *See Eggplant & Ginger, page 83.*
Ginger & Garlic *See Garlic & Ginger, page 113.*
Ginger & Lemon *See Lemon & Ginger, page 298.*

Ginger & Lime A Moscow Mule combines vodka, lime and ginger beer; it can be a lame donkey made with ginger ale. A few drops of Angostura nails a shoe of metallic bitterness to its cumulative kick. Ginger beer is hotter, fuller flavored and cloudy, whereas ginger ale is a clean, amber color, with a subtler flavor that mixes well with dark spirits such as rum and whiskey (the types with an "e": North American or Irish).

Ginger & Mango Mango ginger, or zedoary, is related to neither plant, although, like ginger, it's a rhizome. Native to India and Indonesia, it has a taste that's bitter at first, then sweet, then sour, with a musky, aromatic flavor reminiscent of green mango. It's predominantly used in pickles and curries. As to ginger *and* mango, they combine in a crème brûlée just as nicely as they do with shellfish, and the chef Jean-Georges Vongerichten marries them with foie gras in one of his signature dishes.

Ginger & Melon *See Melon & Ginger, page 273.*

Ginger & Mint Ginger mint has a flavor similar to peppermint, with the slightest hint of ginger. Candies and sodas flavored with ginger *and* mint are popular in America, and a sprig of fresh mint or a slug of mint syrup can pep up insipid ginger ale in the summer.

Ginger & Oily Fish The pickled ginger, or *gari*, that comes with *nigiri* (fish on rice) and sashimi is there to freshen the palate, ensuring you appreciate the fine flavors of each different fresh fish. For the same reason, purists prefer to eat their sushi with chopsticks because the fingers (which are perfectly acceptable to use) might transfer the flavor of one fish to another. If you are going to dip your *nigiri* in soy, dip it fish-side down, making sure you get only a light seasoning of sauce and positioning the *nigiri* so that the fish meets your tongue, so you taste it more. Dip it rice-side down and your soy-drenched nigiri will almost certainly fall apart before it reaches your mouth. In better sushi restaurants the chef will in any case have absolved you of the decision by seasoning his *nigiri* with a soy-based sauce and wasabi as he sees fit. Dipping in extra soy and wasabi will be frowned upon. It's also considered bad form to pop the ginger in your mouth *with* the sushi or scarf it as an appetizer, so do it only when you're sure nobody's looking.

Ginger & Onion *See Onion & Ginger, page 109.*

Ginger & Orange Both spicy and citrusy. In Finland, ground Seville orange powder is used with cinnamon, clove and ginger in a spice mix for gingerbread. Try this sticky ginger and orange cake. Cream 1½ sticks butter with ¾ cup light raw sugar until pale and fluffy. Beat in 2 eggs, one at a time. Grate the zest of a large orange into the mix and sift over 1⅓ cups self-rising flour, 1 tsp baking powder and a pinch of salt, before carefully folding them in. Add 3 tbsp milk and 4 finely chopped nuggets of preserved ginger. Transfer to a greased and lined round 7-in cake pan, smooth the surface and bake at 350°F for 40–50 minutes. Glaze with melted marmalade—½ cup should be enough—and leave to cool.

Ginger & Pork Brought together in a simple Japanese dish called *shogayaki* (fried ginger). *Shogayaki* can also be made with (among other things) beef or squid, but it's most popular by far with pork, which is what you'll get unless the menu indicates otherwise. Marinate ½ lb thinly sliced pork fillet in 2 tbsp

finely grated fresh ginger, 2 tbsp soy sauce and 2 tbsp rice wine for 15 minutes. Working quickly, fry the drained pork in vegetable oil and divide between two plates. Tip the marinade into the pan and heat through before pouring it over the pork. Serve with rice and a bowl of miso soup.

Ginger & Rhubarb The combination of ginger and rhubarb came about because it was considered to be good for the bowels, which may go some way to explaining why they're still paired when, to my mind, the two flavors seem a little annoyed with each other. Chef Jason Atherton serves them pickled with a pressed foie gras and smoked eel terrine and a ginger brioche. Delia Smith combines cooked rhubarb with orange jelly made with ginger beer instead of the usual water. And Andrew Pern, chef-patron of The Star Inn, near Helmsley in North Yorkshire, gives a recipe for rhubarb-ripple ice cream with a local cake called ginger parkin, which you can eat at the inn if you can't be bothered to make it yourself.

Ginger & Tomato For a piquant, gingery tomato sauce, Mrs. Beeton suggests cooking about 2 lb of ripe tomatoes in an earthenware dish at 250°F for 4–5 hours. Let them cool to room temperature, remove the skins and mix the pulp with any juice left in the roasting dish. Add 2 tsp powdered ginger, 2 tsp salt, the cloves of 1 whole garlic bulb, finely chopped, 2 tbsp vinegar and a pinch of cayenne. Pour into bottles and store somewhere cool. It can be eaten immediately but the flavors improve noticeably after a few weeks. It will taste even better if you've powdered your own ginger. Simply grate some fresh onto a baking sheet and leave to dry for 3–4 days, less if outside on a run of sunny days. Keep the dried gratings in a jar and whiz in a spice grinder when needed. You can leave the skin on for an earthier, nuttier flavor, or peel if you like it lighter.

Ginger & Vanilla Add a scoop of vanilla ice cream to ginger ale in place of the usual cola and you'll have what's known as a Boston Cooler. See also Ginger & Cinnamon, page 303.

Ginger & White Fish The citrus notes in fresh ginger make it an obvious partner for fish. Donna Hay recommends lining the bottom of a bamboo steamer with slices of ginger to impart their flavor to steamed fish. As a garnish for other dishes, she suggests a tangle of deep-fried shredded ginger. See also Onion & Ginger, page 109.

Cardamom

Open a jar of cardamom pods and you might be reminded of a vapor rub or sinus-clearing stick. Like bay leaves and rosemary, cardamom contains clear notes of camphor and eucalyptus. As a member of the ginger family, it also has a citrusy, floral quality; depending on their country of origin, cardamoms are likely to

Remove from the heat, strain back into the mug, then add sugar to taste. You could let this cool and use it to make a fragrant banana milkshake.

Cardamom & Coconut *See Coconut & Cardamom, page 280.*
Cardamom & Coffee *See Coffee & Cardamom, page 23.*

Cardamom & Coriander Seed Both spices have distinct citrus notes. If you find cardamom too camphorous for a sweet recipe, "dilute" it by crushing it with a little coriander seed. The coriander's pretty, floral quality plays harmoniously to cardamom's sweeter side.

Cardamom & Ginger *See Ginger & Cardamom, page 302.*

Cardamom & Lamb In Kashmir, cardamom is used to enrich lamb meatballs in a dish called *goshtaba*, for which small pieces of meat are painstakingly pounded with suet, resulting in a super-soft mixture likened to the texture of cashmere. Less time-consuming is *elaichi gosht*, a combination of lamb with an unusually large amount of cardamom. This is Madhur Jaffrey's recipe. Heat 3 tbsp oil in a large pan and add 2 tbsp finely ground cardamom (grind the pods up with the seeds if you can't be bothered to shell them; they add a little fiber to the sauce). Stir once and add 2 lb cubed shoulder of lamb. Continue stirring over a high heat for 2 minutes, then add 2 chopped tomatoes and a small, finely chopped red onion. Stir for another 3 minutes, then add 1½ tsp garam masala, 1 tbsp tomato paste, 1½ tsp salt and 2½ cups water. Simmer with the lid on for 1–1½ hours and add a very generous grinding of black pepper before serving with bread or rice.

Cardamom & Mango Very popular in India. Cardamom's brightness combined with the sourness of yogurt can rescue overripe mango in a lassi. Blend the flesh of 1 mango with 1 full cup yogurt, ½ cup milk, a pinch of ground cardamom and an ice cube or two. Taste for cardamom levels and sweeten with sugar, honey or the rather inauthentic maple syrup, which has a pleasing affinity for the resinous whiff in mango.

Cardamom & Pear *See Pear & Cardamom, page 268.*
Cardamom & Rose *See Rose & Cardamom, page 334.*

Cardamom & Saffron Saffron shares cardamom's yen for all things lush and creamy, and nicely complements its lemon character. Pair them in ice cream, custard and cakes, or use a little of each in savory rice dishes.

Cardamom & Vanilla *See Vanilla & Cardamom, page 339.*
Cardamom & White Chocolate *See White Chocolate & Cardamom, page 343.*

BERRY & BUSH

Rosemary

Sage

Juniper

Thyme

Mint

Black Currant

Blackberry

Rosemary

Rosemary has a eucalyptus character akin to sage, although it contains more pine and floral notes and is sweeter. There are many varieties of rosemary. Among those particularly commended for their flavor are Tuscan Blue, which has a gentle, lemon-pine aroma, Spice Island, which has hints of clove and nutmeg, and Sissinghurst Blue, which has a pronounced smoky character that begs to be thrown on the barbecue. Rosemary tastes fine when it's dried carefully, but it does take on the typically hay-like flavor of dried herbs, and noticeably loses the complexity of fresh. It is a classic partner for lamb and goat cheese, but also makes some excellent matches in sweet dishes with chocolate, oranges and lemons.

Rosemary & Almond See *Almond & Rosemary, page 240.*
Rosemary & Anchovy See *Anchovy & Rosemary, page 161.*

Rosemary & Apricot *Ma'mool,* a plump, stuffed sweetmeat found all over the Middle East and North Africa, usually contains a mix of dates and nuts spiced with cinnamon. The crumbly pastry gives a pleasing contrast to the rich, chewy filling, which can be made with all manner of dried fruit and nut combinations. When dried, apricots lose some of their fruity, lavender aroma and can become rather sour. Rosemary, reminiscent of lavender, restores some of their subtle perfume. Tip 1¾ cups plain flour into a food processor, add 1 stick unsalted butter, cut in small cubes, and whiz into crumbs. Little by little, add 1 tbsp rosewater and 3–4 tbsp milk until a dough forms. Set aside in the fridge. Put ½ cup chopped dried apricots in a pan with ⅔ cup chopped mixed nuts, 1 tbsp finely chopped rosemary, ¼ cup sugar and 4 tbsp water. Cook over a medium heat until the water has evaporated, then stir in ½ cup ground almonds and make sure all is well combined. When the mixture has cooled a little, divide the dough into 20 and roll each piece into a ball. Indent with your thumb to create a bowl and fill with the apricot mixture, easing the pastry over it until it's no longer visible. Place them on a greased baking tray and flatten a little with the tines of a fork. Bake for 20 minutes at 325°F. Once cool, sprinkle with confectioners' sugar.

Rosemary & Butternut Squash See *Butternut Squash & Rosemary, page 227.*
Rosemary & Chestnut See *Chestnut & Rosemary, page 229.*

Rosemary & Chocolate A backdrop of dark chocolate shows off rosemary's cool, evergreen flavors. If this recalls the lovely combination of Chocolate & Cardamom (see page 18), that's because the dominant flavor compound in both rosemary and cardamom is cineole, common to bay leaf too (think how all three make great milk-based desserts). Cineole has woody, eucalyptus, slightly minty notes. In rosemary, these are joined by peppery, camphorous characters, while cardamom takes it in a more citrusy, floral direction—you might say chocolate with rosemary is a wintry alternative to chocolate and cardamom. This recipe for "Little Pots of Chocolate and Rosemary Cream" was conceived by the chef David Wilson. In a heavy-based stainless-steel pan, mix 1¼ cups superfine sugar with 1 cup dry white wine and the juice of ½ lemon. Heat gently until the sugar has dissolved, stirring occasionally. Stir in 2½ cups heavy cream and cook gently, stirring constantly, until thickened. Add 1 stem of rosemary (or 1 tsp dried rosemary) and 6 oz grated dark chocolate. Bring to the boil, stirring until the chocolate has dissolved, then lower the heat and simmer for 20 minutes, until the mixture is dark and thick. Cool, strain into 8 little pots (or even 10, it's very rich), cover and refrigerate before serving.

Rosemary & Garlic See Garlic & Rosemary, page 114.
Rosemary & Goat Cheese See Goat Cheese & Rosemary, page 60.
Rosemary & Grape See Grape & Rosemary, page 249.
Rosemary & Hazelnut See Hazelnut & Rosemary, page 236.

Rosemary & Lamb The bittersweet, sap-green flavor of rosemary alleviates the fattiness of lamb, while its pine-fresh, eucalyptus notes can freshen the meat's gamier qualities. This might explain why rosemary really comes into its own with fatty cuts or older/dry-aged lamb. The Italians pair even the youngest lamb with rosemary, including *abbacchio*, the weeks-old, milk-fed lamb that's roasted whole at Easter. Ground lamb seasoned with rosemary can make good sausages and a rich sauce for tossing with thick ribbons of pasta. Chef Douglas Rodriguez has been known to serve minced lamb and rosemary-seared lamb tenderloin on flatbreads with raisins, pine nuts and goat cheese.

Rosemary & Lemon See Lemon & Rosemary, page 300.

Rosemary & Mushroom At Alinea in Chicago, Grant Achatz makes a *matsutake* mushroom cake and *matsutake* caramel, served with a cube each of rosemary and sherry vinegar jelly, a sprinkling of pine nut salt and a mastic cream. It's an autumnal combination with strong notes of evergreen. Rosemary is an evergreen shrub, as is *Pistacio lentiscus*, of which mastic is the aromatic resin, used among other things as a popular flavoring for chewing gum in Greece. Pine nuts: self-explanatory. The highly prized *matsutake* doesn't have an evergreen flavor but grows under pine trees in Japan; its name translates as "pine mushroom." It has a strong aroma, with a distinct cinnamon note, although the flavor is less remarkable. In Japan the mushrooms are gathered, grilled in the open air over pine fires and eaten with soy sauce. See also Beef & Mushroom, page 49.

Rosemary & Oily Fish When strident rosemary meets rambunctious oily fish such as sardines and mackerel, they knock the hard edges off each other and become, if not exactly refined, then certainly an enchanting couple. Finely chop some rosemary and parsley, then combine them with bread-crumbs and a little grated Parmesan. Spoon this stuffing into your gutted fish, grill them and serve with lemon.

Rosemary & Olive *See Olive & Rosemary, page 174.*

Rosemary & Onion *Socca*, the toasty, nutty chickpea bread (or pancake, depending on which way you look at it) sold at the old market in Nice, is called *farinata* once you cross the Italian border into Liguria, where it's often served scattered with finely chopped rosemary and onion. The *socceurs* or *farinatistas* or whatever they're called have special hot plates for cook-ing, so it's hard to achieve quite the same results at home. No matter: it's still a treat. Whisk together 2 cups chickpea flour (also called gram flour in Indian shops), 1¼ cups hand-hot water, 2 tsp salt and 3 tbsp olive oil and set aside for a few hours. Paint a heavy-based, ovenproof 10-in round (or 8–9-in square) pan with a good amount of olive oil. Heat the pan in the oven at 425°F, then remove and thinly cover it with half the batter. Put it back in the oven for 10–15 minutes: best to keep an eye on it in case it starts to burn. Remove and scatter with very finely chopped rosemary and onion. Eat the first warm while you're making the second.

Rosemary & Orange Orange blossoms became a popular wedding symbol in the nineteenth century, when they were used to decorate cakes, incorpo-rated into bouquets, or their likenesses embroidered on the bride's veil. They signified good luck, innocence, happiness and fertility. Rosemary has had nuptial associations for even longer. Anne of Cleves wore some in her hair at her wedding to Henry VIII, and less grand congregations carried it to signify fidelity and remembrance. Orange blossom and rosemary make a happy couple themselves. Chef Allegra McEvedy gives a recipe for an orange-flavored cake of ground cashews and semolina, sprinkled (once baked) with orange-flower water and topped with a rosemary-flavored syrup. A cake to make for a friend fresh back from honeymoon, to signify that she has no more dress fittings.

Rosemary & Pea *See Pea & Rosemary, page 201.*

Rosemary & Pork There's a great little *osteria* called Il Cinghiale Bianco in Florence. Cross the Ponte Vecchio from the north side of the river, slaloming the Senegalese handbag-hawkers, turn immediately right and it's down the Borgo San Iacopo on your left. (Turn left off the bridge and you'll find a fantastic drinking club off its head with flowers and collections of racy draw-ings on the tables. But that's for another book.) Have the *arista al forno con patate*, magically tender roast pork in a tangy, caramely gravy with roast rose-maried potatoes, the rosemary infusing the meal with a Florentine headiness to give you Stendhal syndrome before you've so much as entered a church. Scrub Stendhal syndrome: it'll give you Stockholm syndrome. You'd happily never leave. *Porchetta*, from Lazio, farther south, is boned pork stuffed with rosemary, garlic and (maybe) fennel and roasted on a spit. It's particularly associated with fairs, where slices are sold sandwiched in chewy white bread, sometimes with broccoli.

Rosemary & Potato *See Potato & Rosemary, page 93.*

Rosemary & Rhubarb Re-creating a restaurant dish at home, says Helen Rennie at culinate.com, is like piecing together a jigsaw puzzle. At Rendezvous, Steve Johnson's restaurant in Massachusetts, Rennie ate a raw rhubarb, dried apricot and dried cranberry compote, but was stuck at first as to how to reconstruct it. The key, it transpired, was to marinate the fruit in honey and rosemary for a couple of days. The flavor of rosemary works well with the sharp, fruity flavors of orange and lemon, and it's easy to see how that would apply to rhubarb too.

Rosemary & Watermelon Rosemary is the barbecuer's faithful friend. Remove the needles by stroking the stalk against their angle of incline and you have a natural, flavor-imparting skewer for cubed meat. Or scent the smoke by throwing a handful of sprigs onto the coals when your meat or veg is nearly cooked. The American food writer Mark Bittman suggests grilling watermelon steaks with rosemary. Cut 2-in-thick "steaks" (including the rind) from a small watermelon. With the tines of a fork, flick out as many seeds as you can without ruining the flesh. Brush the steaks with a mixture of 4 tbsp olive oil, 1 tbsp finely chopped rosemary, salt and pepper, then barbecue for 5 minutes on each side. Serve with lemon wedges. These make a very good accompaniment to barbecued pork.

Sage

A rugged herb, not to all tastes. Some find it too strong, too bitter, or are put off by the medicinal associations of its camphorous-eucalyptus flavor. Fresh sage has lighter, more lemony notes, but loses these when dried, taking on a stronger, musty hay character. Fresh or dried, sage has a particular affinity for dense, sweet-savory foods that benefit from its pronounced flavor and bitter finish—butternut squash, white beans, cooked onions, pork and chicken.

Sage & Anchovy *See Anchovy & Sage, page 161.*

Sage & Apple Sage is rich and domineering, slamming its tankard down and demanding the company of weighty meat dishes. You might think brisk, fruity apple would bestow some freshness on the partnership but it doesn't.

Sage is so doggedly sagey that, for all apple's efforts, a musty darkness prevails. Still, it's a lovely cold-weather combination: use it to furnish poultry with a deep, satisfying stuffing or make a pork stew with sage and apple dumplings. To make the dumplings, put 1½ cups self-rising flour in a bowl with a rounded ½ cup suet, 8–10 finely chopped sage leaves and 1 peeled, cored and finely diced small cooking apple. Gradually mix in enough cold water to make a sticky but still workable dough. Shape it into dumplings with a pair of serving spoons, as you would quenelles. Drop them onto a stew, cover and simmer for 15–20 minutes. Eat in a dank dining room, silent but for the tick-tock of a grandfather clock.

Sage & Bacon A friend was holding a cook-off. On the night, the competitors assembled in her kitchen; we chopped, blended, blanched and julienned. Some time later we were joined by a harried man who smoothed out a scribbled recipe, busied himself for about five minutes, slid a dish into the oven and disappeared into the party with a glass of wine. Of course, he won. And I peeled his recipe off the work surface as a consolation prize. It turned out to be an adapted version of Delia Smith's baked pancetta, leek and sage risotto. The oven method might not result in total textural authenticity, but it does mean you get to hang out with your guests (or the clearly swayable judges at an informal cookery competition that's only supposed to be a bit of fun). Here's how *he* made it. Find a pan that's big enough to hold all the ingredients and can go straight from the stovetop to the oven. Warm some olive oil in it and cook a 8-oz packet of smoked bacon lardons with a fairly finely chopped onion until soft. Add 1 cup risotto rice and stir to coat the grains in the oil, then add 5 tbsp white wine, 2 cups stock, 2 tsp chopped sage and some seasoning. Bring to a simmer and transfer to a 300°F oven. Bake for 20 minutes, remove from the oven, stir in 2 tbsp grated Parmesan and put back in the oven for 15 minutes. Sprinkle grated Parmesan on top and serve.

Sage & Blue Cheese *See Blue Cheese & Sage, page 65.*

Sage & Butternut Squash Try butternut squash raw. It's like French-kissing a scarecrow: straw and damp vegetables. Cooking brings out its sweeter, softer nature, unless it's partnered with sage. Sage's assertive flavor rubs off on the squash, revealing its meaty, virile side—so much so that a sworn carnivore might be won around to this pairing. The Chez Panisse café combines them on a pizza with Asiago cheese, while Russ Parsons gives a recipe for a butternut squash risotto garnished with fried sage leaves and toasted walnuts.

Sage & Chicken *See Chicken & Sage, page 34.*

Sage & Egg Sage is happy in the company of all the classic English-breakfast ingredients, but this Turkish-inspired recipe from the Australian chef Neil Perry puts an exciting twist on the pairing of egg and sage. Crush

1 clove of garlic with 1 tsp salt and mix into ½ cup Greek yogurt with some freshly ground pepper. Set aside and put 4 eggs on to poach. While the eggs are cooking, heat 1 stick unsalted butter in a pan, add 16 sage leaves and cook until the sage is crisp and the butter browned. Spoon a dollop of yogurt onto each plate, top with a poached egg, season with salt, pepper and a pinch of chili powder and finish by drizzling over the butter. Perry suggests some chopped fresh chili on the side. You can always fry the eggs, if you prefer.

Sage & Hard Cheese See *Hard Cheese & Sage, page 70.*
Sage & Juniper See *Juniper & Sage, page 316.*

Sage & Liver A legendary flavor affiliation. Liver gives the fresh pine and cedar notes in sage their due, offsetting the sanguinary flavor shared by both ingredients. But pause a moment to consider their textures. I can't make a chicken liver and sage bruschetta without first brushing the moleskin leaves against my cheek. Cooked liver sometimes takes on a similarly suede-like quality, and together the ingredients make me think of rugged country clothing and damp valleys, or taking a bite of Ted Hughes.

Sage & Onion See *Onion & Sage, page 110.*
Sage & Pineapple See *Pineapple & Sage, page 262.*

Sage & Pork Elizabeth David disliked sage, describing it as having an overpowering dried-blood flavor. The English, she claimed, had become accustomed to the herb through its common pairing with goose and duck, especially in the form of sage and onion stuffing. These days, most English people are more likely to associate the flavor of sage with pork, as it's so often used in sausages. In Italy a pork loin might be cooked slowly in sage- and lemon-scented milk. I've made this at home but, far from my flavor memory transporting me to the rusty-red rooftops and shadowy porticoes of Bologna, the ineradicable sausage association yanks me straight back to the cheerless breakfast room of a lochside B&B in Scotland, scraping at my plate in the deafening silence.

Sage & Prosciutto Salty prosciutto is one of the few ingredients capable of teasing out the fun side of serious sage. In *saltimbocca* (Italian for "jumps in the mouth"), a slice of prosciutto is laid over a flour-dipped escalope of veal, turkey, pork, chicken or even a fillet of flatfish such as flounder, and a brooch of sage is pinned through both with a toothpick. Cook quickly on both sides—sage-side first to flavor the butter. Keep warm while you deglaze the pan with something to match the meat—marsala with veal or chicken, fish stock or white wine with flounder, etc.

Sage & Tomato In Italy, sage is used to season tomato-based dishes like *fagioli all'uccelletto* ("beans little-bird style")—cannellini beans simply cooked with tomato and sage. According to Pellegrino Artusi, the same ingredients were traditionally used to cook the eponymous little birds. The Zuni Café in San Francisco makes a sage pesto, giving the herb a little time in warm olive oil before pounding it with more olive oil, garlic, walnuts and Parmesan. They strongly recommend serving it with grilled or roasted tomato dishes.

Juniper

The flavors of a country estate: juniper is the principal flavoring in gin and is frequently paired with game. While bitter ingredients are commonly balanced with their sweet opposing numbers, juniper is often paired with other bitter flavors such as black currant, tonic water, grapefruit and rare meat. The berries need to be crushed in order to release their thick evergreen flavor.

Juniper & Beef *See Beef & Juniper, page 48.*
Juniper & Black Currant *See Black Currant & Juniper, page 324.*

Juniper & Cabbage These come together most famously in *choucroute garnie*, a sort of carnivore's lucky dip where the prizes are rich, meaty treats: wobbly chunks of pork belly, upright frankfurters, a pale and sinister *boudin blanc*, smoked knee, pickled hock, brined thigh. But the sauerkraut is much more than plastic hay. I love the slender ribbons of pale, barrel-fermented cabbage, scented with the balsamic fragrance of juniper. Its sharpness is what stops the sweet fattiness of the meat overpowering the dish: a cynic in a roomful of corpulent bores.

Juniper & Grapefruit Kingsley Amis recalls enjoying a cocktail called a Salty Dog on a trip to Nashville, Tennessee. Dip the rim of a glass in water, then salt, carefully pour in one part gin to two parts fresh grapefruit juice, add ice and stir. Amis calls gin and tonic "suspect" and "a rather unworthy, mawkish drink, best left to women, youngsters and whisky distillers." He considers gin and water a superior combination, not least because drunk without ice and with just a splash of Malvern water, it will give you a chance to taste the botanical flavorings unimpaired and appreciate the difference between the brands. He also approves of gin with ginger beer and plenty of ice, "one of the great long drinks of our time."

Juniper & Hard Cheese Pecorino ginepro, popular in Italy, is a semi-hard ewe's milk cheese soaked in juniper and balsamic vinegar. It's nutty and tangy, as you'd expect from a pecorino, and the flavor of the juniper works well with the gaminess of the cheese.

Juniper & Lemon Juniper lords it over the other aromatics that lend gin its characteristic mix of flavors—coriander, angelica, anise, fennel seeds, cardamom and orange, among others. Lemon can be one of the more discernible secondary flavors, but then juniper itself has citrusy-lemon notes to it. Which is why, in turn, lemon is such a classic partner for gin—floated in a G&T, juiced in a Tom Collins, in a classic gin fizz (with sugar, soda and ice), or in the mixer, bitter lemon, whose bitterness comes, as with tonic water, from quinine.

Juniper & Olive Flying is tough on the body. After a *lot* of pretzels, a handful of olives, a Bloody Mary, a very small sandwich, two boxes of crunchy snacks of no known national provenance, a glass of champagne, a suspiciously smooth pâté with bread roll, a glass of red, a salad of explosively tough cherry tomatoes, a beef bourguignon whose accompaniment of boiled carrots and mashed potato serves only to emphasize the meat's weird insubstantiality, as if it came from the ghost of a cow, another glass of red, a horrible chocolate mousse, a Bailey's, a doll's-house selection of crackers with silicone cheese, a port (nice), two coffees (not), a snack pack of Jaffa Cakes, a three-quarter-flight selection of sandwich slices, a truly disgraceful cup of tea, a scone with raspberry jam and cream, and, to de-pop the ears on descent, a cherry Life Saver gone misshapen at the bottom of my handbag, I often feel a little brackish arriving at JFK. No better cure for this than a dirty martini. They're like regular martinis, except with brine from the olive jar added; the murky, salty oiliness of the brine brings the gin's aromatics alive. And boy, do they give you an appetite.

Juniper & Orange Gin and orange may have slipped quietly out of fashion but this hardy, wintry pair still finds plenty to do on the plate. They're often used to season strong-flavored meats such as venison and duck, but have an affinity for other bold, bitter ingredients such as cabbage, chicory and dark chocolate. La Sambresse, from the Brootcoorens Brewery in Belgium, is a beer flavored with bitter orange peel and juniper.

Juniper & Pork Juniper's bold freshness cuts through fat. Juniper and pork are classically paired in pâtés and terrines, but you don't have to go to great lengths to enjoy this combination. Elizabeth David recommends mixing crushed dried berries with chopped fennel bulb, salt, olive oil and garlic as a seasoning for pork chops.

Juniper & Prosciutto *See Prosciutto & Juniper, page 170.*
Juniper & Rhubarb *See Rhubarb & Juniper, page 250.*

Juniper & Sage Juniper has a hint of that evergreen aroma, oily but fresh, that clears your sinuses when walking through coniferous forests. The juniper berry is, in fact, the only edible spice from a conifer. Sage also has notes of pine, cedar, pepper and eucalyptus; paired with juniper, it makes an excellent stuffing for duck. Try 1¼ cups fresh breadcrumbs mixed with 2 tbsp chopped sage, 10 crushed juniper berries and some cooked, chopped onion. Season and add a little lemon or orange zest.

Thyme

This chapter, although primarily concerned with the common thyme, *Thymus vulgaris*, also covers some other cultivars, such as lemon thyme and orange thyme. Common thyme is the type you brush past on the mountain trails and coastal paths of the Mediterranean; strong, with a sweet, herbaceous warmth that can tip into smokiness or a medicinal quality. For me, thyme is the essence of the word herbal—almost neutrally so—and forms the backbone of a bouquet garni or *herbes de Provence*. Its bittersweet, aromatic flavor flourishes in slow-cooked tomato sauces, braised meat dishes and bean stews. It also brings a tantalizing hint of lush pasture to dairy, and increasingly turns up in sweet dishes.

Thyme & Bacon See *Bacon & Thyme, page 168.*

Thyme & Beef Delia Smith says she always adds a little thyme to her beef stews. Don't worry if you don't have any fresh—dried can be even better. Thyme is one of the few herbs whose flavor intensifies when dried, in the best cases taking on a distinctly spicy, smoky character that complements its herbal perfume, bringing welcome layers of flavor to a rich meat dish. Orange thyme and caraway thyme are good with roast beef too.

Thyme & Chicken See *Chicken & Thyme, page 35.*

Thyme & Chocolate New best friends. Thomas Keller serves them together at The French Laundry in California's Napa Valley and at Per Se in New York, where, before your eyes, dark lids of chocolate are sprinkled with Maldon sea salt, then anointed with hot olive oil, which melts through the chocolate to reveal the thyme ice cream underneath. At Taillevent, the Parisian restaurant where Keller once worked, they have been known to serve thyme ice cream with a *moelleux au chocolat* (molten chocolate cake), which is somewhat easier to reproduce at home. Taste the thyme custard before its second heating and I think you have a clue as to why this pair seems to work so well. The thyme makes the cream taste as if it came farm-fresh from the churn, more pasture than pasteurized, and the dark chocolate like the freshest milk chocolate in the mouth. To make thyme ice cream, put about 10 leafy thyme sprigs in a pan with 1¼ cups full-cream milk, scald, then leave to cool. Cover and steep overnight in the fridge. The following day, scald the milk and thyme again, then strain. In a bowl, beat 4 egg yolks with ½ cup superfine sugar, then slowly pour the warm, thyme-infused milk onto that mixture, stirring all the time. Transfer to a clean pan and stir over a low heat, without letting it boil, until the mixture is thick enough to coat the back of the spoon. Strain through a fine sieve, stir in 1¼ cups heavy cream, then cool and freeze in the usual way while you get on with making the cakes. You can't go wrong with Galton Blackiston's recipe for melting chocolate puddings, passed on by Delia Smith in several of her books and on her Web site.

Thyme & Cinnamon *See Cinnamon & Thyme, page 215.*
Thyme & Garlic *See Garlic & Thyme, page 114.*

Thyme & Goat Cheese Flavoring cheese with thyme can be traced back to Roman times. The natural thyme flavor in milk from animals that had fed on the herb must have acted as a serving suggestion. Roves des Garrigues, made from the raw milk of the Rove goat in Provence, has a thyme flavor, as do some Greek fetas and Fleur du Maquis, a sheep's cheese from Corsica. (Both *garrigue* and *maquis* roughly translate as "scrubland," which gives you an idea of the sort of thorny, scented stuff these goats have nibbled on.) Bees are crazy for thyme, and thyme honey is highly prized in Greece for its unique, assertive flavor. How perfect that would be drizzled over the cheese.

Thyme & Lamb A classic combination, even if rosemary and mint have a more famous relationship with lamb. In *Papilles et Molécules* ("taste buds and molecules"), François Chartier, a sommelier from Quebec, analyzes food at the molecular level in order to arrive at wine pairings. He notes that thymol is key to the flavor of thyme as well as a constituent of lamb, and that there are red wines from the southern Languedoc in France that share that flavor note and would therefore make an exceptional pairing.

Thyme & Lemon *See Lemon & Thyme, page 301.*
Thyme & Mushroom *See Mushroom & Thyme, page 81.*
Thyme & Oily Fish *See Oily Fish & Thyme, page 157.*

Thyme & Olive These two eke out an existence on the stony, parched soil of the Mediterranean. The feral, fragrant character of thyme contrasts beautifully with heavy, complex olive, somber as a Greek church bell at noon.

Thyme & Onion Thyme adds both depth and freshness to a creamy onion soup. Roughly chop 3 or 4 onions and sweat them in a little butter and oil, with a few sprigs of thyme, until really soft but not brown. Add 2 cups vegetable or chicken stock and 1 cup milk. Season, then bring to the boil and simmer for about 15 minutes. Leave to cool slightly, discard the thyme, liquefy and reheat to serve. Add a swirl of cream and some tiny thyme leaves if the spirit takes you.

Thyme & Orange Thymol, the compound responsible for thyme's distinctive flavor, is also what distinguishes mandarin from other members of the orange family. Inhale the spritz as you're peeling a mandarin and you'll detect a beautiful herbal quality amid the citric zing. Orange and thyme are paired together far less than lemon and thyme but when they are, they often turn up with gamy poultry such as guinea fowl or turkey. The flavors of thyme and orange also occur naturally in orange balsam thyme, which can be hard to find but has a perfumed, musky orange-peel flavor that's popular in West Indian cooking and used to be a standard seasoning for beef. I refer readers who *do* have orange thyme, a 99 lb sea turtle, some truffles and, I imagine, a pot the size of a lifeboat, to a gargantuan footnote in Daniel Defoe's *Robinson Crusoe*, which will enlighten you on how to cook them together.

Thyme & Pork Weatherproof, thankfully. On a trip to the Dordogne it rained all morning, most of the afternoon and torrentially for a large part of the night. When it did let up, it was only for the clouds to accumulate fat raindrops of laughter to weep the moment anyone tiptoed toward the barbecue. Optimistically I had planned to cook garlicky pork sausages and thyme-fried onions to stuff into French bread smeared with Dijon mustard. Royally rained off, I put Plan B into action. Swaddled in layers of shapeless walking clothes, I braved the cold, dank kitchen, attached the gas to the stove and went through a box and a half of soggy matches trying to light the blasted thing. Once I had, I found a cast-iron pot, cleared it of spiders, and put in oil, the sausages, some roughly sliced onions and a jar of sinister-looking beans. From the overgrown garden I plucked a sprig of thyme, vivid green from its soaking, and threw it in the pot with salt, pepper and a long slug of wine. Putting it in the oven, with a prayer to the presiding spirit of butane, I went to sit by the fire that my husband had made from a stack of tatty paperbacks about idealistic English people making an ultimately rewarding hash of moving to France or Tuscany, called things like *Stumbling on Artichokes* or *Nothing Toulouse*. I peeled off a few layers of Gore-Tex and fleece. Gradually the smell of pork, thyme and garlic displaced the fungal reek of long-empty vacation home. Beans have an amplifying effect on the aroma and flavors of food. It started to feel just a little homey. Of course you'd be completely glorifying my dish to call it a cassoulet, but then you'd be glorifying my five days in France to call it a vacation.

Thyme & Shellfish Thyme can be a little boisterous for delicate shellfish, which is more at ease in the company of basil or tarragon. In robust shellfish stews (or Manhattan and New England chowders), however, it furnishes a deliciously spicy, minty-herbal background note. This simple dish will give you a taster. Soften ½ chopped onion in olive oil, then add a can of drained, rinsed cannellini beans and a stem of fresh thyme. Season and leave to simmer very gently while you fry some shrimp in butter and garlic. Remove the herb from the beans and serve them in a neat heap with the shrimp on top.

Thyme & Tomato Thyme with tomato is a sort of tomato-and-oregano-lite. Thyme's flavor comes from a phenolic compound called thymol, which tastes like a softer, friendlier version of carvacrol, the main flavor compound in oregano. Oregano's secondary flavor compound is thymol, thyme's carvacrol, so they have a lot in common. Clearly they can be paired with similar flavors—lamb, goat cheese and garlic, although oregano turns up less frequently in sweet dishes (that said, chef Claude Bosi serves a chilled soup of Jaffa orange and yogurt with oregano ice cream). And it's said that American servicemen, returning home after the Second World War with a taste for pizza made with oregano, created a market for the cultivated herb—until the mid-twentieth century it was grown on only a small scale in the United States.

Thyme & White Fish *See White Fish & Thyme, page 147.*

Mint

Mint is moody. Turns black when you chop it with a knife. In the UK, spearmint is paired with summer produce such as new potatoes, soft fruits, baby carrots and peas, and if you're not careful their delicate flavors can be overwhelmed by its sweet melancholy. Mint really cheers up when it's partnered with strong flavors, as in a richly beefy Vietnamese *pho*, or chargrilled lamb kebabs, feta cheese or dark chocolate. In the Middle East peppermint is sometimes combined with lemon verbena to make a soothing tea. The different flavor of peppermint is attributable in part to its cooling menthol content. Peppermint is grown predominantly for its essential oil, used in candy, ice cream, dental products and the mint-flavored liqueur crème de menthe.

Mint & Anise *See Anise & Mint, page 181.*
Mint & Asparagus *See Asparagus & Mint, page 129.*
Mint & Avocado *See Avocado & Mint, page 197.*
Mint & Basil *See Basil & Mint, page 211.*

Mint & Beef A journalist for *Le Parisien* recently wrote that, as a nation who served boiled beef with mint, the British had no right to comment on anything to do with agricultural policy. On the coach trip of international cuisine, the Brits, so the journalist implied, were the hapless child whose mother had packed him egg sandwiches, while the rest of Europe gagged into their satchels. Leaving aside that we eat our mint with roast lamb, not boiled beef, he might have noted that the Vietnamese eat mint with beef, in their fragrant soups, salads and spring rolls; and that during their Festival of the Holy Spirit, the Portuguese share a communal meal of slowly simmered, spicy beef broth, poured over a thick slice of bread, a chunk of beef, and cabbage garnished with a sprig of mint.

Mint & Black Currant Deep, dark and herbal, this combination has a whiff of one of those cough medicines you actually quite like. Combine them for a black currant and mint turnover. Place a heaped tablespoon of black currant jam and some torn mint leaves on a 4-in square of puff pastry. Moisten the edges with milk, fold the pastry in half on the diagonal and press the edges together to seal. Make a couple of small slits in the top with a knife, then brush with milk and sprinkle with sugar if you fancy that. Bake at 425°F for 12–15 minutes. Don't be tempted to add more jam; it will only escape from the pastry and burn.

Mint & Black Pudding *See Black Pudding & Mint, page 42.*
Mint & Chili *See Chili & Mint, page 206.*

Mint & Chocolate Hell is a milk-chocolate mint crisp. The kind whose flecks of mouthwash-flavored grit the manufacturers hope we'll be too drunk, after dinner, to spit back into the foil. Fudgy, saccharine milk chocolate meets sinus-widening menthol: I've had more appetizing things collect in my dishwasher filter. Mint with bitter dark chocolate, on the other hand, you can feed me till my teeth ache. It was around Christmas 1978 that I realized the potency of the After Eight mint as a symbol of infinity. First, it was *always* after eight, if you thought about it. Second, there was the subtle, if not occult, clue embedded in the name: After 8 . . . After ∞. What came after ∞? Nothing. *Exactly.* Then there was the wafer-thin mint itself. First the delicate snap of dark chocolate, bitter as plum skin. Then soft fondant so sweet your ears start straining back, until the peppermint invades your nasal passages like an inhalation, not so much refreshing your palate as dry-cleaning it, and leaving you fidgety for your next hit of chocolate before you've tongued the last trace of fondant from the roof of your mouth. All seemed powerful arguments for *never stopping eating.* And the packaging smells so good. The crisp foil of a Bendicks chocolate mint isn't a patch on the After Eight's musky black envelope. I could imagine tearing them open, like those fold-out samples you get in magazines, and smearing a hint of Nestlé's No.8 behind each ear.

Mint & Cilantro Cilantro can be slightly soapy, which in combination with max-fresh mint might savor more of the bathroom than the kitchen. Counter this by adding them to a pounded mixture of shallots and chili to make a garden-bright sambal. Loosen the mixture with lime juice or coconut milk. A sambal is essentially a relish, popular in Southeast Asia, and so is used to accompany rice or noodle dishes, simply cooked fish or meat, or spread into sandwiches. In some sambals the ingredients are pounded into a paste; others might have grated fruit or vegetables added, especially pineapple, carrot or cucumber.

Mint & Cinnamon *See Cinnamon & Mint, page 214.*
Mint & Cucumber *See Cucumber & Mint, page 185.*
Mint & Cumin *See Cumin & Mint, page 86.*

Mint & Dill *See Dill & Mint, page 187.*
Mint & Fig *See Fig & Mint, page 332.*

Mint & Garlic Mortal enemies in the breath wars. French chefs keep them apart, whereas their Turkish counterparts stir dried mint and garlic into thick, salted yogurt to serve with roast vegetables. Mint and garlic also feature in this unusual red lentil dhal from Madhur Jaffrey. Cook 2 crushed garlic cloves in 2 tbsp vegetable oil or ghee with ¾ tsp cayenne pepper. When the garlic starts to sizzle, add 1 cup red lentils, ½ tsp turmeric and 3 cups water. Stir, bring to the boil, then simmer until the lentils are tender. Add 3–4 tbsp chopped mint, 3–4 sliced green chilies and 1 tsp salt. Simmer gently while you fry 2 more sliced garlic cloves in 2 tsp vegetable oil until golden. Add these to the lentils, stir and cook, covered, for a minute or two more. See also Globe Artichoke & Mint, page 127.

Mint & Ginger *See Ginger & Mint, page 304.*
Mint & Globe Artichoke *See Globe Artichoke & Mint, page 127.*
Mint & Goat Cheese *See Goat Cheese & Mint, page 59.*
Mint & Lamb *See Lamb & Mint, page 54.*

Mint & Lemon Lacking peppermint's menthol wintriness, spearmint is the warmer, sweeter variant, and will seem even more so in contrast to bitter or sour flavors. Try some in homemade lemonade. Pare the zest off in long strips from at least 2 of 4 lemons, avoiding as much of the white pith as possible. Add to a pan with 1 cup sugar and ¾ cup water and bring to a simmer, stirring until the sugar melts. Remove from the heat and give the lemon zest a bit of a bash to release its oils. Juice the 4 lemons into a pitcher. When the sugar syrup has completely cooled, strain, then mix it with the lemon juice. Add water (sparkling or still) to taste. Add sprigs of mint and allow to infuse a little before serving.

Mint & Lime *See Lime & Mint, page 295.*

Mint & Mango Combine in a fruity raita to serve with a chickpea curry, in a couscous, or with crab in a shredded Vietnamese-style salad. But as a match for mango, mint can't hold a candle to lime—unless, of course, you're in an Indian restaurant with a stack of hot poppadoms and a carousel of pickles. A shard of poppadom with a blob of mango chutney and minty raita is, in this rare case, far superior to the mango chutney/lime pickle combination.

Mint & Melon *See Melon & Mint, page 274.*
Mint & Mushroom *See Mushroom & Mint, page 80.*
Mint & Oily Fish *See Oily Fish & Mint, page 156.*
Mint & Onion *See Onion & Mint, page 109.*

Mint & Orange How can people possibly enjoy this combination? Have they never drunk orange juice after brushing their teeth?

Mint & Parsley *See Parsley & Mint, page 190.*

Mint & Pea Pea with mint tastes like England in June. The pea's flavor is as bright and simple as sunshine, which mint overcasts with its own damp, gloomy take on summer.

Mint & Peanut Mint jelly, or torn-up mint leaves, is sometimes paired with peanut butter in sandwiches. Try it before you decry it. If you're familiar with the chopped mint and roasted peanut garnish served with many Vietnamese dishes, you'll be able to imagine how successful this can be. The Kiwi chef Peter Gordon combines them in a lime-marinated cucumber salad, which, thoroughly drained, might be exceptionally good with peanut butter in a French-bread sandwich, *bành-mì* style. See Cucumber & Carrot, page 183, for more on these wonderful Vietnamese snacks.

Mint & Potato Might turn up in a ravioli in Rome. Mint-flavor potato snacks are popular in India, too, and are sweet and mellow in comparison to many flavors of British crisp. Closer to home, Irish chef Darina Allen adds chopped mint to a potato soup before blending it and garnishing with cream and more mint. But new potatoes simmered with mint and tossed in butter are hard to beat, as noted by George Orwell in his essay "In Defence of English Cooking," in which he proclaims their superiority to the fried potato dishes traditional in most other countries.

Mint & Raspberry *See Raspberry & Mint, page 330.*
Mint & Strawberry *See Strawberry & Mint, page 258.*

Mint & Watermelon For a gorgeous variation on the mojito, make a mint syrup by dissolving 1 cup sugar in 1 cup water over a low heat. Add roughly 20 torn mint leaves (you need to tear them to release their oil) and leave to steep until the syrup has a pleasing mintiness, then strain and chill. Liquefy some watermelon flesh, strain it and pour into a highball glass with a few tablespoons of the syrup, a slug of white rum and some ice.

Black Currant

An evocative combination of fruity, herbal and musky flavors, with a sour taste. Black currants have a distinct hedgerow aroma—the gorgeous, heady fragrance of country lanes on late-summer evenings. To make them properly palatable, sugar or honey must be added in increments in order to reach the optimal sweet spot. Like all dark berries, black currant loves apple, but to see what an exciting flavor it can be, it has to be tried with other bold flavors, such as juniper, peanut and coffee.

Black Currant & Almond Sarah Raven gives a recipe for black currant and almond cake that's as good for pudding as for afternoon tea. If the volume of black currants seems conservative, that's because their flavor is so intense—it would be counterproductive to use more. The nuts are given a boost by almond extract, which gives the cake a marzipan-like flavor. Cream 1¾ sticks butter with 1 cup superfine sugar. When the mixture has gone pale, add 3 eggs, one at a time, beating in well. Fold in 2 cups ground almonds and 1 tsp almond extract. Put the mixture in a greased and lined 10-in loose-bottomed pan and scatter over ½ lb black currants. Cook for 30 minutes at 350°F until golden and firm to the touch. Sift some confectioners' sugar over before serving with cream, crème fraîche or Greek yogurt.

Black Currant & Anise In the days before the alcoholic-beverage industry thought of sugaring hard booze with cordial and flogging it to teenagers, we had to do the work ourselves. We added lime to lager, orange to gin, lemonade to port and black currant to absolutely everything: lager and black, cider and black, Guinness and black, *bitter* and black. (We would have added it to white wine, as we'd heard the French did, in *kir*, and to red in the less famous *communard*, or *cardinale*—which is not so offputting an idea when you think of the black currant notes in Cabernet Sauvignon and Pinot Noir. But wine wasn't sold much in pubs back then.) If you wanted black currant in something daintier than beer, you'd order a Pernod and black, which tasted like those black currant sweets you sucked until the fruity coating cracked and—a little disturbingly, as in a dream of broken teeth—gave access to the soft, slightly salty licorice center. I relive this combination by pairing a tangy black currant sorbet with pastis ice cream. For the pastis ice cream, combine 1 cup heavy cream, 1 cup milk, 4 egg yolks and ¾ cup superfine sugar in a heavy-bottomed saucepan. Cook over a low heat, stirring constantly, until the custard is thicker than cream. Strain the custard immediately into a bowl and stir for a minute or so. When cool, chill in the fridge until good and cold. Add 3 tbsp Pernod or Ricard and freeze according to your usual method. For the black currant sorbet, put 1½ cups water and 1 cup superfine sugar in a pan and heat gently, stirring now and then, until the sugar has dissolved. Add 1 lb black currants and cook gently, covered, for about 5 minutes. Remove from the heat, cool and pass through a fine sieve. Chill, then freeze in the usual way.

Black Currant & Chocolate See Chocolate & Black Currant, page 18.
Black Currant & Coffee See Coffee & Black Currant, page 23.

Black Currant & Juniper Black currant and juniper are both powerfully flavored northern European berries, typically paired to make a sauce for dark game meats. They also make a good flavored jelly to serve with dark meat and mature hard cheese. Or you can infuse gin with black currants and sugar for a potent liqueur. On a softer note, Lindt makes a milk chocolate bar filled with chocolate mousse and a black currant and juniper jelly.

Black Currant & Mint *See Mint & Black Currant, page 321.*

Black Currant & Peanut I was expecting a degree of culture shock when I went to live in America, but what really caught me off-guard were the peanut butter and jelly sandwiches—not *per se*, but because they were made with grape and sometimes even strawberry jam. Who were they kidding, not using black currant? And who was I, a Limey, to complain? Smuckers, manufacturers of America's market-leading Jif peanut butter, clams the PB&J dates from World War II, when GIs were supplied with both products in their rations. In the absence of actual butter, you can see how the peanut variety might have caught on. After the war, the popularity of the sandwich skyrocketed but, without wanting to turn my nose up at something almost as American as apple pie, I couldn't see how sweet strawberry or grape could have been thought a better partner for fatty, salty peanut butter than the sharp and complex black currant. I didn't know that black currants were, in fact, rarely eaten in the States. At the turn of the twentieth century it was discovered that the black currant plant, *Ribes negrum*, was a vector for a disease that affected the white pine tree, which was vital to the booming construction industry. Growing black currants became a crime under federal law, and although enforcement of the ban has since been shifted to state jurisdiction, only a few states have lifted it. So the flavor is much less familiar to the American than to the European palate. Where our purple sweets are almost always black currant flavor, American purple candies are grape—specifically Concord grape, derived from a species native to North America. Although black currant and Concord grape are easily distinguishable as flavors, both share a "catty" quality in common with Sauvignon Blanc, gooseberries and green tea.

Black Currant & Soft Cheese *See Soft Cheese & Black Currant, page 73.*

Blackberry

Store-bought or cultivated blackberries, tall as beehive hairdos, bright as spit-and-polished toecaps, may sometimes be pleasantly sweet but they never, ever have the countervailing intensity of sharpness, mustiness and deep spice that comes of growing in the wild. There are hundreds of different strains of wild blackberry, and a berry picked in one spot may taste quite different from another a few feet away. Look for notes of rose, mint, cedar and clove beyond the generic berry flavors. Some even have a shimmer of tropical fruit. Come late August, when there should be plenty to choose from, treat the bushes like free-sample ladies, and once you've found a juicy, full-flavored strain, denude the bush until your ice-cream carton is full. Black, shiny fruit won't be as sweet as those that have reached the matt blue-black of full ripeness, but then again, they're less likely, having retained their bulbous resistance, to dissolve in your grasp like a teenager's handshake. Although blackberries can add a fruity, spicy flavor to sweet vanilla cakes or sauces for game, they have an overriding affinity for apple—so much so they could almost be monogamous.

Blackberry & Almond *See Almond & Blackberry, page 237.*
Blackberry & Apple *See Apple & Blackberry, page 264.*

Blackberry & Beef Blackberry sauce is often served with duck and venison but can be used on beef and other meats too. Aside from providing a pleasing tartness, the herbaceous, slightly spicy flavor of blackberry complements the meat's natural sweetness.

Blackberry & Goat Cheese *See Goat Cheese & Blackberry, page 57.*
Blackberry & Peach *See Peach & Blackberry, page 277.*

Blackberry & Raspberry From the same genus, blackberry and raspberry share some flavor characteristics, but differ most clearly in blackberry's muskiness and tendency toward delicious cedarwood notes. To confuse matters, there are such things as black raspberries and red blackberries. A surefire way of telling one species from the other is whether or not the core of the fruit stays inside the berry when picked—raspberries come away clean, leaving that characteristic cavity, which I used to love peering into as a child; blackberries retain their woody flower base. Then there are the many raspberry–blackberry crosses. Loganberry, the oldest, is thought to have been an accident. In California in the late 1880s, the lawyer and horticulturist James Harvey Logan, attempting to cross two American varieties of blackberry, accidentally planted them next to an old European raspberry cultivar and the two species got friendly. Like many unplanned offspring, loganberry has had its identity crises, and is given to sourness. But it's more suited to cooking than the raspberry. Further hybrids include the tayberry, tummelberry and boysenberry, which, even more conflicted about its parentage, contains unmistakable strawberry notes.

Blackberry & Vanilla *See Vanilla & Blackberry, page 339.*

Blackberry & White Chocolate Put a small bar of white chocolate in your pocket the next time you go blackberrying on a sunny, late-summer afternoon. When your bag has begun to leak crimson juice, find a warm spot and carefully unwrap the chocolate, which should have melted by then. Select a handful of your best berries, check them for bugs and dip them in the chocolate, safe in the knowledge that even if you were to eat the whole bag you could always pick some more.

FLORAL FRUITY

Raspberry

Fig

Rose

Blueberry

Coriander Seed

Vanilla

White Chocolate

Raspberry

Raspberries have a sweet-sour taste and a fruity, floral (especially violet), leafy flavor; their seeds contribute a warm, woody note. When ripe, they have an intense, perfumed quality and a sweetness that hints at raspberry jam. The ubiquity of raspberry flavor can blind one to its loveliness, but there are few more delightful starts to the day than a spread of good raspberry jam on buttered toast. Raspberry flavor mixes harmoniously with other sweet-sour fruits such as apricot, blackberry and pineapple, as well as some of the lighter herbal flavors. Like strawberries, raspberries have some inherent dairy notes, which make them especially delicious with cream, yogurt and soft cheeses.

Raspberry & Almond Raspberry really benefits from a partnership with the soft, sweet flavor of almond. The nut planes down the fruit's sharper corners but never gets in the way of its sublime flavor. Try them together in an almond panna cotta with raspberry sauce or in a Bakewell tart (an open jam tart smothered in frangipane), made as follows. Line an 8-in tart pan with sweet pastry and spread about 3 tbsp jam over the base. Cream together 1 stick butter and ⅔ cup superfine sugar, then gradually mix in 3 beaten eggs. Add 1 tsp almond extract and 1½ cups ground almonds. Carefully spread this mixture over the jam and bake for about 35 minutes at 200°F, sprinkling 1 tbsp slivered almonds over the top of the tart 15 minutes before the cooking time is up. If you're using bought jam, Wilkin & Son's Tiptree Raspberry with seeds captures the fruit's flavor in a way that'll make you weak at the knees. Use a seedless jam if you prefer not to spend all day digging out your molars with a fingernail, but note it'll lack the woody contribution that the seed makes to the overall flavor, which partners so nicely with the almond.

Raspberry & Apricot Both members of the extended *Rosaceae* family, these two are renowned for their finely perfumed flavors that teeter between sweet and sour. Apricot is a drupe fruit—i.e., it's fleshy, with a seed-containing stone inside. A raspberry is an aggregation of drupelets, each tiny sphere an individual fruit containing its own dinky stone. If you can get your hands on raspberries and apricots in season, at their peak of perfumed sweetness, I can recommend nothing better than to arrange them in a bowl together and serve them at room temperature. Mediocre fruit can be redeemed by baking it, throwing the raspberries in five minutes before the apricots are cooked. Or you could make this very pretty rustic tart. Roll out some bought puff pastry to about ¼ in thick, cut out a 9-in round and place on a baking sheet lined with baking parchment. Scatter ½-in-thick slices of apricot over the pastry, leaving a ½-in margin around the edge, then dot raspberries over and sprinkle with a tablespoon or two of superfine sugar. Bake at 400°F/for about 25 minutes, until the pastry is browned. Remove and glaze with some sieved warmed apricot jam. Cool slightly and serve with a little cream or vanilla ice cream.

Raspberry & Basil Sophie Grigson recommends the partnership of basil and raspberry. She pairs them in an ice cream or suggests that you macerate some torn-up leaves with raspberries, sugar and either a shot of gin or a small squeeze of lemon or lime juice.

Raspberry & Blackberry See Blackberry & Raspberry, page 326.
Raspberry & Chocolate See Chocolate & Raspberry, see page 21.
Raspberry & Coconut See Coconut & Raspberry, see page 282.

Raspberry & Fig Raspberry and fig is a harmonious combination, and "berry jam" is one of the common flavor descriptors for good figs (notably the famed Violette de Bordeaux cultivar), as is honey, after which many figs are named: Autumn Honey, Italian Honey and the slightly less evocative Peter's Honey. Judy Rodgers, of San Francisco's Zuni Café, writes that her favorite dessert is figs dipped in buttermilk, then flour, deep-fried in peanut oil and served with whipped cream, raspberries and lavender honey.

Raspberry & Goat Cheese See Goat Cheese & Raspberry, page 59.

Raspberry & Hazelnut Walking the Dolomites in September, you'll be able to pick enough of both to make a hazelnut cake and the raspberry filling to go in it. Austrian Linzertorte, one of the oldest cake recipes recorded (in 1696), combines hazelnut (or sometimes almond) pastry with a thick filling of jam (often raspberry, but apricot and plum are used too) and a lederhosen criss-cross of pastry on top.

Raspberry & Mint Michel Roux recommends serving berries with a mint-flavored *crème anglaise* (otherwise known as custard). Raspberry and mint will also make a fine dressing, especially on a goat cheese salad—use a table-spoon each of raspberry vinegar and freshly squeezed lemon juice to 4 tbsp extra virgin olive oil. Shake with salt, pepper and 1 tbsp chopped mint until well mixed.

Raspberry & Peach The great French chef Auguste Escoffier created the Peach Melba for Dame Nellie Melba, the Australian opera singer. Apparently the idea came to him as he was writing out the recipe for *pêches cardinal au coulis de framboise*, which the great diva had requested. Peach Melba is virtually the same thing—peaches with a raspberry sauce and a scattering of fresh almonds, if they're in season—but with added vanilla ice cream. See also Peach & Vanilla, page 278.

Raspberry & Pineapple Barratt's Fruit Salads, one of the greatest achievements in the field of British penny candies, combine the flavors of raspberry and pineapple. James Beard wrote about the two fruits' remarkable affinity for each other. As an American born in 1903, he's unlikely ever to have tried a Fruit Salad, but he may well have drunk a "Queen's Favorite" soda—a mixture of pineapple, raspberry and vanilla. The combination also lies at the heart of a French martini, which is made with Chambord, a liqueur containing black raspberries, herbs and honey. Put 3 tbsp vodka, 1 tbsp Chambord and 2 tbsp pineapple juice in a cocktail shaker with ice, shake, then strain into a cold martini glass.

Raspberry & Strawberry *See Strawberry & Raspberry, page 258.*
Raspberry & Vanilla *See Vanilla & Raspberry, page 341.*

Raspberry & White Chocolate Naturally sour raspberries are a great match for the sweet dairy flavor of white chocolate. In London, The Ivy's Scandinavian Iced Berries with Hot White Chocolate Sauce pairs small berries, not long out of the freezer, with a warm sauce of white chocolate melted with heavy cream. The contrasts in taste and temperature are delightful and, its sweetness moderated by the cream, the white chocolate serves to emphasize the berries' sharpness and perfumed flavor rather than, as is so often the case with this pairing, swamping them.

Fig

Fresh figs are sweet, with a light berry flavor overshadowed by the heady perfume that develops when they are properly ripe. They're very difficult to transport at this stage, so for optimal flavor you need either to live near a reliable source or to grow them yourself. Dried figs take on a much more forceful personality, with a bumped-up sweet-sour balance akin to dried apricots. Some dried figs have a sugary, fermented character, sumptuous as fortified wine; others are nuttier. I find thicker-skinned, brown dried figs can have a medicinal quality. Both fresh and dried figs are very compatible with sweet spices and with fatty, salty ingredients like blue cheese and prosciutto.

Fig & Almond Although wasps are too impatient to wait for it, the honey-like drop of nectar at the eye of a fig indicates that it's ripe and ready to eat. It might also serve as a reminder that the honey flavor of figs is a match for almonds; they make for a light, nougat-like combination. If you can't get fresh figs, or the wasps have ruined them all, buy some dried and make the justly famous Spanish *pan de higo*, a sort of dense cake in which figs are ground with spices, seeds and almonds. It's served in slices with Manchego cheese.

Fig & Anise Anise seeds go off like tiny licorice fireworks when paired with sweet, sticky ingredients such as dried figs. Roughly chop 4 or 5 dark dried figs and mix into a tub of cream cheese with a teaspoon of crushed anise seeds and a pinch of salt. Serve on the dark, fibrous crackers you get in health-food stores.

Fig & Blue Cheese *See Blue Cheese & Fig, page 64.*

Fig & Chocolate Some fans of dried figs claim they taste like chocolate. Which sounds a bit like wishful thinking to me. While the best dried figs do enjoy a tantalizing and impressive range of sweet flavors—dark molasses, maple syrup, any number of honeys, maybe a little salted caramel—they're not really the sort of flavors you find in chocolate. The only notes I can perceive in common are the sharp, red-berry flavors identifiable in ripe figs and some cocoa-rich dark chocolate. In the sweet "salami" made for Seggiano in Calabria, chocolate is combined with almonds, apple, pear and lots of dried *dottato* figs, known for their glorious flavor and lack of graininess. If you can get your hands on some, eat it as you might real salami, between slices of white bread. Elsewhere in Calabria, the Italian confectioners Nicola Colavolpe make a fig and chocolate panettone with a sugared hazelnut crust; but a trip to your local Italian deli should furnish a wooden box of figs, stuffed with walnut, orange and lime zest and covered in dark chocolate. Spain's contribution is *rabitos royale*, tiny, chocolate-covered figs filled with brandy and chocolate ganache, while in Seattle a company called The Greek Gods makes a fig and chocolate ice cream.

Fig & Cinnamon *See Cinnamon & Fig, page 213.*

Fig & Goat Cheese Split ripe, vinous figs and fill each one with a teaspoon of goat cheese. Press them back together and bake at 350°F for 10–15 minutes. Eat with your fingers when just cool enough. Some like to wrap a little prosciutto around the fig. Chef Eric Ripert's take on this trio is a goat cheese parfait and bacon ice cream, served with roasted fig and hazelnut in a red wine caramel.

Fig & Hard Cheese *See Hard Cheese & Fig, page 68.*
Fig & Hazelnut *See Hazelnut & Fig, page 235.*

Fig & Liver The ancient Roman precursor to foie gras was called *iecur ficatum*, from *iecur*, meaning liver, and *ficatum*, derived from *ficus*, referring to the figs with which the birds were fed. Many modern European languages have thus erroneously taken *ficatum* as the root of their words for liver—*fegato* in Italian, *hígado* in Spanish, *figado* in Portuguese. (The English word is derived from the West Saxon *libban*, meaning to live, denoting the seriousness with which hard-drinking northern Europeans take the organ.) These days, foie gras geese are fed with a corn and fat preparation, and the figs turn up only at the end, in the form of an accompanying compote or chutney.

Fig & Mint Giorgio Locatelli recommends serving figs with a little mint sorbet and some chopped mint. If you're in Greece or Turkey in late summer, when you tire of plucking warm, ripe figs off the tree and slurping them down like sweet oysters, serve them sliced with your best balsamic and a little fresh mint, or combine them in a salad with feta or grilled halloumi, the squeaky, salty cheese from Cyprus.

Fig & Orange The pomologist Edward Bunyard described dried figs as pungent and cloying, dismissing them as fit only "for youthful palates." There's no denying they are phenomenally sugary but, in their defense, they have a complexity that ensures the sweetness is never monotonous. Orange is often paired with fig, as the acidity of one tempers the sweetness of the other, and they're seasonal partners too. Before Christmas, I make these fig and orange bars to serve with mulled wine. Simmer ½ lb chopped dried mission figs in orange juice with the zest of 1 orange for about 20 minutes, until soft. Stir in 1 tbsp butter, turn off the heat and set aside. In a food processor pulse together ½ cup self-rising flour, ⅓ cup sugar, ¼ tsp bicarbonate of soda and a pinch of salt. Add 1 stick cold butter, cut into ½-in cubes, and pulse until the mixture has a breadcrumb texture. Stir in 1 cup rolled oats. Pat down half this mixture in a buttered 8-in round pan, spoon over the fig mixture (which should be chopped or given a quick whiz in the food processor first) and then spoon the other half of the dry mix on top. Bake at 375°F for 30 minutes, until the topping has turned light brown.

Fig & Prosciutto *See Prosciutto & Fig, page 169.*
Fig & Raspberry *See Raspberry & Fig, page 330.*
Fig & Soft Cheese *See Soft Cheese & Fig, page 73.*

Fig & Vanilla *Tempus fugit.* One day we'll all have dentures, and for all its boozy, floral perfume this seed-riddled twosome will seem as attractive as a poke in the eye with a breadstick. Enjoy them while you can, in fig and vanilla jam, ice cream or crumbly *ma'mool*—make the recipe in Rosemary & Apricot, page 309, substituting vanilla for rosewater in the dough and fig for apricot (also omitting rosemary) in the filling.

Fig & Walnut Dried figs and walnuts are combined in any number of cookies, but I think the idea, recounted by Jane Grigson, of opening up a fresh fig, putting a whole shelled walnut inside it and pinching it back together before leaving it to dry in the sun, sounds much more romantic. In Provence they call this *nougat du pauvre*—poor man's nougat.

Rose

O rose, thou art sickly. So much sugar is needed to balance its natural astringency that the addition of rose can often result in a cloying sweetness. And its floral muskiness, unless you're careful, can recall nothing so much as being pressed to your auntie's perfumed cleavage. Rose needs balancing with bitter ingredients, such as chocolate, coffee and citrus zest, or bitter-edged spices like clove. Alternatively, use rosewater in very modest amounts for an unfathomable background note, like you might vanilla. Most rose is used in the form of rosewater, but dried rose petals and rose jam can be bought in Middle Eastern supermarkets. Crystallized rose petals, which are used mainly for cake decorations, are sold in posh delis. Fresh rose petals from your garden can be used as an ingredient if they have not been sprayed with unappetizing chemicals; a list of suitable species is given in *The Oxford Companion to Food*, edited by Alan Davidson.

Rose & Almond *See Almond & Rose, page 240.*

Rose & Apple A few drops of rosewater give apple juice an exotic edge. Add two fingers of vanilla vodka to make what shall henceforth be known as a Scheherazade.

Rose & Apricot *See Apricot & Rose, page 276.*

Rose & Cardamom Indian *gulab jamun* competes with other syrup-soaked lovelies such as baklava, *loukades* and the rum baba for the title of sweetest dessert in the world. It consists of deep-fried balls of a dough made primarily with milk and shaped to the size of a fruit called the *jamun*. They're typically flavored with cardamom: either a few seeds are placed at their center or the dough has the ground spice mixed into it. *Gulab* means rose, which is the flavor of the syrup in which the balls are steeped. There's a recipe for my ginger syrup version in Ginger & Cardamom, page 302.

Rose & Chicken *See Chicken & Rose, page 34.*

Rose & Chocolate The bitterness of dark chocolate is a better match than milk for sweet rose flavor, even if my lifelong love of Fry's Turkish Delight suggests otherwise. How exotic it was to bite through thick, creamy milk chocolate and sink my teeth into rose-flavored jelly. "Full of Eastern Promise" might have been stretching it a bit, especially as the bar was made in a suburb of Bristol, but I didn't care. On TV, a dark-haired beauty wearing too much eye makeup had only to bite into the bar to be swept onto a white Arab stallion by a chisel-jawed sheikh. I'm still partial, although just as likely these days to indulge my love of this combination by making rose ice cream with milk chocolate pieces. In a heavy-bottomed pan, gently heat 1 cup heavy cream and 1 cup milk with 4 egg yolks and ¾ cup sugar. Stir until the mixture has thickened, then strain into a bowl and stir for a minute or two longer. Cool, chill and stir in 2 tsp rosewater. Check the strength, adding a few drops until you have the right level of rosiness, bearing in mind that the freezing process will knock it back somewhat. Freeze in an ice-cream machine, adding a handful of chocolate chips after the custard has been churned but when it is still soft, before it goes into the freezer.

Rose & Coffee *See Coffee & Rose, page 25.*

Rose & Cucumber *See Cucumber & Rose, page 185.*

Rose & Lemon Nestled side by side in boxes of Turkish Delight, so thoroughly dusted with cornstarch and sugar that you have to hold them up to the light to know which flavor is which. To the palate, they seem to come from very different places on the flavor spectrum, although they both contain the essential oils geraniol, nerol and citronellol. Try to imagine lemon without the citric wince, and you can detect its delicately perfumed, floral notes. Rose and lemon are also paired in a Persian dish called *faludeh*, one of the earliest recorded frozen desserts. Ice would be brought down from the mountains and added to very fine noodles, usually flavored with rose syrup and lemon juice.

Rose & Melon *See Melon & Rose, page 274.*

Rose & Orange *See Orange & Rose, page 290.*

Rose & Saffron Mashti Malone's is a legendary ice-cream joint in Los Angeles, owned by a couple of Iranian brothers. Almost all their ices are made with rosewater. Quite a narrow business proposition, you might think, but in many Middle Eastern cuisines rose is as ubiquitous and versatile a flavor as vanilla. Mashti's signature ice cream pairs rose with cream chips—whole pieces of frozen cream—while other varieties include a magical rosewater saffron with pistachio. Create a similar effect by freezing ⅔ cup heavy cream in a thin layer for a couple of hours, then breaking it into tiny pieces and stirring them into homemade ice cream.

Blueberry

Blueberries get on my nerves. It's not entirely their fault. It's just that all this stuff about antioxidants and phytochemicals gives them an air of piety only increased by that little flared crown they have at the top that looks like the collar of a choirboy's cassock. Plus the white bloom on their skin reminds me of the leprous pallor of an old, sad Kit Kat (although it's supposed to be a sign that the berries haven't been handled too much). The flavor of blueberry is largely in its skin, so the smaller the better. If they're a little lackluster, lure them from the path of virtue by cooking them with sugar and unlocking their aromatic potential.

Blueberry & Almond *See Almond & Blueberry, page 238.*

Blueberry & Apple *See Apple & Blueberry, page 264.*

Blueberry & Blue Cheese Blueberries have a sharpness that makes them especially good paired with blue cheese in salads. Warren Geraghty, the chef at West in Vancouver, serves a disc of cinnamon-buttered toasted brioche under a baked blue-cheese cheesecake, topped with blueberries simmered in port, vanilla and brown sugar.

Blueberry & Cinnamon Cooking blueberries brings out their flavor: add some cinnamon for extra improvement. Marion Cunningham suggests dipping warm blueberry muffins into melted butter, then cinnamon sugar.

Blueberry & Coriander Seed *See Coriander Seed & Blueberry, page 337.*
Blueberry & Lemon *See Lemon & Blueberry, page 297.*

Blueberry & Mushroom Fruit and mushroom is a common and well-liked pairing in the north of Italy. Chef Marc Vetri pairs porcini and blueberries in a lasagna, and is judicious with the berries; the diner should come across a little burst of blueberry only here and there. The fruit flavor contrasts with the mushrooms' meatiness. Blueberry and mushroom risotto, usually made with beef stock, is another popular dish.

Blueberry & Peach *See Peach & Blueberry, page 277.*

Blueberry & Vanilla Good blueberry flavor should have a floral top note and a creamy, fruity finish. Dear old vanilla, friend to all, picks up on this quality and combines with blueberry flavor in the fruit's most redemptive recipe, a blueberry and vanilla cake. Put 1½ sticks softened butter, ⅞ cup superfine sugar, 3 large eggs, 1¾ cups self-rising flour, 1 tsp baking powder, 3 tbsp milk and 2 tsp vanilla extract in a bowl and beat with an electric mixer for 2–3 minutes or until well combined. Fold in 1 cup blueberries. Tip the mixture into a greased and lined deep, round 9-in cake pan and bake at 350°F for 45 minutes or until well risen and golden brown. Cool for 10 minutes, then remove from the pan and leave to cool completely. Beat 8 oz cream cheese with ¼ cup confectioner's sugar and 5 tbsp sour cream. Spread this over the cake and scatter ¾ cup blueberries on top.

Coriander Seed

Coriander seeds have a delicious citrus and balsamic character, not unlike a nice version of those scented wooden balls some people keep in their underwear drawers. They lend a startlingly pretty flavor used as the sole aromatic in cookies or to offset the bitterness of wine when you mull it—after all, their flavor recalls the classic mulling combination of orange, cinnamon and clove. Coriander seeds bring a fragrant, feminine touch to curry powder blends and pastes and to mixed pickling spices. They're also one of the key botanicals in gin. A spare peppermill filled with roasted coriander seeds could easily get you hooked.

Coriander Seed & Apple When the floral perfume of coriander seeds is mixed with sharp, fruity apple, the result is quite apricot-like and especially delightful. Try them paired in an ice cream. Peel, core and chop 1 lb tart apples (e.g., Granny Smiths) and put them in a pan with 2 tbsp lemon juice and ⅓ cup sugar. Cook gently, covered, until soft. Beat in 2 tsp lavender jelly and 1 tsp freshly ground, lightly toasted coriander seeds, then leave to cool. Whip 1¼ cups whipping cream until fairly thick and fold into the apple mix. Freeze in the usual way. Use red currant jelly if you don't have lavender, but lavender and coriander seed have very harmonious flavors.

Coriander Seed & Blueberry Coriander seeds can contain up to 85 percent linalool, a flavor compound with a woody, floral, slightly citrus quality that's a key component of synthesized blueberry flavor. Freshly ground, they can lend a fragrant background note to your home-baked blueberry muffins. Or be more adventurous, like the Ottolenghi deli in London, and make a couscous salad with bell pepper, red onion, wild blueberries, pink peppercorns and coriander seed.

Coriander Seed & Cardamom *See Cardamom & Coriander Seed, page 307.*

Coriander Seed & Cilantro Concentrate hard and you can detect a hint of cilantro flavor in quiet, reliable coriander seed; on the whole they're very different and about as substitutable for each other as sprigs of spearmint are for Tic Tacs.

Coriander Seed & Coffee Morocco is both a major grower and user of coriander. The seeds are used to flavor espresso-style coffee, adding a floral (rose, lavender) and citrus quality. Try adding 1 tsp coriander seed to 6 tbsp coffee beans before grinding and take it from there.

Coriander Seed & Cumin *See Cumin & Coriander Seed, page 85.*

Coriander Seed & Garlic Beauty and the Beast. Coriander seed is such a gentle, pretty flavor that seeing it mixed with lots of garlic, with no more than a dressing of salt and the oil they were softened in, is like catching the head girl in a tattoo parlor. Nonetheless, this seasoning mixture, called *taklia*, is popular in Egypt and Turkey. Soften 3 sliced garlic cloves in some olive oil or butter and pound in a mortar with 1 tsp coriander seeds, a few pinches of salt and (optionally) a pinch of cayenne. Use in spinach or lentil soups or to flavor cream cheese.

Coriander Seed & Goat Cheese Spice experts detect notable differences between the egg-shaped coriander seed of India, which is sweeter, with a creamy quality, and the spicier round seed common to Europe and Morocco. Both have the floral, citrus quality that makes coriander seed a good match for goat cheese.

Coriander Seed & Lemon Harold McGee describes coriander seed as lemony and floral. Others find the citric note more reminiscent of orange. Whichever you think, the seeds pair effectively with lemon. They're often used together to marinate olives, and they also work very well with fish. In the nineteenth century they were frequently joined by cinnamon as a seasoning combination for almond milk or cream puddings. Mrs. Lee gives a recipe for a drink called citronelle ratafia. Combine 2 quarts of brandy with the rinds of a dozen lemons, 5 tbsp coriander seed, ½ oz bruised cinnamon and a sugar syrup made of 4½ cups sugar dissolved in 3 cups water. Leave it for a month, presumably somewhere dark, then strain and bottle.

Coriander Seed & Olive See Olive & Coriander Seed, page 173.
Coriander Seed & Orange See Orange & Coriander Seed, page 289.

Coriander Seed & Pork Coriander seed is an unsung hero of the flavoring world. In combination with other aromatics, it's used in curries, ketchup and pickles; pork products such as frankfurters, mortadella, French sausages and *boudin noir*; and has a rare starring role in *afelia*, a pork stew popular in Cyprus and Greece. Making *afelia* is the perfect way to get to know this spice. Marinate 2 lb cubed pork tenderloin in 1 cup red wine, 2 tbsp crushed coriander seeds and some salt and pepper for 4–24 hours. Drain the meat, reserving the marinade, pat dry, then dredge the pieces in plenty of flour. Brown in olive oil, then add the marinade to the pan with enough water to cover the meat. Cover and cook for 45–60 minutes, removing the lid toward the end of cooking if the sauce needs reducing. Prepare yourself for a waft of what will smell like meat stewed in mulled wine.

Vanilla

A universally popular fragrance and flavor. Vanilla is an orchid whose seedpods are fermented and cured to produce the flavoring. It's native to Mexico but is now grown in Tahiti, Madagascar and Indonesia. Tahitian vanilla is highly prized for its fruity, spicy character, described by some as specifically redolent of cherry and anise. Madagascan vanilla is the type most of us are familiar with; Mexican is spicier, richer and earthier. The Incas used vanilla to flavor chocolate, and this was its exclusive culinary role until Hugh Morgan, Elizabeth I's apothecary, began to advocate its wider application. It's still, of course, vital to the manufacture of most chocolate, but is also the world's favorite flavor of ice cream, an increasingly popular addition to savory dishes, and works beautifully as a background flavor in desserts and liqueurs. Pods tend to give the purest flavor but vanilla extract, powder and lovely seedy pastes are all very workable too. Natural vanilla is expensive, reflecting the labor that goes into its production, and to meet the enormous demand created by the manufacture of cakes, cookies, ice cream, candy and soft and alcoholic drinks, a synthetic form is produced in far greater quantities than vanilla itself—in fact 97 percent of vanilla flavoring consumed is synthetic. Vanillin, the main constituent of vanilla essence and extract, is the character impact compound for vanilla and was first isolated in 1858. In 1874 it became one of the first flavors to be synthesized, using material from coniferous trees. Today vanillin is extracted from clove oil, waste material from the paper and wood-pulp industry and petrochemical products. There are many who wouldn't dream of using vanilla essence, but in blind tests conducted by *Cook's Illustrated*, the panel of cooks and baking experts was unable to detect much difference in custards and sponge cakes made with a range of extracts and essences. While the custard made with vanilla essence was judged to be marginally inferior, the experts preferred the fake stuff for cakes—largely on the grounds of its greater strength of flavor, probably attributable to a higher proportion of vanillin than in the extract.

Vanilla & Almond *See Almond & Vanilla, page 240.*

Vanilla & Anise *See Anise & Vanilla, page 182.*

Vanilla & Apple It's well known that vanilla ice cream—or good vanilla-flavored custard—goes down a treat with apple pie or crumble. But how about grinding vanilla sugar into a fine powder and shaking it over hot, crisp apple fritters? Sift ¾ cup plain flour with a pinch of salt, make a well in the center and drop an egg into it. Beat well and gradually add a mixture of ⅓ cup milk and 3 tbsp iced water to make a smooth, not too runny batter. Leave in the fridge while you peel and core 3 or 4 apples and slice them into ½-in rings. Dip in the batter, making sure they're thoroughly covered, then fry in 1–2 in oil for about 1 minute on each side. Serve sprinkled with the ground vanilla sugar. Cider can be used instead of the milk.

Vanilla & Apricot *See Apricot & Vanilla, page 277.*

Vanilla & Banana *See Banana & Vanilla, page 273.*

Vanilla & Blackberry Spend a late-summer afternoon picking blackberries. Get home while it's still light. Wash and gently stew your berries with sugar, stopping short of making them perfectly sweet. Cool, strain and, as the air turns blue, serve with vanilla ice cream. A little honeycomb crumbled over each makes a great garnish.

Vanilla & Blueberry *See Blueberry & Vanilla, page 336.*

Vanilla & Cardamom Second and third to saffron in the spice price stakes. Luxurious paired in a *crème pâtissière* for a fruit-topped tart (see Cardamom & Apricot, page 306) or in an ice cream, where their sweet, spicy, floral flavors come together very prettily. Many chefs prefer Tahitian vanilla for this sort of recipe, as it's sweeter and fruitier than the Madagascan and Mexican varieties. It also contains less of the usually dominant flavor compound, vanillin, and more floral notes.

Vanilla & Cherry *See Cherry & Vanilla, page 244.*

Vanilla & Chestnut *See Chestnut & Vanilla, page 230.*

Vanilla & Chocolate Both native to Mexico, where the practice of flavoring chocolate with vanilla dates back to Aztec times. Today, most bars of chocolate are flavored with vanilla in some way. The California chocolatiers Scharffen Berger grind whole vanilla pods with their cocoa nibs in accordance with founder John Scharffenberger's belief that chocolate is much enhanced by vanilla. At the other end of the scale, cheap chocolate is often flavored with an overwhelming amount of vanilla essence to compensate for its lack of cocoa flavor. If you're curious as to what chocolate tastes like *without* vanilla, try the bars made by the French chocolatier Bonnat. See also White Chocolate & Chocolate, page 343.

Vanilla & Clove Vanilla and clove, or "baking spice" flavors, are commonly transferred to wine courtesy of the vanillin and eugenol naturally present in the wooden casks used for aging. French oak, with its tight grain, is said to contain these subtler flavors, whereas larger-grained American oak, especially when toasted, has more pronounced coconut and herbaceous notes—see Dill & Coconut, page 187. At the time of writing, a new 300-bottle French oak barrel costs about $700, whereas one made of American oak comes in at roughly half that. To save costs, winemakers sometimes slot new staves into older barrels whose flavor has worn out, or, even more economically, drop wood chips in nets, like giant teabags, into the vast stainless steel tanks in which many modern wines are stored. Many drinkers say the chips impart the sort of one-dimensional vanilla character found in certain sweet Chardonnays, but even high-end winemakers are resorting to, or partially incorporating, this cost-cutting measure.

Vanilla & Coconut *See Coconut & Vanilla, page 282.*

Vanilla & Coffee Spill a fresh espresso over some good vanilla ice cream. Italians call this *affogato*, which a waggish waiter in Rome once told me comes from "Affogato tie my shoelaces." Accident or not, the combination works. The protein in dairy binds to the tannins in coffee, making it less bitter and easier on the palate, although it does dampen the flavor a little too.

Vanilla & Egg *See Egg & Vanilla, page 135.*

Vanilla & Fig *See Fig & Vanilla, page 333.*

Vanilla & Ginger *See Ginger & Vanilla, page 305.*

Vanilla & Hazelnut *See Hazelnut & Vanilla, page 236.*

Vanilla & Nutmeg Elizabeth David says the flavor of fresh bay leaves is reminiscent of nutmeg and vanilla, and notes how well they work in sweet cream. These days it's not unusual to find a bay leaf panna cotta on the menus of smart restaurants. Nutmeg and vanilla come together to make a heady, almost hyper-perfumed sweetness that's particularly irresistible in homemade custard tarts. Forget frou-frou macaroons and pretty cupcakes: custard tarts are deep, honest and intoxicating.

Vanilla & Orange Vanilla and orange meet Stateside in the form of the legendary Creamsicle, the American version of what the Brits call a Mivvi. Both consist of vanilla ice cream on a stick, clad in fruit-flavored ice. You might like to try the more adult combination of vanilla panna cotta with chilled orange segments, or seek out a bottle of Fiori di Sicilia essence, which adds a pronounced floral, orange and vanilla character to meringues and cakes like panettone.

Vanilla & Peach *See Peach & Vanilla, page 278.*
Vanilla & Peanut *See Peanut & Vanilla, page 29.*
Vanilla & Pineapple *See Pineapple & Vanilla, page 263.*

Vanilla & Raspberry Around the age of six, I was always badgering my mother to buy something called Arctic Roll. I loved the ship-in-a-bottle impossibility of a cylinder of raspberry-jam-encircled ice cream enclosed in vanilla sponge *seemingly without a seam.* How *had* Birds Eye done it? Was it a section of a single, vast Arctic Roll, the length of the trans-Ural pipeline, piped full of ice cream by forces of unimaginable scale? And how come you could defrost it without the ice cream leaking out or the cake turning soggy? These days, I like to tell myself I've outgrown my taste for elaborate confections, in favor of very simple, *very* expensive ingredients, simply prepared—fresh raspberries, say, their violet-scented elegance enhanced by a little Chantilly cream, prepared with powdered sugar and vanilla extract and nothing else. And then my husband took me to La Cuisine de Joël Robuchon in Covent Garden and I ordered Le Sucre for dessert. Le Sucre is an iridescent, semi-transparent Christmas bauble of sugar, inflated in much the same way a glass-blower breathes sphericality into brandy balloons, filled with mascarpone and raspberry mousse and steadied on the plate by a ring of pink meringue flecked with gold leaf, on a coulis of mixed berries sprinkled with pistachio dust, next to a perfect quenelle of vanilla ice cream. And I thought, the hell with simplicity, as I shattered the bauble with the back of my spoon. *This* is how to eat.

Vanilla & Rhubarb *See Rhubarb & Vanilla, page 251.*

Vanilla & Shellfish A combination to be approached with great trepidation. It's undoubtedly terrific in the hands of French *nouvelle cuisine* pioneer Alain Senderens, who's credited with pairing the two in the first place. He's said to have developed his dish of lobster in vanilla *beurre blanc* to complement the fine white Burgundies in the cellar of the restaurant where he was cooking at the time; the vanilla in the sauce matched the buttery, sweet, toasted caramel imparted to the wine after aging in oak barrels.

Vanilla & Strawberry *See Strawberry & Vanilla, page 259.*

Vanilla & Tomato When used in moderation, Mexican vanilla, which has a spicy quality, draws out a similar spiciness in tomato-based dishes, while taking the edge off the acidity. Some recommend a touch of Mexican vanilla in a tomatoey chili con carne, but a bolder cook might pair just the two in a soup. Claude Bosi at Hibiscus, in London, takes them in a sweet direction in a dish with fragments of frozen raspberries.

Vanilla & Walnut *See Walnut & Vanilla, page 233.*

White Chocolate

White chocolate is made with cocoa butter, milk, sugar and vanilla, and it's the variation in the quantity and quality of these four ingredients that accounts for what difference exists between the various bars you can buy. Vanilla is almost always by far the strongest flavor. A prominent chocolatier tells me that the possibilities for varying the flavor of white chocolate are so limited that no quality brand would offer more than one pure white chocolate bar in its product range. Cocoa butter has a strong taste but it's not necessarily that pleasant, and nearly all chocolate manufacturers deodorize the butter in their products, even if, like pasteurization, the process suppresses some of the good flavors too. Lightly roasting the beans can get around the problem. That said, the Venezuelan chocolate makers El Rey don't deodorize, and their bar made with 34 percent cocoa butter is often said to be the best tasting on the market. It's less sweet than most white chocolate, and more potently chocolatey. Green and Black's white chocolate is shot through with lots of vanilla seeds and tastes like a sweet but luxurious vanilla ice cream, with hints of strawberry and apricot brandy that recall how white chocolate can benefit from sour balancing flavors.

White Chocolate & Almond White chocolate is so sweet because it lacks the cocoa solids that give milk and especially dark chocolate their balancing bitterness. Almonds, especially roasted, can diffuse its sweetness, as Nestlé may have had in mind when it included them in North America's first white chocolate bar, launched in the 1940s. Combine them yourself and you can both increase the nut content and use your preferred white chocolate. Roast 1 cup blanched almonds in the oven at 350°F for 8–10 minutes, until lightly browned, then set aside. Line a baking tray with baking parchment. Put 7 oz white chocolate in a bowl set over (but not touching) a pan of simmering water. When it's completely melted, stir in the nuts, then spread the mixture over the baking parchment. Bear in mind that you'll be breaking it into chunky pieces, so don't spread it too thin. Let it cool a little, then chill until set. In the States this is known as bark. You could throw some dried sour cherries in too, to give your bark some bite.

White Chocolate & Blackberry *See Blackberry & White Chocolate, page 326.*

White Chocolate & Cardamom The London chocolatiers Rococo make a white chocolate bar infused with cardamom. If you're Danish, cardamom plus the strong vanilla flavor of white chocolate might recall a rich butter cookie. For me, the combination is reminiscent of sweet, spicy, milky Indian desserts like *kheer* and *kulfi*, where the cardamom lifts the suffocating heaviness of white chocolate. Try the pairing for yourself, using the recipe at White Chocolate & Almond, page 342, thoroughly mixing in the finely ground seeds of about ten green cardamom pods. It'll be even lovelier if you substitute roasted, skinned pistachios for the almonds.

White Chocolate & Caviar *See Caviar & White Chocolate, page 152.*

White Chocolate & Chocolate Nearly all chocolate bars, dark, milk and white, are flavored with some form of vanilla; cheap ones heavily so to make up for the essential lack of chocolatiness. Vanilla is white chocolate's primary flavor. Which makes me think that when a decent dark or milk chocolate is combined with white (in a dark chocolate truffle, say, with a white chocolate coating), thereby upping its vanilla content while diluting its cocoa solids, the darker chocolate is essentially being cheapened. It's like adding a blended whisky to your single malt.

White Chocolate & Coconut Go ahead, make a coconut cake and give it an icing of thick white chocolate. But be warned, it's like being buried alive in frosting. Raspberries might be your St. Bernard.

White Chocolate & Coffee To me, the combination of coffee and the rich dairy flavor of white chocolate can be a little too redolent of fudgy *café au lait* made with UHT milk, as it sometimes is in France. UHT is cooked at a high temperature, causing a Maillard reaction in which sweet caramel characteristics, absent from fresh milk, are created. The white chocolate/coffee combination is at its best, I think, in Café Tasse's Blanc Café bar, which contains ground coffee beans that give a pronounced bitter contrast to the white chocolate.

White Chocolate & Lemon Good white chocolate, like Pierre Marcolini or Venchi's Bianco, often has a natural lemon character. Pair white chocolate and lemon in a cake, like Rose Beranbaum's lemon-laced white chocolate cake filled with rich lemon curd and topped with a creamy white chocolate and lemon curd buttercream. She calls it Woody's Lemon Luxury Layer Cake; a mouthful in every sense.

White Chocolate & Olive *See Olive & White Chocolate, page 175.*

White Chocolate & Pineapple White chocolate is a firm friend of red berry fruits, but it also works well with tropical ones. Think how notes of pineapple and passion fruit complement vanilla in an oak-aged Chardonnay. Heady stuff, mind: you wouldn't want too much of it.

White Chocolate & Raspberry *See Raspberry & White Chocolate, page 331.*

White Chocolate & Saffron Artisan du Chocolat makes an elegant pink-gold bar of saffron-flavored white chocolate. It says the white chocolate brings out the hay flavor of saffron. I'd add that the honeyed vanilla flavor of white chocolate shines a light on saffron's floral complexity.

White Chocolate & Strawberry *See Strawberry & White Chocolate, page 259.*

Bibliography

Achatz, Grant. *Alinea*. Ten Speed Press, 2008. See pages 250, 310.

Acton, Eliza. *Modern Cookery for Private Families*. Longman, Brown, Green & Longmans, 1845. See page 143.

Allen, Darina. *Darina Allen's Ballymaloe Cookery Course*. Kyle Cathie, 2001. See page 323.

Allen, Gary. *The Herbalist in the Kitchen*. University of Illinois, 2007. See page 178.

Amis, Kingsley. *Everyday Drinking*. Bloomsbury, 2008. See page 315.

Ansel, David. *The Soup Peddler's Slow and Difficult Soups: Recipes and Reveries*. Ten Speed Press, 2005. See page 84.

Apicius. *Cookery and Dining in Imperial Rome*. Edited and translated by J. Dommers Vehling. Dover, 1977. See pages 114, 139.

Arndt, Alice. *Seasoning Savvy*. Haworth Herbal Press, 1999. See page 22.

Artusi, Pellegrino. *The Art of Eating Well* (1891). Translated by Kyle M. Phillips III. Random House, 1996. See pages 87, 170, 199, 248, 315.

Audot, Louis Eustache. *French Domestic Cookery*. Harper & Brothers, 1846. See page 88.

Baljekar, Mridula. *Real Fast Indian Food*. Metro, 2000. See page 44.

Bayless, Rick. *Rick Bayless's Mexican Kitchen*. Scribner, 1996. See pages 19, 58, 196, 288.

Beard, James. *Theory & Practice of Good Cooking*. Knopf, 1977. See pages 165, 330.

Beeton, Isabella. *Mrs Beeton's Book of Household Management*. S. O. Beeton, 1861. See pages 118, 231, 305.

Beranbaum, Rose. *Rose's Heavenly Cakes*. Wiley, 2009. See page 343.

Bittman, Mark. *How to Cook Everything Vegetarian*. Wiley, 2007. See pages 69, 251, 312.

Blumenthal, Heston. *The Big Fat Duck Cookbook*. Bloomsbury, 2008. See pages 11, 122, 152, 153, 174, 236, 252, 258, 288.

Boswell, James. *A Journey to the Western Islands of Scotland*. J. Pope, 1775. See page 186.

Brillat-Savarin, J. A., & Simpson, L. Francis. *The Handbook of Dining*. Longman, Brown, Green, Longmans & Roberts, 1859. See pages 33, 34, 116.

Bunyard, Edward A. *The Anatomy of Dessert*. Dulau & Co., 1929. See pages 229, 244, 333.

Burbidge, F. W. *The Gardens of the Sun*. John Murray, 1880. See page 65.

Burnett, John. *Plenty and Want: A Social History of Food in England from 1815 to the Present Day*. Nelson, 1966. See page 251.

Byrne, Aiden. *Made in Great Britain*. New Holland, 2008. See page 182.

Campion, Charles. *Fifty Recipes to Stake Your Life On*. Timewell Press, 2004. See page 109.

Cannas, Pulina & Francesconi. *Dairy Goats Feeding & Nutrition*. CABI, 2008. See page 57.

Carême, Marie-Antoine. *L'Art de la Cuisine*. 1833. See page 147.

Carluccio, Antonio. *The Complete Mushroom Book*. Quadrille, 2003. See page 80.

Carmellini, Andrew, & Hyman, Gwen. *Urban Italian*. Bloomsbury, 2008. See page 297.

Castelvetro, Giacomo. *The Fruit, Herbs and Vegetables of Italy* (1614). Translated by Gillian Riley. Viking, 1989. See page 127.

Chartier, François. *Papilles et Molécules*. La Presse, 2009. See page 318.

Chiba, Machiko. *Japanese Dishes for Wine Lovers*. Kodansha International, 2005. See page 123.

Christian, Glynn. *How to Cook Without Recipes*. Portico, 2008. See page 85.

Clark, Sam & Sam. *The Moro Cookbook*. Ebury, 2001. See page 42.

Clifford, Sue, & King, Angela. *The Apple Source Book: Particular Uses for Diverse Apples*. Hodder & Stoughton, 2007. See page 263.

Coates, Peter. *Salmon*. Reaktion, 2006. See page 156.

Cook's Illustrated (www.cooksillustrated.com). See pages 172, 339.

Corrigan, Richard. *The Clatter of Forks and Spoons*. Fourth Estate, 2008. See pages 104, 251.

Cunningham, Marion. *The Breakfast Book*. Knopf, 1987. See page 336.

David, Elizabeth. *A Book of Mediterranean Food*. Lehmann, 1950. See page 94.

David, Elizabeth. *An Omelette and a Glass of Wine*. Penguin, 1986. See pages 255, 275.

David, Elizabeth. *French Provincial Cooking*. Michael Joseph, 1960. See pages 11, 153, 154, 174, 240, 288.

David, Elizabeth. *Italian Food*. Macdonald, 1954. See pages 47, 49, 84, 146, 169, 314, 340.

David, Elizabeth. *Spices, Salt and Aromatics in the English Kitchen*. Penguin, 1970. See pages 265, 316.

Davidson, Alan & Jane. *Dumas on Food*. Folio Society, 1978. See pages 20, 229, 274.

Davidson, Alan. *Mediterranean Seafood*. Penguin, 1972. See page 291.

Davidson, Alan. *North Atlantic Seafood*. Macmillan, 1979. See pages 145, 154.

Davidson, Alan. *The Oxford Companion to Food*. OUP, 1999. See page 133, 334.

de Rovira Sr, Dolf. *Dictionary of Flavors*. Wiley Blackwell, 2008. See page 262.

del Conte, Anna. *The Classic Food of Northern Italy*. Pavilion, 1995. See page 275.

Dolby, Richard. *The Cook's Dictionary and Housekeeper's Directory*. H. Colburn & R. Bentley, 1830. See page 115.

Douglas, Norman. *Venus in the Kitchen*. Heinemann, 1952. See page 39.

Dumas, Alexandre. See Davidson, Alan & Jane.

Dunlop, Fuchsia. *Shark's Fin and Sichuan Pepper*. Ebury, 2008. See pages 54, 109.

Esquire Handbook for Hosts. Edited by P. Howarth. Thorsons, 1999. See page 149.

Farley, John. *The London Art of Cookery*. Fielding, 1783. See page 140.

Fearnley-Whittingstall, Hugh. *River Cottage Every Day*. Bloomsbury, 2009. See page 132.

Fearnley-Whittingstall, Hugh, & Fisher, Nick. *The River Cottage Fish Book*. Bloomsbury, 2007. See pages 1379, 146, 163, 244.

Fearnley-Whittingstall, Hugh. *The River Cottage Meat Book*. Hodder & Stoughton, 2004. See pages 120, 166.

Field, Eugene. *The Writings in Prose and Verse of Eugene Field*. C. Scribner's Sons, 1896. See page 265.

Fisher, M. F. K. *Consider the Oyster*. Duell, Sloan & Pearce, 1941. See page 149.

Floyd, Keith. *Floyd on Britain and Ireland*. BBC, 1988. See page 102.

Gill, A. A. *The Ivy: The Restaurant and its Recipes*. Hodder & Stoughton, 1997. See pages 116, 331.

Gladwin, Peter. *The City of London Cook Book*. Accent, 2006. See page 35.

Glass, Leonie. *Fine Cheese*. Duncan Petersen, 2005. See page 66.

Glasse, Hannah. *The Art of Cookery Made Plain and Easy*. 1747. See pages 54, 96, 143, 154, 156, 161.

Graham, Peter. *Classic Cheese Cookery*. Penguin, 1988. See pages 63, 269.

Graves, Tomás. *Bread and Oil: Majorcan Culture's Last Stand*. Prospect, 2001. See page 276.

Grigson, Jane. *English Food*. Macmillan, 1974.

Grigson, Jane. *Fish Cookery*. Penguin, 1975. See page 142.

Grigson, Jane. *Jane Grigson's Fruit Book*. Michael Joseph, 1982. See pages 245, 266, 333.

Grigson, Jane. *Jane Grigson's Vegetable Book*. Michael Joseph, 1978. See pages 130, 260.

Grigson, Sophie. *Sophie Grigson's Herbs*. BBC, 1999. See page 330.

Harbutt, Juliet. *Cheese: A Complete Guide to over 300 Cheeses of Distinction*. Mitchell Beazley, 1999. See page 299.

Hay, Donna. *Flavours*. Murdoch, 2000. See page 305.

Hay, Donna. *Marie Claire Cooking*. Murdoch, 1997. See page 181.

Henderson, Fergus, & Gellatly, Justin Piers. *Beyond Nose to Tail*. Bloomsbury, 2007. See pages 44, 51.

Henderson, Fergus. *Nose to Tail Eating*. Macmillan, 1999. See pages 132.

Hieatt, Constance B., Hosington, Brenda, & Butler, Sharon. *Pleyn Delit: Medieval Cookery for Modern Cooks*. University of Toronto, 1996. See page 141.

Hill, Tony. *The Spice Lover's Guide to Herbs and Spices*. Wiley, 2005. See page 214.

Hirsch, Dr Alan. *Scentsational Sex.* Element, 1998. See pages 183, 228.

Hollingworth, H. L., & Poffenberger, A. D. *The Sense of Taste.* Moffat Yard & Co., 1917. See page 34.

Hom, Ken. *A Taste of China.* Pavilion, 1990. See page 112.

Hooper, Edward James. *Western Fruit Book.* Moore, Wilstach, Keys & Co., 1857. See page 263.

Hopkinson, Simon, & Bareham, Lindsey. *Roast Chicken and Other Stories.* Ebury, 1994. See page 90.

Hopkinson, Simon, & Bareham, Lindsey. *The Prawn Cocktail Years.* Macmillan, 1997. See page 197.

Jaffrey, Madhur. *Madhur Jaffrey's Quick and Easy Indian Cookery.* BBC, 1993. See page 307.

Jaffrey, Madhur. *Madhur Jaffrey's Ultimate Curry Bible.* Ebury, 2003. See page 322.

James, Kenneth. *Escoffier: The King of Chefs.* Continuum, 2002. See page 278.

Kamp, David. *The United States of Arugula.* Broadway, 2006. See page 153.

Kapoor, Sybil. *Taste: A New Way to Cook.* Mitchell Beazley, 2003. See pages 85, 163.

Katzen, Mollie. *Still Life with Menu Cookbook.* Ten Speed Press, 1994. See page 134.

Kaufelt, Rob, & Thorpe, Liz. *The Murray's Cheese Handbook.* Broadway, 2006. See page 59.

Keller, Thomas. *The French Laundry Cookbook.* Workman, 1999. See pages 24, 152.

Kennedy, Diana. *Recipes from the Regional Cooks of Mexico.* Harper & Row, 1978. See page 196.

Kitchen, Leanne. *Grower's Market: Cooking with Seasonal Produce.* Murdoch, 2006. See page 261.

Lanchester, John. *The Debt to Pleasure.* Picador, 1996. See page 9.

Lang, Jenifer Harvey. *Tastings.* Crown, 1986. See page 65.

Larkcom, Joy. *Oriental Vegetables.* John Murray, 1991. See page 124.

Lawson, Nigella. *Forever Summer.* Chatto & Windus, 2002. See page 26.

Lawson, Nigella. *How to be a Domestic Goddess.* Chatto & Windus, 2000. See page 243.

Lawson, Nigella. *How to Eat.* Chatto & Windus, 1998. See page 72.

Levene, Peter. *Aphrodisiacs.* Blandford, 1985. See page 130.

Lewis, Elisha Jarrett. *The American Sportsman.* Lippincott, Grambo & Co., 1855. See page 217.

Leyel, Mrs C. F., & Hartley, Miss O. *The Gentle Art of Cookery.* Chatto & Windus, 1925. See page 300.

Locatelli, Giorgio. *Made in Italy.* Fourth Estate, 2006. See pages 82, 95, 232, 253, 271, 332.

Luard, Elisabeth. *Truffles.* Frances Lincoln, 2006. See page 115.

Maarse, H. *Volatile Compounds in Foods and Beverages.* CRC Press, 1991.

Mabey, Richard. *The Full English Cassoulet.* Chatto & Windus, 2008. See page 78.

Marinetti. *The Futurist Cookbook* (1932). Translated by Suzanne Brill. Trefoil, 1989. See page 290.

Marsili, Ray. *Sensory-Directed Flavor Analysis*. CRC Press, 2006. See page 49.

McGee, Harold. *McGee on Food and Cooking*. Hodder & Stoughton, 2004. See pages 35, 89, 105, 143, 152, 168, 170, 181, 189, 193, 273, 338.

Michelson, Patricia. *The Cheese Room*. Michael Joseph, 2001. See page 233.

Miller, Mark, with McLauchlan, Andrew. *Flavored Breads*. Ten Speed Press, 1996. See page 101.

Miller, Mark. *Coyote Café*. Ten Speed Press, 2002. See page 182.

Ojakangas, Beatrice A. *Scandinavian Feasts*. University of Minnesota, 2001. See page 161.

Oliver, Jamie. *Jamie's Dinners*. Michael Joseph, 2004. See page 2713.

Olney, Richard. *The French Menu Cookbook*. Collins, 1975. See pages 33, 114, 262.

Parsons, Russ. *How to Pick a Peach*. Houghton Mifflin Harcourt, 2007. See pages 183, 313.

Paston-Williams, Sara. *The National Trust Book of Traditional Puddings*. David & Charles, 1983. See page 266.

Pern, Andrew. *Black Pudding and Foie Gras*. Face, 2008. See pages 42, 305.

Perry, Neil. *The Food I Love*. Murdoch, 2005. See page 313.

Phillips, Henry. *History of Cultivated Vegetables*. Henry Colburn & Co., 1822. See page 146.

Plath, Sylvia. *The Bell Jar*. Heinemann, 1963. See page 31.

Pomés, Leopold. *Teoria i pràctica del pa amb tomàquet*. Tusquets, 1985. See page 255.

Puck, Wolfgang (www.wolfgangpuck.com). See page 163.

Puck, Wolfgang. *Wolfgang Puck's Modern French Cooking for the American Kitchen*. Houghton Mifflin, 1981. See page 151.

Purner, John F. *The $100 Hamburger: A Guide to Pilots' Favorite Fly-in Restaurants*. McGraw-Hill, 1998. See page 68.

Raven, Sarah. *Sarah Raven's Garden Cookbook*. Bloomsbury, 2007. See page 324.

Reboux, Paul. *Book of New French Cooking*. Translated by Elizabeth Lucas Thornton. Butterworth, 1927. See page 225.

Renowden, Gareth. *The Truffle Book*. Limestone Hills, 2005. See page 115.

Robuchon, Joël. *The Complete Robuchon*. Grub Street, 2008. See page 298.

Roden, Claudia. *A New Book of Middle Eastern Food*. Penguin, 1985. See page 287.

Roden, Claudia. *The Book of Jewish Food*. Viking, 1997. See page 53.

Rodgers, Judy. *The Zuni Café Cookbook*. Norton, 2002. See pages 169, 170, 315, 330.

Rose, Evelyn. *The New Complete International Jewish Cookbook*. Robson, 2004. See page 84.

Rosengarten, David. *Taste*. Random House, 1998. See pages 123, 151, 155, 177.

Round, Jeremy. *The Independent Cook*. Barrie & Jenkins, 1988. See page 249.

Roux, Michel. *Eggs*. Quadrille, 2005. See pages 73, 330.

Saint-Ange, Madame E. *La Bonne Cuisine de Madame E. Saint-Ange*. Translated by Paul Aratow. Ten Speed Press, 2005. See page 228.

Saulnier, Louis. *Le Répertoire de La Cuisine*. Barron's Educational Series, 1914.

Saveur Editors. *Saveur Cooks Authentic Italian*. Chronicle, 2008. See page 195.

Schehr, Lawrence R., & Weiss, Allen S. *French Food: on the table, on the page, and in French Culture*. Routledge, 2001. See page 146.

The Silver Spoon. Phaidon, 2005. See page 128.

Slater, Nigel. *Real Fast Food*. Michael Joseph, 1992. See pages 154, 266.

Smith, Delia (www.deliaonline.com). See pages 305, 313, 317.

Smith, Delia. *Delia's How to Cook Book One*. BBC, 1998. See page 220.

Smith, Delia. *Delia Smith's Complete Cookery Course*. BBC, 1982. See page 181.

Smith, Delia. *Delia Smith's Summer Collection*. BBC, 1993. See page 253.

Smith, Delia. *Delia Smith's Winter Collection*. BBC, 1995. See page 243.

Tan, Christopher. *Slurp: Soups to Lap Up and Love*. Marshall Cavendish, 2007. See page 17.

Thompson, David. *Thai Food*. Pavilion, 2002. See pages 26, 79, 163.

Toussaint-Samat, Maguelonne. *A History of Food*. Blackwell, 1992. See page 116.

Uhlemann, Karl. *Uhlemann's Chef's Companion*. Eyre & Spottiswoode, 1953. See page 87.

Vetri, Marc, & Joachim, David. *Il Viaggio di Vetri: A Culinary Journey*. Ten Speed Press, 2008. See page 336.

Waltuck, David, & Friedman, Andrew. *Chanterelle: The Story and Recipes of a Restaurant Classic*. Taunton, 2008. See page 129.

Weinzweig, Ari. *Zingerman's Guide to Good Eating*. Houghton Mifflin Harcourt, 2003. See page 172.

Weiss, E. A. *Spice Crops*. CABI, 2002. See page 57.

Wells, Patricia. *Bistro Cooking*. Kyle Cathie, 1989. See page 155.

Wells, Patricia. *Patricia Wells at Home in Provence*. Scribner, 1996. See page 25.

White, Florence. *Good Things in England*. Jonathan Cape, 1932. See page 157.

Willan, Anne. *Reader's Digest Complete Guide to Cookery*. Dorling Kindersley, 1989. See page 177.

Wolfert, Paula. *The Slow Mediterranean Kitchen*. Wiley, 2003. See page 250.

Wright, John. *Flavor Creation*. Allured, 2004.

Wright, John. *Mushrooms: River Cottage Handbook No. 1*. Bloomsbury, 2007. See page 82.

Wybauw, Jean-Pierre. *Fine Chocolates: Great Experience*. Lannoo, 2006. See page 182.

Zieglar, Herta. *Flavourings: Production, composition, applications*. Wiley-VCH, 2007.

Other

Buttery, Ron G.; Takeoka, Gary R.; Naim, Michael; Rabinowich, Haim; & Nam, Youngla. *Analysis of Furaneol in Tomato Using Dynamic Headspace Sampling with Sodium Sulfate*. J. Agric. Food Chem., 2001, 49 (9) pp. 4349–51. See page 256.

Claps, S.; Sepe, L.; Morone, G.; & Fedele, V. *Differenziazione sensoriale del latte e della caciotta caprina in rapporto al contenuto d'erba della razione*. In: *Proceedings of I formaggi d'alpeggio e loro tracciabilità*. Agenzia Lucana per lo Sviluppo-Associazione Nazionale Formaggi Sotto il Cielo, 2001, pp. 191–9. See page 57.

Kurobayashi, Yoshiko; Katsumi, Yuko; Fujita, Akira; Morimitsu; Yasujiro; & Kubota, Kikue. *Flavor Enhancement of Chicken Broth from Boiled Celery Constituents*. J. Agric. Food. Chem., 2008, 56 (2) pp. 512–16. See page 96.

Simons, Christopher T.; O'Mahony, Michael; & Carstens E. UC Davis. *Taste Suppression Following Lingual Capsaicin Pre-treatment in Humans*. Chemical Senses, 2002, 27 (4) pp. 353–65. See page 204.

Recipe Index

General Index

Pairings Index

389

A Note on the Author

Niki Segnit had not so much as peeled a potato until her early twenties when, almost by accident, she discovered that she loved cooking. As much as she enjoys haute cuisine, she's not likely to attempt to reproduce it at home, preferring to experiment with recipes from domestic kitchens abroad. Her background is in marketing, specializing in food and drink, and she has worked with many famous brands of candy, snacks, baby foods, condiments, dairy products, hard liquors and soft drinks. She lives in central London with her husband.

LAST INDIAN SUMMER

LAST INDIAN SUMMER

THE BLOODY MILK CREEK SIEGE

RICHARD DAVIS

FIVE STAR
A part of Gale, Cengage Learning

GALE
CENGAGE Learning·

Farmington Hills, Mich • San Francisco • New York • Waterville, Maine
Meriden, Conn • Mason, Ohio • Chicago

GALE
CENGAGE Learning

LIBRARY OF CONGRESS CATALOGING-IN-PUBLICATION DATA

Davis, Richard (Richard J.), 1944–
    Last indian summer : the Bloody Milk Creek siege / Richard Davis.
        pages cm
    ISBN 978-1-4328-3039-7 (hardback) — ISBN 1-4328-3039-2 (hardcover) — ISBN 978-1-4328-3034-2 (ebook) — ISBN 1-4328-3034-1 (ebook)
    1. Ute Indians—Wars, 1879—Fiction. 2. White River Massacre, Colo., 1879—Fiction. 3. Frontier and pioneer life—Colorado—Fiction. I. Title.
PS3554.A937658L38 2015
813'.54—dc23   2015008075

First Edition. First Printing: July 2015
Find us on Facebook– https://www.facebook.com/FiveStarCengage
Visit our website– http://www.gale.cengage.com/fivestar/
Contact Five Star™ Publishing at FiveStar@cengage.com

Printed in the United States of America
1 2 3 4 5 6 7 19 18 17 16 15

ACKNOWLEDGMENTS

I would like to thank the publisher of this book, Five Star, for the assistance and encouragement provided in getting this book published. In particular, I would like to thank fellow author Michael Zimmer for introducing me to Acquisitions Editor Tiffany Schofield and Editor Hazel Rumney. I am grateful for their help in leading me patiently through the maze of details required by the publisher. Thanks for not giving up on me.

"THE CAVALRYMEN'S POEM"

Halfway down the trail to Hell,
In a shady meadow green,
Are the Souls of all dead troopers camped,
Near a good old-time canteen.
And this eternal resting place
Is known as Fiddler's Green.

Marching past, straight through to Hell,
The Infantry are seen.
Accompanied by the Engineers,
Artillery and Marines,
For none but the shades of Cavalrymen
Dismount at Fiddler's Green.

Though some go curving down the trail
To seek a warmer scene,
No trooper ever gets to Hell
Ere he's emptied his canteen.
And so rides back to drink again
With friends at Fiddler's Green.

And so when man and horse go down
Beneath the saber keen,
Or in a roaring charge of fierce melee
You stop a bullet clean,
And the hostiles come to get your scalp,
Just empty your canteen,
And put your pistol to your head
And go to Fiddler's Green.

—Anonymous

PROLOGUE

Congressional Record
United States of America
Congress
House of Representatives
Special Committee on Indian Affairs
January 15, 1880

WHEREAS: The Ute Indian Reserve, known as the White River Ute Agency, existed on the White River also called Rio Blanco River, under the administration of Agent Nathan Meeker, an employee of the United States Government Department of the Interior and,

WHEREAS: The White River Ute Indian Tribe, by treaty and agreement and under the sufferance of the Government of the United States of America, resided within the boundaries of said agency and,

WHEREAS: On September 29, 1879, in the State of Colorado the Ute Indian Agency known as the White River Agency under the Governance of the United States Department of the Interior, Bureau of Indian Affairs, was subjected to a revolt by its Ute Tribe residents resulting in the killing of Agent Nathan Meeker and eleven of his male employees and the capture and kidnapping of three white females and two children and,

WHEREAS: On September 29, 1879, three companies of the United States Army under the command of Major Thomas

Tipton Thornburgh were ambushed and attacked by Ute insurgents while on their way to aid the White River Agent, Nathan Meeker, and his staff. This attack resulted in the deaths of fourteen United States Army soldiers, the wounding of forty others, the slaughter of over three hundred cavalry and draft animals and the destruction of over twenty wagons and their supplies and,

WHEREAS: The United States Congress is constitutionally empowered to investigate the origin and nature of conflicts between states, between states and the federal government and between foreign nations, including Indian nations, agencies and reservations,

THEREFORE:

BE IT RESOLVED THAT: The United States House of Representatives hereby creates and empowers a Special Committee on Indian Affairs to hold hearings, subpoena witnesses and hear testimony concerning the origin and nature of those incidents, now termed the Meeker Massacre and the Milk Creek Battle, make a determination of cause and culpability and, if necessary or desirable, refer its findings to the War Department's Inspector General's Office for further investigation, remedies and corrections.

CHAPTER 1

"Alea iacta est." [The die is cast.]
Julius Caesar, January 10, 49 BCE

"He who is prudent and lies in wait for an enemy, who is not, will be victorious. If ignorant of both your enemy and yourself, you are certain to be in peril."
Sun Tzu, *The Art of War*

Captain Joe Lawson watched with clenched teeth as Lieutenant Samuel Cherry and his fifteen-man cavalry squad approached a ridge that ran from close to the right of where he stood upward to the north side of Yellow Jacket Pass. Their horses' hooves kicked up billowing clouds of white alkali dust that curled back over them and powdered their cavalry blue twill like snow.

Before they had ridden a hundred yards from him only the red-and-white guidon of E Company, Third Cavalry, could be seen flapping above the dust at the tip of its ten-foot spiked standard. He tightened up on the reins of his prancing bay mare and patted her withers with his gauntleted hand. "Easy old girl," he murmured deeply. "There'll be plenty o' time for that here shortly." He looked over his shoulder at the rest of his E Company command and then to F Company, Fifth Cavalry, commanded by Captain J. Scott Payne. To his left, Major Thomas Tipton Thornburgh, commander of the White River expedition, silently observed Cherry's detachment fade into the distance. Thornburgh's heavy muttonchop whiskers hid any expression but Lawson felt certain his commanding officer was having deep doubts about his chosen course of action.

Looking to his right Joe's eyes fell on Thornburgh's second in command. Captain Scott Payne's eyes darted back and forth from the squad to Lawson. Scott had earlier confided his deep misgivings about the morning's strategy to Joe but said nothing to the major.

Lieutenant Cherry's cavalry squadron had now slowed to a walk as they approached a line of red-gold-leaved scrub oak. Beyond and above them lay a darkly ominous line of pine, cedar and juniper. Thornburgh ordered a detachment from F Company to form a skirmish line along a similar ridge to the officers' left. The skirmishers' combined positions formed a large "V" with the intersecting end pointed up the incline and into the trees that ultimately intersected at the crest of Yellow Jacket Pass where Indians ran through the clearings among the trees and oak brush.

Lawson ached with a foreboding fostered by more than thirty years of soldiering. All his senses knotted into pinpricks of vigilance. His mare, an experienced warhorse, had stopped dancing and stood, stock-still, legs braced, trembling, head up and ears pricked forward. She knows for sure, Lawson thought.

He sensed an impending explosion. He had felt it for the first time with the Light Brigade at Balaclava even before the bugles had blared "forward." For him time slowed; seconds seemed like minutes, minutes like hours. A mishmash of battle sounds deafened him but he could count each beat of his heart. He could feel the tickle of a soft southerly breeze on his face. He smelled the mare's lathered hide, the sagebrush, the tangy juniper berries and pine needles. The scene proceeded at a crawl, then nearly stopped. He felt as though he could almost count Cherry's horse's strides. He looked up. A vulture that a few minutes ago had been spiraling on the breeze now seemed to hang suspended against the cobalt sky. Even the skittish horses and their anxious riders, his fellow officers, moved as though swimming through molasses. His stomach cramped; breathing nearly ceased, heartbeats drummed in his ears.

There appeared in a clearing on the ridgeline of Yellow Jacket Pass a line of people, perhaps a dozen, evenly spaced, trotting very slowly, each carrying something. As if on signal the line stopped and without breaking formation, knelt down. Thornburgh and his command had come to a standstill, like a tiny audience watching puppets dance across the hilltop clearing.

From his ridgetop position Cherry had seen a band of several dozen warriors five hundred yards away veering to his right. Are they flankers, he wondered; probably not, they seemed in no hurry. Lieutenant Cherry, looking back over his shoulder, removed his hat and waved it as a sign of reassurance.

Thornburgh himself now realized that following Cherry into the assembling Utes threatened his entire command. He told Payne to hold his position on the ridge but not to fire unless fired upon. The major also began to apprehend the mistake he'd made by crossing the creek in force instead of the five-man party demanded by Colorow and Jack. He hoped that he could preempt a catastrophe by a last-minute meeting with Colorow or Jack.

Cherry lowered his hat and someone, Indian or trooper, fired a shot. A bullet hissed past Cherry's head and wounded the trooper behind him, the first casualty of a long battle.

Lawson involuntarily flinched at the sound of the shot, his mare shied and he nearly lost his seat. Within seconds of the first shot the entire tree line blossomed with blue gun smoke, and the late morning air reverberated with the trip hammer blasts from hundreds of repeating rifles. Lawson could see Payne's men firing from a skirmish line, their Springfields adding to the din. The snip, snap and zip of rifle bullets passing among Lawson and his fellow officers followed the clatter of fire.

Lieutenant Cherry's troopers, Lawson saw, were engaged in a fighting withdrawal, apparently under Thornburgh's orders to protect the right flank and not let the Indians get a foothold between them and D Company back at the wagons. Major Thornburgh rode back calmly from his forward position directing the bugle calls while also sending a runner toward Lieutenant Cherry's position. Joe turned and organized E Company, Third Cavalry, into an oblique line aimed between the wagons and flanking Indians. Then he edged forward himself to pick up Cherry's command as they fell back. Joe was keenly aware of the same astonishment he'd felt in so many other battles: Why, with pounds of lead slugs careening within inches of him every minute, had he never once taken a wound? As they retreated, men and horses seized by surprise, deafened by the clamor, blinded and choked by dust and gun smoke ran, reared and collided with each other. It was a cauldron of confusion, a maelstrom of human and equine flesh, leather, metal and blue serge, truly, he thought, the fog of war.

Major Thornburgh, Joe saw, commanded an orderly retreat back across Milk Creek and toward the wagons. But he knew his commanding officer also realized the urgency of collecting his entire command in one defensible position. Companies E and F were in great danger of being outflanked, surrounded and cut off from D Company and the wagons.

"Payne!" Thornburgh yelled. "Take your company west and prevent Cherry from being overrun then retreat back to the wagons and link up with Lieutenant Paddock's D Company that I left as rear guard."

Joe spurred his horse into a full gallop toward the north end of the ridge where his E Company men had dismounted and were fighting afoot. *The Manual of Arms* prescribed fifty feet between individual skirmishers but under battle conditions even the best trained men seeking protection and mutual support often closed ranks and put themselves in greater danger of becoming surrounded. His men were collapsing like an accordion. He intercepted the men leading the horses to the rear. "Turn around and follow me back up," he yelled, "get those men mounted and fight our way out of here." Relieved, he watched as the men didn't hesitate and immediately fell into line behind him.

The Utes had been pouring a withering fire in his direction but he noticed by the sound that the bullets were flying well away from him. The warriors were now so preoccupied with overtaking his men, separating them from Payne's and surrounding them that they were firing on the run, some from the hip. As he and the troopers leading the horses approached the men in the skirmish line he yelled: "Mount up, and keep up screening fire as you fall back."

The last man in line, a very young private, upon seeing Joe, dropped his rifle and ran toward a horse. Joe snatched the reins from the trooper leading the horse and away from the boy attempting to mount it. Joe grabbed him by the back of his shirt and pulled him to the ground. The private looked up at his commanding officer in panic and pain.

"Private! Get to your feet, come to attention and salute." The boy's face registered only confusion. "Now!" screamed Joe unlimbering his revolver and putting the muzzle against the boy's forehead. Training defeated fear. Resolve replaced panic. He bent and picked up his hat and jammed it on his head, came to attention and saluted. Joe returned the salute. "Private, go pick up your carbine, you *will* need it." The boy ran back to where he'd dropped his weapon as Joe unloaded his revolver over the boy's head causing a squad of Utes closer than fifty yards away to run for cover. Once astride his horse the boy took aim and fired his carbine at them while Joe reloaded.

Lawson looked back up the ridge where the Ute were sliding in and out of cover firing as they approached. If, he reasoned, Cherry and Payne had linked up and organized a fighting retreat over this ridge to the northwest they would be in the midst of a shooting footrace to the wagons. He urgently needed to get his company organized into a fighting withdrawal so that the two companies could cover each other's flanks. So far, at least, there had been no noticeable action toward the east. For him, the more pressing action was to back this squad down, link up with the rest of E Company and retreat toward D Company and the wagons. He descended calmly encouraging each man in the line and moving him down a few evenly spaced paces to maintain formation and taking advantage of any cover, tree or rock as they went.

After reforming and again checking for any stragglers he broadened their line of retreat into a large crescent to prevent the Indians from flanking them. He commanded from the center of the crescent but frequently checked the firing to the west where Cherry and Payne were fighting their own rearguard action. With relief he caught sight of the tall lieutenant firing, retreating and reloading and then charging back into the fray. He gave the command to the trooper nearest him to slide in Cherry's direction and then galloped along the crescent repeating the order and making sure each trooper stayed within eyesight of those to his right and left.

Lawson desperately scrutinized the receding ridges and gullies for either the major or Captain Payne. He didn't see Payne.

The captain had had his horse shot out from under him and, unarmed, was in a desperate struggle for his life against a lone warrior. The warrior, armed with a revolver, was almost on top of Payne. But First Sergeant John Dolan had also noticed the captain's plight and circled back and run his charger between the Ute and Captain Payne, firing his handgun at point-blank range. The Indian, though unwounded, stopped in his tracks. By that time, Payne had found his own revolver in the dirt and had begun firing. The Ute fled back up the ravine. A private leading a horse for Captain Payne and Dolan then joined the other two.

Lawson saw a lone trooper riding at an unusually leisurely pace about five hundred yards to the rear of his skirmish line and a hundred or so yards north of the Milk Creek crossing. As Lawson started to leave the line and risk a run to help the trooper, the man threw both arms above his head as if waving or signaling then toppled over backward into the tall sagebrush. Lawson had unknowingly witnessed the major's last moments of life.

Private John Donovan, on his way to the wagons for more ammunition, discovered the body of Major Thornburgh. Donavan waved his hat frantically as he approached Lawson's position behind the skirmishers. He reined in his lathered horse next to Lawson.

"The major's been killed, sir," Donavan reported.

"You're sure it's the major? You're sure he's dead?"

"Yes sir, both . . . shot in the chest, looked like the heart. Excuse me, Captain, I have to get to the wagons for more ammunition. Captain Payne's been wounded and his men are nearly out."

Lawson motioned the private to stop. "How bad is Captain Payne's wound?"

"Left arm and shoulder, sir, but he's still astride his horse fighting and giving orders."

"Carry on, Private."

"Yes, sir."

Lawson, again, started for the point where he'd seen the

major fall but at that moment he also saw a company of Utes charging through the sage to the same spot. His sense of duty won out and he returned to commanding and saving his company.

Joe could hear a sudden, furious but brief amount of fire in the vicinity of a low, north to south ridge maybe half a mile long. He felt a stab of fear. If the Utes had intercepted Payne and Cherry, his company would be their next target. Companies E and F would go down first because they had gone into the field with only forty rounds of ammunition per man. That left D Company, the smallest and weakest of the three, to defend the wagons against a force perhaps three or four times larger. A very bad situation, he thought, that could very quickly get much worse.

He continued to back his troops toward the semicircle of wagons still several hundred yards distant. A trooper appeared on the flank of the hill where the heavy firing had occurred and raced toward him. The man, a sergeant, reined in his heaving mount, saluted and gasped for breath to speak.

"Sergeant Grimes," Lawson yelled above the gunfire in an exaggerated Irish accent, "you're lookin' like you had an Irish Stand Down with the devil."

"Aye, sir that I 'ave. Captain Payne and Lieutenant Cherry are just over that 'ill and a few hundred yards ahead of you gettin' back to the wagons. We came within a wee bit of bein' cut off by the Indians on the other side of that ridge. Captain Payne ordered me and two more to clear the ridge, which we did but only after a hellofa fight. That's when I saw you and reported it to the lieutenant. He told me to inform you that he and the captain and most of their men will be inside the wagon corral directly."

"Thanks, Grimes, until now I didn't know if we had a right flank of army or Indians. How many casualties have you seen?"

Grimes shook his head. "Hard t'say, maybe half a dozen killed and the same wounded. But fights like this always look worse when one's amid 'em. I should be trottin' back."

"Ed, head on back to the wagons and help Paddock with the defenses. Your horse is about to drop dead and I don't want the army's best noncom wanderin' around on foot out here. We'll all be a-needin' your services later on more than Payne and Cherry need you right now." Grimes argued no further but touched the brim of his hat and moved off in the direction of the wagons.

Within twenty minutes Payne, Cherry and their men were within two hundred yards of the wagon perimeter at which point a good three quarters of them had ejected their last spent cartridge. Payne yelled the order to mount the led horses and break for the wagons. He, Cherry and a handful of men with ammunition maintained their positions and fired furiously at Utes, who, bolstered by the slackening fire and mounted race for the wagons, began breaking cover in a full frontal assault.

From the other side, Paddock saw the troops take flight toward the wagons and directed his men to pour concentrated fire at the Indians flooding down the hillside behind them. The single-shot Springfield rifles and carbines with their greater range and accuracy began taking a heavy toll of the pursuing warriors and their horses. Lawson's men were close enough to mount and with the last of their ammunition execute a devastating flank attack. The warriors withdrew to a more heavily wooded hillside leaving a scattering of dead and wounded men and horses.

The men of Companies E and F dismounted once they had reached the wagons corral, collected cartridges from opened boxes and took up positions alongside the D Company men. Lieutenant Paddock, who had gallantly exposed himself in the center of the corral to direct defenses, went down with a hip wound.

As soon as both captains were safely inside the compound, Payne and Lawson met to assess the situation. Payne's blood-soaked left arm dangled uselessly at his side. "Joe," he moaned, "with the major dead, I'm supposed to be in command but I don't think I can function in that capacity with this wound."

"Captain, your wound isn't fatal. I've been wounded worse'n that by my wife with a paring knife. Thornburgh picked you as second in command. I've seen a lot of officers in the last forty years and you're as good as any. I'll be happy to advise you but you need to lead this bunch as the major intended. You make the decisions and I'll back you up."

Payne nodded. "We've got to get out of here. First, we'll turn around and withdraw back to where we left Lieutenant Price and the infantry and then fight our way back to Rawlins for reinforcements."

"Your first bad idea! Even if we could get the men assembled in a line out of here they'd cut us to pieces from the trees. Not a one of us would make it to Price's camp let alone to Rawlins. Make another decision."

"We close ranks and fight here; send for help and wait to be relieved."

"Now you're getting the hang of it. Get Doc Grimes to tend that shoulder while I get things organized otherwise." Joe turned and surveyed the scene surrounding him. Less experienced eyes would have seen a maelstrom of death, destruction, panic, confusion and disaster. One hundred fifty men and nearly three hundred head of horses, mules and oxen seemingly running amok. Officers were shouting orders, buglers transformed them into piercing calls, wounded men and animals screamed piti-fully and dashed blindly back and forth amid a hail of several hundred bullets a minute from the Indians' repeating rifles. An increasing thunder of army rifles and carbines countered them. Small squads of Utes on foot had gotten within thirty yards of the wagons but were stopped by the deadly accuracy of the Springfields. Lawson's experienced eyes saw courage and training winning over fear and stupidity. Officers and noncoms were directing the unloading of wagons to build bulwarks with ration boxes, flour sacks, clothing and bedding. Others directed men and fire.

Lawson checked his watch. It was almost one o'clock. It had taken the men less than two hours to stage a fighting withdrawal against an ambush by overwhelming enemy numbers and secure a strong defense that stopped the attack. As only one who has devoted a lifetime to the profession of arms can feel at such times, he felt exalted and thrilled. But he also knew from his combat experience that this scene, this tiny fortification would be home for quite some time, for some that meant eternity. He looked south hoping to see Major Thornburgh making a mad dash to the wagons on his tall, gray mare, Athena, but saw only men afoot or riding short ponies, Utes flocking in for the kill. Part of the reality of a soldier's life is that one must accept the probability of violent death delivered by the hands of men that one doesn't know, let alone hate. It was a fearful fact. He accepted it when he first joined the Seventeenth Lancers at age fifteen. Joe found it more difficult imagining death suddenly pouncing on some men but not others. Captain Nolan, who, at Balaclava, was the messenger who delivered Lord Raglan's charge order to Cardigan, was the first to fall to the Russian cannons. And General John Reynolds at Gettysburg was downed by a sniper who, in all probability, could barely see him. When a man dons the uniform, he not only relinquishes his freedom to move freely, but hands over his fate to a different set of gods and deliverers. Where, when, why, how and by whom are such determinations made? By what paths did the young major travel to what now seemed an inevitable conclusion?

CHAPTER 2

"Every man thinks meanly of himself for not having been a soldier."

Samuel Johnson

"There's many a boy here today who looks on war as all glory but it is all hell."

General William Tecumseh Sherman

Thomas Tipton Thornburgh's life's course and conclusion were set and sealed on the morning of April 12, 1861, a fate which Joseph Lawson might have forewarned him. But given Tip's single-minded determination to succeed at what he considered to be his life's calling, Lawson's words would have fallen on deaf ears. Actually, he could not remember wanting to be anything but a soldier but on this day he would embark upon that course of action.

On that early spring morning he had awoken long before dawn, long before the larks, finches and robins pulled their heads from beneath their wings and began calling shrilly from the branches of the huge flowing dogwood tree outside the open bedroom window. Wide awake but his eyes still shut to dawn's first light drifting across his bed, he tried to imagine what adventures awaited. But the enormity of it all overwhelmed him. Just trying to comprehend the considerations of time and space were too much let alone all the people, events, fears, joy and sadness. For him, until today, the world, planets and stars revolved around his family and their large farm outside New Market, Tennessee. This morning he would walk out the door of the only home he'd ever known, maybe for the last time, ever. He'd never considered that. A current of fear coursed through his body. He shivered involuntarily. He could change his mind. Mother would be overjoyed. His father would welcome him into his law practice and he could continue indefinitely to fish and hunt with his brother and the dogs. But just as quickly as the thought appeared in his mind, he forced it out again. No, he thought, my life's course is set. Sometime during the night he'd pulled the coverlet up to his chin leaving only his head exposed to the East Tennessee spring chill. Now, the warmth of the sun brought forth an opera, a cacophony of cheeps, whistles, chirps, warbles, peeps and screeches. The massive lump huddled next to him on top of the bedcovers shook, shivered and twitched in response to the birds. Jesse, his rangy, ten-year-old Irish setter, had found the floor too cold and took refuge next to him on the bed. He felt great comfort feeling the flow and ebb of her breathing.

He could hear Georgia building a fire in the iron kitchen cookstove lifting the flat irons and banging them about spitefully to wake the whole house. Georgia had come to the Thornburgh family as payment of legal fees due his father from a Mississippi planter. Though not an abolitionist, his father, Montgomery Thornburgh, a democrat and Tennessee state legislator, had literally no use for slaves or slavery so he took Georgia, freed her and brought her into the Thornburgh family. In return, she nursed the family with her home remedies, cooked, and disciplined the children with an iron hand and willow switch. He heard his parents getting up and dressing and his older brother, Jacob, bounding down the stairs. Tip rolled over onto his back and watched Jesse perform her morning regimen of stretching and yawning.

"Tip!" his father called from the hallway. "Reveille, let's hear your feet hit the floor." Montgomery Thornburgh had attended West Point but instead of graduating he'd left and returned to New Market, Tennessee, to marry Tip's mother and read law with the Jefferson County Attorney. Jake had been born a year after Montgomery had been sworn into the bar and Tip two years later. A man of strong ambitions, Montgomery's law practice flourished and he bought more than a thousand acres of land outside New Market upon which to raise his growing family. Having distinguished himself within the county bar association and become a respected and established country gentleman it seemed to him, as he told his family one Sunday over dinner, that he should pursue a political career. That led him into that Tennessee legislature with high hopes of ultimately climbing the steps of the US House of Representatives and Senate. Being a man of no middling talents that might well have happened except that both his personal political philosophy and Tennessee's ambivalence toward slavery and states' rights came athwart the push and pull of the Southern separatist Democrats and the Northern union Republicans. Though an antislavery humanist, one could not survive politically in Eastern Tennessee as an abolitionist. By subscribing to the Confederate "state-rights" philosophy, he hoped to bridge that philosophical gap. He espoused that the slavery decision should be left to each individual state. A politician so obviously compromised in the milieu of mid-nineteenth-century America was unlikely to sustain a long political career.

Tip hurriedly dressed, washed his face, combed his thick, dark-brown hair and checked his slowly emerging whiskers in the mirror. He'd begun shaving the previous year, but not every day. That particular manly rite of passage he'd eagerly anticipated sharing with his father and brother turned out to be a lengthy regimen of razor honing, soaping and soaking, painful scraping and nicking resulting in a patchy product that made him look like he lost a fight with the cat.

He picked up the leather valise he had carefully packed the night before and set it on the bed. Turning to view the bedroom he'd occupied for as long as he could remember, fear, sadness and remorse again replaced anticipation and excitement. Jesse sat on the bed her ears pricked quizzically and tail flopping against the pillows. His eyes misted slightly. She'd been a present from his father on his seventh birthday. Montgomery had a particular fondness for Irish setters. He had singled her out as being the most promising pup of a litter of eight from his prizewinning bitch, Athena. She had been, for nearly ten years, his constant companion. She escorted him to school in the mornings and from school in the afternoons. Work or play she was at his side all summer long. Every day was a training day and she could respond to at least a dozen hand and voice commands, sometimes even anticipating them. Now, she was ready to begin another joy-filled day following him and Jake as they worked the farm. His eyes burned. The perfect, pastoral world she had known all her life was ending. She didn't know nor would she understand his disappearance. Would she feel abandoned? Betrayed? At least she would live out the remainder of her life well cared for in the quiet seclusion of the farm, as would his mother. The two of them tumbled down the stairs together.

Tip opened the door and ushered Jesse outside for breakfast then seated himself across the table from his father who studied a legal brief in one hand and sipped coffee with the other. Montgomery looked at Tip over the top of his spectacles.

"Better get used to rising earlier and shaving every day," he mumbled. "Olivia, get the boy a cup of coffee."

"Yes, sir," Tip replied contritely. It was part of the morning routine. He couldn't remember his father greeting him with anything but an admonition or reprimand. "But it's not West Point. Most of the enlistees grow beards."

Montgomery put aside the brief and eyed his younger son closely. "Too late for a haircut," he said, ignoring Tip's comment, "but the army will take care of that the first day anyway. Enjoy that coffee." He nodded toward the cup his mother set before him. "You'll probably be getting burned chicory bark instead after today . . . and for some time to come."

"Yes, sir." Tip put the cup to his lips and tested the boiling liquid gently. "Where's Jake?"

"Hitching up the team, he'll take you into town to catch the train to Knoxville where you'll meet up with Colonel Joseph Cooper and his Sixth Tennessee Volunteers. Colonel Cooper and I began West Point together. He's a good man of sound judgment, got excellent grades in strategy and tactics. You can put your complete trust in him, follow his orders to the letter. I'd go with you boys but I'm due in court today."

"Yes, sir."

"Thomas," said his mother seating herself on the chair next to him while Georgia laid a plate piled high with ham, eggs, buttered grits, okra and cornbread, "remember to write . . . as often as you can." Her eyes brimmed with tears, a rare occurrence which made him most uncomfortable.

"I will," he promised laying his hand on her shoulder.

"It's not too late to change your mind. I still say you're too young."

"No, it's not too late but my mind is made up. I never wanted to be anything but a soldier."

Montgomery took off his spectacles and dropped them on the table. "Tip, you know I very much support your decision to make the profession of arms your lifelong career. Had I not met your mother"—he nodded deferentially toward Olivia—"I would have done so myself. Don't misunderstand me. My farm, family and job make me happy. I'm convinced and always will be that I made the right decision. Times change and people change and in a year or two you may be back here. But I encourage you to wait one more year and begin your future profession with your entry into West Point."

Tip handed his mother a carefully ironed and folded linen napkin, a Georgia showpiece, to daub her tears. His mother had been adamantly opposed to his enlistment and even his father only reluctantly accepted it. "I understand and appreciate both your sentiments and advice but as everyone knows that a war—if it's to come at all," he added to soothe his mother, "would last less than a year, maybe a few months, which would mean I'll have lost the opportunity for the battle experience that an officer, even a West Point graduate, needs to further his career. I'm not doing this on a lark. I'm doing it because I believe in the Union, the United States, and because I want to establish myself as a professional soldier. At the end of the war I'll be old enough to apply to the academy. That is what I want to do with my life."

"And a noble goal it is," said his father rising from his chair. He put the document he had been reading into a satchel, took his hat from the hall tree and strode over to Tip who respectfully stood up. "Good luck, son, and may God be with you." He ignored Tip's proffered hand and instead embraced his son tightly. "I love you. I'm proud of you—never forget that." He whispered in Tip's ear.

Tip, surprised and a little embarrassed by Montgomery's emotional candor responded by hugging him all the more tightly. Much to his relief Jake barged through the door announcing the carriage and his father's horse were prepared. Montgomery and Tip stepped back from each other, saluted, and shook hands. Montgomery bent over the table and kissed his wife good-bye and strode out the door.

Tip returned to his breakfast but his appetite had disappeared and he only picked at it until Jake finished his and was ready to leave. His pending departure hung over the silent table like a dark whirlwind sucking the words from everyone's mouth and breath from their lungs. "Well," he said, sliding back his chair and standing, "I guess we should be going . . . don't want to miss the train."

"Just a moment, Thomas," replied Olivia leaving the kitchen and walking into the dining room. After several awkward seconds he walked over to Georgia who he could see was ready to burst into tears. He held out his arms to her and she fairly flung herself into them.

"Oh, Mother Georgia, I will miss you, dear. My earliest and best memories of life are nestling in your arms like a chick under the wings of a hen. You comforted me when I was sad, nursed me when I was sick, reassured me when I was afraid, shared my joys and tears. How can I ever repay you?"

"Mister Tom, you repayin' me and lots more like me what ain't free by walkin' out that door and riskin' your life. No one, man or woman, can ask fo' mo' repayment than that. But I do ask you stop short of dyin' fo' us." She surrendered to sorrow and lapsed into a gasping, soft wail.

Olivia returned to the kitchen. In a show of mock courageousness she held out her hand for him to shake. He recognized that in these last few seconds together the mother-infant bond was forever severed and forever replaced by the eternal mother and son union. He took her hand and felt something pressed into his palm. He grasped it and looked. There were two gleaming gold objects, a Saint Christopher medal and chain and a brand-new twenty-dollar gold coin. He looked at her, shook his head and started to return the gold coin but she folded his fingers back over it. "No, I want you to keep it, not spend it right now.

I want to know that you'll never be completely broke. But, there will come a time and only you will know when that time is, to spend it on something very special." She took the medallion from his hand and held open the chain. He bent forward with sincere reverence. She looped the chain over his head and around his neck. He kissed her gently on each cheek inhaling the same perfume she'd worn for as far back as he could remember. Oh! How badly he wanted to rest safely again in his mother's arms and for her to carry him away from this dreadful act. He loosened his embrace for fear if it lasted another second he might never be able to do so. She had regained her strength. Her eyes were clear and her lips firm. "Good luck, Tom. When shall we meet again?"

He smiled. "When the hurley-burley's done, when the battle's lost and won," he said repeating one of their favorite lines from Shakespeare's *Macbeth*. Tom picked up his valise, donned his hat and walked out the door without looking back. Jake and Jesse were sitting in the buggy waiting for him.

"Jake," said Tip, throwing his valise into the buggy and vaulting onto the seat beside Jesse, "we best hasten on out o' here before I do change my mind. I reckon the direction I'm headed in I'm gonna have a lot more difficult decisions to make in my life but this is the hardest one I've made in the past seventeen." He noticed that his brother didn't reply. He just gave the horse a spat with the reins, the carriage lurched forward.

Jacob leaned against the back of the seat and looked over at Tip. "You and Dad surprised me. In fact, I couldn't have been more surprised if I'd walked in on him and Georgia naked."

Tip scowled back at him. "Now *that* doesn't bear thinking about! Jake, if you'd just witnessed the second coming of Christ you'd find the most distasteful way of announcing it. He surprised me every bit as much as he did you. Tell me, did you ever consider Montgomery Thornburgh, Esquire, had a heart?"

"Yes, but a black one. He is affectionate enough with Mother, who, I suspect, lowered her expectations over the years, and he seems to know when she needs it most, I'll credit him for that. Can't say I ever expected or wanted his love and affection and doubt if you did. After all, we're his sons. Fathers, who only have a limited amount of those emotions, can't be expected to waste them on the male line. That can wait until one's death."

"Fair enough, does that go for brothers as well?"

"Absolutely, I wouldn't be caught dead giving you a hug."

"Right, the thought of my dead brother hugging me is quite repelling."

The brothers laughed and Tip put his arm around Jesse and pulled her over to him. He buried his face in the hollow of her neck to smell the farm dust and the lingering odors of Georgia's breakfast before the crisp air of the Tennessee morning carried them away.

"Tip?" Jake asked in a somber tone, "You feel certain war's comin'?"

"Yeah . . . it's just a matter of how big and how long."

"Which way's Tennessee gonna go?"

"Don't know, we're neither fish nor fowl. We're sure enough below the Mason-Dixon Line, but slavery never took hold up our way, maybe the same reason cotton never did. Father's in a bit of a pickle; a Free Soil politician in a borderline county. Could go either way, could split. What do you figure to do?"

"About what?"

"Your future."

"I'm staying home for the time bein'. Farm and study law with Dad. War comes I'll jump north. I've seen you shoot pocket change out of the air and I don't want to take a chance bein' on the business end of your long gun." Tip laughed, reached around Jesse and patted Jake on the shoulder. "Hey!" Jake responded. "Don't you ever touch me like that in public!" They both laughed. But the thought of either of them aiming a weapon at the other had struck a nerve and Tip wondered how many family members on both sides would be faced with exactly that reality.

"Speakin' of the public—looks like there's a good bit of it in the road up ahead."

Tip looked up to see a cluster of men, horses and carriages around the New Market Post Office. "Looks like. Wonder what's goin' on, not a holiday or county fair."

"Maybe a hangin'," kidded Jacob.

"There you go again, bein' bloody-minded."

"And usually right. Ain't that Cyrus Blankenship?"

Tip looked more closely at a tall, gangly boy walking toward them from the crowd and recognized his former schoolmate. Cyrus and his family had moved to the New Market area when Missouri entered the Union as a slave state. He had been in school for only a couple of years when he had to drop out and work the farm because his father died of typhoid fever. Cyrus jogged the last few yards to the carriage.

"You boys heard the news yet?" He spluttered around a wad of tobacco.

"Howdy to you too, Cyrus," replied Tip, "nary a word."

"Fort Sumter in Charleston Harbor was fired on and surrendered yesterday afternoon. Lincoln's calling up the northern states' militias and the regular army. No doubt about the war now. I envy you, Tip, gettin' in right at the start. 'Old Fuss and Feathers' General Scott'll roll over top Jeff Davis and the rebels and you'll be home to parades before Christmas with medals and all. I wish I was goin' with you." The suddenness of the news took Tip by surprise and both he and Jacob fell silent.

"Fancy that," said Jake quietly staring straight ahead.

"Yeah," murmured Tip thinking of the breakfast debate. "Amazing how much can happen in so little time."

He looked at his ragtag chum, arms and legs akiml buckteeth already stained permanently brown with tobac juice. He couldn't picture him standing at attention with musk at shoulder arms but he could imagine him spread-eagle on th ground with an ounce of lead buried in his chest. "No, Cyrus, he spoke slowly and quietly, "don't envy me. War's not all it' cracked up to be. The price of glory and medals are gore and destruction. You'll be home for Christmas and it is I who will be envying you the warmth of house, home and family. Excuse us, my friend, I have a train to catch."

He held out his hand which Cyrus sandwiched between his two horny, calloused fists. "God bless and take care of ya, Thomas."

The brothers proceeded to the railroad station which was surrounded by horses, carriages and wagons. "Looks like you'll not be without company on this trip. I'd guess about half the men in Jefferson County are here and the rest will be here by the end of the day. You get your ticket, Tip, and I'll wait for you out on the platform."

The boarding platform was awash with stone-faced young men trying to console weeping women, children and parents. Jake felt very ill at ease standing alone with Jesse. It was obvious he wasn't one of the volunteers and drew many scornful stares. He was entering the station to see what was delaying Tip when his brother stepped out of the door.

"Train will be late," explained Tip. "They had to stop and add cars for all the unexpected passengers."

"How late?"

"Don't know. You go on back to the farm so you can get there before dark. Mother and Georgia shouldn't be there alone. Be tactful for a change. Don't just march in the door hootin' and hollerin' that the war's started. She'll hear it soon enough and maybe we spare her a day or two of more intense worry."

33

He held out his hand. "Do *not* embrace me in front of all these people and do *not* shed a tear," he laughed.

"I won't," said Jake. "I'll say good luck and don't forget to duck. When shall we meet again?"

"When the"—he felt a wall of tears filling his eyes—"battle's lost and won." He knelt and buried his face one last time in the hollow of Jesse's copper neck fur, rose, turned his back to them and walked, head and shoulders above almost everyone else into the crowd.

CHAPTER 3

"Love is like war: easy to begin, but very hard to stop."

H. L. Mencken

Josephine Meeker awoke with a start, a sound, an unusual noise, had awakened her. She lay petrified in bed scarcely breathing, listening for it to repeat. Her only answer was the familiar harmonic shrill hum of mosquitoes, frogs croaking, crickets chirping. Like a turtle retreating into its shell, she pulled the quilt over her head. Its comfort from the chill air wafting through the open windows was immediately palpable. Her head had become cold. The cabin room had been desperately hot when she'd gone to bed. She had stripped naked after blowing out the lantern on her bed stand and had lain sweating into the sheets and pillowcases in the dark for what seemed an eternity while listening with increasing anxiety to the distant monotonous tempo of the Ute's thundering war drums.

A constant flow of messengers, once Jack himself, had arrived Saturday and Sunday from Yellow Jacket Pass with the latest information on the approach of the soldiers from Fort Steele. Some of the squaws, including Johnson's wife, Susan, had struck camp and started south across the Roan Plateau toward the Grand River. Most of the younger men had accompanied Jack north on the buffalo hunt forbidden by her father, Nathan Meeker, leaving a small cadre of warriors, adolescent boys and old men to guard the village under the leadership of Quinkent, called Douglas. On the other side, mail carriers Black Wilson and Wilmer Eskridge were kept busy twenty-four hours a day delivering telegrams between Nathan, his bosses at the Department of the Interior and the army command.

No news was good news. Every day, it seemed, brought a new crisis as the troops moved down the 185 miles of the Agency Road from Rawlins, especially since the fight with Johnson.

It'd been almost a week since that pivotal incident. Johnson, a Ute shaman and leader, had confronted Meeker in his office about the agent's recent decision to plow the field where the Utes customarily held their very popular horse races. Meeker, already in a pique about the Utes procrastinating cutting hay until it was past its prime, voiced, in the most undiplomatic terms, his low opinion of the Ute work ethic. Harsher words ensued and her father had bluntly stated that the Indian's sacred ponies should be shot and used for meat over the winter. Johnson replied by jerking the agent off his feet and throwing him bodily out the door of the office over the hitching post and into the dusty road. They parted in opposite directions without further words. The incident prompted another Meeker message in which he described the "attack" by Johnson in exaggerated and inflammatory terms. He demanded through the Department of Interior that the War Department, headed by none other than the Indian-hating war god, William Tecumseh Sherman, send a cavalry force to quell a pending Ute uprising.

Another wave of fear and helplessness washed over Josephine as she appraised the realities of her situation. She was naked and alone in a cold, dark room on the bare edge of civilization surrounded by people who were speedily preparing to kill each other. The loneliness and apprehension nearly choked her. Her life, body and soul had been spent, as her mother's and sisters' and brother's, leashed to her father's idealistic fantasies, irrepressible impulses and chronic indebtedness. Now, nearly in her mid twenties, almost twice the age of Flora Price who had two children, unmarried and childless, she was quickly abandoning hope of any positive changes. Daily, perhaps hourly, she cursed her cowardice for not distancing herself after her college graduation from a family seemingly bent on self-destruction. Rozene, her older sister, dithered endlessly about becoming an undesirable spinster. Ralph, like her father, had a star-crossed career as a journalist, always chasing the next job. Mary had successfully cut her family ties and was living a reasonably normal life but probably only because she gained anonymity. Josephine desperately wanted a man to share her cold bed, one that she could grasp and hold and who would hold her and tell her everything would be fine and he would see to it. But now that had taken on the aura of a childhood fantasy. If she was tormented to near insanity by thoughts of Nathan Meeker and missed opportunities then there was no apt description for the terror that seized her when she thought of the other man in her life and the indescribable chasm that separated them.

Her eyelids finally relaxed and lowered over tear-drained eyes. On the brink of tumbling into comforting darkness she sensed a shadow pass through the silver moonbeam flooding through her open window. But only the chorus of night sounds vibrated through the chill breeze. Then, again, her eyes instantly popped open. She was wide awake, every muscle taught. The zephyrs that had carried the comforting odors of sage, river water, spruce needles and wood smoke were now suffused with the smell of sweat, wet leather, wool and . . . bear grease.

Josephine quickly reached over, grabbed the edge of the bedspread and snatched it across her thin white body. Heart racing, she waited breathlessly for sound or movement. As her oxygen-starved muscles began convulsing she heard a soft, muffled tinkling, a bell.

It was the familiar sound of a tiny silver bell that her grandmother Meeker had given her mother, Arvilla, when Josephine was born. No bigger than the tip of her little finger, her father had later given her a thin silver necklace so she could wear it around her neck. For nearly thirty years she had never taken it off—until they moved to the White River Agency. She had given it away along with her heart.

She exhaled, dizzy with relief. Slipping from beneath the covers, one eye on the window she hurriedly drew on a pair of twill riding breeches and a man's checkered shirt. She dashed to the door but stopped remembering the latch and hinges were rusty and, as her mother said, "Sounded like someone opening the gates of hell." Instead, she went to the open window and cautiously peeked out at the moon-silvered sage landscape. A few aspen and spruce trees had been spared the lumberman's ax when the agency was built and now stood like enormous dark sentinels behind the cabins. The only movement she could see were Ute ponies grazing in the far meadow where the road bridged the river.

She leaned out the window and cupped her hands over her mouth to muffle a whistle but her lips were dry from fear and excitement. She drew a breath but before she made a sound a wraith stepped from the shadow of a tree into the moonlight. She flinched and stifled a squeal with her hands. The form walked quietly toward her vanishing into the shadows of her cabin only to suddenly appear beside her at the window. According to Ute custom neither spoke. The man always speaks first but only after an introductory silence.

"Josephine," he whispered, "I don't scare you." He spoke strained English, each word carefully, haltingly pronounced. At tribal meetings he spoke Ute with great self-confidence as taught by Ouray. It was when they coupled in English that he became inarticulate and insecure. She knew he strained to impress her.

Many Ute leaders, including Nicagaat, were of mixed lineage, often Ute and Apache. His Apache Ute parents died when he was only an infant. Rescued and raised by a Mormon family and educated in white schools in his teens he returned to his mother's Ute band in Southern Colorado.

"Only for a moment, Jack," she whispered back. "I was calmed when I heard the bell. You're taking an awful chance coming here at night, especially now." She looked furtively around the moonlit landscape and darkened cabins for any sign of prying eyes.

Jack appeared unbothered by caution and continued to stare at her. He stood grim, silent and resolute. She motioned him closer and put her hand beside her mouth as she leaned further out the window. He took a step toward her and put his ear against her hand. "Thank you for coming," she whispered. "I'm glad to see you are well." Her hand grazed his cheek.

"Josephine," he replied into her ear. "I had to see you." He put his hand on her shoulder and gently squeezed it. Jose reciprocated and was surprised at the sudden thrill that raced through her body when her fingertips contacted his tawny, muscular shoulders. Seconds seemed like hours, even like an eternity, as she stared into his eyes, their faces only inches apart, so close she could smell the pipe tobacco on his breath and the musky odor of alkali and sweat. "I fear tomorrow we fight at the Yellow Jacket Pass," he finally said. She tightened her grip on his shoulder.

"Why? What's happened?"

"The army may enter the reservation. If so we will stop them. Major Thornburgh has promised he will stop outside the agency boundary and send only five men forward to meet with your father and us. But he's brought three companies of cavalry and one of infantry, many more than needed if he only wants to talk."

"How many?"

"Thornburgh brought nearly two hundred. My friends at Fort Steele tell me that General Sherman has ordered Crook to order Thornburgh to arrest the Ute leaders causing the unrest—myself, Johnson, Douglas and Colorow, at least those four, maybe more. It is said Ouray agrees and will do nothing to stand in their way."

"Jack, let them come, all of them. When they see you don't pose a threat they'll leave. The army does not want to lose men fighting another battle caused by an Indian agent. But once you've killed a soldier, even one, they will never give up until you're beaten."

He nodded. "I know, but remember Black Kettle's Cheyenne at Sand Creek. This land and our ponies are all we have left and now your father wants to take even those from us."

"Jack, it's already bad here, very bad. I'm sure you've heard about Johnson and my father fighting about the ponies. Douglas and Johnson have ordered all the women, children and old men to Ouray's camp in the Southern Uncompahgre. The young men beat the drums and danced until late at night. What will you do?"

"Tell Johnson and Douglas that the army is moving toward the agency and to take those left here to Ouray's protection in the Uncompahgre. They will be taken in at the Los Pinos Agency. I will tell my men your family should not be harmed. You will be safe here."

Josephine's heart sank. She knew that if the situation degenerated into a battle, none of them would be safe from each other. Ouray was a powerful and influential chief, even the whites agreed, but the nation was aflame with fear and hatred after Custer's massacre and even the Father of the Utes would not be able to save them from retribution. "I have no fear for my safety or my family's but I fear for yours and that of all the Utes

whom I have grown to love. Go back to the pass and tell Colorow and your men to allow the troops onto the agency. Come
with them. I will talk to my father and personally promise him
your people will do as he orders in return for a permanent
home on the White River with the pony herd. Otherwise the
Utes will lose the shining mountains forever."

"We are not farmers, Josephine," he said emphatically. "We
are hunters and warriors who need to move about freely. The
Utes did so even before horses. Then the creator gave us the
horse as a spirit companion who roamed the land as we did.
The Utes will not surrender our horses and betray that eternal
bond."

Tension saturated the air between them. Josephine could see
the cords in his throat and cheeks tighten in the reflected
moonlight but his face remained a block of chiseled granite.
She also could feel the enormity of the problems bearing down
on him and they were nearly suffocating. She had never known
a man, any person, upon whom the fate of so many people
rested. She wished Ouray would somehow miraculously appear
and advise him. At stake was the history, culture, traditions and
independence of Ute Tribe, the People of the Shining Mountains, and, for them, any possibility of a future together. A war
would set the whites to destroying the legends and history told
around campfires for hundreds if not thousands of years; the
skills and traditions taught by the Ancient Ones, men and
women, whose names had been lost in time and space. Her life,
she thought with grievous disappointment, would then continue
pretty much as it had for years but the next several hours
represented the crux of his life and those of his kin. Her eyes
filled and overflowed.

"Josephine . . . I go back," Jack said awkwardly. Ute women
did not shed tears except at the death of a close relative. "Colorow and Sowawick will fight if the army enters our land and I

must stand with them." He reached out and took her head in his hands and quickly kissed her eyes and lips.

Her mind reeling from the shock of the Ute chief's bold display of affection she searched wildly for an appropriate farewell. "When do we meet again?" she asked. Jack only shook his head. Stifling a sob she whispered hoarsely: "When the hurly-burley's done. When the battle's lost and won." He dropped his hands to his sides and disappeared into the shadows from which he had arrived.

CHAPTER 4

"Victorious warriors win first and then go to war, while defeated warriors go to war first and then seek to win."

Sun Tzu, *The Art of War*

"War is the business of barbarians."

Napoleon Bonaparte

When Lawson saw a line of horsemen crossing the wagon road they'd just traveled, dismounting and taking up blocking positions among Gordon's abandoned wagons he yelled to no one in particular, "Shit! We're surrounded." He immediately thought of the Seventh Cavalry slaughtered at the Little Big Horn three years before. Every lurid story he'd ever heard from those who had survived such a battle flooded his mind. He fought to stave off the panic that was welling up inside. Joe pulled his watch from his vest pocket; almost twelve noon on the twenty-ninth of September 1879. He closed the watch and read its inscription, "To Joe: All My Love for All Our Lives—Christine." Together he and Sam Cherry sprinted for the breastworks still under construction on the north end of the corral.

From the vantage point of the narrow summit ridge of a mountain less than a mile from the cauldron battle forming next to the Little River, two men stood apart, arms crossed, rifle barrels resting in the crook of their elbows solemnly watching Joe Lawson and his men sprinting to bolster the north end of the corral.

Their garments were eclectic amalgamations of striped, government-issue shirts and trousers, calf-height buckskin boots, and wool blankets tightened around their waists and shoulders. Their coal-black hair, neatly combed and glistening with bear grease fell below their shoulders framing their somber, swarthy faces like monks' mantles.

One man was tall, angular, athletic and bareheaded. His companion, the antithesis—short, fat, rumpled and dirty—wore a comically small, short brimmed plug hat perched on the top of his head. But there was nothing comical about his deeply scarred chestnut face. His wide-spaced eyes were mere slits from years of squinting into the sun. A pug nose, snarling lips and grim mouth marked him as a fighter. He was the end product of generations of fighters.

He turned to the younger warrior and said, "Nicagaat." The younger man looked and Colorow motioned him over with a tilt of his head. The man whites called "Jack" or "Chief Jack" possessed a small, sinewy frame and sharp but darker facial features. A prominent nose jutted over an expressionless, benign mouth. Though not a pugilistic behemoth compared to Colorow, his straight stature and athletic appearance provided him a more regal appearance. His courage, education and intellect served him well after reentry into the tribe. He caught the eye of Ouray who began schooling him in tribal politics and the art of diplomacy with the whites. In time, he joined General Crook at Fort Lincoln as a scout. Crook and his officers thought Jack assertive, never passive, seldom aggressive and quick to learn. He was affable and quickly gained the comradeship of white and Indian alike.

The young chief ignored the gesture and remained aloof and silent. The corpulent Comanche subordinate approached him. Jack bent his head close to Colorow's mouth to hear him over the din of the expanding battle.

"Jack," he said in English, "warn Canalla and Quinkent back at White River that we fight." Nicagaat straightened. The slight did not go unnoticed. Colorow called him by his white name but used Johnson's and Douglas's Ute names. At length he replied in Ute to correct the breech of protocol. "I'll send someone. He can be there after the sun rises."

Colorow nodded, his perpetually glowering, granite-hard visage betrayed no sentiment or sense of urgency. Colorow had joined the tribe as a Munache Ute, a less incorporated band who competed with the Arapaho for the hunting grounds east of the Shining Mountains until the arrival of the whites. Though they spoke the same basic language as the Northern Utes along the White River and the Tabeguache Utes, their dialect contained inflections and words possessing differing shades of meaning. That, along with their adoption of some Arapaho customs and individuals, widened the cultural and territorial gulf between them and the other Ute bands.

In 1869 Colorow's band had been sequestered on a reservation adjacent to the growing city of Denver. Colorow, forever the opportunist, funded by annuity money, gathered his little tribe and sojourned onto Larimer Street on summer evenings to wine, trade and sell. The Indian Bureau caught wind of these antics and sent the Munache into the mountains to live with their Ute brethren. Colorow sullenly accepted these marching orders but persisted in thwarting and confronting the whites at every opportunity. He was neither trusted nor liked by most of the Ute hierarchy. "Then," he added to Jack, "order the rest of the young men and we will drive the whites back to Fort Fred Steele."

"Perhaps," replied Jack, "but Quinkent should move as many women and children south as quickly as possible and ally themselves with Ouray's Tabeguache Band. Some of the men will have to go with them. We can only hold these troops until they're reinforced and then you and I, Johnson and Douglas will be hanged and the rest sent into the desert to starve."

"They *will* return to Fort Steele, replace Meeker and let us live in peace?"

"No!" Jack replied irritably. "The first shot killed that. The threat of battle was our only real weapon. I thought Thornburgh would stop north of Little River as I demanded. He surprised me when he crossed the river onto our agency."

"When Douglas and Johnson learn of this they will destroy the agency and Meeker. You'll take Josephine to Ouray's?" Colorow jumped back when Jack spun around toward him and took his rifle from the crook of his arm.

"No!" he growled fiercely. "I will stay and fight this mistake to the finish. Send someone to the agency to tell them what's happened here. The plow Meeker bought to destroy the ponies' pasture will cost him again." He turned and strode quickly to his horse grazing at the top of the ridge.

CHAPTER 5

"In politics stupidity is not a handicap."

Napoleon Bonaparte

"It is of course well known that the only source of war is politics—the intercourse of governments and peoples. . . . We maintain . . . that war is simply a continuation of political intercourse, with the addition of other means."

Carl von Clausewitz, *On War*

Photographs of Nathan and Arvilla Meeker depict, at a glance, a plain but not completely unattractive couple; he with narrow, bright eyes fixed in a messianic stare and she with an austere, hatchet face and, characteristic of the times, neither were smiling.

The March 18, 1878, *Congressional Record* announced the Senate's confirmation of Nathan Meeker's appointment as agent of the White River Reservation in Northern Colorado. It was to be a fateful appointment for him, for his family and for many others. And it was the culmination of a haphazard, checkered career including social philosopher, journalist, author, political hack, land developer, real-estate salesman and petty bureaucrat—some sequentially, some simultaneously.

Nathan Meeker was born in 1817 on a strip of Ohio land bordering Lake Erie called the Western Reserve. His parents were small farmers of English descent. George Washington once peculiarly honored his grandfather for "contributing" eighteen children to the American Revolution. The family otherwise was indistinguishable from their neighbors or thousands of similar frontier farmers.

Not so Nathan. A precocious child, he taught himself to read and write at the age of five. At twelve he was tutoring his younger friends, writing poems and treatises on love, religion, social justice and injustice, poverty and the evils of indebtedness. Oddly averse to manual labor he lectured constantly to his less literate father, Enoch, how he could grow better crops using "intelligent" farming methods. Predictably, this led to family conflicts, and Nathan, age seventeen, left home.

"I'm moving to New York City," he announced to his parents one morning at breakfast.

"My word," exclaimed his mother who stopped cutting biscuits her voice bereft of either surprise or alarm.

"Whatcha think you're gonna do there?" queried his father after a sip of chicory coffee.

"Write poetry and become a journalist."

"A what?"

"Writer."

Enoch rasped his stubbled, weather-beaten face with a horny fist missing the tips of a few fingers. "Reckon you can eke out a livin'?"

"Yes."

"Whatcha gonna write about?"

"Scientific farming . . . and socialization . . . and other things. The nation is expanding and thousands of square miles of free land lie to the west ready to clear, plow and plant. Emigrants from Europe will be flooding to America to settle and farm. They will bring new ideas to American agriculture. The Indians of the west who now fight us will farm and assimilate and change their way of life. The opportunities for creativity are boundless."

His father grunted. "Good luck to ya. Write if ya find work. Don't expect no help from here; it's all we can do to survive."

"Thanks." After an awkward silence he raised from the table. "I guess I'll pack a few things and then be on my way." Enoch stood with him. Nathan held out his hand but the old man just snorted up some mucous and spat it in a kitchen corner. He turned to his mother.

"I'll fix you some lunch for today. After that I guess you're on your own." She turned back to cutting biscuits with a pint bell jar. He retreated to his bedroom and began stuffing a laundry bag with his newest and most serviceable clothes. He'd buy more when he got to New York and found a good paying job.

When he had completed packing he returned to the kitchen. His father had left for the fields. His mother solemnly handed him a burlap bag of food, enough to last him the rest of the day, and a handful of coins, from her "butter and egg money." Neither spoke. She returned to her housekeeping and Nathan began his lifelong journey chasing fame and fortune across the United States.

He strode into New York City with ineffable romanticism, indestructible optimism and single-minded determination. Those three well-cultivated characteristics were to be for the remainder of his life his crowning glory and crushing defeat.

With the few dollars he'd managed to squirrel away from teaching jobs in Cleveland and Philadelphia he was able to rent a sleeping attic at 116 MacDougal Street. From there he daily haunted the editors of newspapers, magazines and pamphlet printers for writing jobs. His savings dwindled. He finally accepted the most menial tasks to keep body and soul intact: saloon janitor, tutor, fish monger and stagehand. He persisted. The *New York Mirror* finally agreed to publish a series of his poems. His career as a poet was short-lived. Nathan Meeker was not Alfred Tennyson nor was New York, London. Within months he was unemployed and returning to Euclid, Ohio, where he became a traveling salesman for a garment manufacturer. He embraced his new job with characteristic ardor and expectancy. He wrote a sonnet about it.

Nathan, an espoused atheist, through his copious reading became incongruously attracted to humanist philosophy, in particular the social theorists: the Shakers, Mormons, Brook Farm transcendentalists and agrarianism. He vilified what he considered to be the eastern conservative establishment, slavery, conformity and elitism. The theories and practices of socialist Charles Fourier attracted him. Fournier denigrated individualism, capitalism and government. He advocated collectivism, cooperation and utilitarianism. This was a pivotal period in Meeker's life—in more ways than one.

"I did not believe in love-at-first-sight until I laid eyes on Arvilla Smith at a Euclid social occasion," he wrote in his diary. "Never had I seen such a beautiful, ethereal, dainty creature. She was so small and her skin so thin and white as to be nearly transparent. Contrary to my nature, I sidled alongside her, put my arm around her and kissed her tenderly. Her surprise is worth noting."

"Sir!" Arvilla cried loudly. She pulled his hand from her waist and pushed him away with enough force that he backed into a group of guests causing them to spill their drinks. "Who are you and just what do you think you're doing?"

Meeker, oblivious to the havoc he'd just caused offered no apology to anyone. He smiled witlessly at Arvilla. "My name is Nathan Meeker and I was so taken with your beauty that I decided to seize the moment and introduce myself. I saw your eyes following me and knew you would welcome a meeting."

Arvilla was painfully aware that all conversation in the room had ceased. She and this man named Meeker had everyone's attention. "You're much mistaken, Mr. Meeker. I had not noticed you, I was not watching you, and I certainly am not welcoming your attention, particularly in this manner. Our host approaches, I suggest you compose yourself, collect what little dignity you possess and leave quickly before he physically directs you to the door."

Meeker was crestfallen. He had planned to sweep Arvilla off her feet and instead ended up having the rug pulled out from under him. The object of his attention turned her back on him. Onlookers were either stifling laughter or shaking their heads in disapproval. His host stood just out of reach with his head cocked to one side and arms folded across his chest. "You have just time enough for one decision," the man said threateningly, "make the right one."

Meeker turned to Arvilla's back. "Miss Smith," he called to her. She ignored him and fanned her face frantically. "Miss Smith," he tried again. She turned; face reddened and tears welling in her eyes. Her expression was one of furious contempt. "Miss Smith, I do apologize for my unexpected and clumsy approach. I will contact you again but in a more appropriate manner." Her expression turned from anger to open-mouthed surprise. He turned, nodded to the host and departed from the room.

The incident precipitated a number of vain attempts by Meeker to contact her either directly or through acquaintances. Few women can permanently withstand such overt adoration though clumsy, inarticulate and embarrassing. An ignoble man of passion and persistence may often kindle a flame where wealth, nobility and grace in someone less resolute will fail. Plus, Nathan was tall, lean, handsome and well read. Arvilla succumbed, but only if Meeker became a Christian. He willingly joined the Disciples of Christ, also named Campbellites, an austere sect embracing archaic rituals. A more pure form of the faith, he thought. They married in April of 1844. He forsook his traveling sales career and insisted they join one of Fourier's communal "Phalanxes" in Braceville, Ohio, where he became the township's wordsmith, librarian, teacher, philosopher and poet laureate. Two of their children, George and Ralph, were born in the Phalanx.

The commune was largely populated with liberal idealists lacking in the skills and judgment required to sustain it. The doors closed in 1847. The Meeker family, unlike most who owed the Phalanx money, walked away with an uncollectible credit of fifty-six dollars. For Meeker, the eternal optimist, this was strictly business as usual, a mere bump in the road, as it were. Arvilla felt much more discomforted by their penury, instability and insecurity. She confided to friends and relatives that Nathan's seemingly contradictory attitudes troubled her. He was, she said, a man of two minds, not always synchronized. He was a man of great ambition and full of promise but also a wild-eyed idealist always trying to shortcut the path leading to the pot of gold at the end of the rainbow.

"He's not . . . you know . . . I can't think of the word." She queried her father.

"Practical?"

"Pragmatic."

"Has his head in the clouds?"

"Idealistic."

These were portentous observations.

For several years after the failure of the Phalanx Meeker attempted and failed at several small businesses around Euclid. He remained obsessed with and distracted by pretentious dreams of following, almost literally, in the footsteps of such notables as John Charles Freemont and Brigham Young—follow the proverbial rainbow west and build a utopian socialist community.

"I'm inspired," he announced to his family during dinner one evening, "by the achievements of Freemont and Young to write a novel extolling the virtues of the Manifest Destiny doctrine; triumph of white civilization over indigenous savagery benefiting all mankind." He ended this declaration with a flourish of his table knife as though it were Excalibur cutting down the heathen knaves. Arvilla and the children paused their dining only momentarily. Such proclamations had become a regular part of their everyday life.

Meeker, deaf to the collective sigh from his family, embarked wholeheartedly upon *The Adventures of Captain Armstrong*, a Defoian character—part Robinson Crusoe and part Charles Fourier. A man of monumental inspiration and cast-iron will who imbues the naked aborigines with the spirit of Christian cooperation and leads them into a modern era of agrarian industrialization.

"Aren't those contradictory concepts?" inquired Arvilla.

"What?"

"Agrarian and industrialization?"

Meeker, with hardly a pause, declared that they might be considered so amongst the more conventionally minded but that he was bridging that gap with a new genus of wisdom far beyond the comprehension of stodgy academics and illiterates.

Arvilla stoically acknowledged this piece of grandiloquence and went about her commonplace tasks of cooking and housekeeping.

But, Captain Armstrong did capture the attention of the right man, in the right place at the right time, an editor for the *New York Tribune* named Horace Greeley. As Fourier was Nathan Meeker's guiding light, Horace Greeley became his compass leading him and his family, even after his death, inexorably toward their final fates. To him was attributed Manifest Destiny's clarion call: "Go west, young man, go west!" After reading a few pages of Meeker's manuscript Greeley wired him to come to New York to discuss editing and publication.

"I am returning to New York in triumph!" He declared to his astonished wife. "Not to MacDougal Street, no sir, Mr. Greeley has reserved a suite at the Plaza for me near his office."

"It is wonderful indeed," she replied. "Perhaps now we can finally be relieved of some of our indebtedness."

"Yes, of course," he said dismissively, "but the important thing is that now I have entrée to a whole host of newspapers and publishers through which I can expand my social doctrines and influence the course of history throughout America . . . even the world."

"Nathan," she sighed and shook her head.

"What?"

She paused. "It sounds like a wonderful opportunity. The children will be very proud of you." The Meeker/Greeley association was indeed productive and profitable; not as much as Nathan imagined but more than Arvilla expected.

"Meeker's a bit of a weird duck," opined Charles Anderson Dana, Horace Greeley's managing editor.

"Umm . . . perhaps," squeaked his boss, a personified enigma. He appeared physically to be the cross of an Irish leprechaun and a homunculus. Wispy, white hair and beard surrounded his cherubic pink face, giving him nearly albino features. He spoke in a high nasal whisper. But publishers, politicians and princes recognized Greeley to be the titan of printing. He was also a career broker beyond par. His support nearly guaranteed one's success but his condemnation equaled defeat. In Meeker, Greeley had found not only a good writer but a kindred spirit. They were both teetotalers, socialists, Whig Republicans and abolitionists. Like Meeker he was born into poverty and self-educated. "But, Mr. Dana, the same could be said of Karl Marx and Fredrick Engels."

"He has the eyes of a lunatic."

"Ummm . . . Passionate loyalty."

Greeley correctly assessed Meeker's fluid style of writing as a quality well suited for a war correspondent. He sent Nathan to the western theater where new generals named Grant and Sherman were making progress against the Confederacy. Upon his return to New York, Greeley promoted him to editor of agriculture. The Meeker family for the first time emerged from indebtedness. Greeley even hired young Ralph Meeker as a cub reporter. But Horace underestimated Meeker's messianic zeal, the power that collectivism socialist philosophy held over Meeker.

Following the Civil War the *Tribune*'s new agricultural editor was dispatched west to shadow railroad magnate General William Jackson Palmer. One of his first dispatches back to the *Tribune* included a most provocative line: *"The extension of a fine nervous organization is impossible in the Indian, because he is without brain to originate and support it."*

Meeker returned from his western assignment inflamed with renewed enthusiasm for the establishment of a utopian co-operative community in Northeastern Colorado. Arvilla's worst fears were realized when he stepped off the train dressed in boots and blue denim work trousers, with a wad of tobacco in his cheek. "I have found that pot of gold at the end of the rainbow!" he shouted as he hoisted her aloft. Arvilla, in keeping with her character, listened passively on the way home to Nathan's adventures and encounters. When he arrived he didn't bother to unpack before he launched upon the description of his latest grand "plan."

His plan, he said, was absurdly simple and, of course, foolproof. Simply spend their savings on a parcel of land in

Northern Colorado, survey and divide it, sell the smaller parcels to land hungry immigrants and city dwellers who were willing to subscribe to long-term cooperative colonization, his enduring dream. Of course he would be the patriarch of this utopia named Greeley.

The stunned Arvilla protested. The Meeker family now consisted of five children, two of whom were still in school, the tubercular George and Rozene who was "emotionally compromised" from having fallen into a well at a very young age. "Nathan," she implored him, "for once, restrain yourself at least long enough to gain some insight into the number of things that can go wrong with this scheme. Consider your family for a change, what we need and want, instead of building the Empire of Union Colony—Greeley."

He stood in silence pondering the maps spread on the dining-room table. "It needs to be," he said, shaking his right index finger for emphasis, "close to a river so we can divert water for irrigation. Eastern Colorado is much too dry to depend on rainfall alone. But the mountains shed oceans of water all summer. And then we need to be on the railroad for supplies from Cheyenne and Denver and to ship our harvests to them." He dithered on as Arvilla desperate to the point of tears took refuge in her kitchen and her leather-bound volume of John Bunyan's *Pilgrim's Progress*.

Arvilla's fears were well founded. His savings purchased only a fraction of the expected acreage at five dollars per acre rather than the promised ninety cents. Operational costs such as the irrigation system, building, construction and road infrastructure, mounted with great rapidity. The original subscribers who had clamored for shares of the property became themselves indebted and disillusioned. Some simply walked away; others threatened civil actions in fraud. Meeker's angel, Horace Greeley, for whom the town was now named, remained a loyal source of renewed funding. The amounts of his loans, however, decreased inversely to the interest rates charged. Greeley was no great shakes as a businessman but contrasted to Nathan Meeker he compared favorably with J. P. Morgan. In 1870 Greeley granted Nathan one more loan of one thousand dollars to start the *Greeley Tribune* newspaper, a stillborn concept.

Between late 1870 and December of 1872 the dreamer's dreams had turned, once again, to nightmares. Union Colony/ Greeley lay only partly completed along the banks of the Cache-la-Poudre River. Both Meeker and the city treasury were bankrupt. Horace Greeley's health and finances were on the rocks. He was forced to sell his beloved *Tribune* to the rival *Harold* in 1872, his wife died in October of the same year and Greeley followed her slightly more than a month later. Greeley's eleventh-hour divestiture of the *Tribune* meant Nathan and Ralph Meeker's writing incomes ceased abruptly. To everyone's surprise the tiny titan had died nearly impoverished. The administrator of his estate notified Meeker he intended to collect by all means necessary the full amount of the one-thousand-dollar final loan, plus interest. Thus began another vexing job search.

Rutherford B. Hayes had succeeded Greeley's arch enemy, Ulysses Grant, as president, so Meeker contacted, from Colorado, minor Washington power brokers who still held the late Horace Greeley and his acolytes in some esteem. Additionally, Ralph had garnered kudos as a writer and journalist.

Ralph coincidentally stumbled across a new journalistic mission in life: the exposure of corruption within the government's Indian agencies. Ralph's research and reporting piqued his father's interest. From top to bottom the Grant Administration had been rife with corruption not the least of which was the Interior Department's Bureau of Indian Affairs. President Hays appointed a German emigrant and Civil War officer, Carl

Schurz, to be secretary of the interior. Schurz immediately proceeded to reorganize the department and cashiered most Indian agents. This worked to Meeker's benefit. Although Greeley had not been generally popular in Washington, DC, his task taking of Grant and his appointees played reasonably well in the Hays administration.

Meeker therefore aimed toward the position of agent of the White River Agency, a post that had just opened. Undaunted by his Denver attorney's declaration that Meeker would "have a better chance of riding a kite to the moon than procure an agency position," Meeker contacted every politician he could think of including Secretary Schurz. If Nathan Meeker accomplished nothing else notable during the course of his life he did prove, time and again, that unrelenting persistence pays. He unabashedly darkened the doorsteps of friend and foe alike from Denver and Colorado Springs to New York and Washington in pursuit of this new opportunity. Meeker drew fire from family to foreigners for his clumsy leaps of faith, nonsensical business ventures and inept management but not for his unyielding determination. Again his persistence paid off and he acquired the fifteen-hundred-dollar-a-year job on March 18, 1878. His fate along with many others was sealed.

In the spring of 1878 Nathan Meeker moved his family from Greeley, Colorado, to the White River Ute Agency. During the trip he guaranteed Arvilla that after a lifetime of almost constant moving this would be the last.

She nodded and smiled weakly. Mrs. Meeker had heard this too many times before.

CHAPTER 6

"Battle is an orgy of disorder."

General George Patton

"There are five ways of attacking with fire. The first is to burn soldiers in their camp."

Sun Tzu, *The Art of War*

Captain Lawson thought back. He had originally agreed with Major Thornburgh's decision to leave behind a company of the Fourth Infantry Regiment at Fortification Creek nearly seventy miles from the northern border of the White River Reservation. The commanding officer of the White River expedition had met with Chief Jack and his subchief, Sowawick, on September twenty-sixth. Despite the major's assertions to the contrary, the size of the combined Fort Steele and Fort Russell force had unnerved the Ute chiefs who saw the larger force as a blatant invasion threat. The major confided to Lawson that he couldn't imagine a full-blown battle and would placate the Indians by leaving the infantry as a distant reserve. If the cavalry did come under fire he could immediately call up the infantry.

Now, Thornburgh lay dead, the cavalry had ceased to function as an offensive mounted force. Surrounded, with most of their mounts either dead or dying, the horse soldiers dug trenches with knives and mess plates, piled flour sacks and boxes between the parked wagons and fired their blistering hot

59

carbines. Lawson watched as small groups of Ute warriors would assemble and charge the wagon corral both mounted and afoot; meet withering carbine fire and fall back. The Indians never lacked for courage, he thought, but with few exceptions he'd witnessed, they seemed incapable of mounting one single, huge coordinated attack. Thank God! Their tactics were spontaneous and disorganized. He guessed the troopers were outnumbered at least three to one and most of their opponents were armed with repeating rifles. An enveloping surge by the Utes right now would wipe them out like Custer.

By this time, Lawson himself, the only mounted man in the vicinity and obviously an officer, drew intense fire. Bullets whipped the air around him like a swarm of angry hornets and kicked up dozens of dust craters. But he noticed something that caused him to reign in even as the Ute riflemen acquired his range and the fire increased. Squads of men were firing, unloading wagons and unhitching horses but he could see only a handful of wounded. Instead, horses, mules and oxen were dropping by the score in their tracks and traces. The wounded, unmounted cavalry horses crazed and screaming with panic and pain were running blindly through the lines of skirmishers injuring more troopers in their wake than the Ute bullets. The cacophony of this battle, its chaos and confusion, minus the artillery explosions, was as great as he'd experienced at Shiloh or Gettysburg. The Indians were not nearly as numerous as the Confederate troops in those battles but their lever-action repeating rifles effectively multiplied their numbers by a factor of three or four. At this rate, with no trees or other cover to hide them, the three hundred animals that had carried the troops and pulled the wagons from Rawlins would be dead within an hour. The Utes had successfully blocked the troops from entering the reservation but also made retreat impossible.

Joe spurred his horse up to the nearest wall of crates and barrels between two wagons. Now loud claps and thuds accompanied the hiss of the bullets as they struck the barricades. He dismounted and tried to hide his horse against a covered wagon but the panic-stricken animal jerked the reins from his hands and reared. A hailstorm of bullets immediately peppered him, most sending up a plume of dust. The animal simply sat back on his haunches, rolled his eyes and tumbled to one side.

"Cap'n," yelled one of the men behind the barricade of barrels and boxes, "you best come over here and take shelter, sir, or you'll be joinin' the horse." The voice had a familiar Irish accent. Joe bent nearly double and sprinted toward the man motioning to him, First Sergeant Edward Grimes. As he ran he could hear the angry buzz of a swarm of bullets above him. Sure enough, they had his range. When he was still ten feet away from the wall he plunged headfirst at the big Irishman but fell well short. Grimes leaned out, caught him by the back of his shirt and drug him face-first through the alkali dust. Joe, choking, coughing and gasping for breath pushed himself up against a barrel. "I think you'll be needin' this, sir," Grimes said dropping the fiery hot barrel of his Springfield .45-55 carbine in the captain's lap. "I trust ye' have ammunition," said the sergeant, popping up above the barrels, quickly aiming and firing another carbine.

"I do," Lawson replied unsnapping the cover of the cartridge box on his belt. "But only about half the forty rounds we were issued."

Grimes flashed a toothy grin. "The lads have been good enough to separate the ammo boxes from the crackers and canned goods and set them in a convenient place a few feet from your right." Joe felt a great surge of reassurance when he saw the open wooden box and men were scrambling back and forth filling their cartridge boxes. With noncoms like Edward Grimes, he thought, the army doesn't need officers.

"Were you able to get to the major?" asked Grimes.

Joe shook his head, stood up and fired at a plume of gun smoke coming from an oak bush. "The Utes got to him before I could. Captain Payne is in command. I thought I saw Lieutenant Cherry take a bullet."

"No!" yelled Grimes over the rifle fire. "He made it back somehow. Some of the wagons had fallen behind, just below the rise before starting up to the flat land. He tried to get back to them for ammunition but the Utes got his range and he had to turn back. I made it through but could only carry one box back."

Lawson nodded. He was relieved to hear that Cherry survived his close brush with death. "When things are under control here I'm going to take a squad for the major's body."

Grimes was loading and firing faster than two men combined. The man was a machine, a brawny, demented jack-in-the-box. "Wouldn't worry too much about it Captain, we got a heap goin' on right here. If they swarm us we'll be seein' the major on the other side." The machine stopped. Grimes, cursing at the top of his lungs, repeatedly opened and shut the breech of his rifle. "Jammed! The bloody thing's jammed! The case won't extract!" He dug frantically and futilely at the head of the spent cartridge blocking the barrel opening. While Lawson continued to load and fire, Grimes took a knife from its sheath hanging at his belt and began digging and prying at the spent cartridge until the point of the knife broke.

Lawson grabbed the useless rifle and handed his to Grimes. "You're better at this than I am," he said with a grin. He took the carbine from Grimes and turned it over in his hands brushing the grit and grime off the blue steel frame. "It's an eighteen seventy-three!" The two men exchanged grimaces of silent frustration. Lawson looked at the cartridge he was holding. The case was dark brown with tinges of green. He emptied his cartridge pouch and discovered a mixture of shiny brass and dull copper. Grimes picked up a handful of cartridges from the box between them. "Damn," he yelled and threw the casings over the breastwork. "I wonder how many boxes of this old copper crap we have."

Relief troops arriving at the Custer Massacre site on June 26, 1876, discovered innumerable cavalry carbines of the same make and model with broken knife blades and dead soldiers lying next to them. At first the relief troops thought the carbines contained a defect that did not appear until they had been fired many times. The ever-conservative army had apportioned so little ammunition for practice that such a defect could go virtually unnoticed until their first major battle. Instead, the Ordinance Board determined the problem lay in the early soft copper cartridge casings which corroded easily and expanded to the point the extractors couldn't dislodge them from the breech. Lawson knew battles, even wars, were won or lost because of such minutia. An old British colonel had once told him: "Not for a nail the horseshoe was lost, if not for the horseshoe the horse was lost, if not for the horse, the king was lost, if not for the king the battle was lost."

"Sergeant, there are boxes of eighteen seventy-nine infantry rifles in some of the wagons that have cleaning rods on them that can be used to poke out the shell casings if the carbines jam. Take a couple of men and find them, hand one out to each squad for as far as they go. Check all the open ammo boxes and throw anything copper over the rampart. Also, nine-foot-long, quarter-inch wood dowels hold down the canvas curtains on the wagons. Cut the dowels in two for ramrods and bring the canvas into the perimeter for use as stretchers and bedding for the wounded."

"Yes, sir. You reckon they're gonna rush us?"

"No. If that was the plan they would have done so right after the first couple of volleys to kill the draft animals, before we could get the wagons circled in defense. They want us outa here, thing is they're goin' about it the wrong way—we're surrounded and they're killin' our draft animals—we're not goin' anywhere for some time to come. Have you seen the scout, Joe Rankin?"

"I think he's with Lieutenant Cherry. There's your company flag?" Grimes pointed at a red-and-white guidon amidst another clutter of boxes and barrels between two wagons about fifty yards away.

"Any men here from E Company of the Fifth?" Lawson shouted to the men clustered behind the barricade next to him. Two men raised their arms. "Follow me! We're going to get the company reorganized." Each of the men quickly joined Lawson and sprinted in the direction of the guidon.

As they ran, Lawson recognized both Cherry and Rankin, the expedition scout who also owned a stable in Rawlins, firing and reloading carbines and revolvers. The three troopers dove into cover next to the lieutenant and the scout. Lawson unholstered his Colt's revolver and fired a quick shot at a Ute running from one serviceberry bush to another about two hundred yards away. He looked at Cherry. "I thought you were dead for sure when the firin' first started."

"You probably saw Private Firestone. He and another trooper were dropped in the first volleys," said Cherry without looking up from reloading his revolver. Cherry glanced up at Lawson. "Oh! Sorry, sir." He saluted. "I didn't recognize your voice."

"Understandable given the circumstances. We won't bother too much with command protocol at a time like this. Aside from Rankin, are all the men in this position with our company?"

"Yes, sir. There are six helping man the positions on either side of us, that's twelve. So we have twenty from E Company accounted for—counting you and the two you brought."

"Killed and wounded?"

"Only Private Firestone that I know of and I'm not sure which of those categories he occupies presently."

"I'll find out. Ammunition? Water?"

"We're halfway through the case of ammo that Sergeant Grimes brought in from the wagons. At the current rate of fire we'll be out within half an hour or so but several boxes are still inside the lead wagons. Our only water is in the canteens. The men haven't had time to drink but once the firing lets up they'll soak up every drop."

"First, check every box of ammunition you open. Grimes and I found an old box of copper ammo. If you find some, take 'em out to the Utes. Second, don't let 'em drink. Each man gets a swig to wash down the dust. Make sure they know they're not goin' to see any more for quite a while."

"How far is the creek, Captain?"

"Maybe two hundred yards—that and a wall of bullets." Lawson leaned over to the man on the other side of him. "Rankin!"

"Yeah, Captain," the scout said dropping to one knee from his firing position. Joe Rankin, fit and in his early forties, had impressed Joe with his intimate knowledge both of the region and its Indian inhabitants. On the journey south from Rawlins, he and the scout had enjoyed many conversations. Rankin had moved his family west to stake a claim under the Homestead Act. Most recently he'd opened a livery stable and freight company in Rawlins. He had become friends with many of the Northern Utes, Southern Cheyenne and Arapaho, bartered, traded and hunted with them and spoke the three dialects. With them he'd become knowledgeable about every mountain, hill, valley, river and stream between Rawlins and Colorado's Grand River. Now he faced his friend and fellow scout, Nicagaat, with whom he'd served under Crook's command. Even though he had no formal military education or training Joe had taken note of his coolness in the heat of this battle.

"We're surrounded and in deep trouble . . ." Lawson stopped; Rankin's grin broke through the muddy rivulets of sweat pouring down the scout's face. "But I guess you know that." Lawson chuckled and instantly both broke out laughing hysterically. Cherry stopped loading and firing to gaze at them in slack-jawed amazement.

"No shit, . . . sir!" Rankin replied when he had caught his breath. "What do *you* intend to do about it?"

"Send you for help." Lawson chortled.

"Sure thing. I'll catch the next train."

Lawson put a hand on Rankin's shoulder. "Joe, you know the area better than anyone else here. Pick out two or three others, not officers, you know to be good riders so we can increase the chances of at least one getting through. I think Captain Dodge and Colonel Merritt are both in the field so they'll be able to respond quickly. Also, telegraph General Crook at Omaha Barracks informing him of the situation, he can pass it on to Sheridan in Chicago and Sherman in Washington as he sees fit. I'm gonna report back to Captain Payne." He turned back to Lieutenant Cherry. "Sam, you're in charge here until I get back. I'll try to get another case or two of ammunition to you as soon as possible. Tell the boys to sip the water and pick their targets carefully."

"Yes, sir . . . Captain, I would like to mention that Sergeant Dolan came back for me when my horse was shot from under me. I refused his horse when he offered it to me so he dismounted and stood by me until another man ran up leading two horses and we all escaped. Dolan should be awarded the Medal of Honor."

"He's gonna be on a long list. I'll keep it in mind and inform Captain Payne." Lawson sprinted toward what appeared to him to be Payne's company guidon about a hundred yards distant inside another barricade. By now the footing inside the wagon corral was treacherous. He hurdled over dead and dying horses, mules and oxen, tripped over discarded carbines and slipped in muddy pools of blood and guts.

As he approached the south end of the corral Joe was pleased to see Scott Payne standing erect and swinging his good arm back and forth directing men and giving orders. The upper portion of his other arm was swathed in a bloody bandage. Dr. Grimes, the expedition's surgeon, had tailored a towel into a sling for it. Payne appeared the epitome of a commanding officer, unconcerned of the lead bullets splattering small brown clouds of alkali dust around him. "Captain Payne!" Joe shouted and saluted as he drew near. Payne turned and smiled at him and returned the salute as though he was in the officers' mess back at Fort Russell.

"Captain Lawson!" Payne called back. "Glad to see you alive and well."

"Thank you, sir," Joe gasped breathlessly. He was becoming aware of his own burning thirst. "I collected all but two of the men from F Company of the Fifth at the barricades along the northwest side. Lieutenant Cherry and Sergeant Grimes have the north quadrant of the defenses organized and controlled. The Utes tried several squad-sized charges that got within thirty or so yards of the ramparts. All they did was expose themselves to our concentrated fire. They don't seem up to coordinating anything larger. The situation there has stabilized for the time being. Men are entrenching with knives and anything else they can get their hands on. We found some old copper cartridges that jammed Grimes's carbine but I didn't see anyone else having the same problem. There may be some picks and shovels in Gordon's wagons along with more ammo but the Utes have us cut off. Cherry's in charge . . . he says Dolan should be cited for the CMH. The water's becoming a problem. Joe Rankin is also over there. I told him to get a couple of other men and be prepared to run the gauntlet at the earliest opportunity, probably late tonight."

"Thanks, Joe. I'd say we're pretty much stuck here until help arrives. Those braves have food, water and ammunition. All they have to do is hold tight and pick us off until kingdom come . . . or we surrender." Payne noted Lawson's astonished look. "It's always a possibility, Joe, but it won't happen while I'm in command, at least not until the expedition is in very real danger of being completely annihilated. How about you?"

"Sir, I've been in an army since I was fifteen years old. I've never been captured, never been a prisoner and I won't be this time either. If I run out of ammunition I'll eat their faces."

"Water?" Payne struggled to unbutton his field blouse with his good hand. "How can it be so damned hot this time of the year in the mountains?"

Lawson pushed his commander's hand aside and unbuttoned the jacket. "It's about two hundred yards east to the edge of the creek bank and another twenty vertical feet down to the creek. The Utes have a few men dangling off the edge firing with one hand. We can counterattack about dark and force an opening."

"By midnight we'll all be shakin' in our boots and have fond memories of how nice and warm it is right now. No fires for the Utes to shoot at, it'll be cold rations for everyone until we get reinforced."

Joe looked over Payne's shoulder at a new hullabaloo of shouting from the southern barricades. "Uh-oh! Brace yourself; it's going to get a whole lot hotter."

"What?" asked Payne turning toward Lawson's gaze.

"The Utes have just fired the sagebrush and dry grass southwest of the wagons."

Payne turned. "Oh lord! If it gets to even one wagon the rest will burn like a child's paper chain. Some of those wagons may still have ammunition in them. This may be the start of a coordinated attack."

Lawson started toward the southern breastworks. "Once we get a smoke screen," he yelled over his shoulder, "I'll take a detail out and try to clear a firebreak before the flames get here."

Payne nodded. "I'll get some of the men from E Company on the north side to reinforce the south."

"Get 'em prepared but don't send them until I give you a signal or send a runner. This could be a diversion."

As Lawson approached the south end of the wagon corral he could see the men were already retreating from the battlements as the flames and smoke of the fire rolled toward them. Intrepid First Sergeant Dolan stood rifle in hand, against them. "If you don't get the hell back and fight," he bellowed, "I'll shoot ya myself." The word had barely escaped his lips when Joe, within feet of him, heard a loud *thwack*. Straight away he sensed again what he had experienced too often before during the calamity of combat. The immediacy of violent death caused instants to become seconds, seconds seemed like minutes and minutes like eternity. Dolan, arms above his head beseeching heaven, rifle tumbling on the breeze, was lifted slowly from the earth, rested perfectly horizontal in midair and settled, like an autumn leaf, onto the battleground. A fountain of bright red gushed from his chest turning the dust on and around him to rusty mud.

Joe, stunned to paralysis, could only stare at the lifeless form lying at his boot tips. He had known others of Dolan's stature, larger-than-life men, mythical Greek heroes. Warriors by nature who had survived the slaughter cauldrons at Balaclava, Sevastopol, Fredericksburg, Antietam and Gettysburg only to die ignobly from a stray shot, disease or accident. One second they were as immortal as the mountains and then next a heap of blood, flesh and bone. Seconds ago Dolan, the soldier, awaited promotion to sergeant major, the army's highest enlisted rank, now Dolan the casualty laid soaking the soil with his blood. At such times as these, emotions he considered long dead resurfaced through the armor he'd constructed plate by plate over decades. Gradually Lawson, hearing the bullets whizzing around him, pulled himself back to reality. He faced the retreating squads, pointed his revolver at them and yelled: "You heard what the sergeant said. Get back to fightin' or I'll shoot ya myself." The men nearly trampled each other as they flung themselves back to the ramparts, some even firing blindly from the hip into the billowing clouds of smoke.

He paced up and down the line shouting encouragement and threats. Finally, remembering his original mission he yelled, "I need volunteers to come with me out in front of the wagons and clear away brush and grass before the fire gets to the wagons!" To his surprise nearly every man within shouting range raised an arm or stood up. He ran down the line picking every second or third man. "With me," he yelled and vaulted over a low point in the breastworks. He went to his knees hacking at the nearest sagebrush with his sheath knife but its fibrous branches bent like steel rope. He tried pulling it out but his hands merely slipped over the greasy leaves. Joe quickly looked around and saw the rest of the detail having the same problem, then glanced at the flames leaning toward them less than a hundred yards away.

"Stop and follow me!" he shouted and ran toward the blaze. He stopped about ten yards from the wagons and checked his squad. Every man dripped muddy sweat. Their eyes darted from him to the fire. "All of you who have matches start a fire here but make sure you're between it and the wagons. Those of you without matches break off a burning branch when someone else gets one lit and light another one. Keep going until we've lighted a line about a hundred yards long in front of the wagons." Turning he saw the wounded Payne standing with troopers at the breastworks. "Fire between us and over our heads into the smoke; they're right behind it." Payne shouted the order and began firing his revolver. Lead started zipping by in both directions.

Lawson grabbed a clump of dry, brown Indian rice grass, put a match to it and dropped the small torch into the grass at his feet. He grabbed another handful, lit it from the first and did the same thing ten feet away. Noting with satisfaction that his men were doing the same thing for about fifty yards in both directions from him, he dropped back toward the perimeter and tried to shout a command but his parched throat failed him so he fired his revolver in the air. The men, surprised at a shot so close behind them, turned and he motioned them with a sweep of his arm back to the wagons.

When he reached the ramparts he motioned the first defender he saw toward him. When the trooper arrived, Joe squeaked: "Water." He took two short sips from the nearly empty canteen the man handed him, tilted his head back and gargled the mouthful before letting it trickle down his throat. "Now," he shouted, "put it out, every man out in front of the wagons and bring something to dig or beat with."

The men leaped over the barricades with knives, bayonets, spoons and forks, blankets and jackets. "When our fire is about ten yards out we're going to swing out shoulder to shoulder and beat it to death." Joe hoped he sounded more confident than he felt. Lawson grabbed Sergeant Grimes by the arm and pulled him close. "Grimes," he nearly whispered, "my voice is going. Stay with me and holler out my orders as I give them to you."

"Yes, sir."

Lawson paced out about five yards from the wagons. "Move 'em out to here."

"File," bellowed Gaines in his best drill-sergeant rumble.

71

"Step out to this position."

With satisfaction Joe watched as every man stepped off his left foot with parade-ground precision and marched unfalteringly straight into what must have appeared to be the gates of hell. No lack of discipline or courage here. As the column drew abreast of him and Grimes, he whispered, "Column 'alt."

"File . . . 'alt."

"Okay, Grimes, now let's see if we can stop what we started." Joe dropped to his knees and scooped up double hands full of dry adobe soil and pelted the nearest burning bush. Grimes removed his blue twill blouse and flailed away at another. The heat scorched Joe's face and the odor of burning hair drifted up from his goatee. When the flames persisted he turned his back to them, spread his legs and frantically pawed the earth like a dog. It worked! The blaze sickened and died to a smoldering, smoking skeleton. Standing, he looked up and down the line of men fighting fire madly, some successful, others giving ground grudgingly and still others, themselves afire, rolling in the dirt. He found another fire salient, bent over and performed the same maneuver.

The heat pushed them back against the wagons but their efforts were reinforced by a very slight but sudden shift in the breeze from south to north. The flames halted and then receded. He became aware of a new swarm of bullets clipping the canvas covers of the wagons.

"Captain!" Grimes called out. Joe followed the sergeant's point. The Utes were now using the smoke blowing back on them to creep closer to the defenses. He calculated there were nearly a hundred shadowy figures making their way through the seared landscape.

"Back to the wagons," crackled Lawson, "man the barricades." As the men leaped back into the relative safety of the corral he jogged up and down his fire line making certain no one, dead, wounded or injured had been left behind. Satisfied, he vaulted back over the breastworks himself where a thoughtful trooper handed him a nearly full canteen. He nodded thanks and took a sip and returned it quickly so he wouldn't be tempted to drain it.

Struggling to his feet, he first checked the progress of the fires. The Indian fire had reached the area burned by the blaze his men had set and was dying for lack of fuel. The Utes who had been using the smoke for cover dropped back from the hail of lead pelting them from the defenders' fusillades. Sergeant Grimes, his face a mask of dirt, soot and red blisters, appeared at his side.

"Whipped 'em we did sir," he wheezed.

"Aye, and without losing a man—that I could find."

"Captain, it 'pears there's trouble at the other end now. Captain Payne's headed in that direction."

Lawson turned back toward the north. Smoke and fire billowed skyward from outside the wagons. "Take over here Sergeant. Keep pouring it on as long as the men have good targets, check again for casualties. Detail two men to bring ammunition crates to the other end. I'm headed back there." He took a step but stumbled and went to his knees in the dust.

Grimes handed him a canteen. "Drink it all, sir, we can do without that little bit of water but we can't do without you."

"Thanks." This time Joe gulped in a large mouthful and held it in his mouth for as long as he could while he stumbled in the direction of the makeshift command post and Captain Payne.

"You look like something the cat dragged in," Payne said, scrutinizing Lawson from head to toe.

"Thanks," replied Joe, "I sorta feel like you look. What happened out there?" he asked, indicating Gordon's smoking supply wagons fifty yards from the north end of the corralled wagons.

"Some damn fool thought a grass fire would stop the Utes from completing their circle of the wagons. Burn a clean field of fire."

"Didn't work?"

"Not perfectly. Wind shifted and took the fire into Gordon's abandoned supply wagons, burned everything to cinders. The officer should be relieved of command, maybe court-martialed."

"Naw," he said, smiling at Captain Payne, "you been doin' a good job otherwise."

Payne chuckled. "How'd things go at your end?"

"They fired the brush and used it as a smoke screen to rush the defenses. We built a backfire and brought 'em to a stop, one scorched wagon, no casualties to speak of."

"Good work. Joe Rankin reported to me. He says he wants Gordon to go with him tonight; as a teamster he knows the roads and paths almost as well as he. I picked a couple of my company's troopers, who I've noticed are particularly good riders. At least one of the four should get through."

"If not," Lawson joked, "in another couple of weeks someone will miss us and start asking questions." His tone changed. "Speaking of questions, I have one about the smoke rising above the Danforth hills down southwest. You think it might be the agency?"

"Maybe," Payne responded, "but that's twenty-five miles away. The smoke could be another brush fire the Utes started."

"Let's hope so," said Lawson shaking his head. "I hate to be the one to tell you this . . . Dolan's dead. He got hit not two feet away from me."

Payne said nothing. His eyes were locked straight ahead. "Grieves me deeply to hear that," he finally replied in a low, hoarse voice. "He was among the best I've known. But we all know when we don a uniform and take up the profession of arms we hazard such an end."

Lawson reverently bowed his head. "Yes, sir." He then noticed the bottom two buttons of Payne's waistcoat were missing and, in fact, nearly all the bottom front of the vest exposing a bloody second torn shirt. "Captain!" he exclaimed and reached out toward the wound.

"What . . . oh, that. It's not as bad as it looks Joe. I took a ricochet from an iron barrel ring that grazed my stomach. Cut's not a quarter of an inch deep and about three inches long but bled like a stuck pig. Over twenty years in the service without so much as a scratch and then two in one day. I'll have Doc Grimes look at it tonight if things calm down."

Joe felt relieved and pleased; relieved that Payne's wound was not serious and pleased that since the episode earlier when Scott had wanted to relinquish his command he had taken over with authority and leadership.

CHAPTER 7

"Tribal warfare . . . is not more than the sum of individual actions of many warriors acting on their own without direction or coordination."

Gwynne Dyer, *War*

Nicagaat stood on a ridge with his back to the setting sun calmly smoking his pipe watching hell's chaos below more than a mile away. Weary men on both sides fired their rifles and carbines more sporadically than earlier in the day, stalemated. A chill autumn breeze washed the smoke toward the north away from the battleground. He felt both triumph and defeat. His warriors had won a hollow victory by stopping the cavalry from entering their reservation. The Utes would ultimately be defeated.

He heard a horse plodding and puffing its way up the ridge trail behind him. The horse, blowing hard, surmounted the ridge. Jack turned as Colorow slid from the saddle. He walked to Jack but remained behind his left shoulder. Now Jack's despair turned to anger. They stood together in silence. Jack finally took his pipe from between his tightly clenched teeth. "Meeker?" he asked.

"He is dead," replied Colorow with undisguised glee. "All are dead . . . except the women and children."

"Not the women and children?" Jack repeated, careful his voice gave no hint of emotion.

"Canalla and Quinkent have taken the women and children

76

toward the Grand River." Jack said nothing but clenched his teeth on his pipe stem. "I sent a messenger," Colorow said after a long pause, "to tell Douglas and Johnson not to harm them."

Jack nodded. Ute men did not consider rape harmful.

"Look," growled Colorow, "at the piles of dead horses, mules and oxen. By tomorrow they will begin to stink and bring flies; great big fat, blue flies and by the next day the soldiers will have only the fat flies to eat and then they will have to leave and walk back to their forts."

"They will send out riders as quickly as possible," said Jack, "probably tonight, for help. They will escape to the north through the river's narrow cut."

"Then we will be waiting beside the river and kill them."

"No, we will let them go and when help arrives we will lay down our arms. Perhaps we will have bought enough time for Canalla and Quinkent to have reached Ouray's camp. If anyone can protect our women and children it will be Ouray and maybe his friend General Adams. Soon they will sleep in the western darkness. I've sent some of the men back to the agency to collect food saved from the burning buildings and bring it here. When this ends it will be a long, very long time until our bellies are again full, those of us who are still alive. How many of us did not live to see today's sun set?"

"Over twenty," said Colorow solemnly. "Come, our camp is at the sharp bend in the Little River. We will build a big fire to conquer the cold night spirits."

CHAPTER 8

"It's a disagreeable thing—to be whipped."

General William Tecumseh Sherman

Seven men, six in uniform and one civilian, sat covered with a canvas sheet beside a wagon wheel. Under it Scott Payne scribbled intently with his good hand on a pad of paper in the dim light of a sputtering candle held by a corporal while Joe Lawson, another corporal, Joe Rankin and John Gordon studied a wrinkled and stained map.

Lawson tapped the map with his finger. "I suggest the four of you start out together, ride straight to the creek gully and follow it to the north to Stinking Gulch and split up there." Lawson deferred to Rankin's experience and expertise as an area guide and scout. "What do you think, Joe?"

"I partly agree. But they'll be expecting us to sneak north along the Milk, it's the only route we can take, be waiting above on the banks. Plus, the horses' hooves will make a hell of a racket on that river cobble. It'd be like shooting fish in a barrel. Instead, how about the four of us creep south toward the sharp bend where the creek curves the cut in the high ridge. We climb out of the creek bed, ride high on the flank of Yellow Jacket Pass and turn north so that we stay way above the creek and the braves on the east side of the wagons. A couple of miles north we'll drop back down to the Agency Road. We'll split up at Morapos Creek. I'll head back to Rawlins on the road because I know the ranches along the way and can change horses. I'll also alert Lieutenant Price and his men. When we separate, I'll direct Corporal Moquin and Corporal Murphy onto trails which are most likely to lead them to Merritt and Dodge. Gordon, you can take your pick which way to go because you know the area. When I get to Rawlins, I'll wire the captain's message to Forts Steele and Russell, and General Crook in Omaha, anyone else?"

Payne shook his head. "Tell Price my orders to him are to sit it out at Fortification Creek. He doesn't have enough men to help our situation and if they try we'll just end up with more surrounded. General Crook can forward your message as he sees fit. Gentlemen"—Payne raised a page and held it up to the candle Corporal Moquin was holding for him—"here it is. I hope it meets with your approval as both my eyesight and writing hand are giving out.

Milk River, Colorado, September twenty-nine, eighteen seventy-nine, eight thirty p.m.

This command, composed of three companies of cavalry, was met a mile south of Milk River by several hundred Ute Indians who attacked and drove us with great loss back to the wagon train which had parked. It becomes my painful duty to announce the death of Major Thornburgh who fell in harness; the painful but not serious wounding of Lieutenant Paddock and Doctor Grimes. Ten enlisted men were mortally wounded and a wagon master. About twenty-five men and teamsters have been wounded. I am now corralled near water, with three quarters of our animals killed after a desperate fight since twelve o'clock noon. We hold our position at this hour. I shall strengthen it during the night and believe that we can hold out until reinforcements reach us—if they are hurried through. Officers and men behaved with greatest gallantry. I am also slightly wounded in two places.

Payne, Commanding

No one spoke. Finally, Lawson voiced his approval and Payne handed a folded copy of the message to each of the four couriers. "I hope," he said with solemnity, "you have chosen your mounts well. It is upon them—and you—that all our lives may now depend."

"We got the best of the bunch," said Rankin, "but that ain't sayin' much considerin' what they been through. Mine's got a neck wound but it's not too bad and he should carry me to a fresh mount by sunup." He slipped his copy of the message inside his leather jacket.

"Full moon tonight," said Gordon, the teamster, "help keep us on the trails, and help keep the Utes on us."

The breeze that had so violently propelled the afternoon fires had died. In its place an early autumn chill enveloped the dark ramparts of circled wagons. Men clustered together under blankets or pieces of canvas sheets made from the wagon shells. The men huddled under the tarp flinched each time a wounded man's moan or cry of pain, a dying horse's pitiful squeal, the occasional gunshot, pierced the cold darkness.

Sergeant Grimes offered Lawson a chunk of pork fat and a can of beans. Joe shook his head and watched as Grimes feasted with apparent relish and washed the food down with copious swallows of Milk Creek alkali water. Instead Joe nibbled at a brick of hardtack cracker and sipped the brackish water that half a dozen men had risked their lives for earlier in the night before moonrise. He peeked from a corner of the makeshift tent.

The ridges surrounding them were speckled with yellow Indian campfires. Occasionally, in the distance, men's voices rumbled the unintelligible words of a syncopated chant from the fires. Every few minutes came the explosion of a cavalry carbine from one of the duty pickets. Payne had ordered the troopers not to waste ammunition shooting at the fires but the contents of the besieged wagons enticed some of the more daring braves who drew fire in search of battlefield souvenirs.

CHAPTER 9

*The Great Shame: And the Triumph of the Irish in the English
Speaking World*

Thomas Keneally

Joe sat cross-legged with a scratchy, brown army blanket draped over his balding head and drawn around his face. It had the familiar odors of mothballs, mildew and dust; homey smells that he'd grown accustomed to and took comfort from going back more than forty years when at night, as a small boy, he awaited his mother to carry him from where he'd curled up on the hard clay floor of their windowless rock hut and lay him on a canvas mattress ticked with straw. She took such a homespun blanket from her shoulders and tucked it around him while she crooned an ancient Celtic lullaby.

This was the nightly repayment to each of her seven living children for having to abandon them from dawn until dark to work in the potato plots with their father, cook, clean, wash, cut peat, dip candles and nurse the sick and injured. An Irish mother never complained nor would tolerate the same from others, especially her children. Every morning and evening she knelt and thanked God and the Mother Mary for each day of her short, tortuous life on earth, her husband and children.

Hundreds of generations of aboriginal Irish were born into days of unceasing, unending toil and hardship relieved only by death—which usually came blessedly early. The lot of the Irish in life was preordained by God said the Church. The Irish economy was based on the potato and children.

Children were necessary for survival of the parents, family, clan, nature and culture. They were born, lived and died, easily and often. Parental emotional attachment was an unnecessary luxury. Life, from beginning to end, was a constant struggle, a battle. No wonder the Irish were a pugnacious lot. If one wasn't fighting the sparse soil and abundant rocks, it was the weather, disease, injury, the landlord and . . . each other.

Joe actually did not know his age. He was a child of illiterate parents in an illiterate society. His mother may have interrupted her work for an hour to give birth and then returned to it. Dates—even birthdates—were meaningless. The conventional means by which the educated classes of nineteenth-century people measured such things were absent from the life of the Irish peasant: watches, clocks, calendars, rulers, yardsticks. They depended on the Catholic Church to inform them of Sundays, Christmas and Easter. A boy achieved manhood and a girl became a woman when they could walk, lift and carry.

The year the potatoes died, he stood nearly the height of his shovel's length. His father came to the house the middle of the morning holding several potatoes from their withered, moldy stalks like dead rabbits by their ears. He materialized out of the morning fog looking like death himself. Without a word he dropped his burden on the cobblestone walk for all to see, globs of black mush at the ends of the crumpled stems. " 'Tis the blight," he muttered grimly. "We'll not be eatin' again soon 'cept grass, tree bark and mice."

The Irish said that when the four apocalyptic horsemen arose from the abyss, an Englishman guided them to Ireland. The English landholders promised wheat, barley and corn from England which never arrived or in such quantities that every man woman and child in the land of Erin drew a single grain for their allotment. The landlords let their properties lie fallow or sold them for cattle and sheep pasture. The Lawson family joined a lengthening line of evictees wandering aimlessly in search of food or work. Ireland became a refugee nation, a nation of indentured servants, whores, serfs and slaves.

Men like his father, no longer able to support their families by any means, drank themselves to death. Women like his mother sold themselves on the streets of Dublin and Skibbereen for crusts of bread. One fateful day the remnants of the Lawson family, by happenstance, found themselves begging, like hundreds of others, in the Eóghan's Garden section of the town of Limerick. There, three young, drunken British soldiers set upon Joe's sister Sally after she begged for a few coins. They began dragging her toward a narrow alleyway. Joe witnessed the assault from a block away and ran to her aid arriving on the scene after the three had torn off what few rags she wore. The narrow confines of the alley worked to Joe's advantage. He separated the drunks and thrashed each repeatedly until all three lay sprawled on the cobblestones.

At this juncture, their commanding officer, who had also witnessed the assault, arrived at the entrance to the alley and was shocked to see his three Englishmen bettered by an Irish peasant. When they revived, the officer ordered his men to search out a blanket for Sally, empty their pockets to pay for new clothes and return to the barracks straight away to prepare for disciplinary actions.

He turned to Joe. "Saw the incident from out in the street and you in pursuit. I quickened my pace thinking you were in for a bit of a drubbing but from the look of things I needn't have hurried. You accounted very well for yourself—and the young lady."

Joe, still taught and puffing from anger and combat, replied, "I had no such misgivings. Any Irishman is worth at least thrice his weight in Englishmen—armed or not."

The tall, handsome officer smiled. He took a blanket proffered by one of his men and then took from them collectively more than a pound in coins and poured it all into Sally's hands.

Both Joe and Sally expressed surprise. "That's more money than I ever seen in one place at one time, even in the church collection." Sally gasped and nodded. She emptied her hands into a remnant of her shift she pick up from the street. They thanked the officer and nodded curtly to each of the soldiers who he ordered to attention and to apologize.

"Hurry on to Mother with it and say nothing of this to anyone else," Joe yelled at her as she ran down the alley. The soldiers saluted the officer and left.

"Aside from the fact that by the time you reached your sister's side, you had no other choice for a battleground, you did well not to confront the three on the open street. Do you know why?"

Joe answered without hesitation. "In the narrow alley I could fight them one at a time and the other two had no room to surround me and take me down."

"Precisely!" said the officer. "Good thinking, lad." He clapped Joe on the shoulder. "You look as though you'd be the better for something to eat but . . ." He pulled back his hand and after a cursory examination, took a handkerchief from his uniform sleeve and wiped it clean. "Perhaps, it would be better if you collected your family and we all retired to the mess tent at the regimental headquarters. My name's Knox," he said with a smile, "Captain William Knox, of the Seventeenth Lancers."

Joe took his proffered hand and found it surprisingly strong. "I'm Joe Lawson." He felt whipsawed with emotion. He was shaking the hand of the oppressor, an Englishman and army officer, supposedly the personification of aggression and oppression and yet nothing about the captain evoked in him anger or hatred. Joe, tall for his age, looked straight across into the captain's gray eyes. He guessed the young officer to be less than ten years older than Joe. "I'm pleased to meet you, William." He wasn't sure how to address this alien in tight riding breeches, black tunic with gold braid and knee-high polished boots. "I'll thank you for your kindness to my sister and comin' to her aid."

"You're welcome. Come, let's be off, but first collect your mother and siblings. You'll all be in safe 'ands, you've not been conscripted or 'dragooned.' That's my horse over there." William pointed toward a massively tall, sleek animal beautifully caparisoned in oak brown tack and red blanket.

He searched out his mother and siblings on a corner. His mother intently counted out the money Sally had given her, much to the interest of a growing knot of other street beggars. He broke into a run. She looked up as he approached and yelled out: "Joseph, this is more money than I've seen in me whole life." Coins spilled from her hand onto the stone cobbles from which his brothers and sisters snatched the coins back before the mob.

"Are ye daft?" he cried out as he ran up. "Showin' off in front of those who'd kill you for a farthing." His mother stood up and looked about furtively. She pulled up the hem of her grimy, tattered dress and dumped the coins into the fold.

Joe watched the encroaching mob with more fear than he'd felt confronting the three English soldiers. He and his two younger brothers formed a protective cordon around their mother and sisters and began backing into the street. They were more than a match for any dozen or more of the sick, starved, bedraggled scarecrows surrounding them but like wolves drawn to the blood scent of a wounded deer, their numbers doubled and tripled within minutes. Just as he raised his fists to charge the most aggressive man, a particularly fierce-looking black-toothed tough, something knocked him aside by a blow of immense force to the back of his right shoulder. As he fought to keep his feet he could see from the corner of his right eye a towering, dark form, snorting and blowing, clobbering past him toward the dissolving crowd. The beast called out in a familiar voice. "Stand firm, Joe, help has arrived! Be off ye damned rabble else ye run athwart an English saber." The beast sounded like Captain Knox. In an instant the sleek chestnut stallion passed him scattering the remnants of the mob. Those moving too slowly fell before the power of the warhorse or the flat of Captain Knox's blade. Joe quickly turned to his mother her mouth agape at this second rescue from a dark angel whose uniform they'd all learned to dread and hate.

The throng dispersed and William Knox returned to the Lawson family. "Is anyone wounded or injured?" he inquired. His question elicited only shocked silence. Knox looked about him. "Well, I guess that means no."

Joe stepped forward. Standing next to the captain's boot he had to crane his head back as far as it would tilt to see the man towering eight feet above him. "Thanks again to you, Captain, we're safe and sound. We are indebted to you, sir."

"Not t'all." Knox surveyed the street. "But can't leave you here for the vermin to close in again. The steeple yonder tells me there's a church 'neath it. I'm sure the worshipful priest will take you in for minor remuneration. I'll escort you there."

The Lawson family, tattered, filthy and forlorn, fell in behind Captain Knox and the horse he called Thunderer in their entire sartorial splendor. Joe brought up the rear so none of his siblings would fall victim to the street toughs. Also he wanted to be certain that his mother didn't drop any of the coins she still held in the fold of her skirt. Such was the sad situation in Ireland that their procession, a British officer leading a train of starving vagabonds, caused only a few passersby to stop to look.

When they arrived at the church, a massive granite edifice

with a steeple so tall clouds seemed to collide with it, they found that their old slate chapel would have fit into the atrium with space remaining. "Joseph," the captain called, motioning him over. Taking Joe aside Knox spoke softly so the others could not hear. "I'll go in alone and make the arrangements. I don't want anyone, most of all not a priest, to know you have any money on you."

Knox handed Thunderer's reins to Joe and charged the massive iron-sheathed wooden doors opening them like Moses parting the Red Sea. The captain strode straight into the maw of the stone beast whose breath smelled of incense.

No one spoke. Joe shivered slightly from the church beast's dreary cold breath. The Lawson clan waited silently. Thunderer's head drooped; he cocked a rear hoof at ease. Even his ever-vigilant tail swung lethargically at clusters of gnats and flies trying to park on his hind quarters. Joe began to imagine the captain had fallen to silent assassins and how it would look to the constabulary that six beggars were in possession of his horse and handfuls of coins. "Ay!" he thought. "If imagination's a sin, I'm goin' t' hell for sure."

Thunderer suddenly shifted and snapped to attention, head up, ears pricked, legs braced and eyes wide open. Everyone jumped. Captain Knox strode out of the darkness as he had walked in, a soldier of inestimable confidence, a short, portly man in church vestments trailing in his wake. He walked straight to Joe, the others clustered around. Joe handed him the reins.

"Joseph Lawson . . . and family," Knox introduced them with a sweep of his arm. "This is Father O'Bannon, a Dominican brother who will find all of you warm, safe rooms to sleep in, clothes and baths compliments of Her Majesty's Lancers and Lord Lucan. Now, save for Joe and your mother—I want a word with them—the rest of you follow the kindly father."

As the group warily followed the priest toward a monastery

adjoining the church, Joe and his mother approached Knox who took a brown leather pouch from a saddlebag. "Mrs. Lawson, put your money into this bag and know where the bag is at all times. Never let anyone outside your family touch it. Joe, I'll return tomorrow morning."

That said he deftly turned a stirrup, placed a booted foot into it and fairly flew into the saddle. Thunderer remained braced at attention. Captain Knox leaned down from the saddle and said: "I can't promise you a life devoid of pain, deprivation and sorrow but I can promise you and your family the opportunity to be spared lives foreshortened by disease, starvation and separation." Knox saluted with his riding crop. "I'll see you tomorrow morning when you're better fed, rested and clothed." Without so much as the touch of a rein Thunderer spun about on one hoof as lightly as a dancer and clattered down the cobblestones.

"Joseph!" Captain Knox cried out in surprise. "If I'd passed ye on the street I'd not know ye, scrubbed, shorn, and clean as a newborn." Caught with his mouth full of bread, ham and eggs, Joe jumped up from his seat at the breakfast table to greet his benefactor but the captain motioned him back down. "Finish, take your time, slow down and finish your breakfast."

Joe swallowed and washed the mouthful down with a draft of milk. "Yes, sir," he gasped. "We learned that lesson last night. An empty belly doesn't take kindly to sudden gluttony. What of my mum and sisters?"

"Just as uncomfortable as you I'd imagine. Prosperity like all else in life takes some gettin' used to." He paused. "Which brings me to the purpose of my visit this mornin'? From where did you come?"

"South," replied Joe between smaller bites, "County Cork, not far from Skibbereen."

Knox seemed to pause to consider Joe's answer. "What brings you to County Mayo?"

"M' dad's brother, my uncle, had a farm outside Limerick. When we were evicted we hoped to come to Mayo and move in with his family but when we got here they were gone too."

"Where's your father?"

"Dead . . . he took what little money we had, got drunk and fell or jumped into the Shannon."

Knox nodded. "So, when the few coppers from my men's pockets runs out, you'll be back on the streets. You still have the money, I trust?"

Joe reached down and pulled up his trouser leg to show the captain the leather bag dangling from its cord. "Aye," he replied grimly. In the excitement of the last day he hadn't considered this eventuality. He stopped eating and stared at what remained of the bounty on the table. "I suppose so," he replied quietly.

William smiled and patted Joe on the shoulder. "Buck up lad, if you accept my proposal, your life and those of your family are about to change for the better. You see, I've come to Mayo to recruit lads into his lordship's regiment. I've seen that you possess endurance and courage, the first two soldierly requirements. His lordship, George Charles Bingham, the Third Earl of Lucan, has authorized me to provide new, five-year recruits a twenty-pound bounty plus at least twenty pounds per year for each of the five years of your recruitment—or more, depending on promotions plus uniforms and keep. At the end of your enlistment you'll have earned no fewer than ten acres of land. I believe I can persuade the earl to reinstall your family back on their farm to sharecrop for him as rental payment. Of course, they'd not be able to raise potatoes but they wouldn't have to with your pay helping to support them. At the term of your enlistment, the property will be yours. Your family will never again face eviction. What do you say?"

Joe struggled to comprehend the offer. Yesterday at this time he and his family, like thousands of his countrymen, seemed irredeemably impoverished and hoping only for a very short life to release them from the pain of slow starvation. Now, as suddenly as the sun burning a hole through bitterly dark clouds, he glimpsed hope. His mouth gaped but nothing came out.

"Well," said William, "if you can't speak, just nod or shake your head."

Joe nodded. "Y-yes . . . yes," he stammered.

"Very good," William said. "Let's go announce the good fortune to the rest of your family."

"Sir," Joe said, "we are in your debt but how can you be sure Lord Lucan will accept this agreement?"

"I can, because, you see, although I'm called William Knox, my full name is Knox-Bingham. The Knox comes from my Scottish maternal lineage and my estates in Scotland. In all probability, I will be the Fourth Earl of Lucan."

CHAPTER 10

"War is the unfolding of miscalculations."

Barbara Tuchman

"Captain . . . Captain Lawson," rasped a familiar, deep-throated growl. Someone shook him so roughly by his right shoulder he nearly toppled over. "Wake up, sir, the moon is risin' and the couriers are fixin' to leave."

The voice pulled the blanket from over his head. Looking up he saw the sergeant's features in the pale moonlight. "Yes, thank you, Grimes, I must have dozed . . . sorry." He flailed his arms about trying to stand. "Ed," he said reluctantly, "a hand, please, my legs are glued together and my butt's nailed to the ground." Grimes hoisted him and held him upright for a few seconds while the circulation returned to his officer's legs. Joe allowed the sergeant to assist him with a hand under his arm for several paces. "When I was your age," he said to Grimes, "I could leap from the ground to full attention in one smooth move."

"Yes, sir and, I'm thinking, spread your arms and fly to the moon."

"My beard wasn't always gray."

"No, sir, it was pure white. It's the nasty smoke and dust what's turned it gray."

The two men were still bantering when they walked up to Captain Payne and the four couriers. "Judging by you two, I'd say morale remains high," said Payne cheerfully.

92

"Aye, Captain," replied Grimes. "We'll wait for 'em to attack in full force and kill 'em with humor."

"Frankly," said Payne, "I'd rather rely on these four men completing their assigned mission." They turned to the men standing beside their horses. Payne addressed them: "Upon your shoulders rests the lives of many men, perhaps even the very survival of this expedition." He paused thoughtfully. "But you know that. Off you go and good luck."

The four men, Joe Rankin in the lead, walked their horses single file toward a break in the south barricade between the wagons. "What do you think, Joe?" Payne asked Lawson.

"At least one of 'em will make it. Now, it's a matter of time; how long it'll take for 'em to get help and how long it'll take the help to get here. A lot depends on the next couple of hours. If we hear lots of shootin' out there before midnight it'll be a bad sign, if we don't hear any it's likely to mean they got through."

"Either way, you and I won't be getting much sleep."

"I got a few winks earlier so I'll take the first watch. Doc Grimes probably has some laudanum that'll quiet your pain and help you sleep."

"Naw, afraid it'd put me to sleep forever. I'll just find a blanket and curl up next to a flour sack—it'll remind me of my wife."

Joe laughed and walked to a cluster of glowing embers where men were sitting lighting cigars and pipes off each other. "You boys keep your hands cupped around those things, Utes can see in the dark." A nervous chuckle arose from the circle. Better for morale, Joe thought, to take a chance and let 'em smoke some. "Don't smoke 'em all tonight either, ain't likely to be a train with more tobacco through here for several days." Louder laughter.

"Captain, how long you think it'll be before we get help?" came an anonymous voice from within the circle. The voice was very young and high pitched. Joe could hear the boy's fear.

"Depends on how lucky we get. One of those fellows could run into a patrol as early as dawn and we might have an entire cavalry division here before tomorrow evening." Joe could practically feel the men relax at the thought.

"That'd sure enough be a sight," squeaked the voice.

"Sir," asked another, "what if we don't get lucky?"

"Well . . . then it'll take a little longer and you boys can have a picnic or two before that long ride back to Fort Steele."

In spite of the chill mountain air Joe Rankin felt a rivulet of sweat run from his hatband down his sideburns and neck. Milk Creek, as they approached the sharp bend around the bluffs, narrowed and gurgled among rocks that for centuries crashed into the riverbed from the hillsides above. The low water did nothing to drown out the splashing and clattering made by eight boots and sixteen iron-shod horse hooves.

In Rankin's heightened state of awareness, the men and beasts sounded like a herd of buffalo crashing through a creek bed full of buckets and kettles. A man behind him coughed, a horse snorted. He instinctively ducked low and braced for a fusillade of gunfire from the line of woods bordering the creek but if Utes were hiding in the trees, they remained quiet. At this point each step took the couriers further south and west into the Indians encampments. The urge to stampede across the rivulets was nearly overwhelming. His eyes strained in the partial darkness to locate a point on the southern bank that would provide them cover enough to climb to a point where they could switch back north above any Utes positioned along the edge of the stream.

Finally, the moonlight revealed the sharp bend and a patch of willows on the south bank. He motioned Gordon forward to his side. "We'll get out of the creek up there," Rankin whispered in his ear, "get clear of those willows and pick our way through the oak and pine east and then back north." He looked to Gordon for confirmation but the teamster, his eyes fixed toward the inside bank of the river bend, appeared not to be listening. "Joe," he whispered, "we're right on top of 'em—or they're right on top of us."

Rankin looked where Gordon was pointing. Through the willows that defined the separation of river and land he could see a small, flickering campfire. His arm shot up to halt the two following corporals who ducked down. "You see anyone around the fire?" Rankin asked.

"I thought so," replied Gordon, "but it might have been willows or cattails wavin' in the breeze."

"God! If they've seen us we're prime targets out here in the moonlight. If they haven't seen us, they sure as hell will. We gotta make it to that flat water just ahead and hope none of us fill up somebody's sights on the way." Rankin bent low and tugged on the bridle reins. Every hoof step into the shallow, rocky river sounded like a cannonball: *thud, splash, thud, splash!* Worse, the cacophony echoed off the narrow gully walls upstream toward the confluence with Beaver Creek. With every beat of his heart, Rankin expected an explosion signaling the start of a running gun battle. But running which way? The relative safety of the corralled wagons only several hundred yards back was overwhelmingly enticing but Joe also knew that everyone there was depending on them getting at least one man through. He pinned his eyes on the ford and the far bank of the stream and gingerly put one foot in front of the other.

The scout felt a surge of relief as the thicket of tall willows closed around him. He advanced a few yards and turned to check on the others. As each approached him he motioned and whispered, "Keep moving but stop where the willows end."

When the third man passed, Joe fell into line behind him.

There was no escaping the infernal noise. The willows bent and whistled. Men and horses broke through patches of dead, fallen willow that snapped and cracked like gunfire. He felt relieved but also unnerved that they had come this far without a shot fired at them or having to run pell-mell for their lives. In spite of the cold night air, he was soaked with sweat and his heartbeat nearly deafened him.

He suddenly bumped into the tail of the horse in front of him. The column had stopped. The word came back that the first man had reached the edge of the willows. Joe worked his way forward and caught up with Gordon who was kneeling just inside the southern boundary of willows.

"We're stuck," whispered Gordon. Rankin heard the near panic in his voice. He parted the willows and almost within arm's length saw what caused Gordon's alarm. Slightly more than six feet away appeared the towering dark buttress of an adobe cliff. Over centuries the river, in making its sharp turn from east to north had slowly sliced through the low point of the ridge leaving a twenty-foot wall of earth and rock which, at its deepest, seemed to surround the couriers on three sides.

"Christ, Joe, we're trapped like sittin' ducks down here." Gordon's voice changed from a whisper to a high-pitched squeal.

"Shhhh! Calm down."

Joe put his hand reassuringly on John's shoulder. "It's all right. I know this place. I'll take over lead now, you go back and tell the others to keep in tight." He took John's bridle reins and the teamster made his way back through the willows whispering to the men as he went. When he returned, Joe could tell he'd collected his courage. "Okay, we're going to lead the horses a hundred feet or so left. There's a section of collapsed bank with a game trail that angles up it to the top. When we get to that point, it's clear sailin' north behind the Utes around to the

Agency Road. Let's go."

He gave Gordon a pat on the back, led his horse out of the willows and turned quickly to his left. Rankin struggled to contain his own fear. He had to rely strictly on memory of a single experience with this route several years before, during the day. The shadow of the overhanging bluff caused the path to disappear. He kept the willows within reach of his left arm and his right touching the adobe wall. That guideline, plus the fact that neither he nor his horse was tripping over rocks, logs or root snags, indicated that they were on relatively worn ground.

Within about two minutes, he could feel the path begin to incline slightly. As he climbed, the moonlight revealed a narrow path edging upward along the side of a very large loaf of adobe clay. Very quickly he lost touch with the willows to his left. Looking in that direction he could see the tops of the willows which only seconds ago were over his head. He quickened his pace, exhilarated at leaving an obvious death trap and that his recollections ran true. The air changed. The smell of sage, juniper, cedar and pine replaced that of the river and swamp. The incline gradually receded. He stopped to confirm all of his party had made it to level ground and were cloaked in the cover of cedar trees and oak brush.

"Mount up," he whispered to the assembled men. "We're past the most dangerous part, now it's just a long ride." He brought the bridle reins back over his horse's head but discovered when he tried to place his boot into the stirrup, his feet were so swollen and numb from the cold water, they refused to cooperate. After several attempts, one of the cavalry troopers, to Rankin's chagrin, dismounted and cupped his hands aside the horse for Rankin. Joe murmured "thanks" then, putting the midnight moon to his back he slackened the reins and urged his gelding forward letting him pick his way along the trail. They left the sound of Milk Creek trickling between the rocks. Now the only sound was the soft thud of sixteen horse hooves on the packed earth of the trail, and the occasional click of an iron shoe against stone.

Nearly half a mile behind them, three men draped in heavy wool blankets stood holding rifles in the crooks of their arms watching the four shadowy riders pass in and out of the moonlit trees.

"We should have killed them," declared Colorow. "If, as you say, Nicagaat, we are destined to hang anyway then we should take the battle as far as we can, kill as many as we can."

"Perhaps," muttered Jack irritably. "If that's what you want, then *you* lead an attack on the wagons at dawn. We outnumber them at least three to one. Then, if you survive the charge, you can explain your decision to your warriors when the army comes and has you surrounded by ten to one."

Sowawick, one of Ouray's subchiefs, had arrived just after dark. Chipeta, Ouray's wife, had learned from her sister-in-law, Susan, Johnson's wife, of the pending Meeker mutiny and Milk Creek ambush. She sent Sowawick with the message to the one man who could forestall both, Ouray, who was hunting on Grand Mesa. Ouray then told Sowawick to ride to the White River Agency with an order to Douglas and Johnson to avoid a battle.

But, Sowawick said, on his way to the agency he had encountered Douglas and his five captive women and children at a ford on the Grand River. He informed Douglas of Ouray's order.

Douglas had only replied, "Too late." Sowawick rode on to the burned and looted agency where he changed horses from a herd that Douglas had left behind for the men at Milk Creek.

Now, trying to quell the quarrel between Jack and Colorow he said: "Is Meeker's fault."

"Was his fault," corrected Jack. The chief despised Meeker, even beyond his death, yet he knew that the man was not the sole cause of the day's events. In a larger sense he was a symbol of a rapacious and acquisitive white plague that had begun to spread across the land generations ago. No, he despised Meeker, the man, because he'd possessed the power to peacefully control the flow of events leading up to this day's disastrous torrent. "Now we're in the thick of it and it doesn't matter whose fault it is." How, he wondered, could one man, a whining coward like Meeker, cause so much trouble and the deaths of so many? He flinched slightly, remembering Josephine, beautiful, wise and thoughtful Josie. Had she been able to impose her will on her father, everything would have been different. He tightened and shifted his feet thinking about her at the mercy of the vindictive Douglas and his men, especially the love-stricken boy, Persune. Jack turned and walked back to the fire leaving the other two to debate issues.

While Rankin wanted to set the dispatch riders on their separate ways as soon as possible, he couldn't risk it in the dark. Neither of the troopers had any familiarity with the lay of the land and would get hopelessly lost in the dark, if not killed. Knowing that each step took them further and further from what, a few hours ago, he considered certain death, profoundly relieved him. But they all had many a mile to travel and hazards to face before they were literally "out of the woods."

The couriers snaked along a maze of game trails first to the east then north through pine, scrub oak, cedar and juniper trees. Joe stopped when they crossed a low pass into the Morapos Creek drainage and gathered the men around him. "I didn't want to split the group while it's still dark," he said, "but I have to hightail it to Rawlins at this point. The moon's bright enough so you can pick you way along the road until it gets light enough to get your bearings. I suggest two of you follow Morapos Creek to the Yampa River and then east along the Yampa where I think you're more likely to find Colonel Merritt and his men. John, maybe you can cut west toward the Yampa and find Captain Dodge and the Ninth Cavalry. They started north from the Uncompahgre area on the twenty-first, so they should be up here by now. I'll stop at Lieutenant Price's camp on my way and tell him to stay put. Any questions?"

"What if we get lost once we separate?" asked Corporal Murphy.

"There's no such thing as getting lost up here—just extremely inconvenienced. Everybody has a compass. If you get confused, stop, locate north and set a heading and course. If you go directly north long enough, you'll ride into the Yampa and the river will take you to farms, ranches and towns either upstream or downstream. If you're lucky enough to find a creek follow it downstream to the Yampa."

"Tell everyone you run into, at least every white, what's happened and get directions to the next settlement." He studied the men's faces. "Well, that's about all I have in the way of advice. You know how much is riding on at least one of us getting through . . . and it may have to be one of you. So don't stop somewhere for a few drinks and a card game." Joe shook every man's hand and wished him luck and said he hoped to see him again. He pulled out his watch and held it up to the moonlight. It was one o'clock in the morning. Within minutes he caught sight of the double ribbons of wagon tracks which marked the Agency Road and spurred his wounded horse up to a gentle, efficient lope. The cold night air passed through his sweat-soaked garments and around his soggy boots. Dawn couldn't come quickly enough.

CHAPTER 11

"The journey of a thousand miles begins with one step."

Lao Tzu

At six o'clock in the morning, cattle rancher Ben Hewitt and his dogs were in the midst of a venison-and-egg breakfast as they watched the sun peak over the distant northern tip of Colorado's Gore Range.

Hewitt, by nature a solitary man, had become even more so after the death of his wife from pneumonia the previous winter. They'd had no children. After three miscarriages in the Colorado wilderness she kept Ben at arm's length and he gradually despaired himself.

Sometimes they'd go for days without speaking, except to the dogs. Just before Christmas Mrs. Hewitt took to bed with a heavy fever followed by incessant coughing and labored breathing which seemed to rattle the bones of her fleshless body. Hewitt prepared to ride to Rawlins for a doctor but she stopped him. "No!" she pleaded. "I'll be gone before you get back and I don't want to die alone." Together, in silence, they'd waited for death's knock on the door.

The ground being frozen at the time, he sewed her body in a quilt she'd made for their wedding and carried it to the barn and locked it in the tack room until he could bury her in the spring. On the first day of May he dug a grave in her cherished bed of colorful peony blossoms atop the hill behind their dugout

101

beside three other graves with carved wooden planks for markers.

In the dark emptiness following her death, a small wedge of light illuminated for him a dim awareness that he'd somehow mistaken loneliness for love and created for her the living hell he should perhaps have endured by himself. He wore this guilt and shame like a monk's hair shirt.

A distant *click, click, click* roused him from these cold remembrances. Ben recognized it as the sound of a loose horseshoe striking rocks in the dust of the lane. The hoofbeats said the animal was lame. He dumped his breakfast on the floor for the dogs to finish off and stood in the doorway, one hand on his rifle, and watched the bend in the road to see what fool or desperado drove a horse that hard during the night.

The dogs, having made quick work of the biscuits and ham, ran off toward the road barking up a storm. A rider appeared at the apex of the bend and stalled the canine assault by calling out their names. "What the hell," Ben muttered, "is Joe Rankin doing out here at this time o' the morning? Hey Joe," he yelled and waved. The rider waved and reined his horse off the lane toward Ben.

Rankin pulled up a hundred feet short of the cabin and with some effort, slid from the saddle. Ben walked to him. Dirt and dried blood from innumerable small cuts and scratches caked the stubble on Rankin's face. Scrub oak leaves and twigs dripped from his jacket like icicles and matted his hair. Ben had never seen Rankin bareheaded. Joe bareheaded. Ben strode up to him and extended his hand. Joe could offer only a feeble handshake.

"My God!" exclaimed Ben. "You both look plum tuckered out. How long you been ridin'?"

"Most of the night and then some," answered Joe hoarsely. "You have any coffee?"

"Sure, sure. You go on up to the house and help yourself, sugar's in the cupboard. I'll look to your horse. Is that a bullet wound on his neck?" Joe didn't answer. He staggered to the dugout. Ben led the bedraggled horse to the barn and examined the wound. The bullet had gone through the fleshy under part of the throat. Both sides of the wound were crusted with dried blood and nothing poured out when he drank so apparently the animal would survive. Ben clipped off the remainder of the shoe and tossed it into a barrel. After he'd filled the mow with hay, he returned to the house.

Ben walked in on Joe as the guide washed down a moldy biscuit with gulps of cold coffee. He gazed at Ben through slits of red-rimmed eyes. "When you swallow and catch your breath I'd like to hear about it."

Joe nodded. "I gotta get to Rawlins," he gasped between gulps. "There's three companies of cavalry surrounded by three, four hundred Utes this side of Yellow Jacket Pass. They hit us yesterday about noon. The commander, Major Thornburgh, has been killed along with a dozen or so men, probably thirty or forty wounded." He paused for another gulp and inhaled deeply. "Four of us crept out last night about ten o'clock. We split up before daylight and I'm hightailin' it for Rawlins."

Ben nodded. "I guessed they were up to no good. I was over to Peck's Store a few days ago, that's probably why I didn't see your bunch come by on the road. They said the Utes bought over ten thousand rounds of ammunition from him, claimed they was goin' huntin' out east on the Poudre and Platte. Also been hearin' that fancy-pants Meeker got 'em heated up a bit."

"That's a fact." Joe emptied the soot-blackened coffeepot into his tin cup. "I reckon 'bout everybody here'bouts wishes Meeker had done things a little different."

"He don't know how, never did, never will. I lost my shirt to 'im years back on that Union Colony business—snake oil's

103

what it was. He got out o' Greeley, that's what they call it now, by the skin of his teeth. Lots o' folks like me would like to buy in with the Utes and cut a pound o' flesh off 'em after the Indians string 'em up."

Joe silently studied the fat blue fly floating in the bottom of his cup. "Well, be that as it may. I gotta hit the trail again. Can I swap you a dead horse for a live one?"

"Sure, sure, I'm fixin' to take a herd up to Fort Steele. They're out in the big corral. You cut out one you like and I'll put together some food and water for you. Guess you know there's a company of infantry up on Fortification Creek."

"Yeah, that's my next stop, Thornburg left them in reserve. Now they're no good to us ¡ . . not enough of 'em." He started toward the door. "By the way, I lost my hat during the night. You have one I could take?"

"Shore, I'll get one for ya before ya leave."

Joe picked out a short but very muscular red dun with an eel stripe the length of its back and took it to the barn. After he saddled the dun, he checked on the gelding which was asleep on the floor of its stall. "Men and horses are the fuel that feeds the flames of war. Hope I'll be as lucky as you."

Ben handed him a leather bag of food, a canteen, a blanket, and an oversize leather jerkin to replace his torn jacket. "Hung a hat on the gatepost, you can grab it on your way."

"Thanks, thanks for everything." Rankin swung up into the saddle. "What should I do with the horse?"

"Leave her with the army; she was headed there anyway. I'll charge 'em for her when I take the herd up to Steele. Anyway, drop by next time you're through this neck o' the woods and let me know how things turn out." The two men shook hands and Ben opened the gate.

Once Joe was back on the Agency Road, he held the spunky young dun to the same soft, easy lope. He crossed the Yampa

River about midday and continued on the same road twenty miles to Fortification Creek and the camp of Lieutenant Butler Price's Company E, Fourth Infantry. The officer welcomed him into the camp expressing surprise at Joe's appearance. "My God, Rankin, what's happened? You look like you've been dragged for miles."

"Feel like it." Joe tried to dismount but simply fell off his horse. Price and another officer helped him to his unsteady feet. He immediately handed Price Captain Payne's dispatch and orders. "We ran into some trouble with the Utes down at Yellow Jacket Pass. The major's dead and Payne's wounded but still in command. Four of us got out last night. I'm ridin' to Rawlins to wire Fort Steele, Fort Russell and General Crook. My horse is pretty well done in; you got a remount for me?"

Price, while reading the dispatch, replied, "Yes, take mine. Do you think we can expect an attack here?"

"Don't think so. The Indians have plenty on their hands down south—but, if they get the better of the cavalry and decide to ride out, there's not much to stop 'em. They'd killed off more'n half the cavalry horses and draft stock by the time we left and I'd guess they pretty much finished 'em off during the day today. No way can Payne and Lawson form up a counterattack."

"Sergeant." Price turned to a burly noncom. "Saddle my horse for Mr. Rankin while I draft a dispatch for him to take to Rawlins." Price conducted Rankin to his tent where he pointed to a folding chair.

"Thanks," said the scout, "but if it's all the same to you I'll stand for as long as I can; may be many an hour before I have the opportunity again."

"Joe, mind if I make a suggestion?"

"Go ahead."

"You've done your fair share. Let one of my younger men

take it from here. All he would have to do is follow the road to Rawlins."

"I appreciate the offer but I've got to see this thing all the way through. There's still a lot that can happen between here and Rawlins and I'm a whole lot better equipped to deal with most of them than some infantry private. I also feel sorta responsible for the situation."

"How so?"

"Well, Jack and Colorow paid us a couple of visits after we left you. They wanted us to stop where we camped the night before the ambush and send five men from there to the agency boundary. The two of them would meet us there and ride on to the White River to meet with Meeker. The major pretty much agreed. He and I talked about it later and he said he was afraid of separating himself and his officers from the three companies by that much distance. He hoped to find water and adequate grazing for the horses miles before we got to Milk Creek. He passed up a couple of good ones. I should have spoken up but I didn't. I'm sure the Utes thought we'd betrayed them when we crossed the creek with all three companies."

"Perhaps," Price replied thoughtfully, "but he would have also been correct, according to good tactical principles, not to split up his command. I'd like to send one of the men with you to help out . . ." Price held up his hand to deflect Joe's protest. "But I won't because I know that would just slow you down. So, I'll leave you to your devices and wish you good luck."

Joe shook his hand and exited the tent to find a thoroughbred stallion saddled and waiting.

CHAPTER 12

"Cry 'Havoc!' and let slip the dogs of war,
That this foul deed shall smell above the earth.
With carrion men, groaning for burial."

William Shakespeare
Julius Caesar, Act 3, Scene 1

Joe Lawson awoke before dawn shivering uncontrollably. He tightened the thin wool army blanket around him and snuggled up to the side of the shallow revetment Sergeant Grimes had scratched out for him, hoping to absorb some warmth from the dirt. But, like Joe Rankin at about the same time many miles to the north, all he could really do was pray for the sun to hurry above Sleepy Cat Mountain.

Everyone alive inside the defenses had endured a grim night. The unwounded troopers huddled against the mountain cold and listened to the almost constant moans, groans and screams of wounded men, horses and mules. The nocturnal caterwauling merged with horrific nightmares. By dawn, he felt nearly delirious with exhaustion.

He knew the only way to defeat the cold was to jump from his grave-like bed and immediately concentrate on his company's welfare. Many of his troopers were only slightly older than he when Captain Knox had taken Joe under his wing in Ireland and over the sea to a place called Crimea and the first of many wars.

A lifetime of warfare had hardened him, body and soul, to almost any conceivable deprivation or degradation. But too, his body had endured decades of insults; coarsely treated wounds, injuries, disease and starvation until it was little more than hard wood connected by sinew and covered with rawhide. But now, almost forty years later, muscles reacted slowly to his commands and joints fought his demands with pain and rigidity. If he'd stoop to such, age could be a most convenient excuse for ignoring duty and orders.

Knowing he wouldn't be able to warm himself back to sleep, he uncoiled like a snake and began carving out a cupola in the end of his hole so he could at least light a candle to warm his frozen fingers. He looked toward the west where the pale moon disc sat atop the mountain towering over the Ute-occupied hill. He could even see smoke from smoldering Indian campfires floating into the light. But at this hour even the sporadic Ute rifle fire had died out. He enjoyed for a second or two the compelling dream that they had given up and moved back onto the reservation. The sound of a lone shot and "thwack" of the lead striking wagon planking jerked him to present urgencies. He blew out the candle, scrambled from the trench and stood with the blanket around him.

"Uh . . . Cap'n," came a voice from behind him. Joe jumped before he recognized Sergeant Grimes's voice. "Pardon me, sir, I'd leave that blanket behind and wear your hat, the men are a tad jumpy and in this dim light might mistake you for an Indian that wandered in."

Joe laughed. "Thanks, Sergeant," he tossed the blanket into the hole and donned his weather-beaten hat with its gold rope band and captain's bars. "Damn near froze to death last night. We catch a break today, I'll scratch this burrow a bit deeper and find some planks to put over it before night."

"I'll be happy to take care of that, sir."

"Appreciate that but you'll probably have your hands full. Let's go visit Captain Payne."

He and Grimes made their way cautiously to the broken staff and pennant signifying Payne's command post, another, slightly larger "badger hole." The hole was empty. "Cherry! You about someplace?" Joe called out.

"Sir!" A voice came from the darkness followed by a rustling of cloth and the click and clank of brass buckles. In moments, Lieutenant Cherry's statuesque form materialized in the dim light. He straightened to attention and saluted.

Joe returned the salute, pleased to see Cherry had retained his military decorum in spite of the circumstances. Even his blue uniform looked brushed and clean. "At ease, do you know where Captain Payne is?"

"No, sir, sorry."

"Sleep well, Sam?"

"Yes, sir. I passed out right after I stood the first watch and slept like a baby until just now."

Joe heaved a sigh. "Liar."

Cherry shuffled his feet in distress. "Fact is, sir, couldn't sleep at all because of the wounded—both men and horses."

"Don't reckon any of us did. After sunup locate as many of the E Company men as you can and a good place for a command center. I'd like to keep the company together as much as possible. Anything happen on your watch?"

"Not much, sir. I think I spotted a few braves crawling through the sage toward the perimeter. I shot and moved quickly and fired again. Don't know if I hit anything or not. Probably just some Indians on pilfering expeditions. Had to put my greyhound, Rex, down. He'd sought out the creek in the early hours before dawn for a drink. On his way back into the perimeter one of the pickets mistook him for an Indian down on all fours and shot off a paw."

Lawson expressed his sympathy. He considered this an ill portent to the first full day of the Ute siege. "You have anything for breakfast?"

"Enough, sir, I'll scrounge up some more during the day."

"Maybe we can light a fire or two during the day and boil coffee—save some for tonight. Cold coffee's better than none at all."

"Yes, sir, I agree." Cherry saluted. He picked up his adobe laden blanket, gave it a vigorous shaking, then very carefully folded and rolled it.

Joe chuckled and walked off. They may leave West Point, he thought, but the Point never leaves them. As the pink gray light of dawn silhouetted the ridges of Sleepy Cat Mountain he and the sergeant found Captain Payne helping Doctor Grimes tend to the wounded at one of the revetments carved out of the brick-hard soil. The hollow contained eight men less critically wounded; flesh wounds, through-and-through wounds, glancing head wounds, no broken bones. These men, if the situation called for it, could still man a rifle or carbine, or run for safety if need be.

He and Grimes walked up. "Captain Payne," Joe announced in the most military voice he could summon, "Captain Lawson and First Sergeant Grimes are alive and fit for duty." Both came to attention and saluted. Joe, known for his casual manner and quick wit, was, however, a soldier to the marrow of his bones and always conscious that officers and noncoms be diligent in preserving military dignity and discipline before the men, especially the younger ones. If the officers were slack or behaved cowardly so would the men. In crucial battles, if one man, especially an officer, broke ranks everyone would follow.

Payne looked up and smiled. He placed his wounded arm back in its sling and rose stiffly to attention and saluted. "Morning, gentlemen, at ease, looks like a fine day ahead of us, can't

110

see a cloud in the sky yet. I trust you both rested well during the night. What are your plans for the day?"

"Well, sir," replied Lawson, "I asked Lieutenant Cherry to locate the E Company men. I'll assemble them and call roll. Any missing I'll try to locate among the wounded. If I can't find 'em either place, I'll begin to consider the probability they were killed and check among those laid to rest. If you approve, sir, I'll have E Company defend the north-northwest breastworks, D Company the center and F Company the lower south-southwest. The teamsters, outfitters and blacksmiths will cover the east where cover is less and targets fewer. We have to get as much dirt as possible over the dead men and animals as quickly as possible. We'll enlarge the trenches for the wounded ensuring maximum protection from enemy fire and erect some lean-tos out of wagon canvas for shade over them. We didn't take any more casualties during the night—at least not human. The men are pretty sick about the loss of animals, especially their horses, but they've got to get hardened to it because it's only going to get worse. You figure Rankin and his bunch got away?"

"I think so, Captain. We'd have heard a lot more shootin' last night after they left if the Utes had seen 'em. Rankin has the best chance of getting through; he can find help along the way."

"Joe." Payne motioned him away from the wounded men. "I think the Utes know they have a tiger by the tail. Their one plan was to stop us at the agency boundary. They may have hoped that we'd hightail it back to Wyoming but that hope ended when they started killing the animals. They don't know what to do now. We're dug in, have ammunition, some food and water. Their hearts really aren't in the fight enough to accept the losses of a frontal assault. What are our alternatives? Surrender is out of the question . . . at this point anyway. We can't counterattack; the numbers are against us and they've got the good ground. We can't make a run for it and get shot in the back. As I see it, the only feasible alternative is to stay put and wait for help. Tell me, Joe, am I overlooking something?"

"No. Tryin' to back out of here in a fighting withdrawal would be a disaster. We could only take as much ammunition and rations as each man could carry. We'd have to leave the dead and wounded behind and that's not even a consideration. We'd better get used to callin' this 'home' for a while. Speakin' of dead and wounded . . ."

"Yeah," interrupted Payne. "I'm sick at the thought of not being able to retrieve the major's body but it's impossible for a while."

Lawson nodded in agreement. "What the Indians left, the coyotes, buzzards, ants and flies will finish. But something worse prays on my mind."

"Good God! You think it's possible he's still alive out there? You saw him get shot and go down."

Joe looked down and shook his head. "I saw someone I thought was the major shot from his horse. Grimes confirmed the major had been shot. But, the heat of battle and fatigue has muddled my mind enough that as of right now I would not be able to walk directly to the spot. There wasn't another man around him at the time so I don't think anyone else could do so. Sadly, until relieved or reinforced there's nothing we can do for him." Joe snapped to attention and saluted. "Sergeant Grimes and I will tend to the men on the north barricades."

Payne returned the salute. "Carry on, Captain. When you get time, please note your actions and those of your company for my report."

As Joe and the sergeant walked away from Payne and Dr. Grimes, the sun peaked over the mountain ridge to the east and shone on the pair like a giant spotlight. Joe's pause for a moment to bask in its radiance quickly brought a fusillade of rifle fire from the hillside. Bullets hissed and sizzled past them and thumped into the ground. Both sprinted toward a line of troopers returning the Ute shots from inside the breastworks.

They dove for cover. Sergeant Grimes slid over the three candles set into the dirt that the men had been using to try to boil water for coffee. Cries of anger replaced carbine shots. Joe smiled inwardly; American soldiers in battle, he learned, might do without food or water but not coffee and tobacco—and maybe ammunition. "Boys," he spat out a clod of adobe dirt, "since the sergeant and I are responsible for interrupting your morning coffee, I give you permission to build a small fire to boil more water." The men cheered. Joe took out his knife and began digging a small fire pit while the men searched on hands and knees for sticks and scraps of broken crates. Within minutes eager eyes watched a smoke-blackened coffeepot.

The nationalities and names had changed since the days when he joined the lancers, hussars and dragoons similarly urging on teapots outside Balaclava in the Crimea over twenty-five years before but the pleading faces were the same. "Sergeant, take over here, I need to talk with Lieutenant Cherry." Joe stood up. Grimes snapped to attention. The men jumped to their feet and came to attention. A shot rang out from the burned-out supply wagons and the bullet whizzed through their midst. Not a man flinched but all in unison raised their hands in salute. Lawson slowly, steadily returned their salute. "As you were," he said quietly. The men ducked for cover and returned to their coffee-making task. Lawson turned on his heel and strode down the line of wagons.

"Cap'n Lawson's one smart officer," remarked one of the men, "we're damn lucky to have him."

"Yeah," added another, "the captain's the kind of officer if he led you straight into the fires of hell, you'd look forward to the trip."

"First Sergeant," asked a freckle-faced boy Grimes thought

no older than fifteen, "is Captain Lawson West Point?"

Grimes chuckled. "Nay laddie, ee's better'n that, ee's Irish." The men all laughed. Grimes fed the fire a few more sticks as he pondered his story.

"Captain Lawson went to war, real war, not against a few half-starved Indians like what we got here, but against Russians and Cossacks and the Moguls of India when he was a good bit younger than any o' you. And not to mention, he fought in every major battle from Bull Run to Petersburg in our War of Southern Secession."

"If he's that good, why ain't he a general?" asked an older boy with a scraggly goatee.

"Because," shot back Grimes, jamming a burning taper toward the face of the offender, "he's a fighter not a politician, a soldier first and an officer second, a wise man, a just man, a hero, a leader and . . . an Irishman. He took up the Union Jack to buy back the land that Lord Lucan stole from his family. Then, when the great Russian bear came down from the north to eat the God-cursed Turks and make the Mediterranean into a Russian bathtub, he and his regiment, the Seventeenth Lancers, were shipped off to a hellhole known as Balaklava where the devil roasted men slowly in the summer, pickled them with rain in the fall and stiffened them into pillars of ice in the winter to be devoured in the spring. If the Russians didn't kill ya the cholera that swam in the black water would."

Grimes lifted open the top of the coffeepot and dumped in a double handful of coffee mixed with chicory root. To this he added a handful of sugar and replaced the lid.

"Were you there, First Sergeant?" asked the freckle-faced boy.

"Nay, I was barely out of nappies in fifty-four. My father was there in the Eighth Dragoons, part of the Light Brigade; received the Victoria Cross same as Captain Lawson. Ye see, it is one of the last wars in which cavalrymen actually fought toe-to-toe with lances and swords. Horsemanship and a strong arm were of the utmost importance. The lance and sword seldom killed outright; they wounded a man to death rather than killed him. Men carved off each other's noses, ears and scalps; an eyeball 'ere and a jawbone there. Ghastly sights they were, the Wilkinson shaves."

"The Wilkinson shaves?"

"The name o' the company what made cavalry sabers. They also made razors."

The blackened pot came to a rolling boil and spewed mud-colored bubbles out its spout. The small fire hissed in anger and belched forth a mushroom-shaped cloud of steam and white smoke that attracted a crackle of rifle fire from the burned wagons and nearby hillside. The men tucked their backs up against the breastwork being certain the coffeepot first had a level shelf of dirt on which to rest while the grounds steeped to varnish color and consistency.

"Sergeant Grimes," Lawson shouted from down the line of wagons where he was talking to Lieutenant Cherry.

"Sir."

"You're going to upset the WOGs with smoke signals. Bury your fire and conduct the tea party in a more helpful manner."

"Aye, sir," Grimes shouted back as he pushed adobe over the tiny fire pit with his boot.

"WOGs?" questioned one of the men.

"Worthy Oriental Gentlemen," chuckled Grimes. "It's a derogatory Brit term for anyone of color what ain't a Brit . . . 'cept for the Irish for whom they reserve very special derogatory terms."

"Sergeant." Captain Lawson motioned to him.

"Gotta go, lads. Kindly pour two cups for our officer corps. Two of the men handed over their cups and began hunting for others. Grimes set the burning tin aside quickly and donned his leather gauntlets. "Now," he said, "before I stand up, I want you all to return to your firing positions and zero in on the closest wagon in J. C. Davis's burned-out supply wagons which had been following Paddock's D Company. Likely they'll get off one wild round at me; as soon as they do, fill their gun smoke full of holes."

He waited until all the troopers had returned to their positions and were aiming. "Ready," he yelled, "aim . . ." With that he stood up with a cup in each hand. As predicted, several rifles fired from the wagon sending their deadly leaden missiles snapping by him. "Fire!" Nearly twenty .45-caliber Springfield carbines roared to life in near unison. As the men quickly reloaded, the sergeant strolled along the breastworks to Lawson and Cherry who met him with mouths agape.

"Sergeant . . . ," Captain Lawson growled. But Grimes interrupted him by handing him a cup of coffee.

"I know, sir. One more stunt like that and you'll have me stripes . . . again. Captain, these so-called indicia of authority are fair worn to shreds havin' been passed between you and I so many times. If you please, sir, take 'em and keep 'em safe until they can be added to me burial kit."

"Not a chance, Grimes. I'll promote Second Lieutenant Cherry to first sergeant and hold him in that rank until the devil dies of old age." Sam Cherry cast him a surprised look and took the other cup Grimes offered him. Lawson gingerly brought the cup to his lips and blew on the inky mixture before taking a sip. "Uhhh," he gasped and spit to one side. "My God, *that's* awful!" he exclaimed. "You didn't use that creek water did you?"

"No, sir," Grimes replied seriously. "We couldn't afford to use real water sir; we waited until one of the horses stretched out and caught his piss in the pot. If you're not up to it, sir, I'll be more'n happy to drink yours and the lieutenant's."

Lawson waived him off while Cherry studied his cup more intently. "No, no, I'll force it down. Don't want to hurt the boys' feelin's."

Cherry sniffed the cup and feeling his commander's and his subordinate's eyes lying heavily upon him, sipped the brew very cautiously. "Very good, Sergeant, best I've had all day." He put the cup aside. Lawson shook his head. Grimes beamed back and saluted.

The crack of distant rifle shots and the moans of wounded men whose draughts of laudanum had worn off ended the lighthearted camaraderie. The three men tucked themselves in more tightly against the bulwarks. "Break the men into two details," Joe ordered. "I want one maintaining suppressing fire and the other digging graves and trench shelters. Dig right next to the dead animals so we can roll them right into it, we can't be digging a canyon for a mass burial. Get the graves deep enough to cover several inches of dirt over them to stifle the smell. Organize a cemetery for dead troopers at the south end where the ground appears softer. Make every effort to identify the body and leave something with a note in it."

"Like what?" asked Grimes.

"Use a small scrap of paper to write the name and unit on. Roll it up and put it in an empty cartridge case and seal the end with candle wax. Put the case in the man's mouth. And put up some kind of a marker; rock, board or bottle. If you can, scratch the man's name on it as well. Alternate details every other hour."

"What about amputated limbs?" asked Cherry. "I see Dr. Grimes has a pile already."

"Dig one trench for all of 'em. Dig it a bit deeper and bury 'em in layers with a layer of dirt between."

"Latrines?"

"Same."

"Cooking?"

"Cold rations. Pipe smoking only. Ration water, a pint a day per man. We may be able to increase that once we get bucket brigades better organized." Joe looked down at the ground and pulled on the end of his gray goatee. "In lieu of a miracle, Sam, we're in for some pretty rough days. Keep the men busy so they don't have time to dwell on our predicament excessively."

"Yes, sir." Cherry and Grimes saluted and scampered to their tasks.

Joe walked cautiously along the northern arc of wagons inspecting the men at the barricades, checking ammunition, wounded troopers and supplies. As he strode by one of the wagons he saw a man's head silhouetted against the canvas by the early-morning sun. Joe lifted up a section of cloth skirting and peered under it. A boy sat cross-legged in the middle of the rough plank floor bent over scribbling in what appeared to be a ledger book. Looking more closely Joe noted the single gold bars of a second lieutenant. The officer was hatless and Lawson could see no weapon of any kind. The troopers manning the breastworks moved aside and looked at Joe apprehensively as he mounted a flour barrel near the tailgate of the wagon. He pulled aside the canvas flaps covering the rear opening and stepped down to the wagon floor. The young officer looked up and sprang to attention. Joe recognized him as Lieutenant Paddock's assistant. The tall, lanky boy appeared to be in his early teens, too young even to be shaving, and sported a head of thick blond hair. He returned the boy's salute.

"What *are* you doing, Lieutenant?" Joe demanded. "Hiding?"

"Er . . . no sir. I'm taking inventory, Captain."

Joe held out his hand and nodded toward the book. The lieutenant handed it to him, open to the page where he had been writing. Sure enough, the book's pages were carefully lined and lettered with dates, descriptions and numbers.

"Trooper . . ."

"Bliss sir, Tasker Bliss."

"Lieutenant Bliss, *you are* aware, aren't you, that we are under attack and in grave danger of being overrun by several hundred Ute warriors?"

"Yes, sir."

Joe felt fury swelling inside him. "I'm curious why you aren't manning the barricades along with the men of your company?"

"Ah . . . I thought I could better serve our situation by taking stock of our remaining supplies and ammunition . . . assuming we will be here for an extended period such accounting would help rationing." Bliss tucked his chin hard against his neck, puffed out his chest and sucked in his stomach—a West Point plebe's best sign of subordination.

"Who ordered you to make this accounting?" Lawson hissed, his temperature rising with each passing second.

"No one, sir."

"You assumed this task on your own?" asked Lawson.

"Yes sir."

"Lieutenant Bliss, I don't see a rifle, carbine or revolver anywhere in this wagon. Have you fired a shot in the last eighteen or so hours?"

Bliss gulped. "No, sir, my eyesight's not good. Frankly, I'm not much of a shot so I helped build the barricades."

"Yesterday, Lieutenant, I watched as Sergeant Dolan ordered men back to their firing posts under threat of death. He was shot and killed at that instant. I then threatened them in the same manner, and none of them were West Point officers. Now I am about to threaten you to make the same decision. Get out, fight, and take your chances or stand here and be summarily shot by me for desertion."

Bliss, to Lawson's astonishment, began to shake uncontrollably and gasp for breath. Joe had seen officers literally shake in their boots before and during battle but he'd never confronted one sobbing in terror. Bliss looked more like a bullied schoolboy than a cavalry officer. He dimly remembered his own nearly paralyzing fear as his company awaited the order to charge up the Valley of Death. His tone softened.

"At ease, Bliss." He hoped the men outside couldn't hear the dialogue. He handed the officer a handkerchief. "Pull yourself together. Those men out there are every bit as afraid as you. There's no one here that isn't scared nearly out of their wits. Some of them are ten years younger. They need a professional officer, you, to set an example. When this whole thing began yesterday at noon, I wanted more than anything else to turn my horse around, ride all the way home to Kentucky and hide under the bed."

Bliss caught his breath and stared at the captain with wide-eyed amazement.

"You, sir?"

"Damn straight. I've been in more battles and have more wounds than you and your whole family has fingers and toes and I've been puking scared every time . . . including right now. Men're afraid. Sometimes I think we're afraid of everything. Fear is a part of life. What makes one man different from another is how well he controls his fear and does the job. When some of us put on a military uniform we're saying that we are willing to do a job, accomplishing a mission of protecting others who can't do that job themselves . . . and we have the strength and training necessary to steel ourselves against man-killing risks to perform the tasks as duty calls."

Bliss had stopped sobbing. His lips were firmly set and the corners of his mouth straightened. "Yes, sir. I understand. I'm sorry I faltered. I'm ready to take my place beside my men."

Lawson nodded. "Do you have any arms, rifle, revolver or carbine?"

Bliss shook his head. "No sir."

"Follow me, we'll get you a carbine. You may be blind as a bat but just shoot in the general direction as everyone on our side of the breastworks. There'll probably be a hail of bullets when we step out onto the barricade but the two of us will stand there together and you'll calmly check on each of the men below you. Understand?"

"Sir!" Bliss saluted.

Lawson crossed himself and opened the wagon curtain. The second the two officers stepped from the tailgate of the wagon onto the hodgepodge of barrels, crates, flour sacks and boxes the Utes fired a broadside. Bullets filled the air around them sizzling by their heads, thudding into wood and cloth, clanging off barrel bands. Lawson looked down at the astonished audience huddling behind the breastworks. He calmly looked at Bliss who was standing steadfast and expressionless. "You men have ammunition?" Lawson called out. Heads nodded in affirmation. "I can't hear you," he yelled.

"Yes, sir," the men yelled back.

"Then I suggest you fire back at the enemy. This is a battle. You're allowed to do so."

The men fell back into position and began firing furiously. The Ute fire slackened. Lawson jumped over the men's heads to the ground and turned around. Bliss remained standing indecisively on the rampart. "Lieutenant," he shouted, "you best get down from there or you risk being shot." Bliss immediately joined him on the ground.

"Take this." Joe unbuckled his cartridge belt and holster containing his revolver. "Shoot the enemy whenever possible—but mainly just shoot and keep shootin'. Collect your platoon sergeant and constantly check the men for wounds, ammunition, and water and, above all, discipline. It'll do wonders for your own. Excuse me while I go find Sergeant Grimes and rearm myself."

Bliss stiffened to attention and saluted. "Thank you sir for the ass kicking."

Joe smiled. "We all need one once in a while. Return to your post."

CHAPTER 13

"Fatigue makes cowards of us all."

General George S. Patton Jr.

Joe Rankin, during his life in the west, had often spent long hours in the saddle, sometimes as many as eighteen, but nothing approaching forty hours without sleep. The enormity and urgency of the White River expedition's plight urged him on with all the strength he could muster. He forced his blown buckskin into a constant trot not only to maintain speed but to keep him awake. More than once he'd dozed when the horse walked and nearly toppled from the saddle or had been awoken by the sound of the ravenous animal grazing at the side the road.

The road was a blessing. Though the scout had spent years traveling this country his fatigued fogged mind often confused direction and distances and failed to comprehend familiar landmarks. Without the two wagon wheel ruts leading him from one hill to the next, he could easily become lost.

Hunger pangs helped keep him awake. He concentrated on his stomach's growling and aching, eating only morsels of dried bread or jerky from his saddlebags. Price's men had included a small bag of real coffee but had no pot so he stuffed his cheeks with grounds, wetted them with sips of water and sucked the juice.

The fiery stallion upon which he'd departed Price's camp

was now a shambling nag that stumbled over the sticks and rocks in the road. He last stopped to water the animal from the Little Snake River almost twenty miles behind him. Rankin knew that only a near miracle would provide a stream, pond or even puddle at this time of the year.

He hoped for a farm or ranch where he could water the animal or trade him. But dusk was descending. And his senses were so befuddled, he wasn't at all certain that he would even recognize a ranch turnoff. His eyes burned and vision blurred. The mountain autumn colors were a dizzying, shifting kaleidoscope of red, green, yellow and brown. He let the horse slow to a walk and touched the brim of the nearly empty canteen to cracked and bleeding lips. He sipped no more than a teaspoon of the precious liquid and fought against the nearly overwhelming urge to swallow. Instead, he stored it in his coffee-pouched cheeks and let the parched lining of his mouth absorb it.

Joe again spurred the reluctant horse into a stumbling canter but quickly let it slow to a trot and stood up in the stirrups. Even if the horse had been willing, Joe could hardly tolerate the pain of the blisters and rash that had developed on the insides of his thighs and his scrotum. Like many cavalrymen, he had replaced the inseam of his woolen twill trousers with more comfortable gussets of soft cotton cloth, but the expedition's trip south to the agency had worn holes in the cotton exposing his skin to the coarse leather and wood of the McClellan saddle. He would have willingly paid a month's wages for a scrap of sheepskin.

The sun, dimmed to an orange ball, dropped nearly to the saw teeth of the Northern Colorado Rockies. The recollection of last night's deep chill briefly jolted him past pain and weariness. He spurred the horse back to a trot.

His dry, sunburned eyes blinked reflexively. The relief of a single blink was palpable. "My God," he said to his companion in the blackness behind his eyelids, "it gets dark quickly up here at this time of the year. A moment ago the sun was just setting and now it's pitch black." His comment was answered only by the steady clomp of horse's hooves on the dirt road. He fumbled for his watch. The hoofbeats stopped—silence closed in around him. Joe, sensing something amiss, tightened his grip on the reins. Instantly, his arms jerked forward and he somersaulted into the blackness suddenly and painfully transformed into flashes of colored light.

"No," he said, "I'm all right. I must have hit a tree branch in the dark. Now I can't even see the road, must be cloudy, I can't even see the stars. Moon should be up soon, though . . . full moon . . . maybe I'll be able to see the road again or I'll just let the horse have his head. He's from Fort Steele so he knows the way back home. Must have dropped the reins when I got whacked and now it's so blame dark I can't find 'em, can't even see the horse."

The only reply was the jingling of a horse's bit and the crunch of the animal cropping grass. "Damn!" he exclaimed. "How long we been ridin' together and you've hardly said more'n two words. Hell! I can't even remember your name. Help me out a bit here fella, I'm feelin' a bit outa kilter and . . . uh . . . what is your name?"

"Jack Wilson," the darkness replied, "some call me Rabbit Tail. It looks like you landed on your head when you came off. Lucky you didn't hit it on a rock. Take a drink."

Rankin felt something pushing past his cracked lips and a surge of cold on his tongue. He tried to catch his breath but choked, coughed and convulsed. He sat bolt upright and spit the rest of the water and a mass of coffee grounds on the grass. A blaze of white light blinded him. "Aaahhh," he bellowed and covered his eyes with his hands.

"Too much at once," the deep voice cautioned. "And sit up

when you drink. Here, I'll help you up and give you the canteen but sip it."

Joe dropped his hands and grasped the familiar object held in front of him. He took a sip and held the cool water in his mouth a few seconds before swallowing. "Thanks," he croaked. Shading his eyes with his free hand he spied his horse grazing on the brown fall grass. After taking another drink he felt the man's hands wedge into his armpits and lift him quickly and effortlessly from the ground. "Thanks, pard," Joe said.

"No partner, jus' you and me," came the reply. "I found you stretched out on the road and your horse eatin'. Good thing I came along when I did. You were bear bait for sure. What happened?"

"Uh, damned if I know," Joe said. A wave of pain which seemed to come from everywhere at once caused him to catch his breath. Rabbit Tail steadied him with a hand in the armpit. "I remember riding in the dark . . . no . . . wait, it wasn't dark. I was blinking my eyes and then it was suddenly dark. I was talking to my partner in the dark and then something happened and suddenly a lot of lights."

"I think you fell asleep and tumbled off your horse. You're in pretty bad shape. How long you been riding and where you headed?"

"Long time I guess, don't recollect how long. Don't know where I was headed."

"Well, your tracks in the road look like you were headed north, does that help any?

"What's north?"

"State line, Union Pacific Railroad, Rawlins."

"Rawlins, I was headed to Rawlins, but I can't remember why." Joe took another, longer drink from the canteen.

"Well, if you feel fit enough to ride I'll ride along with you and make sure you get there. Or we can camp for the night.

There's a spring up the road a bit."

"Thanks, I'd appreciate your help." Joe handed the canteen back and tried to focus on the indistinct, disembodied voice of Jack Wilson. The cold hand of panic gripped Joe's gut. The man was featureless but his voice revealed a short, sturdy man with a wrinkled, round, brown face. A tall, black reservation hat with a broad, flat brim crowned his head. Two long, black braids dropped from the hat along each side of his head over his shoulders and over his chest. He wore a black drape coat buttoned nearly to the neck but reveling the knots of two bright neckerchiefs, one red and one yellow. "You're Indian?"

The voice answered. "I am a Paiute, a shaman."

Joe's memory suddenly flickered to life; the expedition, ambush, wagons, troops, message, Rawlins. "I have to get to Rawlins quick as possible," he blurted out.

"Yes," replied Jack, "you are on an important mission there?"

"Yes." Joe answered abruptly and held his breath. He pressed his right forearm against his jacket to feel the reassuring bulge of his Colt revolver.

"Then we will go, Joe Rankin, and I will help you arrive in Rawlins as quickly as possible." Wilson turned and disappeared. He returned shortly leading Rankin's horse by the lead rope. He handed it to Joe. The scout relaxed. Evidently Wilson hadn't heard about the Milk Creek fight and Joe wasn't going to chance creating an enemy. He could, under some pretext, part with Wilson safely once they arrived in Rawlins.

"Here's part of your problem, Joe," said Wilson, handing Joe the stallion's bridle reins in two pieces. "Looks like you and your horse had a tussle and broke the reins in half. I can fix that quickly enough. While I'm working on them . . ." He reached inside his coat pocket and retrieved a slender leather envelope. Untying a drawstring, Wilson pulled two short wooden tubes from several in the envelope and handed them to Joe. "These," said the medicine man, "contain the power of life or the power to return life. The Paiute received it from the ancient ones in the south who received it from Quetzalcoatl, who is the creator of all."

Rankin examined the wooden tubes suspiciously. Maybe the Indian was trying to poison him. No . . . if he'd wanted to kill him he would have done so when he had been unconscious. "How do you know my name?" Joe asked the Paiute.

"It's engraved on your saddlebags," the Paiute said.

"Oh, what do I do with these?"

"You open the end and inhale the powder like tobacco smoke, through your mouth or nose. If you use your mouth make sure you breathe it in and not suck on the tube. Either way you'll want to sneeze or cough but you need to hold your breath."

"I'll use my mouth."

"First, drive all the air from your chest, then put the reed in your mouth, hold your head back and breathe in as deeply as you can."

Joe exhaled until he nearly bent double then put the end of the reed well back on his tongue, tightened his lips and drew a deep breath. He felt the tube release its contents into his throat, the feeling of inhaling dust or fine ashes. It felt as though he'd inhaled a campfire flame. He choked and dropped the tube but continued to inhale until his chest felt like it would explode. And it did, in a paroxysm of coughing, sputtering and choking. Waves of nausea whirled through his head. He grabbed the canteen Wovoka held out to him and gulped down a draught of water which quickly retraced its course and exploded from his mouth. He looked at the shaman's silhouette but gained no re-assurance. The regurgitation streamed down his face. He had no breath left to cry out from the pain in his chest and throat. Panic seized him. Did the Paiute poison him?

After a few more excruciating gasps of air the pain subsided, then disappeared, replaced by a sensation of tranquility which gradually changed into a vibrant alertness and physical strength such as he'd never experienced. Within a minute he felt no pain, no fatigue, no hunger or thirst. "My God!" he cried out with both elation and alarm. "I do feel grand. Wilson, your medicine is truly the elixir of life. What is it?"

The shaman had finished splicing the bridle reins and handed them to Joe. He shrugged. "It has no name. It is a very powerful medicine used only to bring the dead or nearly dead back to life. It will not last long. You will need the other one before we get to Rawlins. But any more than two and nothing will bring you back to life."

The Paiute turned and took two page-sized sheets of crinkled, reddish-brown parchment from his saddlebags and a leather bag. "Jerky and dried corn," he said, handing them to Joe. "Eat and drink as you ride. Walk the horse. I have given him some grain but his legs are weak and will not stand a faster pace for very long. I will follow you in case you fall from your horse in the dark."

They mounted. Joe ecstatically discovered that the medicine had also dulled the pain of his saddle sores. His horse too had renewed energy and soon he was in happy rhythm with the animal's quick walk. He looked over his shoulder into the darkness. He couldn't see or hear the Paiute but sensed him following closely.

After hours of almost interminable riding along the moonlit road Joe unbuttoned his jacket and fished in his waistcoat pocket for a match. After three attempts with cold numbed fingers he struck one with his thumbnail and held it up to his pocket watch, two o'clock in the early morning. He'd ridden approximately one hundred forty miles in twenty-five and a half hours, had not slept more than a total of two hours in two days, eaten and drunk little and pissed blood.

Now he stopped at the intersection of the Agency Road and the Union Pacific Railroad tracks, the point at which the White River expedition had disembarked from a train ten days ago and began their ride ill-fated ride south. Now it seemed more like a lifetime—or just a dream. He neck-reined his limping horse onto the road that paralleled the tracks to the Rawlins Station, a mile distant. He could hear the hoofbeats of the Paiute's horse in his wake but neither man spoke.

The shaman had been as good as his word. He'd followed but never interfered. When Joe, while still astride his horse, took the contents of the second reed, the Indian had thoughtfully stationed himself beside him. Now as they were closing in on the telegraph station, Joe's mind was again dimming. The effects of the second reed had lasted only about half as long as the first. In the indefinable distance, Joe saw the crimson twinkle of the station's red lantern. He was too numb to be reassured and instead fixed it in his sight and pressed his knees into the spent mound of flesh, blood and bone he straddled. The horse moved not an iota faster.

The men stopped beneath the lustrous ruby-red lantern that tinted the wood smoke curling skyward from a stovepipe atop the building a ghostly pink. Joe remained frozen on his saddle. His muscles and joints failed him when he tried to dismount. It was as though they had forgotten how to move. He feared to stir his legs back to life, afraid of awakening the savage pain of the blisters and sores.

He gazed dumbly through the red haze at the Indian. The Paiute slipped from his horse with the fluid motion of a young boy and positioned himself with his arms outstretched to the left of Joe's horse and beckoned with the fingers of both hands. Joe dropped his reins and raised himself upward with one hand on the pommel and the other the cantle of the saddle. He slipped his foot out of the right stirrup and over the tail of his horse. But his right hand slipped and his left leg was too weak and he sloughed over into the man's waiting arms. The short man's wiry strength surprised Rankin. He held Joe upright from under the arms until the guide got his feet under him and then gently assisted him toward the station. Joe stumbled up the steps and pushed on the door which creaked open tripping a bell above it.

The agent awakened from a position of repose, chin on his chest and feet on the desk, and leaped to his feet. "Yes," he blurted even before his eyes were open, "what can I do for . . . you? Are you all right?" he exclaimed, jumping from his chair and starting across the room.

Joe motioned to calm the man. "I need," he croaked, "to send an urgent telegram." Unbuttoning his jacket he stiffly reached to its inside pocket. Stunned to discover it empty, he quickly tore open the jacket all the way and checked both sides. Then gazed at the agent in stupefied panic. "Uh . . . uh, I have a dispatch." He turned and stumbled back outside. The Paiute was waiting astride his horse holding the message. He handed it to Joe.

"It came out of your jacket when you fell from your horse. I picked it up and saw your name on the outside."

"You didn't get my name from my saddlebags. You read the letter?"

Wovoka shook his head. "No, I didn't have to. I knew what business you had here in Rawlins."

"You could have left me to die or killed me yourself and delayed help. Why didn't you?"

"Your death would have accomplished nothing, changed nothing. Your death would only have prolonged the inevitable. The Utes are doomed to defeat. Only the quickest arrival of the cavalry will prevent more deaths."

Joe took the offered letter and returned to the telegrapher.

131

"This needs to be sent immediately to General Crook in Omaha, Colonel Merritt at Fort Russell and Captain Bisbee at Fort Steele. I'll wait for confirmation." The agent examined its contents and without another word turned up the wick on the lantern over his desk and began tapping out the message on his key.

Joe went back outside. Only his stallion remained. The Paiute had vanished. The scout checked his horse's hooves for what had sounded like a loose shoe. As he bent to pick up a rear hoof his eyes picked up the distinct trail in the snow-sodden clay of a horse approaching the railroad station, one horse. A search of both sides of the tracks revealed no more tracks either coming or going.

"Mr. Rankin," called the telegrapher from the station deck. "I got your message out and received a confirmation. Is there anything more?"

"No, mission accomplished. I need to get Lieutenant Price's horse to my stable before he dies under me and myself to bed before I die on him."

"Do we here in Rawlins need to be concerned about a general uprising?"

"No, they just wanted to stop us." He returned to his horse, untied the lead rope from the hitching rail and started to mount but found it too painful. Then he tried mounting from the station deck. Finally he patted the stallion's drooping head and ruffled his forelock. "What the hell, it's not very far and I could use the exercise, let's walk."

Chapter 14

Winning's not about who hits the ground first—it's about who's standing last.

Glen McKinney

Joe Rankin laboriously settled Lieutenant Price's stallion into a stall, brushed him down, salved the saddle galls, fed and watered him before collapsing under a blanket on a mound of hay. Rankin, nearly unconscious with fatigue, remained sleepless from the pains that seemed to emanate from every bone, joint and muscle fiber in his body. Near dawn he hunted down a bottle of horse elixir that he'd discovered possessed the same calming effects as laudanum. A few swigs later he plummeted into a black, dreamless sleep.

Unknown to Rankin, Gordon, Moquin, and Murphy had argued after his departure about the best course of action. They had traveled barely a mile further north when Gordon wanted to turn back south in the hope of finding his brother's wagon train that had been trailing Thornburg's column. Murphy disagreed saying that without the cover of darkness such a maneuver would put them right into Ute hands. However, he followed the other two until they reached the point where Rankin had split from them. Without a word, Murphy turned his horse and galloped back north as instructed by Rankin. He ultimately received a citation for gallantry for arriving, after a 170 mile journey of endurance, at Fort Steele, squelching claims from his companions that he'd deserted. Gordon and Moquin probably later wished they had joined him.

About mid morning they topped a small hill about five miles northeast of the battle site discovering to their dismay the cold ashes and dismembered corpses of the supply wagons and crew. Hundreds of empty cartridge cases in and around the scene testified that the men had put up a ferocious fight. Murphy and Gordon saw no Indian bodies.

Another argument ensued. The grief-stricken Gordon insisted on burying the scattered human remains. Moquin, nearly hysterical with panic, pointed out they had no picks or shovels with which to dig graves and that with each passing minute they themselves were at great risk of equally horrible deaths.

"This is shit stupid," spat Moquin, "we're only a few miles from where we started out, wasted a whole day, wore out horses that weren't in good shape to begin with and are probably surrounded by Utes. You might wanna hang about here and end up like your brother but I'm headin' north to the Yampa like Rankin said, it's the only chance we have of staying alive." He mounted his bedraggled horse and reined it back up the road while Gordon only stared from red-rimmed eyes.

"Moquin," he called and the young corporal stopped. "Hang on; we'll have a better chance if we stay together." The two again followed the Agency Road back to the Morapos Trail where in the early dusk they built a small fire concealed between two boulders leaning on each other like huge tent walls. They stretched their blankets as windbreaks over the end openings and curled up under two sweaty saddle blankets next to the fire which they took turns feeding all night. Their exhausted and hobbled horses grazed on the brown fall grass which at least was collecting some moisture from a light skiff of snow that began falling shortly after sundown.

Dawn arrived clear and cold. Gordon and Moquin ravenously finished the scant rations remaining in their saddlebags and slaked their thirst by melting snowballs in tin cups. Later that day the men and their spent horses struggled into Iles's ranch close to the Yampa River and related their odyssey to the owner. Neither courier was capable of continuing the journey. Jimmy Dunn, the teenaged son of a nearby rancher, volunteered to locate Captain Francis Dodge and his Company D of the Ninth Cavalry known to be in the vicinity.

Dunn began his mission bravely enough but alone and barely sixteen years old. The horrific tales told by Gordon and Moquin took their toll on his imagination. On his way to Steamboat, Colorado, the boy encountered Ed Clark, a surveyor from Greeley on his way to survey land for the White River Agency.

"You might want to reconsider this job," said Dunn.

"Why?"

"An army from Fort Steele has been ambushed and surrounded at Milk Creek by thousands of Indians. I was at Iles's ranch when a couple fellas who escaped rode in lookin' for help. They think the agency was probably wiped out."

"God a' mighty, good thing I happened on you or I mighta walked myself right into a real fix. You need any help?"

"Uh . . . yeah, Captain Dodge and a company of the Ninth Cavalry are somewhere around here, maybe over toward Twenty Mile Park and Williams Fork Mountain. I'd appreciate it if you could head over that way . . . I gotta head for home and warn my folks."

"OK," Clark answered apprehensively, "I'll take a look in that direction."

"Thanks," replied Dunn and galloped off in the direction from which he'd ridden.

Clark located the dry bed of Wilderness Creek and followed the trail beside it upstream until it intersected with Twenty Mile Park Road that wound deeper into the dark heavily forested mountains. Within a few miles he too began feeling acutely alone and vulnerable so he wrote a short note hoping that if Dodge was in the area he'd find it. Clark tied the note securely and conspicuously to the top of a tall sage bush beside the road and then hightailed it back east hoping to get to the nearest ranch before dark.

The sun had set and a snow-speckled dusk settled quietly over the Northern Colorado Rockies when Sandy Mellen, a local guide hired by Captain Francis Dodge, reached from atop his horse and snapped off the top twig of a sage bush. He unrolled a piece of paper tied to it and handed it to Dodge who read it aloud: "Thornburgh killed. His men in peril. Rush to their assistance. EEC." Dodge motioned Mellen and a lieutenant over for a consultation.

"We can't ride through to Milk Creek tonight and be fit for a fight. Sandy, you know a place where we can feed the horses and get out of the weather until morning?"

"Yeah, Iles's ranch just a few miles from here." Mellen noticed Dodge's sidelong glance at the troopers in formation behind him. "Right now," he hastily added, "they'll be more'n happy to take in any color soldier you show up with."

Dodge nodded. "Off we go, then."

CHAPTER 15

The chrysalis of a waning moon blurred by snow pellets played hide-and-seek from cloud to cloud as the thirty-five troopers and four civilians of Company D, Ninth Cavalry, rolled up to the same ranch that had been a safe haven to John Gordon and Corporal Murphy many hours earlier. Tom Iles met them with a lantern at the front porch hitching post.

"Welcome Captain. I see Dunn found you."

Captain Dodge extended his hand. "Who?"

"Jimmy Dunn."

"Don't know Dunn. We found a message tied to a sage bush on Twenty Mile trail saying that Thornburgh's men desperately needed help. We couldn't make it that far south before nightfall, so Sandy said we could probably stay here for the night."

"Sure 'nuff," replied Iles as he held the lantern up to the line of troopers that had gathered at the hitching rack. "Uh . . ." he stopped. Tom Iles had been born and raised in the Colorado Mountains. He'd only seen photographs and drawings of Negroes in books and newspapers. Now the sight of more than thirty of them armed and dressed in cavalry blue twill shocked him into silence.

Dodge, recognizing an expression he'd become very familiar with, interjected: "It's all right Mr. Iles. These are the men of Company D, Ninth US Cavalry, an all-Negro regiment, and the finest Indian fighters on the Western frontier. They can ride further, faster and shoot straighter than any white cavalry force

I've ever seen. The Indians call them 'buffalo soldiers' because of their curly black hair."

Iles trying to recover from his shock held the lantern up in front of Dodge. "You're white," he declared.

"All officers are white," replied Dodge. "Only privates and noncommissioned officers are Negro."

"Oh . . . well." Iles motioned with his arm. "You're all welcome. There are a couple of corrals attached to the barn where you can put your horses and fork some hay down to 'em from the loft. There's water a plenty where a creek runs through the corrals. There are also some barrels of cracked corn in the barn you can put in their hay. Feed 'em up because Gordon says there's nothin' to eat down on Milk Creek."

"Gordon?"

"He's one of the couriers that sneaked out in the dark two nights ago. He's in the house. He came in with a corporal who rode out early today headed for Fort Steele. You'll want to talk to him more about the situation. Your men are welcome to sleep in the barn and sheds; we'll send some food out to 'em. You can stay in the house."

"I appreciate your hospitality. I'll get the men settled in and come in to talk to Mr. Gordon. But I'll sleep out here with my company."

Upon entering the house Dodge noted that Gordon held half a glass of whisky. The teamster struggled to rise from the maw of an overstuffed leather chair. With help from Iles and Sandy Mellen he eventually found his footing but then tripped over the edge of a rug sending his drink flying into a corner of the room. "Damn sorry about that," he apologized to the men surrounding him while Mrs. Iles hastened to the corner with a cloth. "It's been a hellish few days and to top it off we came across the pieces of my brother on the way here."

"Condolences," offered Captain Dodge. "I hope we avenge

his death and those of Major Thornburgh and his men. Can you brief me on the situation as you last saw it?"

"I can and will, in so far as I can remember it. A lot happened in a very short time, hell in a very small place, you know."

Iles interrupted. "Gentlemen have a seat at the table. My wife has prepared a splendid dinner and we can talk while we eat."

"Got hit about noon on the twenty-ninth," said Gordon pulling a chair back from the table. "I didn't actually see the first shots, heard 'em, though. My wagons were fifty to a hundred yards behind the cavalry wagons. The major had left a company of cavalry; don't know which one, behind to protect the wagon train. He took two companies and advanced toward Yellow Jacket Pass. I learned later he sent two skirmish lines forward after seeing some Utes on a ridgeline. I was kinda on pins and needles, guess we all were, when we heard a shot to the south toward the pass. I think I jumped. Then, within seconds they was so many shots, I couldn't begin to count 'em. I knew that we were in a heap o' trouble."

Gordon took a deep breath, then continued. "My first thought was to get our wagons hooked up with the ones on the flats ahead of us. But before we could move, we started takin' fire from a butte to our right and the edge of the stream gully to our left. Dropped a couple of my wagon mules right off. We grabbed our rifles, and about a dozen of us laid down a hell of a defensive fire. But even then the Indians from the creek bed were startin' to get in between us and the cavalry wagons. We reloaded and ran in sections to the forward wagons that were circling. I don't think we lost any men but we lost all our draft animals in the first few minutes of the fight."

"Did Major Thornburgh and his two forward companies make it back to the wagons?"

"Most of the companies did . . . the major didn't."

"How many Utes are there?"

"We guessed a couple hundred, some said as many as three or four hundred."

"By the time you left how many casualties were there?"

"Thirteen or fourteen dead, counting civilians, and probably twice that wounded. The Indians mostly had repeating rifles and were puttin' down a solid wall of bullets but aimed mostly at the horses and mules. Almost all the casualties happened in the first two or three hours of the fight."

"What about the White River Agency?"

"Don't know anything for a fact. Tom here tells me it's been wiped out."

"I don't know for a fact, either," Iles interjected. "Just before John and Corporal Murphy arrived a cowboy from down around Grand Junction on his way to Laramie stopped by and said he'd run into a band of Utes with some white women and kids in tow crossing the Grand River just east of the De Beque gorge. He said they weren't of a mind to converse and looked like they were headed toward Grand Mesa. He expected to spend the night at the Ute agency but when he got there it was burned and there was a heap of dead bodies so he didn't linger, cut across country and ended up here."

Dodge shook his head. "Sounds like an even bigger problem than I thought." He turned back to Gordon. "The Utes are well supplied."

"Everything from the agency that they didn't burn. They bought ten thousand rounds of ammunition at Peck's Store before the fight. Not much chance of 'em leaving because they're short of anything."

"What's the best approach, the best way in?"

"Well, you're sure as hell not gonna sneak in. They hold the highest ground around for miles and with lookouts aplenty who'll spot you five, maybe ten miles away. You'll need to get wagons with supplies in there, not that you'll be getting them out again anytime soon. Use the Agency Road, same as Thornburgh. Most likely be fightin' your way in. And, it's not like you're staging a jailbreak. Once in, you won't be fightin' your way out again. Nope, you'll be there a while."

"You're right. No, about all we can hope to accomplish is to bolster the encircled companies until substantial reinforcements arrive. In fact, in the interest of expediency I think we'll run the blockade with pack mules instead of wagons."

Dodge excused himself and left for the barn where the men were huddled around a fire in the ferrier's forge. Mrs. Iles had sent out platters of pork, beef and venison, potatoes, biscuits and coffee. The men were in good spirits. Their creativity and ability to make the best of bad situations always amazed him. A trio, guitar, fiddle and harmonica, accompanied several accomplished singers crooning church hymns and spiritual songs. One of the men handed him a cup of coffee as he approached and called, "Atten-*hut!*" Everyone jumped to their feet and braced.

"As you were, please continue." The men fell back into place. Captain Dodge sipped his coffee contentedly and basked in the warmth of soldierly camaraderie. He recalled General Lee's intentional contradiction: *"It's a good thing that war is so horrible, otherwise we would come to enjoy it too much."* Probably no force on earth, no magnetism known to humans, could bond men this tightly together as does battle and war, the giving and taking of lives.

He basked for nearly an hour before checking his watch and regretfully calling: "Lights out, reveille at zero three hundred." The men without complaint sought out bundles of hay and straw to curl on. Everyone knew that some of their number may be stiff corpses at this same time tomorrow night but for now death's shadow still seemed abstractly remote.

The duty-watch trooper awakened Dodge at three o'clock in the morning. Mrs. Iles started a cook fire and boiled three enormous pots of strong coffee which the men lined up for in the moonlight after they had saddled and harnessed their horses. She handed out meat, bread and potatoes left over from last night's dinner.

Within an hour the company was moving south. John Gordon, much to everyone's surprise, insisted on joining them. Dodge detached eight men and four wagons to inform Lieutenant Price of his plan.

The trek south began. The men rode huddled beneath their greatcoats in silence. The cold morning air reverberated with the rhythmic beat of the horses' hooves, wagon squeaks and groans, clanking of brass harness buckles and steel bits.

Dodge drifted in thought, apprehension and cold. Surely the Ute pickets would see them approaching from some distance. If the Indians numbered as Gordon thought, just a portion of them could organize another ambush and devastate the thirty-nine men of his company. Each man carried two hundred twenty-five carbine cartridges, plus hundreds more on the pack animals, enough for a skirmish but not for a prolonged battle, the wagons with ammunition and supplies sent to Price at some point had to come through as well. A cavalry charge into a mass of dismounted braves would certainly separate the pack mules from the troopers and spell disaster for the relief column.

An hour before dawn Dodge sent a three-man patrol ahead at a gallop under orders to reconnoiter south to where the Agency Road arrived at Milk Creek, or, if fired upon, to retreat back to the column. In no case were they to advance any further than the creek crossing.

Dodge passed the word back through the column for each trooper to check his carbine and revolver then waited breathlessly for a shot caused by nerves and frozen fingers. None came. He reined his horse back south and kneed her up to a faster walk.

As the gray light of morning began to silhouette the mountain peaks around them, Captain Dodge halted the column and galloped back along its length to make certain the pace had not separated the cavalry from the pack train. The horses and mules, fueled with Iles hay and grain during the night, were maintaining a tight formation.

As the sun reflected rose light from the new snow on the peaks to the west John Gordon reined in and pointed south. "Look, here they come."

Dodge's heart fluttered until he realized Gordon had seen the returning scouts. He raised his arm to halt the relief column. The three men approached at a steady, horse-saving lope. Good, thought Dodge, no sign of panic or urgency. The corporal in charge of the scouts pulled his lathered horse up in front of the captain and saluted. "We made it to the crossing, sir. No sign of Indians but we could hear sporadic fire a mile or so further south." The trooper hesitated, his breath smoking in the cold morning light, and looked at John Gordon. "We did come upon the burned wagons Mr. Gordon spoke of."

"Thanks, Corporal. Fall in line." He turned to an officer on his left. "Lieutenant, have the men dismount and stretch their legs for a while." Dodge stiffly descended from his mount and Gordon joined him. "I'd guess we're about five or six miles from the corral, about right?"

"Yes, sir."

"You think they've seen us yet?"

"Captain, you see that biggest anvil-shaped butte off there in the southwest?"

"Yes."

"I don't know what it's called, but the corral is located at the foot of it and about a mile this way from the end of the anvil. I can guarantee you that the Utes have one eye on the corral and one on us, right now."

"And waiting until we get into range."

"Probably."

Dodge pondered this for some time. He motioned for the lieutenant who had returned to join him and Gordon. "We'll continue in this formation, column of twos, until we receive fire or reach the crossing. If we receive fire before the crossing, we'll immediately split the company and bring them forward, one on each side of the pack mules. I'll take command of the right column and you to the left, Lieutenant. Gordon you'll come with me. The troops will be forming a running skirmish line. I want those soldiers to maintain intervals of no more than ten yards. It'll be up to the two officers leading to keep the pack animals in the middle of the skirmish lines until we get to the corral. At that point every man's first duty is to get the mules unhitched and as protected as possible; we'll have to depend on the men already there for screening fire."

CHAPTER 16

Jack, Colorow and Johnson sat cross-legged swathed in blankets around a campfire finishing a breakfast of roast venison, tinned beans, hardtack and coffee when hoofbeats announced the approach of a rider.

A boy no more than twelve years old galloped to the edge of their clearing, slid from his barebacked horse and ran toward them. Jack leaped to his feet and motioned the boy to stop. He walked to the boy while the other men continued to hunker around the fire. The boy, out of breath from excitement and his ride, jabbered animatedly to Jack who asked him a few questions. The others watched the exchange dispassionately. Jack walked back to the fire and packed his pipe then rustled around in the fire for the right coal to light it. After sucking several times he drew deeply on the smoke. A blossom of smoke and condensing breath surrounded his head. He motioned the boy over to sit with the chiefs and then gave him roasted meat, bread and coffee. The boy sat in respectful silence while Jack addressed the others. "The cavalry forces we've been expecting have arrived."

"How many?" asked Colorow.

"He says they were too far away to count but our lookouts thought only twenty or thirty, plus pack mules."

Johnson shook his head. "Too small, they wouldn't send in such a small number only to have them trapped with the rest, maybe a scouting party of a large force."

Jack nodded. "They brought pack mules with supplies and ammunition so the companies can hold out for more reinforcements."

Jack arose, threw off his blankets and picked up his rifle. "Only from the top can we see and decide what to do. He walked toward their hobbled horses grazing in a glade. The other men and the boy scrambled to catch up. Jack bridled his horse, unbuckled the hobbles and mounted bareback. The others followed suit except for fat Colorow who had to mount a saddled horse from the top of a stump. The others, including the boy, started up the trail without him.

The horses were blown and lathered when they reached the summit where at least twenty Ute warriors had collected to watch the distant slow line of troopers plodding like a thin worm on a chalk mark. No one spoke. Jack noted with relief the line numbered only about twenty men. The situation would remain a stalemate.

Colorow arrived as winded as his pony and immediately began dithering in Ute about mounting up and attacking the troop before it reached the river crossing and agency property. "If," he gibbered, "we attack the relief troop from horseback and the surrounded troops on foot at the same time we can destroy them both."

Jack and Johnson exchanged disparaging glances. Jack shook his head no.

Colorow looked pleadingly to the shaman, Johnson, whose face remained as impenetrable as coffee-colored granite. "We can kill them, make them surrender and walk back to Fort Steele."

Johnson shook his head. "At the cost of many Ute warriors. Then they will return with thousands. We've accomplished what we set out to do. We won two battles; the one with Meeker and halting the army from entering the agency. Now we must accept the loss of the war and whatever follows with as little bloodshed and as much dignity as possible."

As they watched, Dodge's troops began crossing Milk Creek in tight formation. His troopers unlimbered their carbines and held them at the ready with the butt resting on their hip. Jack mounted without a word and pointed his pony down a trail that descended the face of the mountain. Johnson and the boy followed leaving Colorow to sulk.

On the road, Sandy Mellen looked up in time to see the riders descending and called Captain Dodge's attention to them. Dodge watched but continued his walking pace which slowed as the pack animals struggled up the steep incline to the first flat bench. The column was now within rifle range of any Ute marksmen in the scrub oak and serviceberry brush that bordered the bench. An oppressive silence settled over the cavalry column. Bits and harnesses clinked, saddle leather squeaked, hooves rumbled. But the black troops, normally jocular and talkative, were ominously silent. Absent too was the meadowlarks' trill that usually accompanied their morning progress from the tops of trailside bushes. Fear and tension became more palpable with each hoofbeat.

The column began ascending another, shallower rise toward the south end of the elliptical bench. To their right began the northern flank of a ridge paralleling, but much lower than the mountain they had been observing. To their left, about two hundred yards from the column, the dry grass and sage disappeared abruptly into the Milk Creek ravine. In spite of the morning chill Dodge felt a rivulet of sweat escape from the leather band of his hat to his neck. He, Gordon and the lieutenant surmounted the low ridge about the same time and reactively halted simultaneously at the scene before them.

Only Gordon, who had ridden from this site two days and half a night earlier, had seen it and even he was shocked. Less than three hundred yards away on the road in front of them lay the burned remains of Gordon's wagons surrounded by the bloated, flyblown bodies of their draft animals. Flanking either side of the charred remains were two separate lines of shallow trenches and breastworks manned by at least two dozen Ute warrior sharpshooters; some watching the besieged corral of cavalry wagons, some apparently sleeping in the morning sun, some smoking pipes or cooking food. Another fifty yards beyond these defensive positions lay the far northern angle of the elliptical shaped wagon corral and its barricades. Smoke from dozens of Ute cooking fires hung over the scene in a low, blue vaporous deck.

The Indians, upon seeing the cavalry troops, rose to their feet with their rifles held across their bodies. Dodge heard behind him the loud crackling of carbine hammers thumbed back. Holding his own carbine aloft he shouted: "Hold fire," which was repeated by the lieutenant and first sergeant. At this point the two lines of entrenched Indians broke and began retreating toward the line of trees and shrubs on the low bluff to the west where Dodge saw more breastworks. He eased the column forward at a slow walk. With each step more Ute warriors stood up, rifles at the ready but not aimed, and backed toward the bluff.

"It's the troops," commented Gordon in a near whisper. "They've never seen buffalo soldiers before."

"Good for morale," replied Dodge quietly. "Let's hope it holds for a while. We're about as exposed as is possible here."

Someone from inside the wagon corral called out and a rolling cheer began that spread through the defensive ellipse from north to south. Men spontaneously abandoned their firing positions and collected to greet the newcomers. The wounded who were able to do so stood and waved or saluted. Their officers, Payne, Lawson, Cherry and others, gathered around the makeshift flagpole crowning Payne's entrenched command post, watched with cautious restraint.

"It's Dodge and the Ninth," said Lawson matter-of-factly, "at least one of our couriers got through."

"No shooting," stated Cherry.

"Everyone's too surprised," Payne answered. "No one expected help, Negro help, to ride in this quickly."

After skirting the burned-out wagons and rotting corpses, Dodge took advantage of the undeclared truce to split his column to the left and right forming a cordon through which passed the pack mules to a point where the defenders were opening a breach in their barricades. Within minutes men, horses and pack mules were inside the corral and the breastwork restored.

Dodge ordered his men to tie the mules inside the wagon corral for protection and unpacking. "If that meets with your approval, sir?" he asked of Payne with a salute.

"It does indeed," Payne replied, returning the salute.

"The men of the Ninth are the whitest black men I've ever seen," Captain Lawson interjected. A volley of shots punctuated his comment and lead whirred through the group.

Payne, Lawson and Cherry stood in placid immobility while the newcomers flinched and ducked. "You'll have to get used to the neighbors," said Payne, "they're a prickly bunch at times. Distribute your men evenly along the western perimeter wherever they can fit. It's from that direction we receive the most fire."

"Captain Payne," Dodge spoke up, "since my men and mounts are fresh I would like to mount a cavalry attack to dislodge the Utes."

Payne glanced at Lawson who shook his head. "With all due respect, Captain Dodge," he replied, "it's not a good idea. We're surrounded. In what direction would you attack? The bluff? It's a steep hill thick with brush that would stop horses dead in their tracks. Toward the creek? You can't get any further than the precipice of the ravine and that would put the troopers with their backs to volleys from the hillside. A charge either north or south invites complete envelopment. Your men are the only troopers here with horses. You couldn't sustain an attack without more support."

Dodge thought about this and then nodded in agreement.

"What's on the pack mules?" asked Payne.

"Some food, blankets, medicine."

"We were hoping for wagons."

"I sent the wagons to Price's company on Fortification. I wasn't sure what we were going to be up against here and I didn't want the Utes taking them."

Payne nodded. "I ordered the wagons you just passed burned for the same reason. Well, we can put all to good use. Captain Lawson will assign a squad to unload and organize when the shooting lets up a bit."

By noon all but seven of Dodge's cavalry horses were dead.

CHAPTER 17

"A cadet will not lie, cheat, steal, or tolerate those who do."

West Point Honor Code

"Mrs. Thornburgh?" asked the man removing his hat. The woman who answered the door looked nothing like the person he had pictured in his mind.

"Yes," replied the woman. She was short and petite with long dark hair and brown eyes. Though small she exuded the essence of a wiry frontier woman; straight and square shouldered. Her sinewy brown hands were devoid of jewelry except for a simple wedding band. Around her tawny neck she wore a small gold cross and chain. Similarly, her face, devoid of any powder, rouge or lipstick was the quintessence of the term "simply beautiful" even with a nearly masculine mouth and lips that appeared to never have experienced a smile.

"My name's Kenneth Church, I'm a reporter for the *Rawlins Daily Times* newspaper. We've just learned that your husband, Major Thornburgh, has been wounded as the result of an encounter with the Indians in Colorado. I'd just like to confirm that report."

"If you're referring, Mr. Church, more accurately, to a recent report that Major Thornburgh has been killed in action, then I can confirm I received such a telegram from General Sherman. It was further confirmed by messages of condolence from the major's more immediate superiors, Generals Sheridan and Crook." She stopped.

Kenneth Church felt dumbfounded, speechless. He detected no trace of emotion in her face, voice or movement. He had arrived at Fort Steele to interview the late commanding officer's widow expecting to meet with a frail, weeping widow draped and veiled in mourning black. Lida wore a dark-red dress caught up at the neck with a silver broach. She looked every bit an officer's wife, crisp and constrained. Collecting his thoughts, he bowed politely and said: "Please allow me this opportunity to add my sincere condolences and those of my employer to those of our government. May I have a few moments of your time . . ."

"For an interview?" she asked in a blunt almost accusatorial tone. "To talk about my husband whose body even two days later has not been recovered?"

"I'm sorry, Mrs. Thornburgh. I intended no disrespect either to you or the major but our readers . . . and the nation, for that matter, want to mourn the passing of its gallant uniformed men who have died in its service. I empathize with your sentiment because as a federal officer during the War of the Rebellion and later under General Crook's command at Fort Lincoln, I too suffered grievously at the deaths of relatives and friends. Your family's loss, as great as it is, is not borne by you alone; over six hundred thousand men did not return to their homes and families in eighteen sixty-five. They sympathize with you and want to share your similar burden, not out of a lurid fascination with violent death but because doing so relieves them, and perhaps you."

Lida considered this. Her face softened slightly. The hint of a smile crossed her lips. "Well spoken, Mr. Church. I rather subscribe to the all-too-human lurid fascination with violence and death theory myself. Either way, more newspapers are sold.

I'm as certain, though, that, as night follows day, I won't be left in peace until my comments are added to Major Thornburgh's eulogy so may the process begin with you. Please, come in." Mrs. Thornburgh stepped aside and swept her left arm in the direction of a living room or parlor where a girl busily dusted the furniture.

"Thank you, I assure you I won't take up much of your time."

"I doubt that, Mr. Church, it's not in the nature of your business. But, on the other hand, time is something which I now possess in abundance." She directed him toward a large overstuffed leather armchair. "Coffee or tea?"

"Coffee, please."

She turned to the girl and in a surprisingly soft and serene voice she asked: "Emily, coffee for two please." She seated herself on a matching leather sofa beneath a large oil painting contained in an elaborate gilt frame of a very tall man in formal military attire standing beside a seated woman, obviously herself—but younger. The man depicted was tall, slender, handsome, clean shaven, had carefully tonsured hair and a casual smile. Mrs. Thornburgh's hair capped by a garland of flowers fell in long curls about her shoulders. Her snow-white dress contrasted distinctly with the dark hair. Elbow-length gloves grasped a bouquet of red roses. No trace of a smile marred her serenely beautiful face and even features. "You and Major Thornburgh?" he asked, indicating the painting.

Without looking up she replied, "Yes, our wedding portrait, it was painted from a photograph. You've never seen the major?"

"I've seen only a few poor-quality photos of him in which he sported a beard."

She picked a smaller frame from its place on an end table and handed it to him. "This was Tip's West Point graduation photograph. I think you can see why I was immediately attracted to him."

"He is . . . was, extraordinarily handsome. He should have remained clean shaven." The man in the photo seemed to be looking directly back at Church. Again, the stylishly cut and combed hair outlined a strong but pleasant face with a jutting cleft chin and the firm lips of what Church saw as a born commander. "He looks every bit the scion of military lineage."

"Indeed he was. His father attended the Academy but ultimately chose family and law. He served in the Tennessee legislature as a Lincoln Republican but not an abolitionist, and died in the Andersonville prison for his efforts."

Church withdrew a notebook from his suit pocket. "I hope you don't mind if I take notes."

"I expect you to . . . and here is the coffee." The maid, Emily, set the glittering silver service on the table in front of the couch and poured two cups. "Cream or sugar, Mr. Church?"

"A bit of both, thank you, Mrs. Thornburgh."

Emily handed him a cup and saucer. He recognized the R. S. Prussia porcelain design, so thin and light it was nearly transparent. The coffee service, the room decor; Church felt as though he'd been miraculously transported to a parlor in New York or Washington. "Are there children?" he asked.

"Three . . . well, two, Bobby who is now seven and Olivia, three. George Washington, our youngest died last March of an unidentified fever. I suspect cholera or typhoid."

"My condolences, I'm sorry to bring up what must be a most painful memory."

"It's quite all right, Mr. Church. Like Tip, I also come from a military family and became aware at a very early age that death may visit suddenly, unexpectedly and frequently. I learned the pomp and ceremony, the sartorial splendor of the uniforms are a very thin, artificial veneer over the hard core of military business, death and destruction.

"I loved my husband very much and I'm certain he felt the

same about me but I also knew that, first and foremost, he was married to the army. General Lee once said: 'A good officer loves the army more than anything else in life. But a good officer must be willing to order the death of that which he loves the most.' I learned, Mr. Church, that when a man dons a uniform he undertakes a very perilous task probably ending only with severe wounds, or death. Did you know that the major was a Southerner by lineage?"

"No. My familiarity with him began only when he was assigned to command Fort Steele. I am bereft of any knowledge of him before that point. A Southerner you say, from what state?"

"Virginia originally. The Thornburghs were members in good standing of the landed aristocracy, owning both plantations and slaves. Tip's grandfather had fought alongside Washington and Light Horse Harry Lee, General Lee's father, during the Revolution. He was the first Thornburgh to eschew slavery, an act for which his family rejected him. He sold his Virginia plantations and moved the family to Eastern Tennessee, near New Market.

"Tip, at age seventeen, against his parents' insistence, attended West Point, joined the Union's Sixth Tennessee Volunteers even as the first shots of the war were being fired at Fort Sumter, South Carolina, in the spring of eighteen sixty-one. He distinguished himself in several battles and rose quickly to the rank of first sergeant before eighteen sixty-three when he applied to and matriculated into West Point. He graduated in the upper third of his class in the spring of eighteen sixty-seven." Lida paused as though she had completed reciting a chapter and was uncertain how to proceed.

Church took advantage of the pause to catch up his notes. It was only then that he became aware of the dead air between them and feared some word or thought might have tipped the scale of some delicately balanced, suppressed emotions that might now come flooding out as tears. He stopped writing, his breath labored, shallow and infrequent. He sipped his cold coffee.

"Had it not been," she continued, "for the major's commitment to his oath to defend the Union and his personal abhorrence of slavery, he would have made a fine Southern cavalier. His father had taught him to respect women, shoot accurately and recognize well-bred horses and Irish setters.

"Tip proposed to me in early eighteen seventy and gave me an engagement ring purchased, he said, with a twenty-dollar gold coin that his mother, Olivia, had given him for a special occasion when he left the farm to join the army in April of sixty-one. We were married in Omaha on the day after Christmas in eighteen seventy. Our wedding was the social event of the season in Omaha. Generals Sherman, Sheridan and Crook attended. The major—he was only a captain at the time—was the lowest-ranking officer attending the reception.

"He transferred to Omaha, was promoted twice and became one of Paymaster General Alford's staff. At that rate he could have looked forward to a sedate, sedentary, safe and secure career all the way up to at least major general—provided he behaved himself according to army politics, nor die without permission." Lida paused and proceeded sardonically, "Well, he did behave himself. Besides being a good officer he had an abundance of political correctness.

"We honeymooned in San Francisco and then moved into a Knob Hill apartment. Tip finally gained command of artillery at the Presidio. That lasted for two glorious years until my father notified us that there was an opening in the Payroll General's Office under his command, but in Texas, not Omaha. It was an opportunity that a career-minded twenty-seven-year-old lieutenant couldn't pass up. My father was willing to promote him to major over one hundred fifty other lieutenants and more than twice that number of captains. I don't know, Mr. Church, if you're familiar with the vexing problem faced by career officers in a very top-heavy army called the Logjam."

"I'm not familiar with the term in that context."

"To begin with . . . military officers, not to mention the enlistees, are ill paid on the theory that the army furnishes most of their daily needs; food, lodging, some clothing, medical care, items for which civilians have to pay. Secondly, officers cannot reliably avail themselves of any pension to speak of until reaching at least thirty years of service and the rank of major general. Most senior officers never attain those ends and simply remain in the ranks until death or dotage. Junior officers may wait as long as ten years between promotions. So, to be jumped two ranks at once was dazzling and we were off to Texas without another thought.

"Life in San Antonio was pleasant enough, though hot, dry and dreary compared to the verdant Bay Area. The major, like a banker, awoke in the morning in a spacious, comfortable house, donned his uniform and went to work. He, or rather his staff, arranged for the transportation of payroll money and kept records of the transactions.

"I had no onerous tasks during the days. My job was to manage a staff of several Mexican servants who performed all the cooking, cleaning and laundry, tend the children and organize our social life. Officers' wives competed viciously to curry favor with the post commander and his family. Rising above the rank of major often depended heavily on such trivia.

"That could have gone on indefinitely, he was almost certain to be promoted to colonel and into the paymaster position on my father's retirement. Father was one of those shrewd and fortunate junior officers who had amassed a small fortune through commercial contacts made while he was in the army.

General Grant should have been so lucky. But," she paused thoughtfully, "there was that damn Carter shooting match." Church interrupted his scribbling and nearly dropped his pencil. He'd never heard a lady of status swear.

Lida smiled reassuringly, as if to say: *It's my house and my story so I'll swear if I feel like it.* "General Crook," she continued, "happened to be one of the spectators at the contest between Tip and a certain Dr. Carver. I think he assumed the title *Doctor.* He publicized himself as the West's best shot and a connoisseur of human liver . . . actually only Sioux liver. Anyway, Tip lost the match by one miss after hours of shooting glass balls out of the air. But General Crook was impressed at Tip's demonstrated marksmanship and when he learned the marksman was in his command Tip had a standing invitation to join the general on his frequent hunting trips, card games and social occasions. His career was made but his fate sealed."

"How so?"

"Tell me, Mr. Church, have you ever given any thought to how a chain of seemingly random events, ranging from miniscule to momentous, that mysteriously connect and contrive, quite beyond anyone's comprehension, to ultimately construct a terminal event?"

"Yes, often it takes months or years of pondering to recognize the interconnectedness but even then never quite understanding the pinprick in time and place, an inch here or a few seconds there perhaps involving an instantaneous decision that brought about a life-changing occasion."

"Precisely," Lida agreed. "If Tip's father had not taught him how to shoot, or at least not how to shoot so well, Crook wouldn't have taken notice of him and posted him to Fort Steele and we would not be sitting here talking about Tip's death at the hands of the Utes." She sighed and her shoulders slumped, the excitement of her epiphany seemed to dim and die. "His career took another jump in seventy-eight when Henry Thomas, the post commander at Fort Steele, wanted a position in the payroll section of the quartermaster corps. General Crook suggested Tip and Thomas trade positions, which would mean a promotion for Tip. Naturally he took it.

"Fort Steele was over ten years old when we arrived, dilapidated and undermanned, morale terrible. One could smell the alcohol a mile away downwind. Fortunately, it was at the intersection of the railroad and Platte River. Tip had building materials shipped in, restored and built new quarters, constructed a pump house on the river and engineered an irrigation system. He convinced General Crook to transfer out entire companies and drilled the new men to exhaustion every day. Within less than a year he had turned the fort around.

"For his first field assignment General Crook ordered him to lead a detachment north on the Platte and intercept a large band of Cheyenne that appeared to be heading toward a crossing. Crook feared they would reunite with the Sioux. The major requested mules for a quick-moving pack train. Instead he received wagons. The Indians crossed the Platte north a day or so ahead of his arrival at the crossing because of the wagons. When he tried to cross the river in pursuit, the wagons sank in the mud, capsized and floated downstream. Tip returned to Fort Steele despondent and wired General Crook. The general did not directly blame him for the mission's failure but his record was blemished nonetheless. Tip felt like he'd let the general down. I think that it was really the first time in his life that he experienced failure, objective, tangible failure. He brooded about it for months. Adding to that was the birth and death of our third child, Washington. Of course, neither event was in any way Tip's fault, but he always demanded much too much from himself and magnified so much as a stumble way out of proportion.

"Last spring he was informed by both Meeker and General Crook of the growing crisis at the White River Agency. I sensed him bracing himself for what he viewed as the mission for redemption, that another Platte River lapse could put an early end to his career. Whether General Crook viewed it in the same light I don't know."

On his way back to Rawlins, Church deliberated on the nature of his article. Before meeting Lida Thornburgh he had set his sights on a story containing a weak and grieving young widow unprepared to face a husbandless life. Instead he'd found a strong individual raised to face the harsh realities of being both a military and frontier spouse. Lida Thornburgh and Arvilla Meeker could hardly have been more different women by nature and upbringing—their husbands more different men. But each had, in her own way, confronted the challenges and dangers of the Western American frontier, and survived.

CHAPTER 18

"The meeting of two personalities is like the contact of two chemical substances: if there is any reaction, both are transformed."

Carl Jung

"You know, Captain," said Thornburgh, "I met Meeker once a couple of months ago."

"That so?" replied Joe Lawson, preoccupied with catching a moth whirring in tight circles around the coal-oil lantern dangling from the center pole of the major's tent. He, Thornburgh and Scott Payne had dinner together the evening of September 28, after which Payne, still suffering respiratory problems from the day's march in clouds of dust and pollen, had excused himself to retire early. "How'd you describe him?" Swish! He took another unsuccessful swipe at the moth.

"He's obstreperous, self-centered and ambitious. His attitude toward the Utes is biased, prejudiced and opinionated—but with oddly paternalistic undertones."

"In other words, an average American man." Swish! Down went the moth right into the major's half-full coffee cup. "Got 'im," declared Lawson peering at his commanding officer over the tops of his rimless spectacles. Without replying, Thornburgh reached across the table and switched his cup for Lawson's.

"How'd this meeting come about?" Joe asked.

"Last August, I traveled to Denver in August to investigate

the purchase of a ram pump to bring irrigation water up to Fort Steele from the Platte River. The fort is dependent upon wells to provide us with water for all purposes; drinking, cooking, bathing and irrigation, in that order of priority. The amount is insufficient to supply enough for the first three uses *and* for irrigating gardens and hay fields. I'm finally putting to work my civil engineering degree for which the nation's taxpayers paid. I am submitting a purchase request for equipment which will not be shipped until sometime before January of eighteen eighty, too late, except for some surveying and designing, to do anything for the remainder of this year. So, in that regard, my journey was successful. But the most interesting part of the journey had nothing to do with pumps, irrigation or the fort."

Joe Lawson sucked on his pipe and blew a cloud of smoke into the throng of insects gathering around the lantern. "Won't have to be too interesting to beat an irrigation story," Lawson said.

"General Crook requested that I schedule a meeting with General Pope, commander of the Los Pinos Agency to inform him in more detail of my irrigation water project. Upon my arrival in Denver, I met with the general who informed me that Nathan Meeker had traveled to Denver a few days earlier to pressure him into sending a cavalry contingent to the agency to subdue the increasingly restive Utes. Meeker intimated to the general that the situation was becoming so hopeless that, after only four months as agent, he was considering resigning." At this, Joe's eyebrows arched and he gathered himself into an upright position in his chair.

"The general said this remark caught him quite by surprise as Meeker had gone to great lengths to acquire the position of agent and had only been at the White River Agency four months. The general told him: 'I'm sure that won't be necessary. I'll order Captain Frank Dodge and D Company of the Ninth

Cavalry north through the Gore Mountain area to investigate the fires you say were set by the Indians and to drop by the agency office for a friendly visit and show the flag. Other troops from Fort Fred Steele and Fort Russell are available if necessary.'

"Pope said that Meeker had been neither placated nor impressed. The general went on to say that Meeker had told him that he had written me innumerable times describing the gravity and urgency of the agency's situation. Saying that I replied infrequently reflected a lackadaisical if not insubordinate attitude and that I had apparently a complete lack of understanding or appreciation of the precarious state of affairs.

"General Pope said that he reminded Meeker that even a post commanding officer is bound only to obey War Department orders, not those from Meeker or the Interior Department. If it was police action he wanted he should inform Secretary Schurz.

"Pope said he told Meeker that he was meeting with me later in the day and would inform me of the concerns but he would not, at this point, order me to the White River Agency. Meeker, he said, then accused him of minimalizing the situation and that he would be contacting our superiors." Thornburgh paused and smiled at Joe. "Pope said that at that point Meeker reiterated the hollow threat to resign his post. The general said he sarcastically urged him to do so.

"General Pope then asked me for my assessment of the situation. " 'General,' I replied, 'he grossly exaggerates both the Northern Ute's combativeness and my indifference. He's been the agent for less than four months, has had no prior contact with any Indians or similar prior experience. He took over the agency expecting the Utes to fall in step like a socialist farmer of his Union Colony. The Utes are horse nomads, have been for centuries, and he's demanded they turn their ponies into plow horses overnight. Of course they've resisted. I understand that I may have to mobilize my command quickly should trouble arise on White River and am preparing the men accordingly.' He asked me about the spate of fires in the area saying that Meeker reported the Utes were setting them. No, I've questioned hunters, trappers, ranchers, teamsters; all white and all saying that the fires have been caused by lightning and travelers. It's been an exceptionally dry summer.

"The general reminded me that Meeker had political allies who are only too quick to back his assertions. He was Horace Greeley's lapdog and is still chummy with Greeley holdovers both in Washington and Denver, many of whom take his ranting as gospel. The Custer Massacre is still a good political card to play, will be for years. Interior Secretary Schurz appears sympathetic, a like-minded socialist agrarian, a wild-eyed one at that, who's not mixing well with nomadic pastoralists.

"He continued, saying that the Bureau of Indian Affairs was rife with corruption and misappropriation. Schurz has set out to right many past wrongs but that he paints with broad strokes policies which are appropriate to one tribe but will fail miserably with another. The previous agent, a trifling political hack, stashed the Ute allotment and sold it on the black market. The Utes nearly starved last winter. Meeker, I conceded to the general, to his credit, has assiduously collected, transported and distributed agency food and supplies to them. He had told Meeker that he may send Frank Dodge and a company of cavalry north from Pagosa Springs to investigate the fires and have him pay a visit to the White River Agency not so much as a show of force but rather a unity of interests.

"Well, Joe, when I arrived at the counter in the lobby to sign for my bill, the only clerk was engaged in a lively argument with a tall, slender, man. I stood listening to the conversation. The guest appeared to be questioning virtually every item on his bill.

Frustrated and anxious, I left the two men to their argument and asked the bell captain to have a bellman collect my suitcase from my room and bring it outside to the cabstand. Turning back to the counter I noticed with relief the contentious customer disappearing into the revolving doors at the front of the hotel. I quickly trotted up to the red-faced clerk and asked for my bill.

"At the station, I discovered my train would be leaving an hour late but my first-class pass allowed immediate access to the club car. The ticket clerk appeared duly impressed with my embossed gold pass and went to considerable lengths relocating other passengers to provide me with what he said was the most comfortable compartment, in the most comfortable car on the east side of the train, one sheltered from the torturous late-afternoon sun.

"The club car, though hot, was considerably cooler than the Union Station lobby. The heavy, plush curtains were drawn over the windows leaving only enough light for the passengers to maneuver around the tables, chairs and booths. As my eyes became accustomed to the faint light produced by the sun leaking in around the velvet window curtains, I could see I was the only patron in the car. The bartender ceased arranging and polishing bottles and glasses to nod in my direction. I found a booth and unbuttoned my wool twill uniform blouse just as the bartender arrived with my beer. He relieved me of my blouse and hat and said that he would put them on the coatrack next to the bar to dry out. Even the suffocating air of the lounge felt refreshing as my sweat-soaked white cotton shirt began to dry. I settled into the padded leather and let the coolness cascade over me over me like a mountain stream. I allowed my thoughts to wander to our spacious house, cool evening breezes off the high prairie rustling the bedroom curtains, Lida lying beside me, the children safely tucked away in their beds and dogs barking at night sounds. Fort Steele, Joe, is not the San Francisco Presidio, but, I'm proud to say, I've turned it around since last spring and it's become a pleasant, quiet post.

"My reverie, however, was shattered by a vaguely familiar voice raised in anger, this time with the bartender who had asked him to produce his first-class ticket. The man took great umbrage at this request and threatened to personally inform the railroad's owner of this discrimination. The barman shrugged indifferently and said that the man could remain in the car but wouldn't be served. I peered over my shoulder. The antagonist stood at the bar with his back to me. Even in the dim light, I recognized the stooped, gaunt frame, thick gray hair, and high-pitched nasal twang of the man complaining to the clerk at the hotel desk.

"I watched the man carefully. Although attired in a conventional suit of current style the legs and sleeves were a good two inches too short making his lanky limbs appear even more awkward and gangly. His thin neck protruded from his shirt collar like a pole from a posthole. He lurched away from the bar and shuffled aimlessly about the car until he caught sight of my uniform blouse and hat. He halted and examined the garment. I wished then I'd hidden it beside me in the booth. The man's gaze shifted directly toward me. With sinking heart I watched him change course and head toward my booth. As the man approached, he thrust out a bony hand which I reluctantly grasped. 'Good afternoon Colonel,' he blustered, 'always a pleasure to meet an officer and gentleman, name's Meeker, Nathan Meeker.'

"Joe! I wouldn't have been more surprised if he'd introduced himself as Satan. I immediately released my grip and retracted my hand as though struck by a snake. It took me more than a few very uncomfortable moments to regain my composure. 'Thank you,' I said, 'for the promotion, but it's Major . . . Major

Thornburgh.'

" 'Thornburgh!' he cried. 'This is a remarkable coincidence. I spoke to General Pope of you and our correspondence this morning in the hotel lobby. May I speak with you for a few minutes?'

"Joe, the man conducted himself as though he'd never heard the terms 'manners' or 'propriety.' He did not wait for a reply and instead of seating himself across the booth from me he forced me aside and seated himself next to me. If I hadn't moved, Joe, I think he would have sat on my lap.

" 'General Pope,' Meeker boasted to me, 'sought me out in the hotel lobby this morning; he readily recognized me and wanted to discuss the distressing turbulence at the White River Agency. I told him it was none too soon to do so; things are, as you know, Thorndike, reaching a boiling point.' Meeker turned and held up his arm. 'Bartender, two beers please!'

"The bartender scowled and I seized the moment to reposition myself to the center of the booth more than an arm's length from the obstreperous Meeker.

" 'It's Thornburgh,' I told him.

" 'What?'

" 'My name is still Thornburgh, not Thorndike and I'm still a major.'

" 'The general,' Meeker continued, without giving any indication he understood my corrections, 'expressed concern when I told him of your delayed replies to some of my letters but I told him there were probably ample reasons for such. You've only recently taken over command of what I understand to be a rather dilapidated post with poor morale, your main concerns and efforts being the restoration and discipline not the dangerously rebellious Utes at the agency.'

"Frankly Joe, Meeker had stunned me speechless with this audacious statement and the only thing that prevented me from immediately taking him to task was Pope's forewarning earlier in the day that he was apt to bait me like this.

"I felt the muscles of my neck tighten against my damp collar and asked him: 'I take it, then, that your Indian constituents still wander from the agency, practically at will?'

" 'Indeed,' Meeker continued, 'as I have informed you and your superiors innumerable times, without military intervention to sanction them I can do very little to make them abide by the department's agency rules. As we both know, the Interior Department has no enforcement section and must rely on the War Department for such. If the agency is to successfully perform its function, accomplish its mission and serve the purpose for which it is intended, I need your support and help. If you are unable or unwilling, then please be kind enough to respond to my letters so that I may acquire it elsewhere, perhaps from Fort Russell.'

"Well, Joe, the thought of General Crook receiving such a message caused my hackles to rise. 'Mr. Meeker,' I told him bluntly, 'with all due respect, I did not intend to belittle your predicament, but nothing so far in your letters has indicated an immediate need for any military intervention. In fact, such an intrusion may be the spark to the torch hole that would cause an explosion. Practically all tribes are now on defensive alert and will battle for the smallest reason. And that reason must be the War Department's discretion. We, not you, are the experts in armed conflict, and it is much more likely that it will be army lives lost in such a conflict, not yours or Mr. Schurz.'

" 'Of course, of course,' replied Meeker, patting me conde-scendingly on the shoulder, 'I don't for a minute . . . even a second, mean to suggest that the army has performed badly. In fact, I would go so far as to suggest that the Bureau of Indian Affairs be absorbed within the War Department as a subordinate agency. I don't have to tell you, I'm sure, that the Bureau of

Indian Affairs over the past decade has gained a reputation for being a veritable cesspool of graft and corruption. But, in order to accomplish an interdepartmental merger effectively, the War Department is going to have to be more conscientious and assertive. No disrespect intended, West Point is a fine school, it turns out good engineers that shine in the eyes of beautiful women, recognize fine wine and cigars and comport themselves gracefully at dinner parties but are sadly lacking in true military proficiency . . . the debacle at the Little Bighorn three years ago is a case on point.'

"I noted that Meeker paused briefly to let that 'point' hang in the hot, dark air whether for effect or sensing that he might have trespassed into forbidden territory, I didn't know.

" 'Not the officers' fault,' Meeker relented, 'given what they have to work with . . . sodden, semiliterate rabble intent upon debauching Indian women at every opportunity—remember the anatomical collections displayed in Denver after Sand Creek?'

" 'General Pope's proposal,' he continued, 'to order a company of Negro cavalry three hundred miles north to the White River Agency is not my idea of a proper show of force. Fort Steele is half the distance away and Fort Russell only a little further by rail. Empty those forts, I say, and send us a force worthy of the name.'

"Lawson, I could barely contain my rage at Meeker's breathless stream of bombastic rhetoric. But control it I did. I sat quietly with my hands clenched together beneath the table.

" 'Mr. Meeker,' I replied calmly, 'I respond only to my commanding officer's orders. Everyone, from General Sherman down the chain of command knows of your problems. I have received no orders from anyone to take any action concerning the White River Agency. It is *your* job to manage and control events within the jurisdiction assigned you by the Bureau of Indian Affairs and Department of the Interior. Invite Secretary

Schurz to the agency and the two of you can lecture the Utes on social theory and the boundless opportunities created by agricultural cooperatives. Of course, I'm only an engineer with little or no sense of social justice.'

" '*Pity the Poor Savage* is the title of one of my well received articles,' he replied, 'social justice is truly my guiding light. But I also recognize the reality that the indigenous race we refer to collectively as the Indians—along with Negroes and Mexicans— are innately incapable of the depth of intelligence or thought necessary to plan and organize their tribe's workforce to be self-sustaining through agriculture and industry, a necessity for the future. White Americans will have to dictate the economic course and heading of this country and the Indians must learn and practice our ways. It's been most pleasant talking to you. Perhaps we could meet in the dining car later this evening to discuss our mutual interests in more detail.'

"I clasped the proffered hand but made no effort to rise and squeezed it until his knuckles crackled. 'Thank you for your company, Mr. Meeker, I'm certain we will be communicating again soon—but not tonight.'

"Meeker seemed poised on the brink of saying more but instead simply nodded and walked out.

"On the sixteenth of September I received the order from General Crook to assemble an expeditionary force and proceed immediately to the White River Ute Agency to assist and support Agent Nathan Meeker and arrest the Ute leaders deemed insubordinate and responsible for the uprising."

Joe sucked on his cold pipe. He took out his pocketknife and reamed the ashes into a tin plate. "A thoroughly disagreeable person and probably never experienced a real fight. I'd bet that if he didn't have the army around to pull his chestnuts out of the fire he'd be a whole lot more motivated to arrive at a diplomatic solution to his problems with the Utes. As it is, he's got everyone in a hellava fix." Lawson stood and stretched. "I'm callin' it quits for the night. Since you outrank me, it's your job to stay awake all night worryin'."

CHAPTER 19

"I intend to make Georgia howl."

General William Tecumseh Sherman

In Washington, DC, General of the Army William Tecumseh Sherman had not received news of the White River expedition's desperate struggle, nor its commander's death, nearly as philosophically as had Lida Thornburgh.

The general gazed at the dome of the Capitol through his reflection in the full-length window of his office on the top floor of the War Department Office Building. The nation's highest-ranking military officer was mad. The image reflected was of an aged, careworn face surrounded by closecropped white hair and beard. The face, once furrowed by a perpetual scowl, belied his anger. He was mad that Thornburgh, an officer with virtually no successful battle command experience, had delivered three companies of cavalry into an ambush; that Thornburgh had been killed. He was mad at Interior Secretary Carl Schurz and the Bureau of Indian Affairs, mad at the president and, most of all, mad at Indians in general and Utes in particular for not having learned their lesson from the consequences of the Little Bighorn battle.

After the 1866 Fetterman Massacre, he had written President Andrew Johnson that, "We must act with vindictive earnestness . . . even to their extermination, men, women and children. During an assault, the soldiers cannot pause to distinguish between male and female, or even discriminate as to age." Sherman observed no difference between annihilation demanded by total war whether against Southern rebels or warring Indians.

Yet he also spoke out against the unfair way speculators and government agents treated the Indians within the reservations. If, as he so steadfastly admonished and advised, President Hays, Congress and Schurz had put BIA administration under the War Department where it belonged troops would have been on hand at the White River Agency to immediately quell any incipient uprising. Instead, Schurz had piecemeal allowed this incompetent agent Meeker to paint himself into an inescapable corner, throw up his hands in helpless despair and again dump the whole reeking mess into the War Department's lap. Sherman had been forced to order the creation of a hodgepodge rescue force tacked together, too little and too late to be effective and marched almost two hundred miles into a catastrophe. Headlines across the nation would read, *Another Custer Massacre*.

Sherman pensively smoothed the telegram with the palms of his hands and looked across his massive mahogany desk at his aide, Major Clarence Pickens. "Major," he growled, "my order in response to General Sheridan's message is brief: 'Take any and all actions necessary within your power to relieve and reinforce the White River expedition as soon as humanly possible . . . copy to General Crook." The major snapped to attention, took the telegram from Sherman and saluted.

"Also, prepare the standard War Department letter of condolence for Mrs. Thornburgh for my personal signature. I knew her father and we may even be distantly related. That telegram should go to the acting commander, Captain Bisbee, I believe, at Fort Steele for delivery to her."

"Yes, sir, anything else."

"No, not for now." Sherman returned the salute and Pickens left.

CHAPTER 20

"In battle it is the cowards who run the most risk; bravery
is a rampart of defense."

Sallust

Joe Lawson watched the sun sink behind the western mountains
at day's end. It too seemed just to collapse with exhaustion. The
subsistence rations and constant exchange of bullets were taking a heavy toll on the men physically and mentally. But worse
still, the overpowering stench of death, screams, cries, whimpers
and moans of the wounded and dying men and animals drained
the defenders. The last stifling breaths of summer had expired
the evening of the twenty-ninth, replaced by early winter's cold
embrace. Rain, sleet, snow and wind had driven the bright fall
leaves and carpeted the hillsides a dusty brown. The courtyard
of the besieged wagon castle resembled a ragged tent city.
Canvas sheets carved into makeshift shelters covered the
trenches behind the barricades.

Lawson could feel morale ebbing. Shaving and bathing were
but distant memories and depressed hopes. Cavalry blue twill
was camouflaged by sun bleaching and layers of dried mud.

Early attempts to use Milk Creek as a continually flowing
sewer ended quickly when the Utes posted sharpshooters on
the far bank. Latrine trenches, "honey pits," were dug, used and
filled. They became so numerous that troopers moved cautiously
from place to place and risked a bullet rather than run incau-

tiously and hazard sinking into a latrine.

Water became more precious. The Milk Creek trickle had been poisoned upstream by Indian excrement and the bodies of dead animals. Virtually no man under siege escaped the ravages of dysentery and diarrhea. Joe even chuckled when, during a lull in the firing, two dozen men jumped up with shovels and barrel staves and resembling a sandbox of cats frantically took to digging, squatting and covering. But Lawson's mirth was quickly squelched at the recollection of the cholera ravages in the Crimea twenty-five years before. A man might exhibit symptoms of the disease in the morning and die before dark. More men and horses succumbed each day outside Sevastopol and Balaclava from disease and cold than from Russian weapons. May God forbid, he thought, I'd have to witness the same hell twice in one life.

The arrival of D Company sent the same signal to the Utes as it did to their victims. More help was on the way. It was only a matter of time. Yes, the Utes had the cavalry surrounded and outnumbered but the Indians were surrounded, with no hope of relief or escape, by an exponentially greater number of whites. Two outrageously contradictory thoughts collided in Lawson's mind. Surrender to the Utes who had nowhere to go themselves and nowhere to take prisoners. They could all sit it out here or at the agency until the entire might of the US Army descended upon them. Or, the besieged could brazenly send an ultimatum of surrender to the Utes pointing out the Indians' dilemma; a complete lack of bargaining resources and no ultimate means of escape. A furious but short-lived exchange of rifle fire snatched him from his reverie, plus his gut was cramping. First things first. He frantically hunted for the barrel stave he'd been using as a shovel.

Doctor Grimes fretted over the increasing numbers of noncombat casualties. He reported to Captain Payne that nearly every trooper to some degree was sick, injured or wounded—sometimes all three. The men under siege at Milk Creek, he thought, suffered no less or more than surrounded soldiers throughout history. The constant threat of attack, starvation, numbing cold and suffocating heat, thirst, inadequate shelter and clothes, diarrhea, stench from tons of putrefying flesh, cries from the wounded and curses from the Utes kept men awake and edgy day and night. Despair and exhaustion were becoming serious problems.

The arrival of the Ninth raised morale for a while but also imposed crowding and sanitary problems, more targets for the Indian sharpshooters and worse stench. The Utes were pragmatic marksmen shooting to wound rather than kill. A man wounded in the leg would require the assistance of two or three others to remove him from the battlefield, clean and dress his wounds, feed, shelter and comfort him and quiet his cries. He remained a constant drain on scarce resources without contributing to the fight. A dead man required nothing.

A wounded horse or mule could cause as much havoc and damage as a volley of well-aimed rifle fire. The panicked animal would usually dash madly about the enclosure striking out at everything and every man in its way until shot dead. Cavalry troopers and Ute warriors shared a love of horses. Their tortured screams damaged morale as much as a dead trooper. Grimes had witnessed more than one wounded horse tearing insanely through the corral until it broke through the thinly protected wood, canvas and dirt covered trenches housing the wounded onto the men whose screams and curses added to the pandemonium. Grimes remembered one particularly poignant incident when a trooper's wounded horse limped to him in his trench appearing to beg for help. The soldier raised his rifle to fire but broke into tears and begged his comrade, "You do it, Charley, I can't." Neither could his friend. First Sergeant Grimes overhearing the exchange walked over, put the muzzle of his revolver behind the animal's ear and fired. For a moment it appeared to the doctor that Sergeant Grimes was going to berate the boys but, instead, holstered his sidearm and walked off.

Doctor Grimes being the expedition's only doctor was on duty twenty-four hours a day. Some of the less seriously ill and wounded assisted the doctor as they were able. The most senior officers assigned others. The farm and ranch boys were the best assistants. Many of them had come from remote homesteads where even a self-trained doctor might be fifty miles away. These boys could staunch bleeding, amputate extremities, cauterize wounds, splint broken limbs, suture cuts and holes and even, although not here, deliver babies. They were also hunters and butchers, these frontier boys, so the most profusely bleeding wounds didn't bother them a tic, even their own. But both the doctor and his assistants were stymied by one horrific result of battle, the one for which none of them had an answer.

Blood poisoning, what the doctor called sepsis or death by inches. More horribly, it was often the wounded man whose nose caught the first sickeningly sweet whiffs of his own irreversibly rotting flesh. As many as a dozen men occupied some of the hospital trenches within the besieged corral. The smell of one man's gangrenous limb could quickly permeate the whole dugout sending nearly everyone into a raving panic. A doctor's first line of defense against this pestilential degradation was carbolic acid which a well-equipped modern hospital might have in abundance. What little Dr. Grimes had been able to bring on short notice was nearly gone after the battle's first twenty-four hours. Its main substitute, alcohol, had disappeared almost as quickly.

The expedition's supply of laudanum, a mixture of opium and alcohol and the most effective painkiller available, was consumed like water until, by the third day, Grimes restricted its use for only the most desperate cases of amputation

The doctor noted in his diary that the arrival of the Ninth Cavalry detachment made a noticeable difference in the spirits of the trapped companies.

The Negro soldiers were, at first, a shock, especially to troopers from the former Confederacy. But battle, he thought, is a potent equalizer. It matters not whether the man next to a soldier firing at the same enemy is black or white—he is a comrade in arms and a potential lifesaver.

The men of D Company were new, fresh and keen to fight. Earlier in the morning a Negro sergeant had climbed atop a wagon tree and shouted, "Give me a Ute to shoot!" His request went unanswered so he sat down on the tongue and began rolling a cigarette. A single shot reverberated across the bluff and a bullet ricocheted off a steel turnbuckle just above his head showering him with lead fragments. The man appeared to fall over backward but instead executed a perfect backward somersault over the breastwork and landed crouched but on his feet back among his black and white brothers in arms amid much applause, laughter and cheering. It was a typical example, he thought, of well-trained and well-led men rising to meet a fear-filled, arduous event, adapt and triumph.

The human mind, Doctor Grimes thought, is the most incredible instrument ever devised or created. It possesses unfathomable abilities to create, even simultaneously, death, pain and destruction or life, health and happiness.

He looked down the hospital trench, now open because a wounded horse had fallen into it. Two men, each with an arm in a sling, were playing gin rummy and laughing uproariously at each failed attempt to shuffle the deck with their good arms. Further down a Negro trooper spooned something from a tin cup into the mouth of a white soldier with both eyes bandaged.

Others smoked, read, talked, dozed in the sun, ate, and wrote in diaries.

The wounded who were able stood night watch so the combatants could catch a few winks of sleep. "I may not be able to shoot," said one man with an eye patch and bandaged right hand, "but I sure can holler real loud." The arrival of D Company had also displaced almost everyone's fear that the Utes at some point would mount a full-scale, overwhelming attack on the defenders. English-speaking Utes continued to yell such threats, which during the first forty-eight hours of the battle had nearly panicked some, but now only drew laughter and replies of "bring it on!"

"How's the medical service maintaining, Doctor Grimes?"

Grimes, recognizing Payne's voice, came to attention and saluted. "Tolerable, sir," he said, noticing that Dodge, Lawson and Cherry accompanied the commanding officer. "Captain Dodge's arrival replenished some supplies of medicines and bandages, especially carbolic acid and chloroform, hoping that the former precludes use of the latter. But we will have to utilize all sparingly in hope of further relief. Along those lines . . . I need to take a look at your scratches." Without approval Grimes slipped the sling over Payne's head and raised his left arm. The captain grimaced and caught his breath but uttered no protest. Grimes withdrew a formidable-looking surgical knife from its case and cut the blood-encrusted bandages from around the wounds. Grabbing Payne by the shoulders he turned the wounds into the sun and kneeled down to examine them closely. Probing with his forefinger, he asked, "That hurt?"

Lawson watched Payne's eyes widen and his jaw muscles tense. "Nope," Payne hissed.

"That's bad . . . means you're dead." Grimes stood up. "The sutures are a bit red and inflamed. Probably a touch of infection. Gotta clean and redress 'em, so stand down for a few." He led Payne to a chair adjacent to the tailgate of a wagon he'd pressed into service for surgery. The other officers stood awkwardly while the doctor worked.

Though Frank Dodge outranked both Payne and Lawson, he'd wisely backed away from ordering a large-scale attack in the first few hours after his company's appearance on site. With that, he and Payne had come to an unspoken understanding that Thornburgh's second in command would happily transpose Dodge's orders regarding D Company into official orders and the new lad on the block would discreetly follow Payne's "recommendations" as though spoken by himself. In turn, Payne would catch Lawson's eye or ear evoking a word from the wise to the wise. All realized the predicament demanded an occasional detour up or down the chain of command while at the same time maintaining discipline and credibility within the ranks below.

Lieutenant Butler Price at the Fortification Creek supply camp wrote in his journal that as of October 2, his command had seen no Indians and in a buoyant burst of confidence stated that with his twenty-nine men he would be able to stand off three hundred.

Within days of the appearance of Kenneth Church's article on the White River expedition and Major Thornburgh's death, Captain Bisbee was besieged by homesteaders pleading for immediate personal protection from the meager number of troops remaining at Fort Fred Steele. Bisbee, in turn, telegraphed General Crook for permission to arm civilians with rifles and ammunition from the army store. Crook gave his permission and the same day wagonloads were on their way to Rawlins for distribution.

On that morning, soon after Dodge's pack train and D Company, Ninth Cavalry, arrived at the scene of the battle,

Colonel Wesley Merritt, sick in bed, was handed a telegram from his commanding officer, General Crook, in Omaha. Merritt unfolded the telegram and began reading as the soldier delivering it saluted and turned to leave. "Stay," ordered the colonel, "this requires an immediate answer." Merritt began scratching his answer on a sheet of notepaper.

Major Thornburgh has been ambushed and killed in Northern Colorado at a Yampa River tributary called Milk Creek. What's left of his force is now besieged there. I'm preparing a rescue force and will take the field immediately on receipt of orders.

"Take this answer to the telegrapher immediately," ordered Merritt, springing from bed, "and send messengers to the regimental adjutant and all the company commanders informing them of a meeting here in my office within the hour. Inform the armorer to prepare rifles and ammunition for shipment for at least five hundred men—the same for field rations for at least a week, upward of fifteen wagons. Contact the railroad in my name and request whatever they can get their hands on quickly. That's all for now, I'll have more for you in two hours."

Seeing the stunned look on his subordinate's face, he smiled reassuringly. "Don't try to do it all yourself. Rustle up some help and start giving orders. This'll be good experience preparing you for a command of your own. Lesson one: Don't waste time thinking. Lesson two: Keep a record of your orders and progress and report to me any problems you may have." He folded the sheet and handed it to the aide. "Bring me whatever maps we have of Northern Colorado in as many different scales as you can find—dismissed."

It was the White River expedition's great good luck that their rescue had been tasked to Colonel George Wesley Merritt, one of the best—if not *the* best—cavalry field commanders in the army. The colonel had been one of three men brevetted all the way from captain to major general during the Civil War, another being George Armstrong Custer, and, by regulation, was still often referred to as "General Merritt." He had been born in New York City, one of eleven children, to relatively affluent parents. His father, an attorney, left his legal practice and moved the family to a farm in upstate New York where he became a state representative and newspaper editor.

A precocious child, both mentally and physically, Wesley Merritt excelled in the classroom, boxing, wrestling and horsemanship. He graduated twenty-first in his West Point Class in 1860. His first duty station was under Colonel John Buford in Utah as a second lieutenant in the Second Dragoons Heavy Cavalry. Buford took a liking to the fit and meticulous new lieutenant and soon made him adjutant in the newly formed Second United States Cavalry.

Merritt distinguished himself during the Civil War as an aggressive yet sensible officer. The erstwhile Custer had led through blind impetuosity, if not insubordination. Merritt achieved the same ends through foresight, discipline, rigorous planning and meticulous organization. He gained a reputation as being one of the army's most stringent disciplinarians.

The garrison at Fort Russell contained Companies A, B, D, F, I and M of the Fifth Cavalry and one company of the Fourth Infantry. Companies D and F were already at Milk Creek. Merritt virtually emptied the fort of cavalry. He mustered five companies, for a total of 328 commissioned officers and enlisted men. He also included forty-two enlisted infantrymen and three commissioned officers from Thornburgh's Fourth Infantry. And that was just the start. More men and officers from other forts and outposts were to follow Merritt in the next few days. Nicagaat's fear-fueled prophecy that the Milk Creek battle would become the focal point of the US Army was, in fact, materializing. Merritt's subordinate commanders were in his office well before the appointed hour, none sat. The colonel unrolled a wall map, studied it for a moment and slashed a red "X" on a spot in Northern Colorado.

"Our objective is to relieve four companies of cavalry under siege by three to four hundred Ute Indians. Haste is imperative but not at the expense of good, sound military principles, organization and operation. We have approximately a hundred seventy statute miles to cover which means at least fifty miles per day. On the way every camp duty will be initiated by a bugler: 'Reveille,' 'Stables,' 'The General,' 'Boots and Saddles,' 'Mount,' and 'Forward.' Each call will sound at exactly the same time each day. We will be marching in columns of fours to minimize dust."

He paused. Then, extending his right hand above his head and pointing his index finger toward the ceiling: "*The* most critical task for the duration of the march will be the care and feeding of the animals. The column will be able to move only as fast as its weakest, slowest animal. With proper foresight, this command will be provided only with saddle stock between the ages of five and eight years old kept fit by first-class feed, water, medical care and exercise. In my experience as a commanding officer I have witnessed well-kept horses and mules endure fifty-mile-a-day marches for up to a week in duration, men less so. Discipline therefore will be severe. No man, officer or enlisted, will be allowed to break ranks mounted, no horse will be watered until all can be watered. Any officer or enlisted man who willfully neglects attention to the smallest details of the march, necessary to the perseveration of the endurance of the men and horses, will be arrested, walk and may face a court-martial. Each company commanding officer will, immediately following this briefing, ensure the procurement of the supplies necessary for his company appropriate to the nature of the mission, the season, and the expected duration of one week at the very most. We will secure a supply line for the expedition if it lasts longer than that. Organization, discipline and attention to detail will transport us to our destination in approximately three days at the most . . . and arrive fully prepared to fight and win a pitched battle. Dismissed!" The assembled officers vanished without question or comment.

Merritt accomplished the rescue preparations in four hours, tasks that had taken Thornburgh nearly a week. By eleven o'clock, chief packer Tom Moore had two special Union Pacific trains loaded with men, animals and supplies. Each train had four engines, two for pulling and two for pushing. At exactly noon the trains departed Cheyenne. They stopped in Laramie to relieve the horses and arrived in Rawlins at five thirty the next morning—almost exactly the same time that Captain Dodge's D Company was riding into the wagon corral at Milk Creek.

Three companies of the Fourth Infantry from Fort Sanders totaling an additional hundred fifty men joined the troops from Fort Russell in Rawlins. Merritt, though pleased with the additional forces, now had to confront the same logistic problems Thornburgh had faced when trying to intercept the errant Cheyenne two years before. Infantry required wagons which were sluggish and complicated river fording.

All companies assembled at the same spot as Thornburgh's troops on August 21. Tom Moore acquired fifty pack mules and loaded them lightly to be led by cavalrymen in case the troopers were separated from the wagons.

During their preparations, Colonel Merritt was introduced to a very haggard Joe Rankin. Realizing Rankin possessed invaluable detailed information not otherwise available the colonel drew him aside.

His first question: "The quickest way there?"

"Take the road to the Snake. There'll be some shortcuts, mostly across river bends and will be obvious. At the Snake, contact Jim Baker, in fact, I'll ride ahead and tell him you're coming. He's an old trapper and guide who has lived here for about fifty years. Pay him to guide you all the way to Milk Creek."

"You won't be going with us?"

"Had enough of it for the time bein'. I got a livery business here that needs lookin' after. I will ride ahead and let Baker know you're on the way. Some here think that the Uncompahgre Utes are riding north to reinforce the bunch at Milk Creek. Myself, I doubt it. They're led by Chief Ouray who's smart and known as a peacekeeper. He's about two hundred hard miles away and won't see any profit in marchin' a force north for several days to reinforce defeat. You heard anything about the White River Agency?"

"No."

"News up here is that the agency was wiped out the same day we were ambushed. The Utes slaughtered all the men and took three women and two children hostage. Douglas is probably in a hurry to hook up with Ouray's band—a lot more than Ouray wants to have 'em. Four couriers, including myself, left Milk Creek in the middle of the night on the thirtieth. I made it to Rawlins early this morning. I just learned that Captain Dodge and a company of the Ninth Cavalry overnighted at the Iles ranch down south last night. Most of 'em headed out for Milk Creek this morning. A few were detached to Lieutenant Price's infantry company that Thornburgh left at Fortification Creek."

"What's the state of Price's detachment?"

"Good, so far as I know. I picked up a fresh horse there. He'd had no contact with the Indians and had plenty of supplies."

"Speaking of supplies, how long do you estimate Payne,

Dodge and their men can last?"

Rankin tugged at his chin and shook his head. "Don't know. That first afternoon we were shootin' and unloadin' wagons to build breastworks and then the other couriers and I left that night. We'd lost all that had been in Gordon's wagons when Captain Payne set fire to them to keep them out of the Indians' hands. My guess is they been on pretty thin rations since then and apparently Dodge's wagons were sent to Price. I'd say they're probably lookin' more favorably at the rotting horses and mules every day."

"My original guess was five days and you've just confirmed that. We're going to have to push hard to get there before they run out completely. You prepare to leave within the hour and ride ahead to Mr. Baker's."

Merritt watched him walk away and remembered a quote from the British biographer, Samuel Johnson: "Every man thinks meanly of himself for not being a soldier." There lurks, he thought, in every man, soldier or not, who's avoided or survived a fight, a battle, a war in which relatives, friends, and comrades were killed or wounded, the lifelong cancer of guilt. The haunting, nagging fear that his survival was purchased by the blood of others, that his presence, actions and courage could have spared the life or lives of husbands and fathers that they might have returned to wives and children and secured more gifted and noble generations. He saw in Rankin the personification of that sentiment.

But he'd witnessed also the fallen, faced with their mortality and totally without blame or rancor, absolve all survivors, friend or enemy, from any earthly blame and thereby release their spirit to take wing and depart a world in which the dust and ashes of victory and defeat merge in equal measure to fall upon those left standing.

Merritt scanned the men, horses and wagons surrounding

him. Shouted orders and yelled replies had ceased, men in the ranks stopped conversing and even the horses stood like statues but snorted their impatience. All eyes were pinned on him, all ears were pricked listening for the bugle calls. Merritt, with a tremendous sense of moment that nearly robbed him of his breath, turned to his bugler. "Sound 'Mount.' " The trooper licked his lips and sealed them against the mouthpiece of his bugle. No more than a few clear, perfect staccato notes had cleared the bell of the instrument when the company buglers echoed it. The colonel noted with satisfaction that the seat of almost every man's trousers hit the saddle at the same time as his. He remained immobile a few seconds savoring these most celebrated martial moments then reined his horse around and pranced through an isle of men to the front of the column. He pulled out his watch. Two minutes to noon. He looked over his shoulder at the perfectly curved column, pulled the pin from his watch and forwarded the minute hand until it precisely covered the hour hand and returned his watch to his vest pocket. "Forward," he ordered the bugler.

"We won't be galloping hell-bent to the rescue," Merritt had told his officers. "Just long, steady days, changes of pace, frequent short breaks, organized watering. Expect to be on the move at least twelve hours per day, maybe as many as sixteen. We need to average at least three miles an hour—that's about as fast as the wagons can push it."

As the column rumbled south Joe Rankin loped by and waved. He disappeared on the other side of a dust cloud that followed him up and over a low hill.

Custer pushed his companies of the Seventh Cavalry excruciatingly hard from the very first day out of Fort Abraham Lincoln. By June 25, 1876, both the men and horses under his command were exhausted. He had given so little care to his cavalry horses on the way that they arrived weak from hunger and actually grazed in the midst of the battle after their riders were killed or wounded. The army had noted this and issued standard operational care for cavalry horses en route to battle. Merritt was not only cognizant of these orders, he had helped draft them.

The conditions of time, distance, terrain, weather and battle scenario dictated marching orders. Each hour of marching was punctuated by a few minutes rest stop. Within each hour the pace varied for a prescribed number of minutes from walk to trot, to canter, even twenty minutes of leading horses.

Merritt divided the fifty-mile-per-day objective into thirds. He planned two-hour rest stops after each of the first two thirds. Horses were unsaddled and allowed to eat, drink, and roll. The men did the same but also busied themselves with grooming their horses, checking for chaffing and galling, replacing or tightening shoes and napping if time allowed. The horses fed, watered and secured for the night on the third stop. He ordered reveille at six thirty a.m.

Because of the late start, the first day's march lasted until nine p.m. and covered an amazing forty-five miles, only five miles short of the hoped for goal of fifty miles for a full day. Few, including Merritt, feared a surprise night attack but wagons were circled, defensive bulwarks built and sentries put on patrol.

Rankin and Jim Baker caught up with the column at day's end. Joe remarked to Baker about the profound difference between Thornburgh's command and Merritt's. The major's column often lurched along from one to four hours at a walking pace punctuated by irregular stops sometimes lasting for more than an hour. On at least one occasion the men were allowed to cut willow fishing poles, catch grasshoppers for bait and spend two hours fishing a nearby stream during which Jack and Colorow rode by chatting amiably with friends and acquaintances.

Rankin, comparing Thornburgh to Merritt, now remembered the White River expedition as a lark, a mission without an objective. Thornburgh's lack of discipline and indecisiveness, he now viewed as catastrophic character flaws.

Merritt approached Rankin. "Thought you were going back to Rawlins from here."

Rankin shook his head. "Thought so but changed my mind. I want to witness first-hand how this affair ends."

Within minutes after sundown Merritt's men, who had been riding in sweat-soaked shirtsleeves, were donning overcoats and caps. The mountain peaks and treetops were outlined by a phosphorescent mass of stars.

A waning moon would not show its nose over the western cliffs until after midnight. Lanterns hung from the wagons but threatened to break and explode into flames with each tooth-chattering crash of iron tire against rock. Some infantrymen riding in the wagons, fearful of such conflagration, disobeyed strict orders and under the cover of darkness abandoned the wagons like sinking ships and trudged behind them. Troopers carried rifles at the ready and every coyote howl in the darkness ended with a chorus of rifle hammers cocking until Merritt, fearful of accidental shots, shouted orders forbidding cocked rifles.

Merritt, mindful of the increasing tension, abandoned his hope of marching until ten thirty and ordered a bivouac an hour earlier near where Jim Baker said there was a creek. Rankin, remembering the fires of the first evening at Milk Creek becoming bright targets, suggested a cold camp. Merritt disagreed.

"Tonight we're going to light up the landscape with fires. If there are any Ute scouts out there I want them to think we're an entire division." He turned to his adjutant. "Cavalry troops water and picket the horses while the infantry and wagon drivers collect wood and build fires. The troopers will fall to and help when the horses are watered and staked. Expand the fires out from the ones in the center and we can see to gather more wood. Once we have fires feed the horses, then the men eat. Squads will stand watch during the night and feed the fires."

He took a lantern from a nearby wagon and strode in the direction of the sound of water tumbling over rocks. The adjutant had the bugler call "Officer's Call." Joe Rankin and Jim Baker rushed to catch up with the colonel who was standing downstream from his horse, lantern in one hand and pissing with the other. Rankin joined him, his first time since noon.

"How you holdin' up, Joe?" inquired the colonel. "You've had quite a go of it the last few days."

"I'm a bit tuckered for sure. I'll curl up next to a fire and sleep well tonight, haven't had that much sleep in a whole week."

The two men backed their horses away from the stream as other officers and troopers stumbled from the darkness toward the lantern. Merritt handed the lantern to a passing officer. He and Joe began picking their way toward a growing clutch of yellow and orange blooms on the hillside.

"You hear directly from Washington?" asked Rankin.

"No, Omaha. But General Crook did extend greetings from General Sherman. Uncle Bill . . . no disrespect to the general of the army, was characteristically blunt and brief. Not that that's a bad thing, it provides us with a great deal of latitude on how we handle this matter tactically. I'm hoping the trip down there will wear some of the edge off the men. I don't want another Sand Creek any more than I want another Little Bighorn."

Rankin remembered the slaughter in 1864 of Black Kettle's Cheyenne encampment in Southern Colorado by Colonel John Chivington and a force of Colorado militia, another link in the ever-lengthening chain of vendettas. "We're all killin' of a fast pace, Colonel."

191

"Yes, we, both red and white, lust for it like nothing else in life, no woman's beauty and love can distract a man whose blood runs hot to drain it from another. It's not likely the Indian wars will end here, not at Milk Creek and not with us. We will subjugate the Indians, force our way upon them and close out this particular chapter in warfare. But we will seek out new enemies whom we will equally love to slaughter and conquer. If I thought that a complete slaughter of the Utes in the coming battle would put an end to this incessant struggle, then, by God, so be it. The white nation we represent is an irrepressible force to which all native people have to subordinate themselves and the quicker they learn that the better for both sides."

The two men joined Jim Baker beside one of the blazing campfires which already numbered in the dozens. They raised their hands against the intensity of the inferno that drove them more than ten yards back. The stockpile of wood feeding it smoked and the pitch sizzled and threatened to ignite itself.

Jim Baker motioned Rankin and Merritt a few steps further and lowered his voice. "Didn't want to talk on the trail today with the other men and officers around, Joe already knows. The White River Agency went to hell the same day as Thornburgh's ambush. All the white men, including Meeker, were killed, the women and children were rounded up and marched south, nobody knows exactly how many or where they are or in what condition."

Merritt considered this in silence for several seconds. "Thanks, Mr. Baker . . . and for your discretion. Joe and I were just discussing that the men were in a vengeful mood and this would certainly inflame their anger. There is nothing we can do for the captive women and children at this point but it will be an issue of the utmost importance once we've secured the Milk Creek battle site and unarmed the Utes. Gentlemen, I shall leave you to care for your animals and forage for yourselves. I must be off to tend to my command . . . a task I will happily do afoot."

Rankin turned to Baker. "Jim, got anything for saddle sores? I grew a crop ridin' to Rawlins."

"Sure do." The trapper opened the buckles to one of his saddlebags and pulled out a small jar. "A mixture of mink oil, wool grease, coal tar and menthol. Smell will scare off a starvin' grizzly but works like a charm. Got it from an old Paiute medicine man couple of years ago."

Joe took the jar from Baker. "Paiute, ya say? Remember a name?"

Baker paused and looked at the ground. "Nope, don't know that he told me."

"Wear a big, black, flat-brimmed reservation hat?"

"That's him. You know him?"

Joe thought for a minute. "Not really, might have met him once but can't say for sure."

Rankin returned to the fire which had transformed itself into a circular carpet of glowing coals. He felt the saddle pad he'd laid across his saddle and found it warm and dry. The heavy wool rug doubled as a saddle blanket and tolerable bivouac mattress when unfolded inside out. He and Baker sat wordlessly looking into the embers and nibbling on cold smoked bacon, hardtack and tinned beans. "Wish I had some coffee," he mumbled. Baker answered by handing him a half-full canteen, which he opened and sniffed. His wish was rewarded by the bitter odor of "Black Joe," a soldier's mixture of coffee and chicory bark. "Thanks. This'll do nicely."

"Take a swig but leave some for breakfast," Baker replied. "I daresay we'll have no more until this fracas is over with."

Joe filled his mouth and handed the canteen back to Baker. He sloshed it back and forth from cheek to cheek letting a few drops trickle down his throat with each pass until, even after his mouth was empty, he savored the flavor lingering on his teeth and in his saliva. He chased this down with a big gulp of wood-smoke-tinged night air.

A sentry came by and fed the dying embers a few fragrant juniper branches which snapped and crackled and smoked with resistance but finally succumbed to the nibbling yellow flames. Joe stretched out on the saddle pad and pulled the two layers of wool army blanket up to his chin and then, as the night air bit his ears, over his head. Inside his cocoon he could see the light of the flickering flames seeping through moth holes in the wool blankets. In the shimmering light he saw the quick dance of many bare legs. His eyes closed and the darkness swallowed him until he felt the chill of a breeze on his face which pried them open to see a hand had gently pulled back the edge of the blanket. A dark figure silhouetted against the fire glow and cinder spark knelt beside him. He sensed, rather than saw the Paiute. He felt no fear. The shaman's other hand gently brushed his eyes closed and replaced the blanket over his head. He again surrendered to the tranquil blackness.

As Rankin, Baker and most of the expedition's troopers drifted off to sleep, Merritt and his officers were weighing choices in his command tent under the day-like beams of several lanterns.

"My main concern," repeated Merritt, "is getting there before their rations and ammunition run out—if they already haven't."

"Sir," said one of the officers, "I understand and agree with your sense of urgency but even if we're as much as a day late, the men won't be dropping from starvation."

"Probably not," countered Merritt, "but the condition of some of the wounded already weak from blood loss might be gravely aggravated. Too, the temptation to surrender if supplies run out and they don't realize help is just around the corner may turn the tide against them. Or, the Utes may become emboldened enough to rush the wagons in force. No, I think our best bet to secure a positive outcome in this conflict is to push our force, the fitter, better-fed and better-rested troops, as hard as we dare. Now, with that settled, let me direct your attention to the maps on the table."

The colonel once again opened the debate over whether it was better to separate at least some, perhaps most, of the cavalry from the wagons. The wagons, were, of course, road bound and might be slowed by rough road sections, water courses, sandbars and alkali pans. On the other hand the wagon train would have to rely solely for protection on the infantry troops some of them were carrying. Shortly before midnight the meeting adjourned but with many officers finding reasons to continue discussions within the nearly summer-like temperature of the tent and lanterns. Merritt finally dismissed them and directly an orderly unfolded a canvas cot upon which he stretched out without removing a thread of clothing or his boots. The orderly took his hat and handed him two blankets one of which he folded to make a pillow. "Leave one of the lanterns on low," ordered the colonel, "and tell the officer of the guard to have the duty watch wake me at five thirty."

"Yes, sir." The orderly cranked down the lamp wick and hung the lantern from the ridgepole of the tent.

My God, thought Merritt, I've spent at least ninety percent of my adult life's nights sleeping alone on bunks, cots, or on the ground in barracks, tents, or wagons. Most men sleep in beds in bedrooms next to a female mate from early manhood until one or the other dies. Before he died, General Lee said that his only regret was that he had gotten a military education. I wonder how many others feel the same . . . Grant, Sherman, Sheridan, and Crook. In his last living moments is that how Thornburgh felt? We who wear the blue twill and gold trim serve a nation bent on conquest no different than Roman legionnaires and their predecessors. Thousands left home and family and simply vanished into the far dark corners of an avaricious empire. Almost certainly some of the men serving under him will die lonely deaths at the hands of heartless enemies and be buried in unmarked graves, leaving no marks themselves for which to be remembered.

Merritt closed his eyes and turned his thoughts to being nearly buried into the crevasses of a feather mattress under a down comforter next to the warm, fragrant body of his wife listening to her deep breaths. He too floated off into blackness.

Throughout the night squads of cloaked sentinels roamed the bivouac stoking the campfires and lighting their clay pipes from the embers. The squads trudged back and forth, up and down with the collars of their woolen capes pulled over their heads so they appeared in the moonlight to be cloisters of monks on an endless, aimless journey of penitence.

Captain Joe Lawson, less than one hundred miles to the south, gazed from his dugout at the same moon and pondered the fate of the White River expedition. Hearing the approaching clomp of booted feet he sat up and peered over the edge of his dugout. He could make out Captain Payne's form plodding toward him. Payne's body was truncated at the waist by a low hanging deck of smoldering campfire smoke so that his legs appeared to step woodenly like a puppet directed by hands and arms made invisible by his cavalry cloak. The wagon corral scoured clean of every bush and blade of grass was all the more sinister by human and animal burial mounds, unburied horses and mules and the infernal stench of death and rot.

Joe gathered his emaciated legs beneath him, stood to attention and saluted. Payne returned with a slight wave. "Joe," he panted as he sat himself on a discarded saddle, "got a minute?"

"Make it quick. I have a dinner invitation from General

Sherman that I have to dress for." Payne bent over laughing until it turned to a choking cough. "S'not funny," Lawson replied with mock offense. "The General takes grave umbrage at tardiness."

"God, Joe! You're a savior. I don't know what we'd do without your sense of humor."

"Be left to laugh at yourselves. Much funnier. The Lancers take great pride in maintaining a sense of humor in the worst situations. They consider laughter a very serious matter."

"Speakin' o' dinner Captain Lawson—we're just about out—and breakfast and everything else."

Once the elation at the arrival of Dodge's light company had subsided the realization that their situation was no better than it had been yesterday came home to roost. Not only was it not any better, it was becoming worse because Dodge's men were now consuming the last of their own rations and tomorrow would begin on what precious little the original three companies had left.

"Yes, sir, that we are. Halved rations three or four times in the last four days and we're still out."

"Any ideas?"

"Same as you."

"Morale ain't real good, we go to killin' what horses we got left for food, it's gonna take another serious drop."

"That's so. I told General Crook on the starvation march in the winter of seventy-seven, I'd rather eat one of my brothers than kill my horse and eat 'im."

"As I recall, you did eat horses."

"Them that was already dead."

"Uhh," Payne groaned. "Sounds like poison for sure."

"Nope, I subsisted for more'n a week once in the South African bush on meat so rotten it would gag a maggot. Learned from the Bushmen to boil it first for at least half an hour and then roast it through and through. Gotta eat it all the first time around though, it'll kill you if you try that recipe a second time."

"We don't even have any firewood left. Boxes and barrels and wagon slats are about all gone. There's probably not enough to boil carrion let alone roast it. And speakin' of boiled water—if we can't boil it, we can't drink it, either, because the bastards have turned it into a privy and slaughterhouse flume upstream."

Both officers looked up at the sound of another pair of approaching boots. The outline of Captain Dodge appeared in the moonlight. Both straightened to attention and saluted. "Heard you two palavering out here and thought you'd come up with a secret to get us out of this fix."

"We were comparin' recipes for horse biscuits, gonna bake a mess of 'em to take with us when we sneak off tonight and leave you holdin' the bag," joked Lawson.

"And damn sorry you'll be tomorrow when the Utes are dinin' on your carcasses and we'll be feasting on the catered dinner Merritt's bringin' in."

The officers fell silent. "He might be here in the morning," offered Lawson.

Payne and Dodge murmured agreement. Lawson produced the stub of a candle, struck a match with his thumbnail and lit the wick and stuck the candle in the dirt piled to the side of his dugout.

"If we're gonna attempt a breakout," asserted Payne, "it has to be soon—before we're all too weak—the next twenty-four hours." He looked at Dodge and Lawson hoping for confirmation but neither spoke. "Have to be at night," he continued, "during the day they'd pour fire down on us from the ridges. We could put what horses and mules we have left into harness, load up the best wagons we have left to transport the wounded. The able-bodied men will form columns on either side of the wagons. If they close with us. . . ."

"If they close with us," interrupted Dodge, "it'll be another Custer Massacre. They outnumber us four . . . maybe five to one?"

"It *may* be a massacre," Payne replied testily, "but I'd rather die fighting than waste away here of starvation and disease until they walk in and do the same thing."

Dodge stood up. "Merritt or someone else *will* be here before this time tomorrow night! I guarantee it."

"You can't guarantee it. You can't guarantee anything," Payne shouted and struggled to stand but in his weakened state and without both arms to push himself up, fell back on his wounded side and yelped. Dodge extended a hand but Payne swore at him and slapped it away. Dodge recoiled with surprise and drew a deep breath.

Before he could reply, a third voice spoke up.

"Could ye officers and gentlemen speak a bit louder, please? I think there may be some dead lads who may not have heard ye." Without turning, Captain Lawson recognized Sergeant Grimes. He grinned. He'd seen, during his military career, more than one general humbled by the thundering sarcasm of an Irish sergeant, mere captains were no match at all. Grimes continued: "God knows it would work wonders for morale to hear their fine, upstandin', noble leaders squabbling over our predicament; it'll put hunger, thirst and fear of a violent death right out of their minds." He snapped to attention and saluted. "Sirs."

Payne had reclaimed his feet and swatted at the dust on his blood-and-sweat-encrusted garments. "Thank you, Sergeant, your comments and advice are always welcome. Could you provide us with a report on the state of our food stores?"

Without hesitating, Grimes spoke: "If you don't include horses and mules—alive and dead—and wood shavings—we don't have any—food supplies. Now you didn't ask me about water but I'll include it in the report anyway. If you don't include the river which has become an open sewer, we don't have any of that either. We do have ammunition but none that's edible or drinkable, sir."

"Sarge," Lawson finally spoke up. "Captain Payne is in command. Captain Dodge and myself may offer advice but he makes the decisions and gives the orders. I, personally, and I think the other officers present value your knowledge and experience. With the permission of my commanding officer I would like to ask you for your opinions about our situation and what you would consider doing." He looked at Payne who hesitated at this surprise breach of protocol. He nodded. "You have permission to speak freely, Grimes."

Dodge said nothing.

"Truce," muttered Grimes. The comment drew shocked silence.

Grimes continued. "First, a cease-fire. Since neither of us trusts the other, everyone remains armed. Our sick and wounded get a free pass out immediately. The remainder gets a free pass to collect wood and water. We kill one or two of the remaining animals to sustain us until help arrives. We know and they know, help is just up the road a piece and when it gets here the help'll be in the mood to do some serious killin'. So in return, we guarantee that the leaders will not be summarily shot or hung. Ouray will return the captive women and children to the agency who will also oversee the safety of the Ute women and children. Upon the arrival of our relief the Utes will lay down their weapons and the men can return to the agency as well. The leaders will have to answer to justice. Draw the enemy's attention to two accomplishments: They've rid themselves of Meeker and stopped us from entering the reservation and in acknowledgement, we surrender."

After a long silence Lawson spoke: " 'Tis a grand idea, grandly presented, one worthy of the genius of the Irish mind. And it won't work."

Payne said nothing but looked questioningly toward Lawson.

"Because it's reasonable and rational and it depends on humans behaving so under the most unreasonable and irrational of circumstances imaginable, *war*. Plus, no officer who surrenders a command of United States soldiers to an Indian tribe will ever again be promoted. Captain Payne, you're among the wounded, twice so in fact, and you're in command. You could save your life and the lives of a lot of men *if* the Utes would accept such a plan as the sergeant suggests, but, honor dictates, you're not about to take such action. Captain Dodge, you're the senior officer here. You even hinted at taking command and leading an attack. I'm sure that issue will be revisited in a court of inquiry once this is over, even if none of us are alive. I have little doubt you would not entertain, even for one second, the term 'surrender' construed in its most liberal sense, to save the nation let alone a few cavalry companies and perhaps you would be correct."

Lawson continued. "Last January or February one hundred fifty British infantrymen in the Zulu Kingdom successfully defended themselves in a situation much like ours against three or four thousand Zulu warriors until relieved. They were able to do so because they maintained organization and discipline." Lawson paused to allow the others time to consider this. "Besides," he continued, "I doubt very much that of all the courses ever taught at West Point since eighteen-oh-two, not one of them was entitled, *How to Properly Surrender*. If something's not in the Point curricula then it really doesn't exist. So, it looks like we'll be here until the buzzard belches."

"Well," said Sergeant Grimes shaking his head, "it was a thought, probably the result of being sober too long." The officers laughed and went their separate ways leaving the two Irishmen to ponder their fates. "I knew," ruminated Grimes, "that dressin' up like a soldier could certainly foreshorten one's life at the hands of a stranger, went ahead and did it anyway. Can't believe we both lived through four years of the big 'un."

"True enough," Lawson agreed. "Any regrets?"

"Wish I'd married and had a litter. How 'bout you?"

"Wish I'd spent more time with my wife and litter. After I joined the Kentucky Cavalry in sixty-two I probably didn't see 'em more'n eight ten times afore she died and the kids scattered. Army's a damn poor substitute for good wife and family."

Nicagaat, Ute war chief and General Crook's favorite scout, stared through the flames of his campfire at a mound of agency-allocated blankets surrounded by empty ration tins, gnawed bones and whiskey bottles.

The blankets ballooned and deflated rhythmically each accompanied by a low-pitched growling wheeze. Colorow snored the same as he appeared and behaved; fat, guttural, snarling one minute, whining the next. Jack imagined drawing his knife, leaping the fire, throwing the blankets aside and quickly cutting his fat throat. So heated was his anger that he could hear the point of his knife pop through the flaccid skin and fat, slice with sickle sharpness from ear to ear and listen to the man slosh in his own blood, piss and excrement. Instead, he stirred the fire and watched a shower of sparks and cinders land on the blankets.

Before burning the agency, the Ute warriors had emptied the barns, cabins and sheds of all food supplies; flour, meal, sugar, tinned fruit and vegetables, candy and whiskey. The men had gone on days-long eating binges until most were continuously sick. The hillsides surrounding the besieged corral were covered with befouled abandoned campsites. No one moved after dark anymore for fear of falling into an uncovered sewer filled with vomit and liquid excrement. Water also had become a problem. While the Utes had access to fresh water above the point where they had nearly dammed the stream with rotting animal carcasses they lacked buckets, kettles and other means of transporting it up the ridges to distant firing lines. Dehydrated men became listless and then nearly useless. The bravado of the successful ambush and nearly complete victory on the twenty-ninth had all but disappeared. They no longer boasted and beamed victoriously. Jack had overheard bits and pieces of hushed conversations about abandoning the siege and following Johnson and Douglas south. Each morning he saw fewer familiar faces.

The warriors possessed abundance of ammunition but to little avail. Rifles captured or bought over the years but never maintained stopped firing. Very few of the Ute men had the knowledge, skills or tools to repair them. Pistols, never a decisive weapon, were mostly decayed Civil War remnants that required unavailable powder and ball. Jack had watched these useless pieces of wood, iron and brass pile up at the abandoned camps. Often the warriors whose weapons failed simply walked away from the battle. Unlike the white army which deterred desertion by summary execution, most Indian tribes might, at worst, banish them. Ute warriors who had spontaneously joined the insurrection expecting to rid the tribe of the onerous Meeker and his employees and halt the cavalry from entering the agency, felt their mission accomplished. Most knew by the first days of October that a decisive, conclusive victory by a single overwhelming attack was impossible.

A wraithlike figure materialized from the blackness behind the sleeping Colorow. Jack's hand tightened on the stock of the Winchester rifle he cradled on his crossed legs. He thumbed back the hammer and curled his index finger over the trigger.

He relaxed as the figure approached the outer corona of the fire.

The short, stout man's long black hair dropped in front of his shoulders and down his chest. His face was neither young nor old but very dark and lined. The reflected flames from the campfire turned his eyes a demonic red. He wore a long, nearly white, brain-tanned doeskin tunic, fringed and trimmed with beads. Jack saw no weapons. In his right hand he held a high-crowned, broad-brimmed black hat. He walked slowly, purposefully around the prostrate Colorow into the fire's periphery.

Jack laid his rifle aside and stood. "Welcome, Grandfather," he spoke softly and solemnly, "come, share the fire and food."

The man shook his head but stepped closer. "Nicagaat, I can only pause briefly. To their great harm, the Lakota again grow impatient. Unlike you, their young men have misunderstood the message of the Ghost Dance and Shirt. They are not iron shields and armor against the American bullets. Instead they are intended to shield and protect the red man from the white indignities through dignity, forbearance and endurance." The man moved to position himself between Jack and the fire so that the Ute could see only a silhouette. His words were slow, deliberate and deep.

"The Ute's time here on Little River is now measured in hours, not days. A storm approaches that will forever extinguish the Ute campfires along the White River. Temper your blows against the American soldiers and know that the whites will remember every death and exact revenge. Let go those warriors who want to leave, they will serve little purpose from now until the end."

"What shall we do at the end?"

"Lay aside all your weapons, stand calmly and do what the Americans say."

"What of our women and children who have moved south?"

"Your men will join them at the new Los Pinos Agency in the Uncompahgre. You, Douglas, Johnson and Colorow are accused already of crimes against the Americans. I can tell you no more."

"What have become of the white captives?"

"They live, but white men too eagerly imagine the worst when their women are seized. The Meeker daughter you long for suffered no less or more than the others."

"The war, our war, will not end here," Jack declared.

"Wars will not end so long as men look and act differently, as long as one man has something another wants. War is as natural to men as flight is to the eagle. Kill no more, your battle has ended. Colonel Merritt with hundreds is camped less than two days from here." The Paiute turned and glided toward the darkness.

"Grandfather," called Jack, "let me come with you."

The man stopped and half turned. "Another time, another place. I will take your hand and together we will pass through the shining mountains." He drifted into the darkness.

CHAPTER 21

"The first virtue in a soldier is endurance of fatigue; courage is only the second virtue."

Napoleon Bonaparte

Sometime during the night, Rankin awoke shivering violently. The sentries feeding the numerous fires had burned all the wood stores. He arose and found a shovel. He paved the smoldering coals with a thin layer of dirt and on that spread his damp saddle blanket. He folded the two top blankets to double thickness and curled up in a knot beneath them.

But it seemed as though his eyelids had only just closed when the first notes of the colonel's bugler called "Reveille," repeated immediately by the company buglers along the hillside. He heard in the darkness men cursing and scrambling from their blankets and for their rifles. Lantern wicks flared. Cooks and orderlies banged coffeepots and skillets, horses stomped nervously at their picket stakes.

Rankin slowly pried himself from the comforts of his now-warm bed. Stiff, aching joints and bruised muscles at every curve and bend of his body reminded him of his monumentally long ride. Incredibly the nearly unendurable pain from the saddle sores inside his thighs and along his butt was nearly gone. No white doctor with all his balms, unguents and lotions had ever been able to heal them as quickly.

In minutes the lanterns illuminated neat ranks and files of

troopers answering roll call. Merritt, already astride his horse, awaited the results. Sergeants reported to company commanders and commanders shouted reports to the colonel who then order the men to their horses and the infantry to their wagons. The animals were fed generous amounts of corn, oats and barley mixed with molasses while being brushed vigorously. Hooves and shoes were checked, animals blanketed, saddled and bridled. Only when a sergeant had inspected each horse, man and equipment were the troopers allowed to line up at the mess tent. While the troopers ate, the teamsters fed and harnessed the draft animals, ate, packed up, and took their seats. The infantrymen having no horses to care for helped the teamsters and cooks and boarded their wagons for another day of bone-breaking travel. Cramped like sardines in a tin, the foot soldiers padded their seats with blankets and sagged back against the canvas wagon sheets to catch moments of restless napping. Every two hours the infantrymen were ordered out the tailgates of the moving wagons to march or dogtrot in formation alongside them, always with rifles held ready.

Merritt's concern that the wagons would slow the march proved unfounded. The roads for the most part were dry and solid. In only two places did the cavalry have to rope onto the wagons to help them through the ruts and holes. The weather accommodated the column. Rain and snow showers lasted an hour's duration and served the troops well by settling the dust without turning the trail into bottomless muck. Colorado experiences two monsoon seasons, spring and late summer. The 1879 fall monsoon had been light and ended early so water in streams and rivers was at low ebb, clear and clean.

At precisely two o'clock in the afternoon of October 4, Merritt ordered the column to halt at a small, unnamed stream. Jim Baker and a team of scouts rode ahead reconnoitering the road for at least two miles to the south. Their orders: If they encountered the enemy in any number or were fired upon to return immediately to the wagons and not engage them. At two miles they were to split the team in two and return in separate semicircles to the column.

Horses were unsaddled, unbridled and unharnessed, watered, hobbled and set to grazing under guard in a nearby glade. Men ate and settled in formation onto the warm, fragrant grass to nap, their rifles always within easy reach.

Merritt deployed scouts while he and his officers continually paraded around the perimeter of sleeping men and grazing horses. A sudden and unexpected wave of angst swept over him. The meadow with his men in repose across it jolted his memory of interludes preceding Civil War battles fifteen years before. But then, instead of a few hundred dozing, hundreds or even thousands lay dozing from exhaustion but praying fearfully they'd live to see the sun set and moon rise. He addressed his officers.

"It was a cold, short night. Today will be longer, tonight shorter. Dismount and rest in the sun briefly. Keep your halters in hand and your rifles next to you. I'll call assembly in about an hour." He saluted the group and touched his horse's sides with his spurs.

As he trotted off, one of the infantry officers who had joined the expedition at Rawlins turned to a cavalry officer from Fort Russell. "Doesn't he ever sleep or eat?"

The cavalry officer smiled. "Very little, and he's ill. He climbed from his sickbed and mounted his horse. He harbors the odd notion that officers lead by example. He believes that since officers are better paid, fed and housed than enlisted men, by God, they'd better sacrifice more in the field as repayment. If the men under them are uncomfortable, afraid and suffer then a good officer should sacrifice his comfort to relieve their discomfort."

At precisely four o'clock the procession began again to the bugle's blare. Merritt increased the pace to make up some lost time. Before sunset they reached Lieutenant Price's supply encampment. Rankin immediately sought out Price and informed him that his prize stallion enjoyed lodging and resting at the livery stable in Rawlins. "He's alive and tuckered out but much less the worse for wear than I am." Merritt delayed only long enough for Price and his men to saddle and harness and then headed the growing column southwest toward the setting sun.

Joe Lawson was almost thankful for the nearly bursting bladder and diarrhea attack that sent him scrambling from his dugout. Physical necessities among the besieged had long ago replaced privacy, modesty and sanitation, especially among the ill and wounded. No such word as "privacy" in the vocabulary of these soldiers. He grabbed the shovel from the dirt rim of the hole and headed toward the designated latrine area. After only a couple more steps he quickly bent and shoveled furiously. Squatting above the shallow depression he gasped with pain and relief, nearly oblivious to the men walking by who looked deferentially straight ahead. He turned back to his burrow. Sergeant Grimes stood waiting with a brick of hardtack in one hand and a tin cup of chicory in the other. Lawson washed up with a handful of adobe and wiped his hands on his trouser legs. He took his breakfast from Grimes.

"It's cold, sir," said Grimes.

"Yes, I'm freezing."

"I meant the coffee is cold, there's hardly a stick of wood left 'cept for the wagon boxes we're usin' as breastworks. Took the wood wheels off the last of 'em yesterday. In fact, that's the last of the chicory as well. A water detail ventured out this mornin', nary a shot fired. I guess the Utes are a bit weary of this affair as well. After they strained the water through a wagon sheet Doc Grimes added his last pint o' his medical liquor to the barrel, said it would kill some of the wee beasties givin' us trouble."

"Appears we're only about one step away from the sawdust and clay pies we dined on as children in Ireland. Mum would be disappointed about my lack of progress."

Grimes laughed. "They wasn't so bad if ye had a wee bit o' wild honey or fresh fried blood and grease t' mix 'em with."

Lawson smiled at their mutual recollections. They conjured up a lot of painful memories but also the satisfaction of having survived desperate hardships. Ireland was an inhospitable land begetting fierce, durable generations who embraced donnybrooks, music and poetry alike. As the two of them approached the troopers, the men jumped to attention and saluted. He returned the salute. "You boys don't need to be wastin' energy squabbling among yourselves. What's the problem here?" It appeared to be troops from the Ninth on one side and the Third on the other.

"One o' you boys better answer the captain real quick or you'll all spend the rest of the day digging latrines," boomed Grimes.

"Sir!" spoke up one of Dodge's men. "We was talkin' about shootin' one of those turkey buzzards on that dead horse out there and eatin' it. We was just figgerin' out who was gonna do the shootin' and who was gonna fetch it."

Lawson looked at Grimes and laughed. "I have to hand it to you, boys, I'd never thought of those birds as food. Let's see . . . that horse is about a hundred yards from the wagons and thirty from the Utes. We once had a man who could easily hit one with a Springfield carbine but Major Thornburgh's . . . indisposed. Even if we could hunt up another marksman of that lethality someone would have to sprint out and back, expose himself to concentrated fire for . . . say, thirty seconds, to at least one hundred repeating rifles in the hands of experienced shooters firing at the rate of a round every five seconds. That's upward of six hundred rounds. Ain't none of you fast enough to fetch the bird and make it back here before the Utes turned you into buzzard bait same as the horse. Ponder it a while. In the meantime, any of you wastes so much as a shot on those poor, harmless birds I'll have the lot of you hanged or shot for insubordination. Good day."

He returned their salute but only got a step away when the crack of a shot came from somewhere behind the tree line. The first shot Lawson had heard since before dark last night. He turned in time to see the buzzards take flight from the carcass and instinctively ducked. "Down!" he shouted and dove for the breastworks. The young troopers scrambled for carbines resting against the barricade. A volley followed the first shot. He could hear Payne and Dodge behind him shouting orders and troops flying to obey them. He quickly checked the men around him. They had quickly sorted out the rifles and positioned themselves at regular intervals sitting or kneeling against the barrier. Grimes was on his left, cool as always. Despite the volume of fire, which seemed to come down the hill and taper off when it reached the Ute trenches outside the tree line, he heard no bullets whizzing past nor saw any dirt kicked up. He took off his hat and drew his revolver from its holster. He popped his head above the dirt-filled barrels providing him cover and just as quickly dropped back down.

"See anything?" asked Grimes.

"Something's moving out there, coming this way."

"They're chargin' the wagons," cried one of the young Negro troopers.

"Shut up," yelled Lawson, "they're not chargin' the wagons. You'll start a panic." He could see some of the wide-eyed boys on the verge of dropping their rifles and running. He himself started to hunker down but decades of training and experience took command of his body.

He could hear what sounded like hoofbeats within yards of the corral. He cocked his revolver and stood up. Almost instantly a familiar dark form took to the air outside the breastworks a few feet from him. He had only time to point the revolver into the dark mass coming down on top of him and pull the trigger. The .45 exploded at the same time Lawson yelled "damn" and went down on his back beneath a spray of blood and in a cloud of dust. *Whoom!* Something incredibly heavy landed on his chest popping the air from his lungs as though they'd exploded. He gasped for air but only inhaled a mouthful of alkali dust. "I shot myself," he thought, "after all the years, wars and battles I shot myself to death." He listened helplessly to the pandemonium and chaos around him. Men and boys yelled unintelligibly and ran to and fro. Gunfire exploded all around him and he recognized Payne, Dodge, Cherry and Grimes yelling, "Cease fire!" The pain of suffocation overcame him. "Oh, God, please let me faint or die," he prayed. Layers of darkness covered his eyes like curtains drawn over windows. But with the darkness the pain disappeared and he surrendered to the twilight.

Someone, Doctor Grimes, kept yelling at him, "Captain Lawson, Captain Lawson, can you hear me?"

The voice echoed distantly in the darkness. Lawson tried to yell back but couldn't answer. He wanted to tell him to stop yelling because it was too painful to answer.

Then the captain felt someone roll him on his back and a finger dug the mud from his mouth. He retched, coughed and heaved but, with great relief, drew in a deep breath of mountain air. His comrades lifted the great weight smothering him. "Can you breathe all right?" asked the doctor. "Does anything feel broken?"

Joe sat up and vomited. He moved his arms, legs and head.

He spat out another wad of mud and looked at his red soaked field blouse. "Is that my blood?" He gasped. "What happened?"

"Mostly deer blood. You got clipped on the head—a small cut but head wounds bleed excessively."

"Clipped by what?" Joe demanded.

"Here, take a drink, wash your mouth out." Grimes handed him a canteen. Joe took a swig and rolled it around in his mouth and spat it out.

Grimes laughed, "After nearly five days of shootin' you finally hit something . . . a four-point buck. The Utes evidently jumped him and two does, that's what they were shootin' at, not us. The animals came down out of the trees right into the Indian trenches which caused more shootin', panicked and headed for the nearest shelter they could see, the wagons. By the worst luck they picked your spot to jump over just as you popped up."

"I came up just as something left the ground. I thought it was a Ute jumping his horse over the breastworks. I couldn't aim. I just pulled up and fired into it . . . felt like a train hit me."

"You won't be getting any medals for being wounded in action by a deer." Lawson looked over his shoulder to see Payne, Dodge and Cherry behind him. He struggled unsteadily to his feet and wobbled so badly that some of the men had to catch him under the arms to keep him from collapsing. He managed to only wave an arm as a salute.

"Cap'n Payne, if I go to my grave without another medal of any kind, I'll die a happy man."

"By a miracle," said Dodge, "the men killed and retrieved the two does that were with your buck without shooting each other in the process. We'll eat well enough now until help arrives."

"Raw?"

"Maybe not, we'll take the floorboards, tailgates and seats out of the wagons. Since we got no horses to ride we'll strip the leather off the saddle trees and use them; handles off the picks, shovels and axes, stocks off broken rifles. One of the water details found dry wood from a beaver lodge higher up on the creek."

"Draw fire?"

"Not yet. Rankin scouted up as high as where the couriers crossed the river and found an Indian fire pit but no sign of 'em otherwise."

"Strange," added Payne, "as much as we were movin' around in here since shootin' the deer, we haven't attracted any interest from the other side. Maybe they know something we don't."

"M'be," replied Lawson. "But remember the first rule in any fight is 'defend yourself at all times.' Man the barricades with at least fifty percent of your force. The men'll still have to eat in shifts." He looked deferentially at Payne. "If that meets with your approval, sir."

"It does. Sergeant Grimes would you please help Captain Lawson to the folding chair at the command post, so he won't be in the way." Grimes nodded.

"Oh," added Lawson, "one other thing . . . several smaller fires. One big fire with lots of men around it is still liable to attract attention . . . and . . ."

Payne nodded. "Thank you, Captain Lawson, that will be all for now."

"Don't forget to assign men to serve the wounded and sick first."

Grimes gently put a hand on Lawson's shoulder and directed him toward the improvised flagpole that represented the location of the command post.

Lawson turned to Grimes: "That wasn't just a lucky shot, Ed, that buck was a victim of quick thinking and lightning-like reflexes."

"Aye, sir, and I'll do m' best to make sure that's the accurate story that'll be handed down for decades, centuries." Out of the corner of his eye he could see Lawson's lips part and curl at the corners into a broad grin.

"Just so, lad, just so."

Wesley Merritt did not know about the surrounded troops' stroke of good fortune, but even if he had it would not have diminished his sense of urgency. The thought of even one death, man or animal, in Payne's command which might be prevented by him arriving one minute, even one second earlier, tore at his conscience and propelled his command forward under intense pressure. He believed with steadfast confidence that the outcome of other battles and momentous events earlier in his life were the culmination of a sequential causality known only to God. A meticulous organizer, Merritt was no less painstaking in the execution of every minute tactical detail.

He cleansed his mind of anything and everything, family, friends, career, food and sleep that did not directly apply to accomplishing this mission. His every order went to serve that purpose. He watered, fed and rested himself and the men under his command only to serve that purpose.

"Baker!" the colonel yelled through cupped hands toward a dim light more than a hundred yards ahead. The light kept moving. "Baker!" he tried again, louder—nothing. Turning to the shivering young private next to him, he said, "Assembly." The private pulled his brass bugle up from where it swung tethered to his saddle pommel and the mouthpiece from its nest next to his stomach inside several layers of clothing. He attached the piece to the horn and played several clear bars of the call. The light stopped and reversed direction. Merritt nodded his approval to the boy who smiled and dropped the piece on its cord back under his shirt.

In the rough, darkened terrain the horses were allowed to

walk at their own pace with heads down somehow sensing the road's rocks and snags and picking their way among them. Merritt kept the column moving at a steady if sluggish pace. The last time he'd checked his watch it was nine o'clock, at least an hour ago.

He halted the column as Baker pulled up abreast of him. "Mr. Baker, you're getting too far ahead of us again."

Baker leaned over with his hand cupped to his ear.

"I said," Merritt yelled, "slow down. We need your light."

Baker nodded. "About half a mile to a stream," he yelled back. "Better stop there, no water to speak of till we get to Milk Creek."

"Any firewood?"

"Not much. Some old beaver lodges. Won't be much more in the way of wood from here on either."

"Press on." Merritt had been hoping for more, like last night. Good fires can make a world of difference in troop morale under cold, dark, dangerous conditions. But at least the men and animals were well-fed and, aside from the mountain cold, the weather had been most favorable. Except for brief snow flurries and icy showers, the men had been spared the sudden and engulfing blizzards for which these mountains were infamous at this time of the year. Waist-deep snow and subzero temperatures could be fiercer enemies than any Indian or soldier.

Morale and resolve remained strong. But the singing and nearly unbridled enthusiasm of that first day now seemed remote. Men and animals slogged silently, fixedly forward; determined yeoman-like manner but no longer headstrong knights to the rescue. Most were young, in their twenties and thirties, lots of immigrants and first-generation Americans—Irish, English, Polish, German and Scandinavian used to hardship and deprivation. They were well-trained and disciplined and had earned their spurs in the Indian Wars of the last decade.

Slackers and misfits deserted and those remaining were hardened veterans who could cast an unflinching eye down the sights of a Springfield in the heat of battle.

Indian warriors were every bit the troopers' equal in courage and endurance. Their shortcomings were organization and discipline, the ability to snatch victory from the jaws of defeat at the last minute of a desperate struggle for survival. They couldn't comprehend the European strategies of "total war" and "annihilation" and battles that cost tens of thousands of lives.

The Indian tribes of pre-Columbian America had no word for "war" in the European sense of the term. Raiding, often without deadly violence, for women, horses and hunting ground across fluctuating boundaries was common, stylized and ritualistic. A warrior's proudest possession and sometimes his only battle weapon was a "coup stick." Its use measured warrior's skill and courage, not by the number of men he killed but how many warriors he closed with and touched with the stick—to "count coup" and escape alive, unwounded or captured. They learned their "art" of war from the tribal elders and male relatives far from four cloistered years in a castle on the Hudson River.

No true soldier, Merritt thought, could ever engage Indians in battle, live and return without more respect for them than he'd had before the fight. White soldiers were tied to miles-long wagon trains bearing food, water, blankets, medical supplies, arms, ammunition, infantry, horse feed, horseshoes, spare parts, the list was practically endless. Custer's 1874 Black Hills expedition trailed a train of seventy-four wagons. By contrast, an Indian war party consisted of a couple dozen warriors each equipped with antiquated rifles, bow, arrows, knife, tomahawk or club. There was no uniform, no order of battle. Their ponies were short, mongrelized animals of tremendous endurance and a source of food when necessary. Often they rode a horse until it died from exhaustion and then ate it. A blanket or buffalo skin doubled as coat and sleeping cover. They lived off the land; rock hard in winter and sizzling hot in summer. Freshly killed or dried meat was their staple. They drank from streams, lakes, rivers, swamps and mud puddles—or not at all. They could ride nearly a hundred miles in twenty-four hours and successfully fight a pitched battle.

Merritt could see Jim Baker ahead had stopped and set his lantern on a nearby boulder and he could hear the sound of water rushing over rocks. The bugler sounded "Officer's Call" assembling Merritt's company commanders. "Lanterns forward along the stream," he ordered. "Water your horses by company, four minutes per company. Horses that won't drink in four minutes will have to wait until morning. Teamsters, unhitch your teams and water them last. Leave the wagons parked in the road. Infantry, collect wood and build three separate fires then everyone fall in to picket and feed the horses. Guard-duty schedule is reversed from last night. Make haste, tonight's rest will be even shorter than the last."

Bugles blared, orders were shouted, men and animals fell into formation. Watering became a predicament. Unlike the previous night they watered where the road crossed the stream and the slope was gentle enough for the horses to drink a few at a time. Cavalry regulations declared that horses be led rather than ridden to water. Now troopers fearful of slipping and falling into the icy water or even soaking their boots and socks shied from leading the horses into the stream. Officers keeping careful track of the four minutes turned the often unwatered horses back into the pack bunching up behind them. Lanterns usually used to find wood flickered up and down the banks of the stream to find additional watering points which delayed fire building for an additional hour. Merritt checked his watch with increasing anxiety. At this rate his troops and animals would get less than four hours' rest before the two-thirty morning reveilles.

Horses, having drunk or not, were picketed, brushed, blanketed and fed copiously. In spite of exhaustion both officers and sergeants meticulously conducted inspections.

Finally, after the horses, the men received attention as conditions would allow. Each trooper, guide and teamster was given a lukewarm helping of bacon, beans and hardtack. They ate in silence huddled around fires or over clusters of candles and then took their turns at wood gathering. An icy southwesterly wind began sometime after ten o'clock. It drove like needles through their blankets and coats and into skin, muscle and bone. On the other side men scattered from the furnace-like heat, burning coals and cinders and choking smoke. Infantry wagon floors became layered with men seeking shelter from the ravaging wind.

Merritt crawled beneath a wagon instead of his tent and shrouded himself in blankets and spare wagon canvas to repel the wind. He'd warmed a couple of rocks in a fire and rested his cold boots on them. They would hold their heat for at least an hour, time for his feet to warm up until time to ride. Sentries had one-hour instead of two-hour duty watches. He calculated his schedule should bring the relief column into the Milk Creek site sometime after dawn. He fell asleep listening to the clinking of frozen brass harness buckles in the breeze and trying to imagine them to be the wind chimes on the front porch of his house in Laramie on a hot summer night.

Joe Lawson lay in his dugout with a stomach uncomfortably full of venison. Its rumbling and gurgling portended a night of sudden awakenings and maddening scrambles for his pre-dug latrine. His feet were freezing and he knew they wouldn't get any warmer during the night. And, his entire body ached from the collision with the buck earlier in the day. Every man in camp had fallen to and stripped the carcasses bare. Only bone, antler and hide remained above ground—and not for long. The meat not consumed immediately was cooked and stored or left out to freeze overnight.

"Joe, you awake?" Lawson recognized Payne's voice and fought to unwind himself from his blankets.

"Yes, sir."

"Relax, stay as you are."

"Thanks, Cap'n."

"How you feelin'?"

"I'm a bit stove up, but nothin' a shot of Bushmills wouldn't cure."

"I reckon you'll have to go uncured then. Think we'll have any visitors today?"

"Mebee . . . you know, 'cept for the deer fracas, I ain't heard more'n a dozen shots from the other side in the last twenty-four hours. I'm thinking they're thinking this is a lost cause."

"When you get up and about this mornin', why don't you go over and knock on the door and find out?"

"You get me a pair o' crutches and I'll do it. Those boys wouldn't shoot an old man on crutches, would they?"

"Course not. They have the same notions about chivalry and fightin' fair that we do." Payne paused for the irony to have its effect. "Joe, I'm certain that we're gonna get helped out of this mess in the next few days. The healthy boys will be all right on a few boiled bones until then. I'm worried about some of the sick and badly wounded. Some of them might not last a few more days, especially if the weather takes a turn. The army's more inclined to sit around talkin' about gettin' something done than actually doin' it . . . remember Fredericksburg and the like?"

"Yeah, try not to, though."

"What you think about sendin' someone up the road tonight t' see if there's any sign of help on the way."

"Have to send 'em afoot, not a horse left that'd get any more'n a mile before droppin' dead."

"I'll talk to Dodge about it."

"How you two doin'?"

"Okay. Funny thing is, I don't think he really wants to be in command any more than me. The army's just taught him to feel like he *should* want it. Hell, when the major named me second in command I didn't think for a minute that it'd really happen."

"Well, you've done a pretty fair job of flyin' for a bird with one wing. How's it feelin'?"

"Okay, I think. The doctor looks at it twice a day, takes a sniff and says it's okay. Now'n again I get close enough to one of the wounded who isn't okay, get a whiff and damn near scares me crazy for a few seconds."

"Yeah, hearin' him removing someone's limb without chloroform just about sets me to screamin' myself."

"The boys do see to each other real good, though. Some of 'em took the coals left over from the cook fires last night and spread 'em over the dirt on the hospital dugouts. Ones inside said it felt like a summer evening after about an hour."

Payne paused thoughtfully. "Battle and war are made even more terrible because men love 'em so much. I love my wife and children but no more than being afield with the men. I left home for the Academy when I was seventeen years old. Even then all I wanted in life was to be a soldier, everything else was second. Didn't care whether I died at seventeen or seventy, from battle wounds or old age so long as I was in uniform." Payne, apparently hoping for affirmation or confirmation, stopped. "Joe . . . Joe, you okay?" He took off his hat, knelt down and poked his head inside Lawson's dugout. He was greeted by the sound of heavy breathing. He regained his feet, saluted and walked toward dawn's first gray light in the east.

CHAPTER 22

Wesley Merritt awoke from dreamless sleep to panicked alertness. Something had happened? Something hadn't happened? Something was wrong. He fumbled through the layers and folds of wool blankets for his vest's watch pocket. It was empty. He found the chain and followed it down to the case while unrolling himself from the bedding. He found his match case in the other pocket and retrieved a match that he struck on the floorboard of the wagon above him. It flared but he struggled to focus his eyes on the watch face and hands.

First: relief; it appeared to be twelve fifteen. Next: anger and panic. It was three o'clock! "Officer of the guard!" he yelled from beneath the wagon. "Officer of the guard!" he screamed as he rolled out onto the frosted grass.

Within seconds men scrambled from under and inside of wagons. Horses galloped toward him in the darkness. Sergeants yelled, "To arms!" Bugles stuttered and blared "Assembly."

"Horse!" shouted the general to his orderly, who saluted and began brushing the animal furiously. A captain galloped up to Merritt, dismounted and came to attention. "Sir." He saluted. "Captain Keady, officer of the guard, reporting."

Merritt strode up to him and returned the salute. "Captain, I left a specific order for reveille at two thirty, did that order not get transmitted to you?"

"No, sir, it did not."

Merritt softened. "Very well, I'll seek out the officer responsible for this error later. For now, return to your company and have them mounted and ready to move in half an hour." The officer saluted, mounted and rode into the twilight created by lanterns, small fires, smoke and fog. The orderly trotted up with his horse. "See to your horse," ordered the general, "and get something to eat and drink if you can and be in formation in half an hour. Do you know where I might find the scout, Mr. Baker."

"He and Mr. Rankin are warming up over yonder," said the boy, indicating a blossoming fire not far from the streambed.

Baker and Rankin, still bleary eyed from cold and exhaustion, rotated themselves like meat on a spit edging as close to the fire as they dared when Merritt trotted up to them out of the darkness. He jumped down and joined them at the fire.

"How much further?"

"Four to five hours without stopping," answered Baker. "We should be there before dawn."

"Starry sky," added Rankin. "Should be a clear day."

"We cross Milk Creek a couple of miles before the wagon corral, that right?"

"Right. Then it's up a hill to the first bench. We'll be most vulnerable from the river crossing to the plateau. The horses and mules in harness will find the going tough and have to slow down. The road hillsides so no room to maneuver. Send the cavalry up first to secure the plateau and the infantry to protect the flanks, especially to the right where the terrain ascends the mountain slope and offers good cover."

Merritt nodded in agreement. "You'll be scouting ahead?"

"As soon as it begins to get light. Won't do anyone any good for me or Joe to go wanderin' around in the dark."

Merritt had noticed in the time he'd been talking to Baker and Rankin that the troops had saddled, bridled and harnessed the livestock and put on nose bags of grain. Each cavalry trooper had fallen into company formation behind a color-bearer next to a lantern carrier. Infantry companies stood at attention, port arms. He looked to his bugler. "Mount." The bugle notes echoed out through the darkness followed by commands of "Prepare to mount!" and "Mount!" Troopers vaulted aboard their horses and infantry filed into wagons. Again he nodded to the bugler who sounded "March." Merritt looked at his watch and shook his head in disappointment. It had taken them nearly a half hour to depart, but, he thought with pleasure, he'd planned for a two-thirty reveille and three-thirty departure—they were on time.

Jack, in a sense, envied fat, old Colorow. Nothing interfered with the Comanche's sleeping or eating. The young Ute chief had done neither since the shaman's visit. He had spent his time traveling from fire to fire, even man to man, explaining that the end was near.

"No more fighting and killing. We will not surrender. But at my signal you will simply put down your weapons and walk behind me. Or, you can leave now. But you cannot go home, our home, the agency, no longer exists. Take your weapons, food and horses and journey south to the Uncompahgre and Tabeguache, join with Ouray. Say nothing about what happened here . . . forever."

Some warriors who may have had relatives at the Los Pinos Agency or were from there originally, silently arose and with their eyes to the ground departed. Others squatted motionless, staring into the fire. Jack recognized many of them as Northern Utes whose families had fled south with the captive white women and children. They now had no homes, family or possessions to return to. Like all armies defeated in a distant land they had only each other and scant possessions to which to cling.

Jack knew that by the end of the day, he would no longer be free to come and go as he pleased. The white soldiers and politicians would tell him where, when and how to go, maybe for as long as he lived. They would feed or starve him as they saw fit. He would toil and rest at their sufferance. He knew all this because he'd seen it as the northern and eastern tribes had dropped one by one like leaves from the trees in the autumn.

By the time Jack had found his way through the dark back to the camp he shared with Colorow, the fire was a mere wisp of silvery smoke in the moonlight but the mountain of blankets rose and fell in rhythm with Colorow's snoring. Jack rekindled the fire from its last dying embers. He slipped the loop of a leather bag from his saddle horn and emptied its meager contents onto a folded blanket. He picked up a rock-hard egg-shaped ball and began gnawing listlessly on a pemmican of dried and powdered jerky, corn, leached acorns, flour, salt and honey. Colorow stirred beneath his beaver lodge of blankets, coughed, moaned and farted. Had he been awakened by the smell of food or the sound of him eating?

Jack tossed a few larger logs on the fire. Slowly, a mass of greasy black hair emerged from beneath the blankets like the head of some monstrous, crossbred tortoise. Similarly, a ham-size hand with black nails emerged and brushed back the hair from a sagging dark face. Colorow gazed through the fire at Jack and characteristically extended a begging hand. Jack nibbled at his pemmican. The fat, dirty fingers beckoned. Jack tossed the greasy ball through the flames, it bounced off the bones surrounding Colorow's bedding and scuttled like a small animal into the hollow where his neck met the blankets. Colorow retrieved it, his lips curling in a disappointed grimace when he saw what it was but gnawed on it like a grotesque dog with a bone. "It's over?" he queried, nearly inaudibly from under a mouthful of pemmican.

"Soon," Nicagaat replied at length.

"We won."

"What?"

"We won."

"We won *what*?" Jack spat.

"We kept them from the agency, from our land."

"Our land," he declared, "never *belonged* to us, *to anyone* and now not to the whites. Like those before us, we borrowed it, passed through it, camped on it, we ate from it but it was not ours to buy, sell or give away. Land that the People of the Shining Mountains will never enjoy again will someday be taken from the whites. Someday no humans will be left to fight over it."

Colorow grunted and continued to scratch away on his pemmican like a beaver on a wood knot. The cold darkness descended upon Nicagaat and his spirit shrank and ached from an alien sensation.

The emotion *fear* is so antithetical to Ute survival and existence that their language has no exact word for it. Ute children learn, by instinct and experience, to control people, animals and conditions in order to survive, but the emotion fear is like the air without form or substance and therefore nonexistent. Now, though Jack fed the fire more branches, the flames were without warmth or comfort. The icy cold blade of fear, something unutterable and nearly beyond his comprehension, had buried itself in his breast.

Colorow, like a turtle, had retreated beneath his blankets.

Jack struggled to keep his eyes open. He looked to the east for a hint of sunlight but the mountain peaks were outlined with stars. He turned his cold back to the heat of the embers and pulled the edge of the blanket over his head and body. Beneath it he tucked his legs up against his chest and rested his head on his knees. Sleep arrived sporadically, a few seconds or minutes at a time and then departed just as quickly. Dreams and consciousness merged and separated irregularly like two snakes fighting and copulating. Wovoka and Colorow taunted him; Josephine and Ouray scolded him; Meeker sneered condescendingly.

He awoke with a cry. Sweating, he threw his blanket aside. The morning sun struck him full force and he quickly brought his hands to his eyes. To his left Small River trickled and gurgled its light-footed way through a myriad of tiny runnels and channels. Wrens, sparrows, and camp robbers chirped, squeaked and trilled from willow and pine trees. Peeking through his fingers he saw a small, mixed gang of crows and magpies ravishing the boneyard surrounding Colorow's greasy blankets. The fat chief and his horse were gone.

Jack uncurled, stretched and walked to the creek where he relieved himself and drank. Returning to camp he disdained the pemmican remnants and returned them to their bag. This morning he had no appetite. He thought it an ill omen that the crowd of crows and magpies had grown so bold from their unimpeded scrounging that his presence within a few feet warranted only the occasional glance. He listlessly stirred the cold, charred nodules of last night's fire with a willow hoping to find a yet glowing ember but raised only a cloud of soot.

Through the spruce and pine he could see his hobbled pony grazing the dry grass turned golden straw by the morning sun. The pony stopped grazing and raised its head, ears pricked. Jack listened and looked for someone approaching but, except for the birds, the forest remained silent and motionless. Then he heard the first notes of a very distant but familiar bugle call, a call he hadn't heard since his days as a scout for General Crook. Named "Officer's Call" it heralded the approach of friendly forces. He threw aside the willow stick, picked up his revolver and slipped it beneath his belt. On his way to the horse meadow he picked up his rifle from against a tree. A bandolier of rifle ammunition dangled from a broken branch above the rifle. He picked it up, looked at it and then replaced it. In the meadow he unhobbled his horse and led it back to the camp where he'd dropped his saddle. His spirit glided into a heightened state of awareness capturing sights, smells, sounds and emotions as though he sensed each may be the last. He dutifully prepared himself and disciplined his mind for the momentous day ahead of him. But for a moment, only a moment, his slipped and allowed Josephine Meeker to enter his thoughts.

Merritt shifted more frequently from one cheek to the other on his saddle. His toes were numb and he had to keep changing the bridle reins from one hand to the other and drawing his free hand inside his overcoat sleeve to warm his gloves.

His eyes were riveted on the ridgeline of a slope illuminated by the morning sun across the thin, shallow Milk Creek rivulet. He turned and checked over his shoulder the long line of wagons and cavalry parked behind him. The clouds of breath rising from men and horses merged with the vapor from the stream trickling around ice-edged rocks. He'd seen no sign of Baker and Rankin. But no shots either. He signaled his bugler to play "Forward," and the bugler began fishing his mouthpiece from beneath his coat. But then Merritt glimpsed movement in the sunshine moving its way down the slope of the ridge. He stopped the bugler with an open hand. Another figure emerged and the general recognized his two scouts. He turned to an officer on his right. "Major, bring the two most forward companies of cavalry up here prepared for deployment over that ridgeline one hundred yards either side of the road. Support the cavalry with two companies of infantry on the flanks. But wait for my order to disperse." The major saluted and rode rearward as Rankin and Baker trotted up.

"All's clear, Colonel, at least to the top of the first bluff. No Indians, no shots, no smoke," puffed Rankin as he and Baker reigned in.

"Very good, thanks. Before sending the wagons up, I'm deploying two companies of cavalry along the road and infantry to the rear. They will fall into line behind the wagons as they pass. The three companies of cavalry now in the forefront will spearhead the advance into the cauldron beyond. At no time will we falter or stop. We have the strength to force the Utes out of their trenches and off the high ground and we're going to use it immediately."

The young major arrived at a trot with a company of cavalry either side of the wagons. Merritt waved him forward across the creek. He saluted and slowed to a walk as he crossed the creek slackening his rein to allow his horse to lower its head and pick its way about the slick, round rock which could easily claim an ankle or cannon bone.

When his entire detachment had forded the stream, he sped to the top of the knoll at a gallop with his companies in tow. Merritt watched as the major cleared the ridge now aglow in the sun and split his detachment to either side of the road. He then signaled to the infantry company commanders to march their men forward. He watched approvingly as the infantrymen double-timed up the trail to the plateau and dispersed evenly right and left.

As the last of the infantry disappeared from view, the colonel signaled the wagons forward and reined his horse aside to watch. The wagons lurched and swayed and rolled over the river rock. Some listed dangerously close to capsizing in the streambed, only to hit a stone on the other side and list in the opposite direction. Any wagon or wagons lost at this point would be left behind with their horses for possible retrieval later. Like Thorn-burgh nearly a week earlier, they *were* passing the point of no return. And no man, animal or piece of equipment would impede their surge forward.

One by one the teamsters lashed their titanic teams which dug their massive hooves deep into the rocky soil and strained their harnesses to the breaking point. Heads and necks bowed, veins and tendons stood out in dramatic relief through hide and hair. Elephantine muscles bunched at the shoulders and hips like boulders. And, one by one the heavily laden wagons nearly took wing as they topped the hill. Merritt, Baker, Rankin and the remainder of headquarters command followed closely.

At the top Merritt viewed a military textbook illustration of battlefield formations. The leading cavalry companies and wagons had run the gauntlet of the two protective cavalry formations on either side of the road. Those men had formed skirmish lines nearly a half mile long and were turned about with carbines ready to face any foe arising on either side. Infantry companies, in columns of twos, flanked the rear of the cavalry out to within fifty yards of the scrub oak, pine and juniper of the hillside. One column knelt, the second stood behind them all with rifles aimed and ready to fire at anything that moved. To the left the other company of infantry had taken up positions near the edge of the table land overlooking the river but not so close as to be exposed to fire from directly below. The general nodded his approval and galloped forward to join Baker and Rankin.

Rankin pointed as Merritt rode up. "Yonder's the remains of the wagons we burned the first afternoon so the Utes couldn't loot the supplies and use them as close cover." Merritt could see the tops of some blackened poles a few hundred yards distant over a low knoll. "The north end of the defenses is about fifty yards beyond the top of that knoll."

The colonel's first impulse was to collect his forces and charge triumphantly over the crest of the knoll but experience had tempered such a brazen move with caution. Unnerved and weary men in the bastion might easily mistake his cavalry for an all-out Indian assault resulting in friendly casualties. He motioned to his bugler.

"Sound 'Officer's Call,' " he instructed. The boy retrieved his mouthpiece and slipped it into the brass tube. He trotted ahead a few yards and began. The bugle's high notes wafted through the early-morning air and echoed off the nearby slopes. When he'd finished, the bugler rested the bell of the instrument on his hip and waited. The only response was the billowing breath of the horses in the lead group. The bugler looked back over his shoulder. "Again," said Merritt. The boy put the bugle to his lips again. The rising sun glinted off its highly polished golden surface. The notes sang out followed by the echo, which instead of fading, cried out even louder. Several seconds elapsed before the refrain repeated itself. Merritt turned his horse, stood high in his stirrups and yelled, "Forward!" The call flowed from man to man in the column and like a great wave. The troops and wagons surged forward. For the first time in days, a large smile rippled over Merritt's numbed face. The scouts and officers around him were laughing and cheering. His shoulders slumped from being relieved of their great weight. But almost as suddenly he tensed wondering what sort of spectacle would greet them when they surmounted the low ridge a few hundred yards distant.

Merritt ordered no galloping charge, no wild melee of knights in blue serge with sabers drawn prepared to hack down every last remaining insurgent Ute warrior in repayment for their grossly miscalculated attack on the United States Government's army. In keeping with his character he maintained a cautious walking pace so the trailing infantry companies protecting their flanks could easily keep pace.

As he drew abreast of the burned-out wagons, the Milk Creek

bastion came into full view—and smell. Besieged men climbed to the tops of barricades and stood cheering with rifles and carbines raised high. Still fearing a surprise Indian flanking attack, Merritt stopped the column and watched carefully as the infantry dashed up the ridge to their right and dropped into firing positions behind cover at the top. The companies on the left swept forward covering the Milk Creek gully until the stench of the bloated, maggot-infested bodies of the slaughtered cavalry and draft animals stopped them dead in their tracks.

From over the barricades between two of the wagons climbed three men assisted by nearby troopers. A fourth man carrying the guidon of D Company, Third Cavalry, on a makeshift pole hopped over after them. The man in front carried his left arm in front of him in a bloodstained sling. The gaunt figure to his right limped noticeably and leaned on a cane made from a broken branch. The third man was much younger. Merritt recognized him as Captain Dodge. He guessed the other two must be Payne and Lawson although they resembled photos the colonel had seen of Union soldiers released from the Confederate prison at Andersonville. The bloodstained, filthy dark rags which hung from their skeletal frames, except for the brass buttons, gold trim and gold shoulder bars, bore no resemblance to uniforms. The hatless Captain Payne's fine, dark-blond hair plastered to his head with dried blood and mud. Bloody bandages barely concealed gaping rips in his trouser legs. The top of one of his normally meticulous high-topped cavalry boots fell in shreds down to his foot's instep.

Merritt had only seen Lawson in photographs. He'd appeared as a moderately tall man held bolt upright by a sparse, wiry frame, a very prominent, carefully trimmed goatee and a regulation cavalry hat. The man limping toward him looked like a scarecrow—that the crows had taken to task. The colonel dismounted and walked forward to greet the approaching offi-

cers who stopped about ten feet away as Payne called out, "Ten-
hut." The bedraggled group stiffened to attention as did Merritt
and slowly brought their grimy hands up in salute.

Merritt returned it and ordered, "At ease." With his eyes
brimming and arms open, he hurried forward and embraced
Payne.

"Colonel," Payne sobbed into Merritt's ear, "I am *so* glad to
see you. I've dreamed of this moment for five days—a hundred
forty-two hours, to be exact."

"Captain, I wish I could have gotten here more quickly but it
was faster than trying to build a rail line up here from Rawlins."
The two men laughed beneath the trails of tears washing ar-
royos of dirt down their cheeks. He held Payne out at arm's
length and examined him. "You truly look like something the
cat dragged in. I have wagons loaded with every kind of food,
clothing, medicine and shelter—except the one thing that you
men need the most—a bath tub."

"A bar of soap and the creek will do fine for right now,
Colonel. Allow me to introduce you to my most able second in
command, Captain Joseph Lawson." Joe saluted and reached
out a heavily bandaged hand which Merritt artfully avoided by
grasping his wrist.

"The much-fabled Joseph Lawson. It's an honor meeting
you, Captain. Your good friends, Generals Crook, Sheridan and
Sherman, asked that I make it a personal point to seek you out,
dead or alive, and bestow upon you their personal thanks for
serving above and beyond—again."

"We'd have been much poorer," added Payne, "without Joe's
sound advice and remarkable Irish humor. It was Captain Lawson's chance encounter with an attacking stag that brought us
back to life yesterday, an action worthy of an award of some
kind. And, of course, I don't have to introduce you to our first
savior, Captain Dodge. The arrival of D Company of the Ninth

came at a most critical time when our morale was flagging."

"My compliments to you, Captain Dodge," said Merritt, offering his hand, "and my thanks for enduring the hardships of a forced march to aid our comrades in arms."

"And," said Payne, "last but not least, my color sergeant, First Sergeant Edward Grimes. His courage, leadership and gallantry were without parallel during the entire course of the engagement. There is no award high enough to reward this soldier for his devotion to duty." The sergeant stepped forward and saluted but surprised Merritt when he continued to within an arm's length of the colonel. He unbuckled a haversack slung over his left shoulder and took from it a large tin which he presented to the officer.

"The lads inside ask that I give you this can of peaches they've been saving for a special occasion. They decided this lifesaving was about as special as an occasion can get. And their deepest thanks."

Merritt took the can and examined it for a moment, then handed it back. "Sergeant, return it to the men with my thanks and tell them we'll all share a bit of it tonight."

At that moment a cacophony of rifle fire interrupted the gathering and all heads turned in the direction of the hillside to the west where the infantry had taken up positions.

The colonel ordered Payne and his officers back to the wagon corral while he mounted and led a cavalry company to a skirmish line from the base of the hill occupied by the infantry to a point south where the mountain the Indians had occupied for the previous five days dropped down toward Milk Creek. They jumped their horses across the nearest line of Indian trenches all of which were now empty. The cavalry skirmishers received no fire from the trees and oak brush on the hill and without targets did not fire aimlessly. The colonel ascended the infantry hill to seek out the unit's commander, Lieutenant Price.

By the time Colonel Merritt found Price, the shooting had stopped. He noted that the infantrymen held tight to their cover with their rifles seeking out targets hidden amid the tangle of brush and trees on the hillside to the west. Price, having seen Merritt from the crest of the occupied hill, trotted down to him.

"What's happening, Lieutenant?" inquired Merritt.

Price pointed toward the distant slope with the barrel of his carbine. "The men encountered a small Ute party coming downslope about two hundred yards out and engaged them. The Indians scattered and fired back. I think the number and accuracy of the infantry rifles surprised them and they retreated uphill."

"Any casualties?"

"None on our side. Do you want us to take the hill and pursue them?"

"No, too steep and high. You have good cover here and support from the cavalry if need be, maintain your position."

When the colonel and his cavalry detachment arrived at the corral, the first wagon of his train was clearing the top of the bluff above the charred wreckage. He directed them to cross the burned strip of land to the south of the barricades and then turn back north and form a protective curtain fifty feet out on the west side of what was left of the fortification wagons .

"Unhitch the horses," he ordered his adjutant, "but leave them in harness and picket them between the two lines of wagons. Position two companies in defensive positions on the outer wagon wall. I'll take the remainder inside the corral to relieve the defenders. Set up a field kitchen between the wagon lines at the far north end. Detach a company of strong-stomached men to search among the dead animals south of the corral for troopers' bodies. Leave any Indian bodies where they lay. Don't bother dealing with the dead animals. I'll be moving everyone a couple of miles north before evening anyway. The carrion eaters will have this place clean down to the bones before the end of the month." He looked up at the squadrons of vultures spiraling overhead. "Probably the end of the week."

Joe Lawson and Scott Payne, along with the two doctors, separated the dead and wounded from the revetments serving as hospitals. The dirt and planks had been removed from the tops of those trenches to more easily extricate and segregate the men. Some pulled blankets over their heads or covered their eyes with their hands. It was the first time since the afternoon of September 29 that they had seen the sun. Merritt retched when he rode up and caught full force the stench of gangrenous flesh and human excrement. He signaled Captain Lawson over. "Joe, I need to assemble a party to retrieve Major Thornburgh's body. Who knows best where he's located?"

"Probably Lieutenant Cherry. I think he was closer than any of us."

Merritt sent Lieutenants Cherry and Price, scout Sandy Mellen and a small party of troopers to retrieve the body. The troopers, though relieved at being detached from the onerous business of digging graves and dismantling breastworks amid the stench of tons of putrefying flesh, were nonetheless nearly petrified to be separated by nearly a mile from the rest of the battalion. They clustered tightly together until after crossing Milk Creek to the south near the point where Rankin and the other couriers crossed the first night of the battle. Price and Mellen conferred with Cherry who indicated a spot at the base of the "V" ridges where on the morning of the twenty-ninth, the Utes first engaged Thornburgh's skirmishers. "I think I saw him fall about midway between the bases of those two bluffs—give or take a hundred yards," he told Price.

"This isn't going to be a pleasant experience, he's been restin' there nearly six days—" the lieutenant replied.

"He'll be overripe for sure," said Mellen.

The comment drew a reprimanding frown from Cherry. "Careful how you speak. And maybe all the shooting kept the coyotes and bears at bay. I suggest we fan out about two hundred yards this side of the spot where I think I last saw him and walk up to the base of the triangle. It may be necessary to go back and forth a couple of times. Some of that sage is pretty tall and thick."

"We'll be a bit exposed out there," commented Mellen. "Could be the Utes will aim some parting shots our way."

"If we draw fire," Price addressed the remainder of the detachment, "return fire and drop back slowly. The colonel will reinforce us if he hears rifle fire." He paused. "Let's get to it."

The squad followed Cherry single file along a game trail until they reached the designated point of separation. Cherry had planned to use himself as the center point of the southbound line of search. Each man would then separate himself fifty feet from the two next to him and everyone proceed in a straight line toward the base between the two bluffs. It quickly became apparent that this plan was nearly impossible because of the height and thickness of the brush. The maze of game trails that meandered and crisscrossed through the brush forced each man to follow them repetitively.

The first sweep turned up nothing and Price ordered an about-face. "Don't follow the same paths," he yelled, "follow only those that don't have fresh hoof marks." The squad had almost returned to their original starting point when one of the troopers yelled, "Lieutenant," and dismounted. Price called a halt and waited while the trooper disappeared into the tall brush. A few seconds later he reappeared with his carbine pointed skyward its muzzle decorated with some object. "Cap," yelled the trooper.

"Dismount and form a semicircle," Price ordered, "with me at the center and close in on that mark." The men proceeded to do so but without further success.

"Spread out a hundred feet either side of me, keep walking north. His hat probably came off when he was shot. He and the horse probably carried on for a few yards before the major fell off."

They had begun traversing very slowly when Cherry, who was on the far right flank made an abrupt right turn perpendicular to the squad's line of travel. Price, irritated, dropped back to call the lieutenant back into formation when Cherry raised an arm and yelled, "Here!"

Price felt both relief and a stab of apprehension. He noted that no one in the squad hurried to Cherry's position. Most of the men were veterans of at least one Indian battle and knew the enemy never displayed any respect or sanctity for the wounded and dead. They were almost certain to witness a dismal spectacle, especially after five days at the mercy of animals, insects and weather.

"How bad?" Price asked the young lieutenant as he approached.

Cherry shook his head. "This was my first fight. I saw some of the other men who had been shot and wounded or killed but the major is the first that I've seen . . . maltreated."

Other members of the party dismounted and walked solemnly toward Major Thornburgh's remains. As they approached, they removed their hats and caps. "Not as bad as I expected," declared Price. The veterans nodded in agreement. Sandy Mellen began a story of an officer he'd observed after the Sioux women had finished with him. But Price and Cherry both raised their hands and silenced him. He stopped in mid sentence. "Couple you men go back and bring up a wagon," ordered Price.

"Better bring up a stretcher and canvas as well," said Cherry. "We'll never get a wagon through this sage. We'll have to load him on a stretcher and carry him out."

Price agreed and picked two men for the duty. "Report directly to Colonel Merritt and inform Captains Payne, Lawson and Dodge if you can find them." The men mounted and trotted off north as best they could through the maze. "You other men split up and ride near the top of those two hills. Price indicated the knolls to the south where the battle began almost a week earlier. "Don't ride the ridgeline and make targets of yourselves; dismount and walk just high enough to peek over and take a look at the higher ground to the west."

The two messengers Price had dispatched to inform Colonel Merritt and procure a wagon happened across Captain Payne in the care of Dr. Grimes in the bathing area that seconded as the doctor's surgery. Captain Payne lay half naked on two planks bridging two folding tables. Dr. Grimes bent over him with knife and scissors delicately removing the blood-and-dirt-encrusted bandages and clothing. They paused. Neither had ever seen a wounded man, let alone an officer, a near-naked officer.

Payne immediately pegged them as two of Merritt's men; they were mounted and their uniforms and tack looked to be in parade-ground condition. "Well?" he asked.

The two men turned toward him and saluted. "Sir," said one, "we're reporting back to Colonel Merritt that we found the body of Major Thornburgh. Lieutenant Price said we should also inform you, Captain Lawson and Captain Dodge."

"Grim business I'm sure," growled Payne. "In what condition did you find the major?"

"Neither of us saw him up close," answered the other trooper. "I was more than ten feet away looking from the bottoms of his feet toward his face and head. The brush, other men and horses obscured his arms. I saw he was naked and the body was swollen and crusted nearly black. His face and head were black with blood and, I think, gunpowder burns. I thought it looked like he'd been shot several times."

Payne nodded. "Scalped?" he asked.

"I don't know sir. I couldn't see the top of his head but . . ."

"But?"

"Well, sir, I couldn't tell for sure because of all the blood on the body but . . . I think . . . you know"—he pointed down toward his crotch—"they was gone."

"Customary." Payne nodded solemnly. "Thank you. Go report to Colonel Merritt."

The two troopers galloped off glad to be out from under the captain's questioning.

They slowed to a walk as they approached the wagons. One turned to the other. "Lord, Will! You could get us into a peck o' trouble lying to a captain like that."

"The hell you say! I didn't lie to him about anything."

"You know damn well he was scalped and I could see plain as day they cut out his heart or guts."

"Damn your miserable hide. I didn't see any o' that. I saw him for about ten seconds from the bottoms of his feet up and you were standin' within a yard of me. You're the one who'll be skinned alive for storyin'."

"If I'm gonna be flayed for lyin', you're gonna be drawn and quartered. I wasn't standin' next to you. I rode up from the other side lookin' down at his head. I never dismounted. I was lookin' across from you—well almost."

The two boys walked in silence for a while. Finally, Will spoke. "God almighty, Tom, we were just lookin' at the same damn thing and each of us saw something almost completely different. We keep goin' like this, one of us will say he was burned at the stake and the other will say he jumped up and ran toward Mexico like a jackrabbit."

"Yeah, guess we better not tell the colonel two different

stories. Let's just say we didn't get close enough to get a clear view. The colonel will see him when we bring him in and see what he sees."

"Okay, we better lather up these horses a bit before we hit camp or they'll think the major's in bad shape and we didn't want to hurry back and talk about it."

That evening Sam Cherry wrote Lida Thornburgh about the recovery of her husband's body. He told her that the major had apparently died quickly of a bullet wound to the chest and head while gallantly leading his troops in a fighting withdrawal back to the relative safety of the wagons. He mentioned neither mutilations nor the photograph of an unidentified Indian that lay beneath the major's hands. Cherry drafted an almost identical letter to Captain Bisbee, the acting post commander. Early the next morning a dispatch rider, newspaper reporters and an escort were sent to Fort Steele.

Joe Lawson, like Payne, had stumbled to the makeshift hospital on the creek bank where he'd bathed, shaved and treated his festering wounds. From Merritt's supply wagons, he refitted himself in new clothing and boots, sans rank and insignia.

He heard a commotion amongst a squad of troopers dismantling one of the barricades. He walked up in time to see two or three troopers pry a body loose from beneath scraps of cloth and mounds of dirt. In the chaos of the afternoon of September 29, men were frantically constructing breastworks from anything they could find, including the dead bodies of their comrades. Joe recognized the body that had tumbled from beneath the dirt to be that of Charlie Lowry, one of Thornburgh's scouts. To the amazement of all the surrounding troops the corpse sat up. Two men assisted Lowry to his feet. He hacked and coughed as Merritt's physician, Dr. Kimball, began examining him. "What's a matter boys?" he asked as someone handed him a cup of coffee. Lowry took a gulp and within seconds his legs crumpled and he sank back to the ground as though some invisible force had suddenly robbed him of all his bones. Kimball searched for a pulse in his wrist and neck. After several seconds the doctor pronounced him dead, for the second time. He had reposed with a bullet in his brain, in an unconscious or semiconscious state, through freezing nights, without food or water, for nearly six days.

"I been soldiering for almost forty years," said Lawson to a trooper standing next to him, "and never seen anything like that." He walked to Merritt's command post and told the colonel of the incident.

"My God!" replied Merritt incredulously, "I can't begin to imagine the hell on earth that man went through over the past six days, literally buried alive."

Lawson replied that Lowry's response—"What's a matter boys?"—led him to think the trooper was probably unconscious or delirious most of the time and blessedly unaware of his predicament.

Merritt nodded. "While you and Payne were off dallying in the baths like a couple of Roman emperors"—he winked—"the Utes raised the white flag and I rode out to meet with them. By incredible coincidence an employee of the Los Pinos Agency named Brady representing Chief Ouray arrived with a message ordering the White River Utes to cease hostilities. Jack followed Brady out but didn't volunteer a comment. You remember Jack, don't you? Handsome man and smart, he was with us and Crook at Rosebud and Slim Buttes, good scout. He was all dressed up in that buckskin suit and hat that the general had given him, along with the silver medal that the president gave him when he visited Washington. Well, anyway, I went over to Jack and offered him my hand. He took it and asked about my health and family. We conversed for several minutes; Brady stood off to one side and listened along with some of the Utes. I didn't see Colorow, Johnson or Douglas. My guess is they headed south with the women and children and a majority of the men. We talked about the old days—probably happier days for both of us and parted on good personal terms.

"I'm going to set up camp and command post a mile or two up river to get away from this stink. There's grass enough to last us several days."

Lawson swayed slightly and shook his head remembering that Thornburgh's expedition had passed the same location on their way to cross Milk Creek but had not stopped as the Utes had demanded because the grass appeared too sparse.

"Cherry will be my battalion adjutant and command F Company, Fifth. You'll remain in charge of E Company, Third. But I'm adding Payne's F Company, Fifth. Both of you will receive commendations, decorations and promotions accordingly when we get this mess cleaned up and get back to the fort."

"Yes, sir, thank you. What about the agency?"

"That's next. Tomorrow you and I, provided you're up to it, will take a detachment south. I'll leave Dodge in command. We'll provision ourselves for three or four days. I don't want to be encumbered with wagons if we have to move quickly. Brady, the man who delivered the message from Ouray at Los Pinos, said he'd come through the agency and observed that hardly a building was left standing and he could see a number of bodies scattered about. One of the Utes told him that the band with the white captive women and children were headed toward the Grand River and probably up the north side of Grand Mesa. General Adams, the Los Pinos agent, is gathering a search party. For now we'll leave that business up to them, there's more than enough to do here. Once we're finished at the agency, we'll pursue Douglas and Johnson from this side of the mesa. Do you feel up to going, Joe? You've been through a lot."

"Yes, sir, I'd like to go. If I can get a decent night's sleep tonight and stay well fed for the next few days I'll be fine. It'll be good to straddle a horse again."

"Good, prepare to ride out at eight tomorrow morning. According to the stories I'm hearing, I doubt if a doctor can help anyone there but I'm thinking of taking Grimes."

"Better take Kimball. Grimes is wounded and exhausted, it's a wonder he's still walking."

"Kimball it is. Who else?"

"Jim Baker, Joe Rankin and a squad who know how to ride and shoot. There may be some Utes between us and the agency."

"I'm ordering you to be missing in action for the rest of the day. Get some food and rest."

"No argument there, sir." Joe saluted, turned on his heel and aimed himself in the direction of the nearest mess line. To the west of the line of wagons he saw another line forming at the Ute trenches which reminded him that he hadn't had a normal hot meal in over two weeks and his stomach certainly would rebel if suddenly attacked by solid food. Plus, his gut still cramped from the poisoned river water. He ate one portion cautiously and part of a second before rebuffing further temptation.

A jolly quartermaster sergeant supplied him with a folding cot, two blankets and a pillow. Since the camp was moving no tents had been erected. Joe, from force of habit, sought out his old dugout and set up the cot next to it. He stretched out on the cot under the blankets in the early-afternoon sun; uniform, boots and overcoat. Within minutes he was conscious of an almost forgotten sensation—he was too hot! But by that time he was also too sleepy to care. He fell into a deep, dark pit of silent tranquility.

His nearly bursting bladder awakened him to darkness. He

poked his head out from beneath the blankets and rocked from side to side thinking he still occupied the ditch that had been his home for nearly a week. Instead he rolled off the cot into the dust.

"Steady, Captain," he heard a reassuring voice. "Y' don't need t' be tumbling out of bed, sir. Y' visited the ground quite enough the last few days."

"Sergeant Grimes?" he asked in bewilderment.

"Yes, sir."

"Where the blazes am I . . . are we?" Even in the darkness, Lawson could sense the hulking Irishman straddling him. Unseen hands scooped him up by the armpits and seated him back on the cot.

"Still at the old camp, sir. I talked the colonel into letting you sleep rather than roust you out when they moved down the road. So we just built a tent over you, and I stayed behind to cover your needs." Grimes struck a match and lit a lamp which revealed the comforting canvas walls and another cot. "Anything you be needin', Captain?"

"Yes, a trip outside." Lawson launched himself toward the slit door fumbling with his buttons on the way. In spite of the chill, Joe stood and admired the carpet of brilliant stars ending on top of the black ridgeline to the west and absorbed the palpable relief knowing he could do so without risking his life.

"Captain," Grimes called out from inside the tent, "are you all right?"

"Yeah, Ed. I'm out here enjoying a good piss . . . enjoyin' life."

"Yes, sir! But I'd prefer you come in here and enjoy life. You come down with an episode and don't make officer's call in the mornin', the colonel will bust me so low I'll be saluting privates."

Joe returned and sat on the edge of his cot. He reached for his watch and remembered that he'd turned in his waistcoat along with the remainder of his old uniform and the watch was in the pocket. "Damn!" he exploded. "Ed, what did they do with our old gear?"

"Burned 'em." Grimes replied calmly, observing the panicky look on Lawson's face. "All 'cept this." He pulled the gold disk from his pocket and dangled it from its chain in front of Lawson who took it and clasped it to his chest.

Grimes had the horses saddled and ready the next morning before awakening the captain. Lawson was pleasantly surprised that one good meal and fourteen hours of sleep had so quickly revived him. "Sergeant, will you be accompanying us today?"

"No, sir, have to work today making sure the lads are fed, bathed, shaved and clothed. Weapons need repair and cleaning, horses and tack supplied. I've a notion what you'll be seein' and I can do without."

"Just so," Lawson agreed. "Together, you and I have seen enough death and destruction during our lives to populate a small city in the hereafter."

CHAPTER 23

"So may all who engage in such lawless conspiracies perish."

Scipio Africanus the Younger

Merritt and Lawson stood their horses athwart the top of Yellow Jacket Pass and surveyed the thin ribbon of trail winding down through Coal Creek Canyon disappearing into the smoky haze of the distant White River floodplain. The wide double-track wagon road they'd been traveling tapered into an eroded trail barely wide enough to accommodate a wagon. Oak brush, mountain mahogany, serviceberry and cedar hemmed it from both sides like thick walls. The trail sliced through the brush one to two hundred feet above the late season trickle of unnamed creeks. "Steep," commented Merritt.

"And narrow," added Lawson. "If the Utes had let us get this far a week ago they'd have wiped us out for sure. No way to maneuver wagons for defense down there."

"One Ute with a Winchester could stop us cold."

"One Ute with a slingshot could stop an entire regiment cold."

Merritt spurred his horse forward and brought his legs upward in the stirrups and leaned back against the saddle cantle to help the animal's balance on descent. Lawson and the others followed suit.

Lawson noticed the red trail soil had been ground to thick

dust and the rocks broken and scarred from the hooves of hundreds of ponies winding their way back and forth from the plundered agency supplying their comrades at Milk Creek. The once pronounced double track with a centerline of foliage was now a "V" shaped single track on which the horses carefully picked their way to avoid turning their ankles.

The morning sun quickly warmed the canyon. Men held the reins in one hand and constantly unbuttoned overcoats and waistcoats with the other. Birds took flight fluttering and scrambling from one bush to the next looking for dropped or discarded morsels.

Larger more ominous birds glided northward on the heated updrafts from the canyons and hills looking for another sort of meal. A constant swarm of black shadows traversed the trail and bushes. Merritt and Lawson, by far the most widely traveled men of the party, remarked that even the thousands of gulls inhabiting coastal fishing villages they had visited over the decades did not match the numbers of buzzards they could see nearly from horizon to horizon. Merritt spoke over his shoulder to Lawson. "Now that the shooting has stopped and the battlefield's empty, these boys and the coyotes will have it down to a boneyard in a matter of days."

Lawson slackened his reins and let his experienced horse drop his head and pick its footing among the rocks and deadfall. He leaned back and let his body follow the animal's roll and pitch. "Puts life in a different perspective, don't it?" he said to Merritt in front. "Doesn't matter if you're a general or private, rich or poor, short or tall, smart or dumb, man or horse, we all pretty much taste the same to a buzzard, coyote or worm."

"Not you, Captain," replied a voice behind him, "you'll taste like an old leather strap or weathered pine plank."

Joe put his hand on his horse's rolling flank and looked over his shoulder at the line of enlistees behind him. "That's pine plank, *sir*, to all of you."

"Sir!" echoed several of the men.

The road remained monotonously the same grade mile after mile. Merritt tried giving the horses and men breaks by dismounting and walking until several of the men slipped on loose rock and fell beneath the hooves of the horse behind them. Every few miles the track descended to the stream, crossed over and down the opposite hillside. The men dismounted at these crossings, watered the horses and themselves, smoked and adjusted tack.

During one of the stops, a shot from deep in the oak brush reminded everyone that they were still in battle and sent them scrambling for their carbines and defensible positions. After several very tense minutes and no further fire, Colonel Merritt gave the order to stand down and ordered the men to separate by at least twenty feet. The younger men who had earlier chatted with near abandon and even attempted a few marching songs now proceeded in silence. Even Lawson flinched a few times when a horse's hoof clipped a stone with a loud pop.

Above, battle-wise buzzards stopped their northward migration and circled the patrol. A few had even landed in trees along the path like Roman spectators lustily waiting for the first death in the coliseum. The men watched them with disgust and apprehension as they rode past but no one dared to curse these morbid prophets nor moved to scare them away.

The sun was well into its downward slide toward the crests of the western mountains when the colonel finally signaled a halt at another stream crossing and gathered the men around him. "How far to the agency?" he asked Baker.

"About four hours. This trail's a heap worse and slower than I thought. It's about two hours down to where it opens up and flattens out. They's a couple old ranch shacks down there—what had fireplaces if they haven't burned or fallen down. We could hole up there and make it into the agency tomorrow mornin'. Ain't nothin' there that won't keep one more night?"

Merritt nodded in agreement and slapped his bridle reins against his leg. Lawson could tell from his commanding officer's expression and his own long experience with the colonel that he was not pleased about this unintended bivouac, it thwarted his self-imposed schedule.

"Wesley Merritt's greatest strengths," General Crook had once confided to Lawson, "are his unflagging attention to detail and organization. But like a lot of individuals with similar strengths, he lacks creativity and adaptability. If something goes awry with his plan, he has difficulty improvising. He falls in love with and gets married to one and only one course of action."

Joe could sense this failing now. Thornburgh's vacillations still unnerved him. While he still felt strongly about supporting his commanding officers he suffered recriminations for not having more strongly contested the major's plodding indecisions which had led the entire expedition onto the reservation. Now, he felt similar misgivings about Merritt's inflexibility and determination.

Except for Baker and Rankin, everyone in the patrol was in unfamiliar and hostile territory. The bodies of the agency victims—almost certainly there were no survivors at the plundered agency—would be almost impossible to find in the dark and only slightly more scavenged by carrion eaters by tomorrow morning. The pragmatic course of action would be to settle into a convenient camp before dark and proceed at dawn.

"Plus," added Baker, "if I recall correctly, they's some corrals left so we won't have to hobble or picket the horses tonight."

Merritt nodded. "All right, let's hurry up and get out of this bottleneck before dark. Jim, you ride ahead and find the place, get a fire started and we'll locate the smoke. If we don't show up by dark come back up the trail."

Sunset approached as the troops cleared the last of the gorge's steep escarpments and entered upon a broadening alluvial plain bisected by low rolling hills. The narrow drainage path now also flattened and returned to more of a true road. While a relief for both horses and men, it also exposed them to ambush from the darkened cedars. During the afternoon, Joe had begun feeling the effects of the last two weeks' deprivations and injuries. He struggled to stay awake in the saddle. Removing his feet from the stirrups and leaning back on the cantle forced him awake to keep from falling off.

Merritt called for a stop but not dismount. Joe rode up beside him. The colonel inhaled deeply. "Smell anything?" he asked.

Joe sniffed. "Smoke—it's too early for Baker to have made it to the cabins." He then noticed a thin, blue stratus hovering over the hilltops and up against the mountains. "Don't see where it's comin' from—could be Indian camps."

Joe Rankin volunteered to ride to the summit of a small hill a mile away on the right of the road. Lawson provided him a small telescope and cautioned him not to crest the ridge and make a target of himself. While the squad watched, Rankin galloped to within a few hundred feet of the crest, dismounted and climbed near the crest on foot and then crawled to the top. He lay looking southwest for several minutes and then scooted back down to his horse. He galloped back to Merritt and Lawson.

"I wouldn't count on shelter for the night," he reported breathlessly. "The fire's out, just smoke now. Chimney is about the only thing left standin'."

"Any sign of life? Baker? Indians?"

"No, sir, neither." Rankin handed the telescope back to Lawson.

"How's the ground there?"

"Flat; someone cleared the trees and brush for several hundred yards, that's probably why the fire didn't spread. Looks like some of the corrals might still be there."

"We'll move in before we lose the sun. Joe, send out a couple of flankers with orders not to fire until targets are verified. Baker's still out there somewhere, maybe even agency survivors."

The flankers galloped about half a mile either side of the road but remained in sight. Merritt moved the patrol forward at a trot. When they crested the hill, he could see a distant funnel of smoke. Through his telescope he could make out what appeared to be the remains of a fireplace and chimney and the dotted outline of a pole fence, not much more. No sign of life. He signaled forward. As they drew near, Merritt could see no buildings remaining anywhere, only smoking cinders, ashes and the chimney, not a person in sight. But looking down he could see the soil was well-trampled with hoofprints and footprints. One of the men called out, "Colonel!" He looked in the direction the trooper was pointing. The figure of a mounted man emerged from the cedars outside the pole corral. Merritt could see it was one of the flankers and another man riding double behind him—Jim Baker!

Baker slid off the rump of the horse and walked over to Merritt, Lawson and Rankin. He had no rifle; his holster was empty and his hat missing. "Colonel, sorry to be late in getting back to you but, as you can see, difficulties arose."

Merritt laughed, "I'd say so, yes. What happened?"

"I pulled in here early in the afternoon and found the place in pretty good shape. I built a fire in the fireplace to warm up. I sat here a bit and heard horses outside—thought they was you. When I opened the door I came face-to-face with Piah, one of Douglas's men and about a dozen other Utes that I didn't recognize. They didn't do me any harm, in fact, we had a bite to eat together, and *then* they stole my horse and saddle, weapons and hat and set fire to the house, barn and privy. They left and I lit out for the woods in case they changed their minds and decided I had too much hair."

"This means supper will be late?" Lawson inquired.

"It do."

"Think they'll be back?"

"Don't think so. If they'd really wanted to have another go-round they'd have killed me and waited to ambush you. They appeared to be more interested in heading south. But that ain't saying there's not another bunch lurkin' around out there who'd do us harm tonight."

Merritt ordered one of the pack mules forward and its load distributed among the others. The men created a saddle from the pack frame by adding three blankets for padding and a hackamore bridle rigged from odds and ends of leather latigo. Baker climbed aboard and, after a few minutes of snorting and crow-hopping, proclaimed the animal broken to ride.

Upon reaching their degraded destination, Merritt ordered the horses corralled and fed, wood gathered, fire started and a tent erected. "We only have one squad tent so you'll be shoulder to shoulder during the night. Any man with delicate sensibilities can sleep outside. Keep your weapons and ammunition close by at all times. Two-man watches for two-hour intervals from eight o'clock until sunup. Rank has privileges; I and Captain Lawson will take the first watch. No moon, so keep your ears pricked. Pay attention to the horses: they'll hear, see or smell something long before you."

The scarcity of wood precluded a large fire. The original homesteaders had cleared trees for cabins, barn and fence for several hundred yards' radius around the ranch house. Every man in the squad hustled into the deepening twilight to scrounge for twigs and branches. Rankin happened upon a stack of fence poles probably intended for expanding or building a new corral but never used. The men celebrated as though it was stack of gold only to discover that no one had thought to bring an ax or saw. After building the fire the poles protruded from it like the spokes of part of a wheel. As the poles burned down they were simply pushed further into the fire. Spirits rose with the addition of a hot dinner, even if it was field rations of beans, frankfurters, hardtack and coffee.

Joe Lawson, in spite of his long night's sleep, felt himself succumbing to drowsiness and drank more coffee. He felt as though he'd lived another entire life during the past two weeks.

Merritt's men, Rankin and Baker, participants in the "lightning march," were also falling silent, victims of full stomachs and previous days of little rest and sleep. It took only one of them to stand and "call it a night" as a signal to the others to follow suit. Soon only Lawson and the indefatigable Merritt were left to tend the fire. The colonel turned the wick up on a lamp and declared he was going out to have a look around.

"Good idea," said Joe, "I heard a Rocky Mountain barking spider out there a bit ago. You come across one, holler out and we'll catch 'im and have 'im for breakfast."

Merritt just shook his head.

Inside the tent, Captain Lawson waited until the rest of the men had settled under their blankets before retiring near the slit door. He knew that he'd be up at least once during the night to relieve himself and wanted to avoid tripping on two or three men on the way out. He was already cold and the chill seemed to burden his muscles and joints excessively. Every muscle, tendon, bone and ligament in his body felt bruised, broken, pulled or torn. From his coat pocket, he pulled a half-empty bottle of laudanum Doc Grimes had given him and examined its contents closely. Maybe three shots left. One tonight, one at the agency and one on the trail back to Milk Creek. He took a gulp of the stinging liquid and smiled as a soft, gray mist enveloped him even before he burrowed beneath the blankets but lying on his side with his torso resting on one arm so if he nodded off the fall would awaken him. The frozen, rocky ground softened. The air warmed and the cacophony of six snoring men became a soothing drone. He blinked his tortured eyes furiously while staring into the glowing embers.

Joe awoke from the blackness of a deep sleep to the sound of a crackling fire, clatter of tin cups and the low murmur of men's voices. He opened his eyes to an empty sunlit tent. Casting his blankets aside he poked his head through the slit door. The men, including Merritt, were huddled around a breakfast fire drinking coffee and thawing rations. Merritt, smiling, sat staring at him. Joe jumped to his feet at attention and saluted. "I beg the colonel's pardon for oversleeping and missing my share of last night's watch."

"Acknowledged. Your punishment will be singed fingers and a cup of the worst coffee I've ever tasted. Even I saw no reason to rouse you from such a restorative sleep. Approach and be seated."

Joe shuffled toward the fire and someone handed him a scorching tin cup filled to the brim with something the color and constancy of mud, not coffee. "Do I drink it or cut it up and eat it?" he asked.

"Hurry up and drink it," laughed Baker, "or it's gonna eat holes in that cup."

Joe took a sip that scalded his lips and tongue. "Whoa!" Joe exclaimed. "It bit back. Hand me a frozen canteen to kill it." One of the men handed him a canteen and he diluted the coffee with ice chips. He sat down and accepted some warm ham and beans from the men and settled in as contented as an old dog. A cold, rosy morning in the mountains, blazing campfire, warm cavalry wool twill, shared food and comradeship, he was right at home.

He had loved his wife and sometimes longed for her, the

comfortable house and familiar store in Kentucky. But they had occupied only a few years of the nearly half century he'd devoted to various armies and wars. He never feared death or wounds. They'd been constant companions from the day he was born in a windowless stone hut in Ireland. He'd never been alone. First were his mother and siblings, and then the army, his wife and two children who perished before the American war, then the American army. His wife succumbed to pneumonia. Privacy and individualism were curious, alien concepts. Peace and tranquility had been little more than fleeting respites between battles and wars. "War," General Sherman once confided to him, "for me *is* life. Everything else is just waiting."

"All right," said Merritt rising to his feet. "We're burnin' daylight . . ."

"Time to piss on the fire and call in the dogs," interrupted Joe.

Merritt chuckled. "Fold the tent, police the area and be ready to mount in thirty minutes. Captain Lawson, Jim Baker and Joe Rankin stand down. We need to talk before I give marching orders." The four men clustered around the dying fire.

"Baker," asked Merritt, "how far are we from the agency?"

"About five to seven miles mostly brush and sparse timber lowering into sage, grass and cedar."

Merritt nodded. "I don't expect an ambush or shooting match do you?"

"No, I think they got killin' out of 'em for a good long while to come. They didn't kill me when they had the chance. I think they're headed south toward the Uncompahgre and whatever safety Ouray can offer them."

"Anyone else of a different opinion?" asked Merritt. No one spoke. "I'm pretty certain what we're going to find today—a welcoming committee of body parts and mutilated, rotting corpses or bones, depending on the efficiency of our feathered friends. I think most of us have extensive experience with such. Let's wrap this up today, return to Milk Creek tomorrow. Unless there's a compelling reason to do so we're not camping overnight at the agency. I think every one of us has had our fill of reeking death." The morning sun was slicing the frost from the grass when the detachment rode south at a canter slightly after eight o'clock in the morning.

Merritt, to save time, followed the road and decided not to deploy flankers into the thickening brush and trees. Within an hour the squad had nearly covered the distance between their last encampment and the agency. As they approached the White River, a lone mounted figure emerged from the trees onto the road several hundred yards distant leading another horse. The colonel raised his hand to stop his men. He uncased his binoculars which revealed the rider was not an Indian but little else. Jim Baker eased up alongside Merritt.

"Mind if I take a look, Colonel? Something looks familiar." Merritt handed him the binoculars. Baker adjusted the focus back and forth then exclaimed: "I thought so. That's my horse he's leadin'."

"You sure?" asked Merritt with a surprised tone.

"Yep, I recognize the rider. It's Black Wilson, the mail carrier. He carries the mail back and forth between the agency and Peck's Store to Rawlins." He handed the glasses back to Merritt and nudged his horse forward at a canter with the rest of the men close on his heels.

As they rode up to Wilson, Jim Baker yelled out, "Horse thief! Lynch 'im." The man exhibited no fear as the troops reined in around him. He calmly held out the bridle reins of the led horse to Baker as he pulled up.

Merritt muttered, "Good God!" to Lawson who knew immediately what he meant. Always the immaculate, fastidious West Pointer, Merritt had, at times, expressed a deprecating view of Jim Baker. The quintessential mountain man dressed day after day in greased buckskins, a long but well-combed black beard, shoulder-length hair. Baker's hands, blackened by the sun and weather, grimy from months if not years of grease, blood and dirt, dangled from his sleeves like reptilian feet. He seemed not to notice or care that even the hardiest of soldiers and scouts shared no campfire seat with him and consistently positioned themselves upwind from him. But this new chap had taken Baker's less desirable but more salient features to nearly indescribable lengths.

Even Lawson, who had experienced some of the most ragtag examples of humankind that Asia and Africa and America's frontier west had to offer, was appalled at the figure straddling bareback a white pony. Both man and horse merged into a formidable automaton constructed of dirt, bone, brambles, hide and hair. This truly, he thought, was the earthly embodiment of illustrations he'd seen of the Grim Reaper.

The man's greasy hair clung to his head like a tarnished silver helmet covering his ears to his neck where it was gathered in back in one long, white braid. He wore no hat and his forehead glistened as though painted with black lacquer. His eyes hid in dark caves below bushy eyebrows giving him the appearance of some primordial mammal. He possessed few other facial features, save for his sunburned nose. A thick beard cascaded to his chest. He spoke with a deep, guttural voice and an indefinable accent. His garments mirrored Baker's buckskins except he wore a crude sheepskin vest. Both men laughed and exchanged greetings. "Where'd you find him, Black?" asked Baker.

"Grand Valley," growled Wilson. "Piah came into town yesterday with half a dozen braves. I'd holed up there since hell broke loose at the agency. They stopped at the store to buy food and I recognized your horse. I ask Piah how he came by him and he said he bought him off you the day before. I knew he was lyin' through his teeth but bein' outnumbered I didn't push it. I rounded up a dozen or so men armed to the teeth and persuaded Piah to swap your horse for a safe ticket out of town, one of 'em still has your rifle, though."

"I'll manage. I'm just glad to get my horse back." Baker introduced him to Merritt, Rankin and Lawson and the troopers.

Wilson surprised everyone by walking his horse alongside Merritt and offering his hand. Merritt hesitated and then took it. The mountain man twisted the colonel's hand palm-down and leaned over to examine the enormous West Point ring. Merritt recoiled but said nothing. "What year?" growled Wilson.

"What?" Merritt responded.

"Year you graduated?"

The colonel appearing perplexed answered: "Sixty."

Wilson released Merritt's hand. He pulled his beard aside with his left hand and with his right pulled forth a leather thong looped around his neck from inside his vest and greasy jerkin. He held it up so the sun twinkled brightly from a swinging circular ball. "Eighteen sixty-one," he muttered. "Same as Custer."

"Good God!" Merritt exclaimed. "I remember that class . . ."

"Because," Wilson finished, "we were the only class in the history of the Point to graduate a year ahead of time—the army had a dire need of officers for cannon fodder."

"Yes! Precisely. What . . . ?"

"Happened, that I ended up like this?" Wilson swept his hand downward like an actor taking a bow. "Well, to begin with, I was in the fray from the First Bull Run all the way to Petersburg. Almost everyone I knew in the army in the spring of eighteen sixty-one had been killed by sixty-five. By that time I'd had a veritable herd of horses shot out from under me; lost an entire company at the edge of the cornfield at Antietam. The day after Lee surrendered I took my company onto a plantation south of Richmond to serve notice to some irregulars that the war was over. They ambushed us and most of my face got shot away. Army dismissed me for bein' too ugly. So I drifted west and heard Major Thornburgh was in command at Fort Steele. I was his commanding officer when he was a sergeant major with the Tennessee Volunteers. He got me the job of delivering mail, something I could do without too much human contact. Heard Thornburgh lost his command at Milk Creek?"

"No, in fact they defended themselves admirably and probably didn't lose as many men as the Utes. Unfortunately, Major Thornburgh fell in the first few minutes of the battle. I'm sorry to learn of your ill fortune Mr. Wilson," sympathized Merritt.

"Could have been worse. I could have been Custer or Thornburgh."

"Indeed. Did you know Meeker and his workers?"

"All of 'em. They're scattered all around what's left of the agency."

"Would you mind accompanying us there and help identify the remains?"

"I'll do that but we'll have to move slowly. This pony's about played out."

"How'd you come by that bag o' bones Black?" asked Baker. "Anytime I ever saw you deliverin' mail, you were on first-class horses."

"Man ridin' a mule with a packsaddle hasn't got much to brag about. Well, I kinda forgot to mention I swapped my horse to Piah for yours and got this one as part of the deal."

"Saddle up Baker's mule now that he has a horse," injected Merritt. "I'll see to it you get another first-class mount when we get back to Milk Creek."

"Agreed," said Wilson.

The squad continued their way south. To Lawson's surprise, Colonel Merritt dropped back and fell into step beside Wilson. The two men were soon talking and laughing over stories from their West Point days.

Within an hour they spied in the distance the blackened, skeletal remains of buildings and, where the road curved toward them, the first victim, still the object of interest among an aviary of crows and magpies, the vultures evidently having abandoned the carcass for better pickings at Milk Creek. "Colonel," yelled Lawson, "I think we have our first body." Merritt broke off his conversation with Wilson and galloped forward. One of the troopers sighted his carbine at the flock of birds.

"No shots!" ordered Merritt. "We don't need to be drawing the neighbors' attention to us." He motioned Wilson forward.

"Frank Dresser," mumbled Wilson indicating the body. "He was from Greeley, only about fifteen years old, so homesick he'd write his mother every day. Josephine Meeker looked after him like a mother hen, took lunch to him every day. Like a kitten, he was harmless, helpless and hopeless." He urged his pony forward. Merritt followed subordinately in his tracks while Lawson held back the others. As the two men approached the black feathered harpies shrieked their annoyance and spread their black wings ominously but eventually retreated and perched upon the bony branches of dead trees to await their next opportunity.

Merritt had braced himself, expecting to see a partly eaten, partly rotted corpse, but was surprised instead to view a random scattering of bones and scraps of cloth, but no skull. It was a sight so innocuous his horse would have walked through them if he hadn't reined in the animal. "Good Lord! There's nothing left. How did you know who it was?"

"I first saw the bodies within a day or so after the slaughter," Wilson replied. "Frank was the only man here with curly blond hair—what there was left of it. He also still had part of a face at that time. I reckon some critter carried off his skull." Merritt signaled the rest of the party forward and gave two troopers instructions to dig a shallow grave but cover it well with heavy rocks and a note in a bottle identifying the victim for later reinterment elsewhere. Doctor Kimball, noting each in his notebook, carefully separated the bones from identifiable personal effects to send to his parents in Greeley.

The remainder of the party proceeded through the ghostly ruins of burned buildings, scattered empty barrels, sacks, cans and bottles. The fleeing Utes had taken everything else edible or deemed useful and transportable. Abandoned cast-iron soap cauldrons and farm equipment were surrounded by the fragments of heavier rocks that had been used in futile attempts to destroy the much-hated white-American trappings.

In the middle of the road, the Utes, it was obvious, had taken a special interest in putting a particular carcass on display. All the men dismounted and removed their hats as they approached the miraculously intact but bloated body of Nathan Meeker. His wrists and ankles had been bound spread-eagle to deeply driven iron stakes. Some brave had taken special pains to skin his scull from the eyebrows and ears to the nape of his neck. Equally as gruesome, a large, conical stake driven into his mouth pinned his head to the road. His eyes were wide open and bulging through a crown of gore. "I can only hope," said Merritt solemnly, "that he was dead or at least unconscious at the time."

"Afraid not," observed Wilson. "You can see his wrists cut to the bone from trying to pull loose from the stakes. Here lies a much-hated man."

The detachment established a routine. Black Wilson would lead the party to each victim. The bodies were found in varying stages of disintegration. He and Dr. Kimball collected what limbs and bones they could find. Scalps were missing, severed or mutilated genitals were usually found in mouths, heads and faces bludgeoned beyond recognition. These were not victims of a dispassionate, lightning strike for freedom by subjugated men and women but rather the end product of years, decades of torment, injustice, anger, frustration and hatred. Rotting flesh, bones, clothing fragments and what few personal items the Utes had ignored or discarded were buried and marked for identification.

By day's end they had accounted for Nathan Meeker, Frank Dresser, Harry Dresser, George Eaton, Wilmer Eskridge, Carl Goldstein, Julius Moore, William Post, Shadrack Price, Fred Shepard and Arthur Thompson.

No women or children were found. An unspoken question hung like a pall beneath late-afternoon leaden skies.

As befitting such an occasion, a cold southwesterly wind mixed with sleet began pelting the small cluster of men and horses. Merritt ordered his silent band to prepare to return to their previous night's camp.

No one moved.

"Or," Wilson spoke up, "we could move up to the horse pasture next to the river where the horses could be hobbled and graze on decent grass during the night."

Merritt surprisingly agreed, "Yeah, or we could do that."

The men hobbled and cared for their horses and erected the tent close to where an eddy in the river had piled up literally tons of driftwood. After dinner, despite the chill, there was no shortage of men volunteering to stand watch and feed the roaring fire. Lawson watched the men huddling together under blankets and clasping their carbines closely. He felt that he could just as well be watching a group of prehistoric hunters fending off cold, enemies and malevolent spirits. Humans instinctively sought out heat, light and each other when the gloom of night threatened to devour them. Even the men inside the tent maintained the blazing, fuming wick of a lantern all night long. He and Merritt abandoned the smoke-filled tent about midnight rather than order the light extinguished. So fear-filled and sharp-edged were the men that the officers chose the cold rather than risking a mutiny.

The next day's journey back to the Milk Creek encampment was ponderously slow and silent. The previous evening's wind and snow pellets portended an early winter storm that drove the men deeply into their coats and blankets. Wilson gratefully accepted the generosity of his new companions who gave him shirts, a jacket and two blankets. The coiled and deeply furrowed path back to Milk Creek proved no easier to negotiate on the return. The horses having struggled with the narrow, rock-sewn journey to the White River Valley now had to contend with wet, slippery clay footing. Merritt, much against his urgent nature, yielded to frequent stops during which the men tore strips of cloth from their spare clothing to bandage and protect their horses' bleeding knees. The troop uttered an audible, collective sigh of relief late in the afternoon when the party surmounted the crest of Yellow Jacket Pass. Seemingly in celebration of that event the sleet ceased, clouds parted and the bright sun birthed a collective cheerfulness the men had not experienced for three days.

Army wranglers herded horses and mules from the railhead at Rawlins south to the battle site to replace those killed during the siege. Over three hundred cavalry and draft animals had been killed, most of them in the first twenty-four hours after the ambush. Colonel Merritt let Black Wilson take first pick of the new arrivals and provided him with a new saddle and tack. "Thanks for your help," he said holding out his hand.

Wilson, sporting short hair, trimmed beard and a well-scrubbed appearance, took the colonel's hand. "It's been nice meetin' up with you, Colonel. Every once in a while the Point graduates a human, from what I've seen you're one of 'em. I hope our paths cross again someday."

"I hope so too . . . what is your real name?"

"Wilson Black. When I signed on with the mail service I put my last name first just like the army taught me. I've been Black Wilson ever since."

Captain Dodge and the men of the Ninth Cavalry departed Milk Creek nearly casualty free. They rode to Rawlins and traveled by rail from there to Denver and then to their post at the Los Pinos Agency.

Captain Payne's wounds prevented him from riding back to Rawlins. Instead, he was forced to endure a bumpy wagon ride directly back to Fort D. A. Russell. He never again engaged in combat but remained behind a desk in the army until his death several years later.

Captain Lawson, Lieutenant Cherry and Lieutenant Price led what remained of the Third and Fifth Cavalry regiments and Fourth Infantry back to Rawlins where they caught the train back to their respective forts.

Merritt received orders to pursue Douglas's band with the captive white women and children. Earlier information that they had crossed the Grand River and followed the Uncompahgre River south to the Los Pinos Agency now appeared to be incorrect. From local ranchers and farmers the colonel learned that Douglas, whether by design or confusion, had led the group east paralleling the river and crossed it just west of the town of Rifle. If their objective remained the Uncompahgre Valley, a direct course over the nearly eleven-thousand-foot-high Grand Mesa would be the shortest but most difficult route. One or two trails led directly over the summit and into the valley. Otherwise the north slope of the plateau was a maze of game trails through nearly impenetrable forest and brush. Only a guide intimately familiar with the area would be able to stay on the right trail.

Yes, thought Merritt, the Utes had occupied Western Colorado for centuries but he doubted that the reclusive Northern White River Utes possessed much familiarity with the territory south of the Grand River. He examined the hopelessly limited and inaccurate map given him by Jim Baker and scratched his head. The agency massacre had erupted suddenly and spontaneously. Probably the same was true of the white women's capture and the Utes' flight from the burning agency. Seized by panic the addled and alcoholic Douglas wandered off in the most logical direction, away from the battle. He really didn't know where he was leading the band—just fleeing. Merritt decided to take the most direct route to where Douglas was last seen crossing the river and try to pick up his trail on the other side. He did not yet know of General Adams's rescue efforts and his journey north from Los Pinos.

CHAPTER 24

"A wind has blown the rain away and blown the sky away
and all the leaves away, and the trees stand. I think, I too,
have known autumn too long."

E. E. Cummings

On October 2, 1879, Chief Ouray, "The Arrow," was a dying
man. His once artfully tailored buckskins pouched over a
bloated frame. Diagnosed with dropsy, the once-powerful warrior and Ute chief now could barely walk unassisted. That morning, his sister, called Susan, the wife of Northern Ute's shaman,
Johnson, had brought news of the massacre at the Meeker
agency. An hour later, Yanko told him of the ambush at Milk
Creek. Sick and depressed, Ouray told her he wanted to end his
life.

He had been born in the 1830s, the son of an Apache father
and Uncompahgre Ute mother who loaned him to local
Mexican and Anglo families for various jobs. This proved
ultimately to be a boon to the precocious Ouray who soon spoke
four languages: Ute, Apache, English and Spanish. He also possessed the physical foundations of a great warrior; short and
sturdy but muscular and athletic. He was naturally quiet and
contemplative but when roused to anger, exhibited dominating
fierceness, an opponent feared by white and Indian alike. All
these qualities combined as he matured to propel him at an
early age into leadership roles. Gradually, his mother's band the

xenophobic Tabeguache Utes accepted him as a developing chief. His first wife, Black Water, died young leaving young Ouray with three children.

It wasn't long before a chubby but most beautiful Tabeguache woman, ten years younger, caught his eye. Chipeta, eventually acknowledged as "Queen of the Utes," became not only his wife but his closest, wisest and most-trusted confidante during the most difficult period of his life.

As Ouray's prominence grew within the band becoming more commonly referred to as the Uncompahgre Utes, he was thrust, because of his linguistic abilities and recognized wisdom, forward as a chief representing that particular subtribe. He was invited to attend the negotiations resulting in the ill-fated 1868 treaty in which virtually all of Western Colorado and much of Northern Colorado were ceded to the Utes forever. His name appeared first among those of his fellow tribesmen on the treaty. Among the whites he quickly gained the titular title of Chief of the Utes.

The ink on the treaty had barely dried before opportunistic white miners and settlers violated it. Ouray, frustrated and angry, demanded and received an audience with President Chester Arthur, a Harvard graduate and former Civil War general, who emphatically stated that Ouray was "one of most intelligent men I have ever conversed with."

A new treaty, the so-called Brunot Treaty of 1873, was drafted and broken. Ouray recognized the futility of a few thousand Utes waging a successful war with the United States Government. "Dealing with the whites," he said, "is like being a bull buffalo shot full of arrows; you eventually just roll over and give up." By the evening of October 2, 1879, he felt like doing just that.

However, the next day he was spurred to act when he learned that Merritt was assembling a large force to break the Milk

Creek siege. With the approval of the Uncompahgre agent, he dispatched his brother-in-law Sapovanero and agency employee Joe Brady with a cease-fire order to the battle site. He also learned that Douglas and Johnson were headed south with the captive white women. This news was particularly alarming. Indian men, while not completely indifferent to the sexual torment of their women by a foe, more or less accepted it as a by-product of battle. The rape of one's wife or female relative would be repaid in kind at some future time. It was not something, except in extraordinary circumstances, that led to the ostracization of the victim and all-out retribution against the offending side. But the white man considered this personal violation cause for the most extreme acts of revenge, especially if the rapists had darker skin. The whites would use the violations to wreak havoc equally on all Utes.

Exactly that was paramount in the minds of a group of white men, the powerful and the powers to be in Colorado. Carl Schwanbeck, a German emigrant, married to the daughter of Colorado's territorial governor, Edward McCook, took the name General Charles Adams when McCook appointed him commander of a small militia force. The name stuck.

Adams, a tall, muscular Teuton and carnivorous politician, had gained some measure of personal notoriety by once physically throwing three-hundred-pound Chief Colorow bodily out of Governor McCook's office. He later spent four years as the White River Indian agent, a position from which he conspicuously profited. When, on the fourth of October, Adams heard the news of the Meeker Massacre and Milk Creek battle, he imagined the worst when white females fell into the clutches of rapacious racial inferiors. But, his motives for saving the white women and children centered upon an opportunity for political gain and personal aggrandizement.

He wasted no time. Secretary Schurz, prompted by Adams,

ennobled him with a title: Interior Department Special Agent. Adams immediately contacted Governor Pitkin and Senator Teller to tell them he would be collecting a troop of militia to rescue the women. Ever on the lookout for more publicity, Adams curiously and incongruently included a visiting German diplomat, Count August Donhoff of the German legation in Washington, DC.

Douglas probably did not know of—perhaps could not comprehend—the forces closing in on him from both north and south. Though Western Colorado was sparsely populated in 1879, the days when raiders, Indian or white, could disappear into the vast wilderness of the west were long gone. Farmers, ranchers, travelers, trappers, traders and the army created an enormous network utilizing railroads and telegraph to track desperadoes and marauders. Josephine Meeker, watching Douglas constantly tippling a whiskey bottle and wandering aimlessly along the spiderweb of game trails, knew that the addled chief was virtually lost. She told her mother and Flora Ella Price to take heart because they would eventually blunder into a rescue party.

CHAPTER 25

"Most people's lives contain a brief passage so dramatic as to make the rest of their days seem like sideshows, trifling and unremembered."

Marshall Sprague, *Massacre: The Tragedy at White River*

"Men are by nature merely indifferent to one another; but women are by nature enemies."

Arthur Schopenhauer

During the day Douglas's band twisted and turned aimlessly. In the evenings by the campfires, Josephine Meeker busied herself measuring, cutting and sewing army blankets into warmer clothes for the other women and the children. She had abundant time to reflect on the chain of people and events that had inexorably swept her to this singularly remarkable point in her life. Most of all, Josephine remembered Monday the twenty-ninth of September like a series of very vivid photographs and paintings.

She had been unable to sleep after Jack's midnight visit; the allure of a forbidden tryst, an excitement she'd never before experienced with a man, the heart-stopping thrill of a brief kiss. The surprise of a moist afterglow had kept her awake until the first dawn birds pulled their heads from beneath a wing and chirped. Had anyone seen them? Would she have to answer to her mother or, worse, her father? Would she have to submit to the embarrassment of a physical examination to prove she'd maintained her innocence? It was all too much. Though her eyelids closed briefly, the pictures still raced through her mind's eye. The cocks crowed to the setting moon as she finally dozed off.

"Josephine!" Her leaden lids snapped open immediately with her mother's first call. Panic! Was that an accusatory tone? A remonstration? Her first fleeting thought was to roll over, fall from her bed to her knees and beg God and Arvilla for forgiveness. She threw aside the covers and sat up so quickly she nearly fainted. She steadied herself with a hand on the sheet. Her palm sank into a deep, slippery wetness causing her to nearly faint.

"Gracious!" exclaimed Arvilla Delight Meeker. "I didn't mean to startle you so. All you all right? It's unlike you to sleep so late."

Josie, her mind moving like cold tar, nodded. "Yes, mother, I'm fine. I just had a very restless night."

"What with all the drumming and chanting and dancing, I'm not surprised," said Arvilla crossing the room to close the window. "It's freezing in here. The night air carries bad ethers that cause discomfort, illness and nightmares. I've always thought part of the cause for your sister Mary's peculiarities was her nocturnal circumambulations. I didn't sleep well either, thinking about Major Thornburgh's arrival today and what problems it might cause."

"Yes," replied Josephine, ". . . most disturbing. How is Father this morning?"

"Tired and cranky. The events of the last few days, especially that violent incident with Johnson, have had their way with him. He put on his best possible face for my benefit when he got up this morning but he stooped and shuffled about the room like a man half again his age. This job is wearing on him dreadfully and we've been here less than a year. I would have to think we've earned enough to pay off that doggedly burdensome Horace Greeley debt so he could resign and we could be out of here before winter. Surely your brother Ralph and he could collaborate on some writing projects and we could return to New York."

"Mother, Ralph is no better off than us. His collaborations with Father have served only to burn political bridges putting us in our current predicament. Any more collaboration and we'll all be in a workhouse or debtor's prison—if such still exist. No, I'm afraid, in lieu of divine intervention, we will face more than one winter here."

Arvilla ignored the remark. "I started the breakfasts for you but you need to get dressed before the children get here." Arvilla muttered something inaudible and walked from the bedroom.

Josephine tentatively examined her damp palm and moist spot on the sheet. She quickly washed her hands in the basin on the dresser, smelled her hands and nodded, satisfied she'd extinguished the musky odor, and pulled off her nightgown.

The Meeker family, except for Josie, shunned nudity. She was almost certain that her parents had never seen each other naked. She wondered if they had closed their eyes when bathing her and her siblings as infants. Oddly, though, when the girls reached puberty, Nathan subjected them to periodic "checkups" to ensure they were developing on schedule and blemish free. Father, she remembered, excused Ralph from such examinations. If Arvilla knew of these inspections she said nothing to the girls.

Each weekday morning the agency women combined efforts to provide the Ute children with huge breakfasts, perhaps their only meal of the day, as enticement to them and their parents to leave the children in school for the rest of the day.

"At least we won't have to feed Colorow today," Arvilla called from the kitchen. The children were often accompanied by the voracious and corpulent Colorow, who ate but then instead of staying for class, loaded himself with as much leftover bacon, eggs and biscuits as he could carry.

"Oh, why not?" asked Josie.

"Flora Price said she saw him riding out before dawn and he told her that he and Jack were riding up to meet the cavalry."

Josie stopped lacing her shoes. She felt the stab of cold fear in her stomach. "Nothing else?" Arvilla looked at her. "Did he say anything else?"

"No, at least Flora didn't say so. Why?"

"Just wondering. Everyone, including me, is a bit edgy. It could make for trouble if the army were to enter the agency unannounced."

"Well you're right about being edgy. But Father has assured all the Ute leaders that won't happen without his permission and Major Thornburgh is just coming to investigate complaints. Nothing to worry about. He's a West Point graduate and an expert at handling such controversial matters."

Josephine rolled her eyes at her mother's naïveté. "General Lee and General Custer were also West Point graduates."

"Yes dear, that's right, they were." She exhibited no recognition of Josie's ironic point. How incredibly alike her parents were, Josie thought, both intelligent, single-minded, and obdurate.

With her mother out of the room, her thoughts turned to Jack again. She went to the window and looked at the coppice of trees illuminated gold by the morning sun from whose protective shadows Nicagaat had emerged into the moonlight last night. The remembrance again sent that unfamiliar thrill through the length of her body. She turned quickly from the window fearful of the excitement that surged toward her loins. But what emotions? She had not done or said anything wrong, embarrassing or sinful. She was a proper teacher and Jack a devoted student. Unlike Persune, the smitten young warrior with a wife and two children, who embarrassed her with obsequious fawning, Jack was a paragon of self-control and discretion; oddly, qualities her father respected in men.

At eight o'clock breakfast was ready and the women stood by to serve the expected throng of women and children standing in line at the door. At eight thirty they paced the floor silently. Flora Price had told them her husband, Shad, had loaded his rifle before going off to plow and had left it in the kitchen corner—just in case. Josephine thought it unwelcome information. Flora fed her two children, Johnnie and May, breakfast and teased them by telling them that the Utes had offered her high prices for them.

By ten o'clock it was obvious something was wrong. The women silently ate as much of the excessive breakfast as they could. They salvaged the remainder for the evening dinner, and the scraps were put outside for the dogs, cats and jays.

About noon, Josephine announced that she was going down-river to the Ute camp to find out what had happened. Arvilla argued against it and finally stood in front of the door with her arms folded while Flora tried to distract herself by playing with the children. Josephine finally prevailed by saying that she would only go to Art Thompson who was working on the roof of one of the agency's storage sheds.

Josephine found Art sitting cross-legged beside the shed taking his lunch break.

"Art, none of the children or women showed up for breakfast or class this morning. Have you seen any of them?"

"Nope, in fact, I haven't seen any Utes this morning. I guess they all stayed up too late last night. No . . . I take that back, a man came through on a lathered horse about an hour ago."

"He say anything?"

"Nope, just waved when he went by."

"Which way was he coming from and goin' to?"

"From upstream to downstream."

Josephine's heart skipped a beat. She walked quickly toward her father's agency office. In her haste she stumbled, tripped and nearly fell in the middle of the road. Thoughts, imaginings and fears collided inside her head with such a din that it blinded and deafened her.

A man yelled from somewhere behind her but she couldn't understand the words or recognize the voice over the clamor in her head. Recovering from the fall, she felt seized by such panic that she abruptly changed course and pulling up her dress above her knees ran pell-mell for her cabin and the comfort of the women inside.

Her mother met her at the door. "Josie! What's wrong? What . . . ?"

Josie shouldered Arvilla aside in the doorway but then turned, reached back, grabbed her arm and jerked her inside with such a burst of strength the older woman flew across the room striking a wall with such force pictures fell from their nails to the floor. Arvilla lay motionless; her mouth gaped in astonishment, forming a mute cry. Josie slammed the door closed and pinned the locking bolt in place. Flora sprang from her chair to help Arvilla but collided with Josie both of them going to the floor beside Arvilla. Several seconds of silence followed as all three lay gasping for breath. Josie found her feet first and stood half-crouching leaning with her hands on her legs. She raised her head and stared at the others. She heard in the background the jabbering of the children. "It's begun," she wheezed almost inaudibly.

"What?" Arvilla asked with genuine bewilderment.

Josie swallowed and cleared her throat. "It's begun." She ran to the windows, closing inside-reinforced shutters with their peepholes and firing loops. The other women chased after her. Flora began lighting candles and lamps in the darkening room. Josie stopped and leaned on the mantelpiece. "The army's crossed into the reservation," she gasped.

"How do you know that?" asked Arvilla.

"Art Thompson saw a lone Ute rider on a spent horse riding from the north . . . he was headed southwest toward the village to tell Douglas." Her mother looked at her incredulously shaking her head. "Trust me," growled Josie, "I know things you don't."

"We have to warn your father," said Arvilla hastening toward the door. "He'll know what to do. He'll talk to Douglas and settle this matter."

Josie grabbed her arm. "Not this time he won't, it's too late."

"Josie," interjected Flora, "do you have any weapons here?"

"No, just in Father's office."

"Then I have to get Shad's rifle from my cabin . . . now!"

Josie nodded. Flora was a farm girl who had grown up with firearms and knew how to shoot. "All right, and bring as much ammunition as you can carry."

Flora glanced at her infant children sitting in the middle of the floor silently transfixed by the commotion around them and simply said, "I'll be right back, behave."

She ran into the kitchen. Josie heard the door open and close. Both she and Arvilla knelt on the floor and occupied themselves with the children. She listened intently for gunfire or war whoops but she could only hear her own heart keeping time with the clock ticking on the mantle. It was one thirty.

Josie and Arvilla both flinched at the sound of the back door opening. They looked at each other. The door closed. Josie felt as though she was suffocating as she heard footsteps across the kitchen floor, one being the clomp of a man's boots. Flora Price entered their room holding a lever-action rifle. Josie nearly fainted with relief. But then she gasped as a male form entered the shadowy room behind Flora, Frank Dresser. He was unarmed, his eyes wide and wild with fear, his forehead beaded with sweat. He gasped out: "Douglas and a lot of Ute men in war paint just walked to the edge of town." The whites had become used to calling the agency settlement "town" because it handily distinguished their settlement from the Indian encampment and it was the closest facsimile to a "town" for many a mile.

"Are they armed?" asked Josie.

"Yeah," Frank answered. "At least some of them. They were trying to hide their rifles alongside their legs."

"Josephine," Arvilla insisted, "we've got to get to Nathan's office and warn him. He'll take command and settle this fracas." Her daughter shook her head in disagreement. She knew that from the day Arvilla married, she had completely subordinated herself, and later her children, to the authoritarian Nathan Meeker; where, when and how they would live, schools, friends, occupations, dress, manners and behavior. Her mother was lost without his guidance.

"Father's too busy," Josephine remonstrated bluntly but felt a stab of guilt. She considered her father dead and scalped.

Flora, like Frank Dresser, was still in her teens and now adrift upon a tidal wave of events quite beyond her comprehension, let alone, control. Josephine recognized it now fell to her to take matters in hand and assume leadership, but still her first instinct was to seek out her father.

She led the group to the front door of the cabin. Frank Dresser had to help her lift the heavy iron drop bar from its tight cradles on either side of the door frame. The group huddled together in a tight knot and started down the road to

Nathan's office. Less than one hundred yards away stood Douglas and his painted warriors their rifles and revolvers now in plain view. To get to Meeker's office the tiny group had to walk toward the war party. The Utes took umbrage. They fanned out across the street and raised their rifles ominously. Josie stopped and held out her arms to halt her companions. Without an order, the warriors aimed their rifles. For Josie breathless seconds ticked by insensibly slowly—years, decades, eternity. She saw the billowing blue smoke seemingly hours before she heard the shot followed by the women's screams. But Josie's eyes followed the direction of the smoke plume and watched Arthur Thompson clutch his chest and topple from the roof where he had been working.

"Back to the house!" she yelled, scooping up May Price with one arm and spinning her mother around with the other. Arvilla tried to shake her off. She pirouetted plaintively squeaking out, "Nathan" as she tried to bolt for his office. Dresser, seeing the predicament, ran back and grabbed Arvilla's other arm and the two of them pulled her, screaming toward the cabin. Somewhere behind Josephine could hear a man pleading: "Please, don't kill me! Please, don't kill me! I haven't done anything to you." It sounded like William Post, a quiet, nondescript worker Nathan had hired away from the Greeley colony. A shot ended his pitiful begging.

May Price, who she was carrying over her shoulder, screamed at the top of her voice. Josephine didn't turn around but with Dresser's help pulled and dragged her mother into the cabin and slammed the heavy door shut. The iron bolt which before she'd barley been able to lift now felt light as a balsa-wood plank as she dropped it into the cradles.

All, except for the caterwauling children, lay strewn about the dimly lit room gasping for breath. Arvilla crawled to the dining table, snatched her precious *Pilgrim's Progress* from it and curled protectively around it beneath the table.

Flora Price, still clutching her husband's rifle, drew her wailing children to her breast. "Shhhh, shhhh," she hissed comfortingly, "it's all right, it's all right," she sobbed. "Daddy will be here soon and we'll all go back home for dinner."

No, Josie thought, it's not all right, Daddy's never coming home to dinner again, nothing is ever going to be the same again. Even through the thick log walls and window shutters she could hear the din of slaughter. Blasts from the Indian rifles and revolvers echoed up the road like rolling thunder. The screams of white victims merged with the whoops of the enraged Indian victors. Suddenly, she realized that it was only a matter of time before the warriors reached their cabin. The front door and window shutters were closed and bolted, but Flora and Frank had come in the back and she hadn't heard that door being locked.

Josie jumped over Frank's prostrate body and ran into the kitchen. Her heart skipped a beat at the sight of the wide open door but sighed with relief seeing the kitchen empty. She tried pulling the heavy half-timber barrier door shut but its thick iron hinges stuck on rust.

"Frank!" she yelled and ran outside to push on the door. "Help me." Wedging her way between the door and outside wall of the cabin she lifted her dress around her waist and brought her feet up against the door and pushed with all her might. At that moment Frank peeked around the corner of the door and stood, mouth agape, staring at her bare legs and up her open dress. "Pull on the damn door," she screamed. He ducked back and in a few moments the door creaked and groaned into place. She stopped pushing and ran inside just before it closed to help Frank pull it shut and drop the iron bar into the latches. Frank started to mumble an apology. "Save it," she said waving her hands, "we've got more important things to think about." She stopped as the figure of a man flitted by one of the kitchen windows. "Frank, get the rifle while I shutter the windows."

Frank sprinted into the living room. Josie slammed shut one pair of window shutters and bolted them. As she ran around the kitchen table to the other window, a gruesomely painted face appeared with its nose almost against the glass. She and the Ute stared at each other in silence. The warrior's hand came up to shade his eyes from the afternoon sun.

"Frank!" Josie yelled.

"Here," came the reply and the sound of a cartridge jacked into the rifle's chamber. Then silence. "It's my friend, Jata," cried Frank.

"Shoot!" screamed Josie.

"He's Johnson's brother."

"Shoot *now!*"

A white hot volcano erupted over her right shoulder, the concussion blowing her head aside and the sparks burning her cheek, neck and hair. In the same instant the window glass shattered and the devilish face disappeared in a cloud of blood, bone and brain. Ears ringing and still groggy from the blast she jumped forward and slammed the heavy shutters and locked them in place. Turning, she stared straight into a rifle barrel. The shocked Frank Dresser hadn't moved since firing the shot. Josie instinctively ducked but recovered and pushed down the barrel very gently. "It's all right, Frank," she said as softly and calmly as possible. "You did well."

"He was my friend," the boy whispered, tears filled his wide eyes.

"I know, but that's all changed, as of right now the Utes are enemies and we're fighting for our lives." Josie took the rifle from him and led him back into the dining room. Her mother remained curled up around her book murmuring the Lord's Prayer over and over. Flora, clutching her children to her breast, cast Josephine an expression of helplessness.

The gunfire outside subsided then stopped and Josie imagined a scene of scalping, maiming and pillaging. That their redoubt had thus far been spared astonished her, but with sinking heart she knew this relative security and tranquility would not last long.

Turning to Frank who was now sitting on a kitchen chair with his head in his hands, elbows resting on his knees, she put her hand gently on his shoulder. He looked up at her, tears still dripping from his jaws.

"Frank," she said softly, "I need you. I need you to grow up right now and become the man your mother would want you to be. I need you to grow up instantly and help us live. Do you understand me? Look at me and tell me you understand?" He looked directly into her eyes and nodded.

"Go to the washbasin in the kitchen, fill it with clean water and wash your face and comb your hair then come back in here, we have to talk about what we're going to do." She collapsed onto a chair. May Price wriggled loose from her mother and ran to Josie who picked her up and stroked her curly blonde hair. My God! she thought. It hasn't been twenty-four hours since I spoke with Jack last night but it seems like I've lived a lifetime since then—someone else's lifetime.

"Josie, what *are* we going to do?"

"I don't know exactly, Flora. We'll talk about it."

"Do you think Shad's still alive?"

As much as she wanted to assuage the big farm girl's fear, the facts glaringly militated against an affirmative answer. "Maybe, he was out plowing alone when the shooting started, if he'd run for the river brush and hidden when he heard the shots." The answer seemed to quiet the girl.

"Why haven't they come after us?" Frank asked as he reentered the room.

"They're too busy looting, drinking and burning and they know we can't get away without them seeing us."

"Eventually then they will try to break in and if they can't, they'll burn us out."

"They'll burn us out . . . or in," Josie said matter-of-factly.

Flora sobbed and Frank's eyes began filling again. "Maybe we should surrender," he suggested. "They won't kill women and children."

"No, they'll kill you, rape us and sell the children. Our only chance is to wait for night and try to make it to the river in the dark."

"What if they come for us before dark?"

Josie thought for a moment then walked to one of the shuttered windows and peaked out the loophole. Thick, black smoke blanketed the road in front of the cabin, so dense, in fact, that she could not see the buildings on the other side. "We may have our opportunity right now!" she exclaimed as she ran to the kitchen followed by Frank. Peering from the hole in the kitchen shutters she could only make out the dimmest outlines of the trees less than one hundred feet away. Looking down she saw the prostrate form of the young Jata. He must have been alone. "Frank, how far do you reckon it to be from here to the milk house?"

"Maybe two hundred feet."

"Go out and get Jata's rifle and revolver and ammunition. First chance we get we're going to run across the road to the milk house while the smoke is still thick enough to hide us. Most, if not all the Utes, will be down the road in the opposite direction looting the warehouses. If we wait too long, the fires will have burned themselves out." She started for the dining room but Frank didn't move. His wide, reddened eyes stared at the kitchen door.

"Frank," said Josie in a calm but firm voice, "I told you that we needed a man. Now's the time. You have a few minutes to grow up. He's dead. You shot him in the face. He can't harm you and right now there's no one else out there. I'll help you unlatch the door." Taking him by the arm she led him across the room and taking one end of the iron bar herself, looked at Frank and nodded toward the other end. He shuffled over and lifted the bar. "Don't bother locking the door when you come back," she said, "there won't be anyone here when they walk in." Also, she thought, it'll keep you from losing your nerve and running.

Josie knelt beside Flora and the children. She put her arm around the girl's trembling shoulders and pulled her in close. "I have a plan to get us all out alive but we have to move from this house to the milk house across the road. We have to move quickly while the smoke will hide us." The girl nodded. Frank entered the room holding a Winchester rifle and a revolver. Josie motioned him over. "Frank and I will open the door. He'll carry Johnnie and lead the way and you'll be following him with May. I'll help Mother. Once we run, look straight ahead, don't stop for anything. Don't stop for anything or anyone even if they're alive and crying for help. If we're lucky, they'll think we escaped and will start searching the area around the agency. Tonight, before the moon rises we'll run for the river, hide in the willows and follow it downstream. Go to the door and get ready."

Josie walked quickly and deliberately to her mother. Always a short, slight woman, Arvilla now appeared no larger than a rag doll. She reached down and shook Arvilla's shoulder. "Mother," she demanded, "we have to go now, we have to move to another, safer place." Arvilla surprised her by uncurling and asking where. "The agency milk house," she replied, "it's made of adobe and won't burn. But we have to hurry so the smoke will hide our escape from here."

"Nathan's here?"

"No." Josephine thought for a moment. "Father will meet us there later . . . he's talking to Douglas right now." God forgive me for lying, she thought, but right now I'd lie about anything if it will speed things along and save us. She helped Arvilla, still tightly clutching *Pilgrim's Progress* to her breast, to her feet and then to the door where the others were waiting. Flora and Frank each carried a child and bag of ammunition in one arm and a rifle in the other hand.

Josie struggled with the door's iron brace but finally cracked the door open to peek out. The smoke was dense, strong and acrid. It burned her eyes and throat. She could see no further than the center of the road and only the late afternoon's occasional slanting ray provided her time and direction perception. She stepped back and herded the others forward; Frank and May, Flora and Johnnie and her mother. Frank stumbled uncertainly out into the dark cloud but at least he kept moving forward, maybe because Flora was nearly stepping on his heels so as not to lose sight of him. Josie grabbed her mother's arm half lifting and half pulling her down the steps and onto the road. While she stared fixedly at Flora's back, she remained mindful of their heading and listened intently for sounds of nearby Utes. She heard only the crackling of distant flames and the crash of falling roof and wall timbers.

Josephine had imagined the worst, tripping over the corpse of either Shad Price or her father. Detritus of battle covered the road but they encountered no bodies. During the first few seconds of crossing, she had been vaguely aware only of the chokingly thick smoke but then, with each step, the burning in her throat and tightness in her chest increased. Her legs cramped with fatigue and her pace slowed to a walk. It now felt as though she was dragging Arvilla but didn't stop to look. Suddenly, she could see the dark form of a building above Frank and within seconds he disappeared into it; then Flora and May.

The setting sun now blazed into her watering eyes. She was both relieved and horrified. The wind had shifted blowing their smoke screen north. But she was to the milk house door. Frank held the door open and she dived in on top of Flora and pulling Arvilla down with her. He slammed the door shut. Except for the whimpering children the group stood stock-still in the darkness gasping for breath. Josie pulled Arvilla to her. "Are you all right, Mother?" she wheezed.

"Yes, dear," she replied clearly, "but it's very dark, I can't see. Is your father here?"

"No, later . . . he'll be along later."

"Where are we?"

"The milk house."

"Why?"

Josephine's answered with a spasm of coughing that set her throat and lungs afire. "Just rest your head in my lap while I catch my breath. Flora, I can hear you but I can't see you, are you and the children all right?"

"Yes," croaked the girl, "I'm still choking from that foul smoke. Johnnie and May are both here. Frank, where are you?"

"By the door. I can see out through a crack. We made it just in the nick of time. The wind's come up and is blowing the smoke away. Another few seconds and they probably would have seen us."

"Can we light a candle or lantern?" asked Flora.

"No," replied Josephine, "I don't have any matches. Maybe our eyes will adapt the longer we're in here."

"How long will we be here?"

"Until it's nearly dark outside. We'll sneak out then and hide nearby until the moon rises and we can see well enough to make it to the river. Frank! What's that racket? What are you doing?"

"I found a table over here," replied Frank. "There's no lock on the door so I'm going to move it over there."

"No! Be quiet, one of them might be standing right outside. And, if someone pushes on the door and something's blocking it they'll know we're in here. When we can see a little better, we'll hide back in the milking stalls. I hope when they see we're gone, they think that we hightailed it north to find Major Thornburgh's cavalry—nobody would be fool enough to stick around here."

"Headin' north after dark sounds better to me. If the major crossed onto the agency this mornin', he's had all day to travel south so he can't be that far away."

Josie pondered this. "Maybe, but if the small band that's here head north hunting us and run into the troops they'll turn back right into us. I still think we best head south."

"I agree," replied Flora.

Arvilla didn't answer. Josie stroked her mother's hair and felt her sobbing quietly. Frank stopped moving furniture. Josie's eyes were growing accustomed to the dark and she could now see his outline standing in the shafts of afternoon sunlight seeping through the cracks and chinks in the doorway. Johnnie and May Price whimpered for water and food. None of them had eaten since breakfast.

As their eyes adapted, Josie moved everyone into one of the plank-enclosed stalls in the back of the house close to the rear doors split by cracks through which the afternoon sun poured. The cows' bodies lay where the Utes slaughtered and butchered them in the pasture adjacent to the milk house. Frank found cups and the small party silently slaked their thirst and soothed burning throats from milk cans lining the wall where a spring bathed them with cold water. Josie thought ruefully about the uneaten breakfasts intended for the Ute children now wasting away in the cabin. The cold, raw milk calmed the Price children and afterward they napped fitfully.

No one spoke. They didn't have to. There were only three items occupying each mind; their survival, the Ute warriors and the dead whites.

Frank left the confines of the stall and returned to his vigil at the crack in the front door.

"Josephine," he called out to her in a hoarse whisper.

She joined him at the door. "They're at the cabin," he whispered.

Peering through a knothole, she saw half a dozen Ute men, led by Douglas and Persune, approaching the cabin cautiously, rifles at the ready. Douglas mounted the front steps and pushed at the half-open door with the barrel of his rifle. He disappeared inside. In less than a minute he reappeared, excited and gesturing. The others followed him back inside, this time for several minutes. Finally, three of the men ran from in back of the house joining the others who'd come out the front door. Their arms were loaded with food and clothing. They talked animatedly while smoke began boiling from the open door. Douglas appeared to give an order and the men ran into the lengthening shadows of tall sagebrush to the north. The old chief, however, sidled toward a group of men assembling a pile of booty in the middle of the road next to a body. He took a pint bottle from his jacket pocket, sucked it dry, tossed it disdainfully onto the body and laughed.

Josie pulled Frank away from the door crack. "I was right," she said excitedly. "They think we escaped and will stop looking for us here. Now we just have to wait until dark." She looked at Frank. The boy's eyes glistened with fright, his face contorted with terror. "Frank," she whispered, "take the lead; Flora with May and then Mother will follow you and I'll bring up the rear with Johnnie." Frank nodded.

As the sun dropped behind the western peaks, Josephine began stirring the refugees to prepare for escape. "We'll wait for the darkness to cover our run for the river but we have to time it so there's still light enough we won't get lost or separated." Arvilla pleasantly surprised her by awakening alert and appeared ready to run on her own. Josephine grasped one rifle and handed Frank the other.

Josie sucked in a breath and held it. She slowly opened the door and peeked to her right toward the road through a thick coppice of trees and tall brush. No men moving in the deepening gloom. In the other direction a path led south along the rails of the pole fence of the cattle corral toward an open pasture and then into the river willows. No one was in sight. She stepped out pulling Frank with her. "Stand here until I give the word," she ordered. His eyes gleamed. "I'm depending on you to lead the way!" she hissed. He nodded again. "Follow the path along the fence, keep the fence on your left, understand?"

"Yes," he croaked, "keep the fence on my left." His voice sounded distant, mechanical.

"Once you're past the end of the fence run straight ahead until you hit the willows then stop and wait for us to regroup." She hurried back inside, picked up May and handed her to Flora. "Follow Frank like when we crossed the road. Put your eyes on his back and don't look right or left." Flora too nodded and walked through the door. Josie turned to her mother who seemed to be rising to the challenge. "I know," Arvilla said firmly, "I'll follow Flora and not lose her." They embraced tightly and Arvilla also slipped quietly through the door. As soon as all were in line outside, Josie, cradling Johnnie in her right arm and the rifle in her left hand, whisper-yelled, "Go!" The six proceeded at a trot down the path.

Josie saw Frank come abreast of the midpoint in the fence line to his left. He straightened up and lengthened his stride immediately more than doubling the gap between him and Flora. Josie cursed under her breath and accelerated to a full run to keep up but had to slow for Arvilla who was quickly weakening. She could see Flora, as instructed, still following Frank, who, to her horror had fixated on the fence so keenly he made an abrupt left turn and followed it east after the southernmost post instead of running straight for the willows. Flora, fifty feet behind with May dutifully turned east with him. Josephine, now breathless, tried to yell, "Stop, wrong way!" but couldn't. Utter despair immobilized her. To her left, she watched Flora and Arvilla obediently trudging after Frank toward certain capture, enslavement and possibly hideous death. To her right, she saw, less than two hundred feet distant, thick willows, safety and escape. She stumbled straight ahead impervious to Johnnie's cries, her heart beating in her head like a bass drum and the racket of the Ute warriors who had begun a scalp dance around the torched Meeker cabin. She walked to the corner post of the fence, Frank's point of detour, dropped the rifle and began trotting in pursuit of her companions.

Frank, by the time he reached the second corner post dimly realized he'd erred. He slowed and looked back over his shoulders into the dusk to see two people running behind him. Instead of two women he saw two Ute warriors gaining on him and redoubled his efforts. When he emerged from the sagebrush at the edge of the road he realized that he had turned one hundred eighty degrees and stopped stock-still near the front of the milk house where he could see the Ute men drinking and dancing in front of the still-flaming cabin. Now, he thought, I'm surrounded! Drowning in panic he dropped his rifle and bolted across the street toward the sage and cedars he could see in the firelight.

One of the sodden dancers stopped. From the corner of his eye he caught movement in the sage along the road. When his eyes focused he could see a white man, his form illuminated red by the fire's flames. He screamed fearfully certain the white man he saw dart into the road was the malevolent spirit of one of the dead Meeker employees. Panic-stricken, he turned to run and crashed into the dancers following him. The dancing stopped and chaos reigned as he yelled his alarm to the others and pointed at Frank Dresser who picked that moment to dash across the road where he also attracted the attention of a party of mounted Ute warriors returning to the agency with news from Milk Creek. With whoops of excitement at finding new prey they charged after him.

The last Josie saw of the shy, fear-filled blond youth from Greeley, he was disappearing into the tall sage on the north edge of the agency, the Ute horsemen in hot pursuit.

Flora, May and Arvilla had reached the edge of the road when the braves caught sight of Frank and galloped after him. Josie caught up to the women and let Johnnie slip to the ground. All three were speechless with exhaustion, confusion and shock. Josie's fury at Frank's cowardice turned to alarm as a man's shout down the road to the west jolted her back to the urgency created by their own exposure. She looked. Her heart stopped. One of the Ute dancers had spotted them. He began running toward them firing his rifle. The others followed. Bullets hissed and sizzled through the air above their heads and into the ground at their feet. No time to plan. No directions given. The women snatched up the children. Terror drove them in the direction opposite their pursuers, northeast and onto the grassy expanse of the horse racing track where, even in dim light, they made nearly perfect targets.

What had begun as a hope-filled sprint from the milk house had deteriorated into a lumbering walk. Dangling Johnnie in her right hand and pulling Arvilla with her left, Josie staggered behind terror-stricken Flora Price carrying a nearly unconscious May over her right shoulder like a sack of potatoes. "They'll rape us, rape us to death . . . take turns raping us," she blathered, as she stumbled across the cobbled ground her dead husband had been plowing in the morning.

Josie heard, from close behind them, a rifle shot and almost simultaneously her mother uttered an anguished moan and fell to the ground pulling Josie down with her. Johnnie screamed into her ear as she fell on top of him. In seconds the Ute men surrounded them aiming their rifles like a circular firing squad. She was too exhausted to care. One of the men, almost covered from head to toe in war paint, approached Arvilla and held out his hand. "You all right, missus?" he asked politely.

"Yes, sir," Arvilla replied. She took his hand. "I think the bullet only grazed my leg. Thank you."

Josephine regained her feet but bent over with her hands on her knees, breathless, spent. She expected any instant to hear a gunshot and feel the impact of a bullet or hear the crunch of a club splintering her skull, neither bullet nor blow came. The men too stood in silence sucking in air. She looked for Johnnie and for a panicked moment thought one of the warriors had already made off with him but then saw one of the men leading Flora and the two children to the group surrounding her. Relief and dread simultaneously swept over her. The apprehension and ordeal of the last eighteen hours had ended and they were still alive. But did they, the women, now face what the dime novelists termed "a fate worse than death"? As if in answer to her question, the young Persune stepped toward her from the darkness and looped a plundered gold watch and chain over her head and around her neck thus silently announcing to all that he claimed her as his personal prize. Another, very short man, standing on his tiptoes, performed the same ritual with Flora Price and then stood beside her. In another setting, thought Josie, the scene would have been comical. The blonde, six-foot-tall Flora claimed as a slave by a man twice her age and nearly half her height.

Chief Douglas walked to Arvilla who stood her ground. Twenty-four hours earlier she had reprimanded the chief for not sending his son to school more often. "Yes, yes," Douglas had agreed, "heap sorry, will do." Then, she remembered, the boy had come to the house mid-morning, not for class, but for matches so he could smoke. As he left with the matches, Arvilla had shaken her head and muttered prophetically, "More likely, I'll wager, he'll burn the place down."

Now, less than twelve hours later, Josie watched Douglas take Arvilla by the arm and rather roughly steer her back toward a flame that towered above the trees around it and washed away the darkness. The conflagration engulfed the cabin where they'd spent the morning in the naive expectation that her father or the cavalry would step in any minute to quell the rebellious Utes and all would proceed peacefully through the remainder of the Indian summer. Now, even if they were to survive captivity, rape and torture, Josephine would never again savor those bright leaves and cool dusks, family feasts and blazing hearths. This day was a point of eternal departure for all involved. For white and Ute alike, this *was the last Indian summer.*

The three women were marched down the road illuminated by other Ute warriors holding lanterns, torches and whiskey bottles shouting their triumphal approval and thrusting their hips obscenely. Any sign of revulsion by the women to these gestures brought peals of laughter.

The crowd quieted when Douglas raised his hand and spoke to them in Ute, his speech culminating in a grand procession west past the charred skeletons of the agency's homes, barracks, warehouses, suttler's store, blacksmith shop and Meeker's office. The small throng at times parted and flowed like river water around bodies lying in the road. Persune took his prize, Josephine, by the arm and gently guided her around the corpses lying inside large circles of shiny black mud. She made no attempt to look at them. Douglas, she noticed, took sadistic pleasure in pointing out each one to her mother and commenting briefly on each. They passed by the booty heaped in the middle of the road that they had collected during the massacre. No one seemed interested in it now.

The warriors herded their captives west out of the former agency toward the Ute encampment on the Rio Blanco. Josie could see a mile or so distant, near a bend in the river, the evening cook fires illuminating the conical tepees and ridgepole tents. Gradually she could see shadows and silhouettes of women and children swathed in blankets moving among the fires, but no warriors. She'd prayed that Jack would be there. He probably neither would nor could order their release but his presence would certainly serve to cool Persune's ardor.

Then, she realized the agency had been attacked and the men all killed by a handful of young braves. Jack, Colorow and Johnson along with the rest of the men from the camp were battling Major Thornburgh's cavalry. The courier that had arrived shortly after noon had surely brought the news to Douglas, no wonder she'd had ominous premonitions when she heard about him. Whether by plan or opportunity, Douglas knew the time had come to sack the agency and hold the women and children captives as poker chips.

The Ute women and children regarded the captives entering the camp with surprise and arrogance. These same whites who had once doled out food, blankets and cloth at their whim now lived or died at the Utes' sufferance. One of Persune's wives walked to Josie and studied her head to toe with unblinking eyes and an impenetrable expression. She then assumed a position of coruler on her husband's right side. Two other women followed suit. Josie knew then that she was the slave of four owners, three of whom would dedicate their lives to making her as miserable as possible.

Douglas dragged Arvilla, known by all to be the hated Meeker's wife, into the firelight and addressed the band in Ute. At first the group seemed cheered by his speech. Then they listened implacably, a few shook their heads and others seemed to study the ground at their feet. Someone need not understand Ute, Josie thought, to realize he had concluded his speech with news of an immediate escape from their homeland. When he finished, the band walked purposefully toward the tepees and tents. Josie looked to Persune for directions.

"You come," he demanded, "help pack. We have to go far *now!*" He motioned his wives away and walked to Douglas. Josie followed the women to a large, canvas-covered tepee and helped them collect their cooking utensils, pots and pans and place them in old army footlockers.

Persune's domineering first wife directed her to roll up the bedding inside the tepee. She opened the storm flap and poked her head inside. A dim flame from a lamp with its wick turned low exposed a jumble of blankets and skins arranged in a rough circle around the circumference of the lodge. In the center where the lamp rested on a flat-topped rock lay the remnants of a fire pit filled with charcoal and partially burned bones.

The first breath she took inside the tepee nearly strangled her. In spite of the chilled evening air the malodorous blast that stopped her short was warm, nearly swampy and suffused with a mixture of burned flesh and bone, human sweat and waste from unwashed blankets and clothing, rawhides and canvas waterproofed with rancid animal fat. The stench of months or even years accumulation of menstrual blood on clothing and bedding repulsed her. She stood paralyzed with repugnance.

Suddenly the scene went dark and lights danced behind her eyes. The next second she lay amidst the filth, her mind screaming in pain and confusion but her eyes focused incredulously upon the gold ring on her right hand that twinkled brightly in the firelight. One of the wives had delivered a strong blow to the side of Josephine's head with something. A void separated the pain and immobility from conscious thought. What had happened? Was happening? Someone was yelling. A woman was shouting.

Josie rolled on her side and looked up to see the dim image of First Wife in the firelight preparing to strike again. She held a large wooden mallet and stepped forward. Josie leaped to her feet and raised her arms in defense. But a hand reached out of the darkness and clasped the other woman's arm preempting the clout. Persune took the mallet and brushed the woman aside but turned to Josie and growled, "You make hurry, Josephine, no lazy!" Then, pulling a crusty handkerchief from his pocket he reached out and wiped her nose and mouth. The rag came away bloody. He handed it to her and left while admonishing First Wife gruffly in Ute.

Head aching and ears ringing, Josie hastened to comply. She folded the reeking blankets and stacked the rancid skins only to have one of the other wives come in and slap her and motion that they needed to be rolled and tied, not folded.

The little cook fires were stoked to bonfires and packhorses led into the light. Josie caught sight of her mother and Flora who had not been spared the same indignity and drudgery. They helped pack the horses while enduring curses and blows from other women. But Johnnie and May were allowed to run about freely while eating lengths of jerky and bricks of hardtack. Watching them Josie remembered she hadn't eaten any solid food since breakfast, only cups of milk while hiding. She sighed. It would probably be quite some time, if ever, before she would again see the like of meals to which she'd become accustomed over the last twenty years.

The village had packed with amazing speed. Articles too big, cumbersome or heavy for a packhorse or travois were abandoned or burned. The camp's wagons remained empty. That told her they would be moving fast over rough terrain; but to where? Not north toward the approaching cavalry. To the east lay ranges of towering peaks which would be nearly impenetrable with winter coming. The west was nearly as formidable for the opposite reason; deep canyons and high desert with little water or game. They must be heading south toward the relatively sheltered Uncompahgre Valley and the protection of Chief Ouray.

The men collected riding horses for themselves, the women and children and finally, the captives. The Price family shared one saddled horse, a sturdy gelding. Flora, an experienced farm girl, ignored the offer of help from her diminutive possessor, mounted confidently with one child in front and one behind. She placed May athwart the horse's withers in front of the McClellan cavalry saddle and Johnnie behind her. Josephine's heart sank when Douglas handed Arvilla the reins of an aged, barebacked nag with pronounced ribs and backbone. Risking another blow or lecture she strode bravely up to the still-sodden chief.

"Douglas," she addressed him as firmly as her fear would allow, "you can't make her ride like that, she's sixty years old, frail, hasn't eaten and has endured much today. If you force her, you'll have a dead white woman captive to answer for later."

She surprised Douglas. No Ute woman dared address a warrior in such a manner, much less a chief. He scanned the assemblage to see if anyone had understood that the captive, a virtual slave, was demanding something. He then directed a man to fold a blanket and put it on the horse, and then directed Arvilla to mount, a near impossibility since she was still clutching *Pilgrim's Progress*. Josie led her mother to the horse and tried to relieve her of the book but she jerked away and gripped it tighter. "I'll give it back to you once you're mounted," Josie said calmly. Arvilla handed her the book which she laid on the ground, then, cupping her locked fingers together she bent over and made a stirrup for her mother. Once she had Arvilla aboard she picked up the book and handed it up to her while the Utes looked on with bewildered expressions.

She walked back to where Persune stood with a saddled horse. He handed her the reins then dropped to his hands and knees beside the horse. Surprised, she stood and looked at the dozens of dark copper faces gleaming in the firelight. None expressed the slightest emotion—except Persune's three wives who glowered darkly at the servile display. She drew one of the reins over the horse's neck and grasping the pommel hopped up on the brave's back and slipped her right leg over the saddle. He arose without looking at her and mounted his horse. Douglas, on the other hand, climbed aboard behind her mother pulling her tightly back against him while leading his own horse. Persune motioned to Josie and they rode forward to join Douglas and Arvilla at the head of the throng. This, thought Josie, is not going to turn out well.

Josie stoked the campfire and drew the candle closer. Her fingers ached from pushing and drawing the inadequately small needle through the heavy wool felt of the blanket but when she saw Persune emerge from the tent and motion to her she redoubled her efforts and concentrated on her work.

He walked into the firelight and she could ignore him no longer and looked up at the firm but almost boyishly handsome face. He was lighter skinned than Jack and most Utes with an almost aquiline nose, widely spaced black eyes and distinctly curved lips. His face was blemish-free, save for a knife scar on his upper right lip. But he possessed none of Jack's confidence, sagacity, or leadership. And by excessively struggling to emulate the Ute chief he invariably reduced himself to a pathetic sycophant within the tribe. Saying nothing he motioned her toward the tent. She stood and followed; bitter and reluctant, but resigned.

First Wife was waiting outside and opened the tent flap with a sardonic leer and followed her while Persune remained outside. Josephine having become accustomed to the powerful odors of a Ute tent welcomed the humid warmth.

The interior of the tepee was aflame with light from the cooking fire in the center, candles and lamps. She stood quiet and silent as First Wife unbuttoned Josie's bodice and another wife unbuckled the belt holding up the men's canvas trousers Josie now wore because the oak brush on the trails south had quickly shredded her skirt. When finished they all stared with unabashed curiosity at her white slender body and blonde hair as though wondering wherein lay the attraction to Persune or trying to find an embarrassing blemish. The blankets covering Persune's buffalo robe sleeping pad were folded back and Persune was ushered inside. Then the three clustered along the opposite tepee wall to observe, as was the Ute custom, another debasement of the enemy.

CHAPTER 26

"Women always excel men in that sort of wisdom which comes from experience. To be a woman is in itself a terrible experience."

H. L. Mencken

Josie awoke the following morning sensing something indescribably different before she even opened her eyes. She was cold. When she summoned the courage to move her right hand she discovered the reason she was cold. She was alone. When she rolled over and opened her eyes to the sunlight filtering through the canvas walls she saw the beds of all three of Persune's wives were also empty. For the first time since her capture she was truly alone.

The morning sounds she had become accustomed to waking her were missing; no children running, playing, yelling and crying, no birds, no wood being broken for cooking fires. She hurriedly dressed but instead of donning the filthy canvas trousers Persune had given her the night of her capture, she put on the dress and jacket she'd made from blankets. She opened the flap of the tent and was surprised to find Persune standing like a statue next to the opening. So too did Johnson, whose tent was next to Persune's. The entire camp stared fixedly up the pathway through the small village of tepee tents at a group of thirty or so riders coming toward them. The two leaders were exceptionally robust in a most Anglo-Saxon way. They rode straight and tall in their saddles compared to their Indian followers. Their horses were at least two or three hands taller than those of their companions. Their leader wore a wide-brimmed hat with a low flat crown and his companion a bright green hunting cap the type of which she'd only seen in photographs and illustrations.

From the corner of her right eye she saw three forms emerge from Johnson's tent and walk toward her. Persune's three wives parked themselves directly in front of her.

The leader beneath the black hat rode to within a hundred feet of them. The huge man reined in his horse and dismounted. Josie now recognized General Adams and experienced such relief that her knees nearly buckled. She gathered herself and stepped forth resolutely pushing the Ute women aside. She wanted to burst out crying and throw herself upon him but curbed her emotions and instead held out her hand and said: "Mr. Adams, I'm so glad to see you."

Adams took her hand and removed his hat with the other. He greeted her with undisguised emotion, "My poor child . . ." His voice nearly broke. After an awkward silence he asked, "Are the others all right?"

She nodded. "Quite well . . . considering."

"Please excuse my being direct but can you tell me who killed your father and the other men?"

"No, none of us, the women, actually witnessed the killings. We were hiding at the time."

"How were you treated?"

Josephine looked over her shoulder at Persune and the three women. "Better than I expected, especially since we got to this campsite."

Adams paused. He too looked directly at Persune. "What I mean is, please excuse this indelicate question. Did they, the Ute men, commit any . . . ah . . . physical indignities?"

She continued to stare at the boy warrior for several seconds.

He and his squaws remained stone-faced. For the past week she'd hated him and his women with helpless homicidal fury hidden behind clenched teeth. But now she saw again a beaten human incapable of comprehending the enormous events of the last two weeks nor the havoc that would now descend upon himself and his tribe. The hatred dissolved and was replaced by sorrow. She turned and, as though reading from a script said: "Oh, no, Mr. Adams nothing like that."

"I'm relieved to hear that," he said. His voice too lacked conviction. "It will expedite your journey home."

As they spoke, Douglas and Ouray's sister, Susan, emerged from the woods where they had been hiding Arvilla, Flora and her children. Adams expressed surprise when he saw Johnnie trotting along happily chewing on a strip of jerky and May skipping in step with Susan. After inquiring about their health, he remarked to Josie he thought Arvilla incredibly wan. A normally thin woman she now appeared almost skeletal. Flora appeared tan and plump as though she'd been on a holiday.

He posed the same delicate question to them as he had to Josie. They each gave him almost identical answers. Unconvinced, he decided to pursue the matter no further.

Adams and his cohort were in for more surprises, however. For one thing, he'd thought their arrival would simply notify the Utes that they had now *lost* the rebellion in every sense of the term. He hadn't considered the fact that his party, excluding the Uncompahgre Ute volunteers, was outnumbered at least three to one and that Douglas and Johnson might make another fight of it.

But the chiefs knew Merritt, though at least fifty miles away on the Roan Plateau, commanded a considerable force of cavalry with their minds set on retribution for Milk Creek, the agency massacre and the female captives. Douglas had not only his original five captive bargaining chips but now, conceivably, also Adams's miniscule rescue party.

During five hours of bickering and wrangling Adams sickened. His original ebullience faded into anger and discouragement. The sun lay low on the western horizon and the southwest breeze augured a cold night when old Sapovanero rose in obvious frustration with Douglas and Johnson and stated that *he* personally represented Chief Ouray, *The Arrow*. "Today is Tuesday," he said angrily. "If the women do not arrive safely at the Los Pinos Agency by the end of the day Friday, Ouray will send a war party against you that will drive you into Merritt's forces and you will all be killed. If you release the captives to us, General Adams will stop Merritt."

Douglas frowned skeptically as he pondered this. He looked at Adams. "You can do this?" he asked. "You can stop Merritt?"

Adams's confidence returned. "I represent President Arthur," he started boldly. "I will telegraph Secretary Schurz to advise the president to tell General Sherman to order Merritt to halt." He stopped abruptly and seeming to realize his threat weakened with each phrase, he folded his arms, clenched his teeth and nodded with all the authority he could muster.

Douglas looked at Johnson who simply dipped his chin and walked back to his tent. He then stepped up to Adams who towered over him by at least a foot and held out his hand. "Douglas heap happy to give you damn women. You send messenger to Chief Arthur now."

"Tomorrow morning, Chief Douglas. Tonight we eat and rest." He then walked to where Josephine, Arvilla and Flora, prohibited from attending the negotiations, stood anxiously awaiting news. "You are free!" he announced simply. He had expected jubilation, screams of joy and relief, crying and effusive appreciation. Instead, all three women stood motionless, quiet and staring. Finally, Josephine Meeker extended her left arm around her mother's shoulders and firmly embracing her lay her head on Arvilla's right shoulder. Arvilla clenched a book tightly to her breast and laid her head on Josephine's. Flora Ellen Price knelt, opened her arms wide and drew both children to her. Josephine remained dry-eyed and silent, thinking first of Jack's fate and her own future. A tremor of fear coursed through her body as she realized that no event in the remainder of her life would be as momentous, as important, as frightening, nor as exhilarating as what she had experienced during the last twenty-four days. The days of the last Indian summer.

EPILOGUE

"History is written by the victors."

Winston Churchill

September 30, 1909

Dear Mrs. Thornburgh:

I hope you receive this letter. I contacted your daughter in Omaha explaining who I am and our personal connection nearly forty years ago when I interviewed you at Fort Fred Steele in the aftermath of the Milk Creek ambush and battle. At the end of the interview you asked that I keep you informed of events and personalities involved in both the Milk Creek incident and the associated Meeker Massacre. Alas, I have been remiss in doing so and beg your forgiveness.

I remembered my promise to you when I received an invitation to the Town of Meeker's fortieth memorial to that sad day, September 29, 1879. I immediately contacted Mrs. Charles Adams, "General" Adams's wife, Martha, who is now living in Manitou Springs, Colorado, a small town in the foothills just west of Colorado Springs. She had received a similar invitation but could not attend because of ill health. During that exchange of mail and telegrams she received a letter from the late Chief Ouray's widow, Chipeta, saying that she too had received an invitation but was not up to the journey. She suggested, however, that the two of them get together for a visit in Manitou Springs where Chipeta could seek out relief from her arthritis in the hot springs. Mrs. Adams remembered the article I had written for the Rawlins, Wyoming, newspaper and several follow-ups on personalities and events in the ensuing years. I, of course, jumped at the chance to exchange remembrances and anecdotes. Now, a day later, I've just finished my article and began this letter to you.

Captain Scott Payne, Major Thornburgh's second in command and the officer responsible for directing the defenses at Milk Creek, after Colonel Merritt's arrival, was transported by wagon to the railhead at Rawlins and from there to Omaha Barracks Hospital. Captain Payne was promoted and awarded a Distinguished Service Citation. Due to ill health and his battle wounds, the War Department provided him a desk job in Omaha until his death a few years later.

Captain Dodge returned to Fort Russell and a Distinguished Service award. He also was transferred to Omaha where he was assigned to the paymaster general's command for the remainder of his military career.

Captain Joe Lawson distinguished himself in both the British and American armies during more than forty years of military service. Interesting, since he was otherwise a man apparently devoid of pretensions and ambitions. The man was honesty and modesty personified. He was on a first-name basis with Generals Grant and Sherman, Hancock, Merritt, Miles and Crook; an honor few, if any, officers of his rank could claim. Yet he never used those connections for purposes of self-promotion. He retired a year after the Milk Creek Battle and returned to his family, farm and store in Franklin County, Kentucky. He was shot and killed a few years later during an argument with a former Confederate officer.

Lieutenant Samuel Cherry, a most promising young officer, died of wounds suffered in a gun battle with a gang of bank robbers he'd been ordered to arrest and return for trial. He was less than thirty years old.

Wesley Merritt eventually attained the rank of major general. His promotions were accelerated by his "lightning march" from Fort Russell to Milk Creek without losing a man or animal. He attended and then taught at General Sherman's Command and Staff College. When he retired, he returned his family to New York City where he'd grown up.

Some military historians claim that proportionately there were more Medals of Honor awarded for intrepidity and gallantry at Milk Creek than any other single battle since Congress approved the medal during the Civil War. Eleven men received the medal, and fifteen Certificates of Merit were awarded to others. Lawson originally recommended twenty-five of E Company's forty-nine troopers receive the medal, but General Sherman, thinking fifty percent of one unit was too many, rejected all but four. Lawson himself, along with all the other prominent officers received no awards or medals because in the years following the Civil War, Congress considered outstanding gallantry and intrepidity part of an officer's routine duties.

Thirteen army troops and civilian workers were killed at Milk Creek and forty-two wounded. Over three hundred horses, mules and oxen were killed. Eleven white men died at the White River Agency. Nicagaat testified at the House committee hearing that approximately twenty Utes died and an unknown number wounded.

Providence treated the luckless Meeker clan no better after Nathan's death than before. Josephine, the star of the drama, had, as the result of the hearings, received a five-hundred-dollar-a-year federal annuity as compensation for her captivity and attracted Colorado Senator Henry Teller's attention. He offered her a secretarial job on his staff in Washington and arranged convenient living quarters. If, as many suspected, he was motivated by romantic fantasies they were doomed to failure. Teller had an almost homicidal hatred for Indians in general and Utes in particular. Josephine steadfastly denied that she had been ravished while in captivity but then subtly cultivated everyone's suspicions when, at the behest of her sister, Rozene, her mother and brother, she went on a speaking tour which included dramatizations of submission to Persune. To make matters worse the addled Arvilla let slip in interviews that she had observed her daughter with Jack in the willows along the Rio Blanco. Teller took her along with the rest of his staff when appointed Interior Secretary but otherwise distanced himself from her.

In December 1882, Josephine contracted pneumonia, allegedly as the result of a drunken assignation in a snowdrift behind her boardinghouse. She managed to leave her sickbed long enough to telegraph her brother Ralph in New York City who hurried to her side. His presence had no effect on the outcome of the disease. Josephine Meeker succumbed shortly before Christmas 1882 at the age of twenty-five. She died as she might have otherwise lived, in near anonymity, save for that fate-filled month in September and October three years earlier.

Except for a letter found in Josephine's room, Flora Ellen Price disappeared so quickly after testifying at the Washington hearing, that some newspapers reported that she had committed suicide as the result of the rape controversies. Josephine intimated to someone that the letter's tone was depressing and gave no hint as to her location, health or the condition of her children. At this time neither Johnnie nor Mary, now adults, has ever stepped forward to counter suspicions.

Rozene Meeker capitalized as much as possible on her sister's and mother's short-lived fame. Letters from Rozene, Arvilla and Ralph were found in Josephine's room urging her to advance her [their] finances by dramatizing her traumatic captivity. In one letter Arvilla describes how she and Rozene were living solely on bread and milk and asking Josephine to send money. I made numerous fruitless attempts to learn of their respective fates after Josephine's death. They apparently joined Flora Ellen in spiritual anonymity. Ralph, however, continued to enjoy a modest writing and journalistic career until his death several years ago.

Only Mary, of Nathan and Arvilla's four children who survived to adulthood, and who had distanced herself from the family in late adolescence, produced children; when and where I don't know.

Practically no written records exist concerning the fates of most of the Utes participating in the Milk Creek Battle and Meeker Massacre, only the principal chiefs. Ouray struggled and lived long enough to guide the others to a peaceful capitulation to the white authorities. Two commissions, one in Colorado and one in the United States House of Representatives, declared the Utes exculpated mainly because both incidents had transpired on Indian property, the rough equivalent of foreign land, and were therefore considered to be a war between two independent nations. Additionally, no one could accurately identify any single, specific killer of any of the White River Agency victims.

Federal officers removed Douglas, enroute to the House committee hearing, from his train and transported him instead to Leavenworth Federal Prison. He remained incarcerated there for nearly a year. When he was released as being "insane," newspapers reported that he returned to Colorado.

Similarly, Johnson returned from Washington and rejoined the Utes in Utah.

Jack returned and moved north to the Rosebud River area where he died at the hands of American cavalry when caught with contraband arms and whiskey.

Ouray, "The Arrow," died from dropsy in August 1880, at age forty-two. His death saddened both whites and Utes. He hated and condemned the whites for their callous maniacal acquisitiveness. He mourned the deaths of friends and relatives who fought to preserve the old ways. But, during his brief life, he had witnessed the rapidity of change. Wisdom and courage won out over anger and pain. He accepted and preached the inevitability of white rule. The best the Utes, all Indians, could hope for was accommodation and preservation without subjugation.

Thankfully he did not live to witness nor endure the exodus forced upon the tribe in September of 1881. Chipeta returned to live on a new reservation created in Southern Colorado. She abandoned almost all white accouterments and sequestered herself in a brush hut at the head of an adobe arroyo. Chief Ignacio and his family care for her.

Now, forty years later, few . . . very few, even remember—let alone were directly connected to—the events which transpired on Monday, September 29, 1879, at a remote Indian reservation and an obscure stream in Northern Colorado. Those events directly, or even indirectly, affected a comparative miniscule number of people. But those touched felt not a gentle breeze wafting over their skin and clothes but rather the iron fists of history and destiny which changed, shortened or ended their lives.

Nathan Meeker, the icon of that infamous day, was an intelligent and gifted man blinded by ambition, self-importance, bigotry, and paternalism. He has become the causal focal point of those events. But Meeker was only the end link in a long chain of broken treaties, acquisitive land-grabbers, unscrupulous Indian agents and rapacious frontiersmen. Meeker, compared to his predecessors, was honest and compassionate, if condescending. He assiduously righted many wrongs; distributed food, blankets and other annuities that previous agents had plundered for black-market sale. But patience was not one of Meeker's virtues. All his life he divined tasks and threw himself headlong into them and expected the same of family, associates and, most of all, *"Pity the Poor Indian"* . . . subordinates.

Sincerely,
Kenneth Church

ABOUT THE AUTHOR

Richard Davis is a third-generation Western Coloradoan. He has written for, among others, *American Way Magazine, Iron Man, Wild West, True West, Denver Post, Rocky Mountain News, Olathe Messenger* and *Rocky Mountain Mason*. Western Reflections published his first western fiction book, *A Man to Cross Rivers With*, in 2000. He is a member of Western Writers of America.

Richard attended Colorado State University and law school in Chicago. He worked for federal, state and local governments and taught political science. In 2009 he retired to his ancestral homestead, the Two Bar K (=K) homestead in Olathe, Colorado, where he and his blue heeler, Jesse, live in the same house in which he grew up.